T0140086

// Advances in Intelligent Systems and Computing

Volume 695

The series "Advances in Intelligent Systems and Computing" contains publications on theory, applications, and design methods of Intelligent Systems and Intelligent Computing. Virtually all disciplines such as engineering, natural sciences, computer and information science, ICT, economics, business, e-commerce, environment, healthcare, life science are covered. The list of topics spans all the areas of modern intelligent systems and computing such as: computational intelligence, soft computing including neural networks, fuzzy systems, evolutionary computing and the fusion of these paradigms, social intelligence, ambient intelligence, computational neuroscience, artificial life, virtual worlds and society, cognitive science and systems, Perception and Vision, DNA and immune based systems, self-organizing and adaptive systems, e-Learning and teaching, human-centered and human-centric computing, recommender systems, intelligent control, robotics and mechatronics including human-machine teaming, knowledge-based paradigms, learning paradigms, machine ethics, intelligent data analysis, knowledge management, intelligent agents, intelligent decision making and support, intelligent network security, trust management, interactive entertainment, Web intelligence and multimedia.
The publications within "Advances in Intelligent Systems and Computing" are primarily proceedings of important conferences, symposia and congresses. They cover significant recent developments in the field, both of a foundational and applicable character. An important characteristic feature of the series is the short publication time and world-wide distribution. This permits a rapid and broad dissemination of research results.

More information about this series at http://www.springer.com/series/11156

Vikrant Bhateja · Carlos A. Coello Coello
Suresh Chandra Satapathy
Prasant Kumar Pattnaik
Editors

Intelligent Engineering Informatics

Proceedings of the 6th International Conference on FICTA

Editors
Vikrant Bhateja
Department of Electronics
and Communication Engineering
SRMGPC
Lucknow, Uttar Pradesh
India

Suresh Chandra Satapathy
Department of Computer Science
and Engineering
PVP Siddhartha Institute of Technology
Vijayawada, Andhra Pradesh
India

Carlos A. Coello Coello
Departamento de Computación
CINVESTAV-IPN
Mexico City
Mexico

Prasant Kumar Pattnaik
School of Computer Engineering
KIIT University
Bhubaneswar, Odisha
India

ISSN 2194-5357 ISSN 2194-5365 (electronic)
Advances in Intelligent Systems and Computing
ISBN 978-981-10-7565-0 ISBN 978-981-10-7566-7 (eBook)
https://doi.org/10.1007/978-981-10-7566-7

Library of Congress Control Number: 2018930368

This Springer imprint is published by the registered company Springer Nature Singapore Pte Ltd.
The registered company address is: 152 Beach Road, #21-01/04 Gateway East, Singapore 189721, Singapore

Preface

This book is a collection of high-quality peer-reviewed research papers presented at the 6th International Conference on Frontiers of Intelligent Computing: Theory and Applications (FICTA-2017) held at School of Computer Applications, KIIT University, Bhubaneswar, Odisha, India, during October 14–15, 2017.

The idea of this conference series was conceived by few eminent professors and researchers from premier institutions of India. The first three editions of this conference: FICTA-2012, FICTA-2013, and FICTA-2014 were organized by Bhubaneswar Engineering College (BEC), Bhubaneswar, Odisha, India. Due to its popularity and wide visibilities in the entire country as well as abroad, the fourth edition FICTA-2015 has been organized by the prestigious NIT Durgapur, WB, India. The fifth edition FICTA-2016 was organized by KIIT University, Bhubaneswar, Odisha, India. All papers of past FICTA editions are published by Springer AISC series. Presently, FICTA-2017 is the sixth edition of this conference series which aims to bring together researchers, scientists, engineers, and practitioners to exchange and share their theories, methodologies, new ideas, experiences, applications in all areas of intelligent computing theories, and applications to various engineering disciplines like Computer Science, Electronics, Electrical, Mechanical, Bio-Medical Engineering.

FICTA-2017 had received a good number of submissions from the different areas relating to decision sciences, intelligent computing and its applications. These papers have undergone a rigorous peer review process with the help of our program committee members and external reviewers (from the country as well as abroad). The review process has been very crucial with minimum 2 reviews each, and in many cases 3–5 reviews along with due checks on similarity and content overlap as well. FICTA-2017 witnessed more than 300 papers including the main track as well as special sessions. The conference featured seven special sessions in various cutting-edge technologies of specialized focus which were organized and chaired by eminent professors. The total toll of papers received, included submissions received cross country along with 7 overseas countries. Out of this pool, only 131 papers were given acceptance and segregated as two different volumes for publication under the proceedings. This volume consists of 67 papers from diverse areas of Intelligent Engineering Informatics.

The conference featured many distinguished keynote addresses by eminent speakers like: Dr. Siba K. Udgata (University of Hyderabad, Telangana, India) on Intelligent and Soft Sensor for Environment Monitoring; Dr. Goutam Sanyal (NIT Durgapur, WB, India) on Vision-based Bio-metric Features; Ms. Kamiya Khatter (Sr. Editorial Assistant, Springer Nature, India) on Author Services and Tools. These keynote lectures embraced a huge toll of an audience of students, faculties, budding researchers, as well as delegates.

We thank the General Chairs: Prof. Samaresh Mishra, Prof. Veena Goswami, KIIT University, Bhubaneswar, India, and Prof. Suresh Chandra Satapathy, PVPSIT, Vijayawada, India, for providing valuable guidelines and inspiration to overcome various difficulties in the process of organizing this conference.

We extend our heartfelt thanks to the Honorary Chairs of this conference: Dr. B. K. Panigrahi, IIT Delhi, and Dr. Swagatam Das, ISI, Kolkata, for being with us from the beginning to the end of this conference, and without their support this conference could never have been successful.

We would also like to thank School of Computer Applications and Computer Engineering, KIIT University, Bhubaneswar, who jointly came forward to support us to organize sixth edition of this conference series. We are amazed to note the enthusiasm of all faculty, staff, and students of KIIT to organize the conference in such a professional way. Involvements of faculty co-ordinators and student volunteer are praise worthy in every respect. We are confident that in the future too we would like to organize many more international level conferences in this beautiful campus. We would also like to thank our sponsors for providing all the support and financial assistance.

We take this opportunity to thank authors of all submitted papers for their hard work, adherence to the deadlines, and patience with the review process. The quality of a refereed volume depends mainly on the expertise and dedication of the reviewers. We are indebted to the program committee members and external reviewers who not only produced excellent reviews but also did these in short time frames. We would also like to thank the participants of this conference, who have participated in the conference above all hardships. Finally, we would like to thank all the volunteers who spent tireless efforts in meeting the deadlines and arranging every detail to make sure that the conference can run smoothly. All the efforts are worth and would please us all, if the readers of this proceedings and participants of this conference found the papers and conference inspiring and enjoyable. Our sincere thanks to all press print and electronic media for their excellent coverage of this conference.

We take this opportunity to thank all Keynote Speakers, Track and Special Session Chairs for their excellent support to make FICTA-2017 a grand success.

Lucknow, India Dr. Vikrant Bhateja
Mexico City, Mexico Dr. Carlos A. Coello Coello
Vijayawada, India Dr. Suresh Chandra Satapathy
Bhubaneswar, India Dr. Prasant Kumar Pattnaik

Organization

Chief Patron

Achyuta Samanta, KISS and KIIT University

Patron

H. Mohanty, KIIT University

Advisory Committee

Sasmita Samanta, KIIT University
Ganga Bishnu Mund, KIIT University
Samaresh Mishra, KIIT University

Honorary Chairs

Swagatam Das, ISI, Kolkata
B. K. Panigrahi, IIT Delhi

General Chairs

Veena Goswami, KIIT University
Suresh Chandra Satapathy, PVPSIT, Vijayawada

Convener

Sachi Nandan Mohanty, KIIT University
Satya Ranjan Dash, KIIT University

Organizing Chairs

Sidharth Swarup Routaray, KIIT University
Manas Mukul, KIIT University

Publication Chair

Vikrant Bhateja, SRMGPC, Lucknow

Steering Committee

Suresh Chandra Satapathy, PVPSIT, Vijayawada
Vikrant Bhateja, SRMGPC, Lucknow
Siba K. Udgata, UoH, Hyderabad
Manas Kumar Sanyal, University of Kalyani
Nilanjan Dey, TICT, Kolkata
B. N. Biswal, BEC, Bhubaneswar

Editorial Board

Suresh Chandra Satapathy, PVPSIT, Vijayawada, India
Vikrant Bhateja, SRMGPC, Lucknow (UP), India
Dr. Jnyana Ranjan Mohanty, KIIT University
Prasant Kumar Pattnaik, KIIT University, Bhubaneswar, India
João Manuel R. S. Tavares, Universidade do Porto (FEUP), Porto, Portugal
Carlos Artemio Coello Coello, CINVESTAV-IPN, Mexico City, Mexico

Transport and Hospitality Chairs

Ramakant Parida, KIIT University
K. Singh, KIIT University
B. B. Dash, KIIT University

Session Management Chairs

Chinmay Mishra, KIIT University
Sudhanshu Sekhar Patra, KIIT University

Registration Chairs

Ajaya Jena, KIIT University
Utpal Dey, KIIT University
P. S. Pattanayak, KIIT University
Prachi Viyajeeta, KIIT University

Publicity Chairs

J. K. Mandal, University of Kalyani, Kolkata
Himanshu Das, KIIT University
R. K. Barik, KIIT University

Workshop Chairs

Manoj Mishra, KIIT University
Manas Kumar Rath, KIIT University

Track Chairs

Machine Learning Applications: Steven L. Fernandez, The University of Alabama, Birmingham, USA
Image Processing and Pattern Recognition: V. N. Manjunath Aradhya, SJCE, Mysore, India
Signals, Communication, and Microelectronics: A. K. Pandey, MIET, Meerut (UP), India
Data Engineering: M. Ramakrishna Murty, ANITS, Visakhapatnam, India

Special Session Chairs

SS01: Computational Intelligence to Ecological Computing through Data Sciences: Tanupriya Choudhury and Praveen Kumar, Amity University, UP, India.

SS02: Advances in Camera Based Document Recognition: V. N. Manjunath Aradhya and B. S. Harish, SJCE, Mysore, India.
SS03: Applications of Computational Intelligence in Education and Academics: Viral Nagori, GLS University, Ahmedabad, India.
SS04: Modern Intelligent Computing, Human Values & Professional Ethics for Engineering & Management: Hardeep Singh and B. P. Singh, FCET, Ferozepur, Punjab.
SS05: Mathematical Modelling and Optimization: Deepika Garg, G. D. Goenka University, India, and Ozen Ozer, Kırklareli Üniversitesiy, Turkey.
SS06: Computer Vision & Image Processing: Synh Viet-Uyen Ha, Vietnam National University, Vietnam.
SS07: Data Mining Applications in Network Security: Vinutha H. P., BIET, Karnataka, India, and Sagar B. M., RVCE, Bangalore, Karnataka, India.

Technical Program Committee/International Reviewer Board

A. Govardhan, India
Aarti Singh, India
Almoataz Youssef Abdelaziz, Egypt
Amira A. Ashour, Egypt
Amulya Ratna Swain, India
Ankur Singh Bist, India
Athanasios V. Vasilakos, Athens
Banani Saha, India
Bhabani Shankar Prasad Mishra, India
B. Tirumala Rao, India
Carlos A. Coello, Mexico
Charan S. G., India
Chirag Arora, India
Chilukuri K. Mohan, USA
Chung Le, Vietnam
Dac-Nhuong Le, Vietnam
Delin Luo, China
Hai Bin Duan, China
Hai V. Pham, Vietnam
Heitor Silvério Lopes, Brazil
Igor Belykh, Russia
J. V. R. Murthy, India
K. Parsopoulos, Greece
Kamble Vaibhav Venkatrao, India
Kailash C. Patidar, South Africa
Koushik Majumder, India
Lalitha Bhaskari, India
Jeng-Shyang Pan, Taiwan

Juan Luis Fernández Martínez, California
Le Hoang Son, Vietnam
Leandro Dos Santos Coelho, Brazil
L. Perkin, USA
Lingfeng Wang, China
M. A. Abido, Saudi Arabia
Maurice Clerc, France
Meftah Boudjelal, Algeria
Monideepa Roy, India
Mukul Misra, India
Naeem Hanoon, Malaysia
Nikhil Bhargava, India
Oscar Castillo, Mexico
P. S. Avadhani, India
Rafael Stubs Parpinelli, Brazil
Ravi Subban, India
Roderich Gross, England
Saeid Nahavandi, Australia
Sankhadeep Chatterjee, India
Sanjay Sengupta, India
Santosh Kumar Swain, India
Saman Halgamuge, India
Sayan Chakraborty, India
Shabana Urooj, India
S. G. Ponnambalam, Malaysia
Srinivas Kota, Nebraska
Srinivas Sethi, India
Sumanth Yenduri, USA
Suberna Kumar, India
T. R. Dash, Cambodia
Vipin Tyagi, India
Vimal Mishra, India
Walid Barhoumi, Tunisia
X. Z. Gao, Finland
Ying Tan, China
Zong Woo Geem, USA
Monika Jain, India
Rahul Saxena, India
Vaishali Mohite, India
And many more…

Contents

About the Editors

Vikrant Bhateja is an Associate Professor in the Department of ECE, SRMGPC, Lucknow, and also the Head of Academics and Quality Control at the same college. His areas of research include digital image and video processing, computer vision, medical imaging, machine learning, pattern analysis and recognition. He has authored more than 120 publications in various international journals and conference proceedings. He is an associate editor for the International Journal of Synthetic Emotions (IJSE) and International Journal of Ambient Computing and Intelligence (IJACI).

Carlos A. Coello Coello received his Ph.D. from Tulane University (USA) in 1996. His research has mainly focused on the design of new multi-objective optimization algorithms based on bio-inspired metaheuristics. He has authored over 450 publications and is currently a Full Professor in the Computer Science Department at CINVESTAV-IPN, Mexico City, Mexico. An Associate Editor of the IEEE Transactions on Evolutionary Computation, he has received several prestigious awards including the 2013 IEEE Kiyo Tomiyasu Award "for pioneering contributions to single- and multi-objective optimization techniques using bio-inspired metaheuristics." His current research interests include evolutionary multi-objective optimization and constraint-handling techniques for evolutionary algorithms.

Suresh Chandra Satapathy, Ph.D. is currently working as Professor and Head of the Department of CSE, PVPSIT, Vijayawada, India. He was the National Chairman Div-V (Educational and Research) of the Computer Society of India from 2015 to 2017. A Senior Member of IEEE, he has been instrumental in organizing more than 18 international conferences in India and has edited more than 30 books as a corresponding editor. He is highly active in research in the areas of swarm intelligence, machine learning, and data mining. He has developed a new optimization algorithm known as social group optimization (SGO) and authored more than 100 publications in reputed journals and conference proceedings. Currently, he serves on the editorial board of the journals IGI Global, Inderscience, and Growing Science.

Prasant Kumar Pattnaik, Ph.D. (Computer Science), Fellow IETE, Senior Member IEEE, is a Professor at the School of Computer Engineering, KIIT University, Bhubaneswar, India. With more than a decade of teaching and research experience, he has published a number of research papers in peer-reviewed international journals and conferences. His areas of specialization include mobile computing, cloud computing, brain–computer interface, and privacy preservation.

Assessment of Integrity Auditing Protocols for Outsourced Big Data

Ajeet Ram Pathak, Manjusha Pandey and Siddharth Rautaray

Abstract Due to overwhelming advancement in sensor networks and communication technology, Internet of Things is taking shape to help make our lives smarter. Devices working on Internet of Things are responsible for generating multifarious and mammoth data. In addition to this, academia, business firms, etc., add vast amount of data to the pool of storage systems. This data are big data. Cloud computing provides paramount solution to store and process this big data through database outsourcing, thus reducing capital expenditure and operational costs. As big data are hosted by third party service providers in the cloud, security of such data becomes one of the significant concerns of data owners. The untrustworthy nature of service providers does not guarantee the security of data and computation results. This necessitates the owners to audit the integrity of data. Therefore, integrity auditing becomes the part and parcel of outsourced big data. Maintaining confidentiality, privacy, and trust for such big data are *sine qua non* for seamless and secure execution of big data-related applications. This paper gives a thorough analysis of integrity auditing protocols applied for big data residing in cloud environment.

Keywords Big data · Database outsourcing · Third party auditing Provable data possession · Integrity auditing · Proof of retrievability

A. R. Pathak (✉) · M. Pandey · S. Rautaray
School of Computer Engineering, Kalinga Institute of Industrial Technology University, (KIIT), Bhubaneswar, Odisha, India
e-mail: ajeet.pathak44@gmail.com

M. Pandey
e-mail: manjushapandey82@gmail.com

S. Rautaray
e-mail: sr.rgpv@gmail.com

V. Bhateja et al. (eds.), *Intelligent Engineering Informatics*, Advances in Intelligent Systems and Computing 695, https://doi.org/10.1007/978-981-10-7566-7_1

1 Introduction

Due to wide availability of sensor devices and proliferation of sensor, and communication technology, the Internet of Things (IoT) is enriching our lives with intelligent devices and services. IoT is producing massive data at unprecedented velocity, known as big data. Big data are represented by 4Vs as volume, velocity, variety, and veracity. Focusing on volume aspect of big data, every sector ranging from academia, industries, scientific research domains, and social media is contributing to big data. The size of big data is getting doubled every 2 years, and it is anticipated to reach 44 zettabytes in 2020 [1].

Storing and processing such big data are major problem. Cloud computing provides efficient way to store and process big data at large scale. It works on the principle of pay-as-you-go and follows the SPI (SaaS, PaaS, IaaS) model. Outsourcing allows data owner to upload data on cloud and alleviates burden of data management.

Database outsourcing refers to outsourcing the database on cloud for managing, storing, and administering the data by cloud service provider (CSP) and let data owners focus on business logic [2]. Due to powerful resources and services offered by cloud, data owners opt to host data at cloud's site. This reduces costs associated with in-house database management such as database storage, maintenance, and personnel management and leads to reduction in both capital expenditure and operational expenditure. But this way of outsourcing big data imposes the following challenges.

- Dishonest cloud service provider: CSP may discard the rarely accessed data and still claim that availability of data in cloud for maintaining reputation.
- Chances of data loss in cloud: Irrespective of maintaining reliability measure to secure the data, possibility of data loss in cloud cannot be denied [3].

Due to these challenges, security of big data becomes major issue. Data security is commonly understood in terms of confidentiality, integrity, and availability (CIA) triad. Focusing on integrity aspect, integrity refers to assurance that data remain intact, consistent, and accurate over its complete lifecycle and guarantees that data are accessed, modified by trustworthy (authorized) entities only. To apply quality checks on the services provided by cloud, auditing is required. It is divided into internal auditing and external auditing. Internal auditing is performed to check practices followed by cloud services to achieve objectives related to service level agreement. These audits are performed internal to the cloud environment for evaluation of processes carried out by service. On the other hand, external auditing is done by entities external to the cloud environment. For example, data owner or third-party auditor audits quality of services through available interfaces.

External integrity auditing is a procedure of checking whether the data uploaded by data owner on cloud are residing in reliable or trustworthy environment supported by cloud. Integrity auditing aims to check data hosted by CSP are being operated in fair, trustworthy, and honest way. The intent of this paper is to assess external integrity auditing protocols for outsourced big data.

The contents of the paper are portrayed as follows. Section 2 enunciates verification methods of integrity auditing. The state-of-the-art integrity auditing approaches for outsourced big data along with the thematic taxonomy of protocols are discussed in Sect. 3. The paper is concluded in Sect. 4.

2 Verification Methods of Integrity Auditing

The previous works [4, 5] well explain database outsourcing model and general requirements for achieving security. According to number of entities involved in the process of verification, integrity auditing protocols fall into two categories, viz. private verification, public verification. Integrity auditing methods generally follow challenge–response protocol. Idealistic features of these protocols are as follows.

- The challenge–response protocol should use small challenges and responses to reduce the overall complexity.
- The challenges should be generated in random and unpredictable manner in order to curb the responder from precomputing and storing challenges instead of storing actual data.
- The protocol should be enough sensitive to detect minor changes, data corruptions, and shuffling done in data blocks to timely identify misbehaving cloud server.

In case of auditing protocols employing private verification, verification rights of data hosted by cloud are given to data owner/clients only. This is also known as two-party auditing. The procedure of *challenge–response verification* occurs directly between data owner and cloud server as shown in Fig. 1a. Auditing protocols based on private verification cannot fully utilize cloud computing services due to the following reasons. Client may possess resource-constrained devices which have less capability to withstand computation and communication overheads. For verification process, client should be equipped with Internet connectivity. This condition cannot be satisfied when client stays in remote area with no Internet connection. Due to these reasons, auditing protocols based on public verification are chosen. Public verification lets anyone to check the possession of data without acquiring the private parameters of data owner. So, data owner delegates task of integrity auditing to a third-party auditor without compromising the private information. This method of delegating auditing task to a third-party auditor is known as delegated verification.

The working of third-party auditing protocols with delegated verification is shown in Fig. 1b. It alleviates the computational overheads of clients in two-party auditing by delegating task of verification to trusted third-party auditor. This method is more efficient than two-party auditing. Initially, data owner uploads data on cloud server. In the setup phase, data owner generates the metadata, tags required for auditing and delegates the task of auditing to third-party auditor.

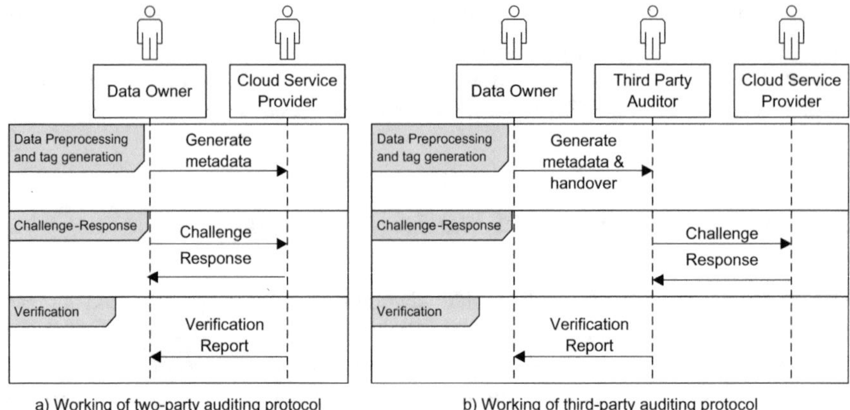

Fig. 1 Verification methods of integrity auditing

Third-party auditor and CSP follow *challenge–response* protocol. Finally, third-party auditor returns auditing report to data owner stating whether data are corrupted, missing, or intact.

3 Integrity Auditing Approaches for Outsourced Big Data

The taxonomy of integrity auditing protocols is depicted in Fig. 2. Table 1 shows the parameter-wise comparison of integrity auditing protocols on the basis of technique used, updating mode, mode of verification, security model, possession guarantee, batch auditing, achieved security requirement, and security assumption.

Different security techniques are used for integrity auditing such as homomorphic verifiable tags (HVTs), bilinear maps and pairings, hash index hierarchies, victor commitment, Boneh–Lynn–Shacham (BLS) signatures. Apart from this, authenticated data structures are also used for supporting the dynamic updates like balanced update trees, authenticated skip lists, Merkle hash trees (MHTs). Auditing models follow security assumptions for achieving cryptographic security. This assumption includes computational Diffie–Hellman (CDH) and strong Diffie–Hellman (SDH) problem, message authentication codes (MACs), Rivest, Shamir, and Adleman (RSA) algorithm, knowledge of exponent assumption (KEA), discrete logarithmic and linear problems, and collision-resistant hash functions.

Integrity auditing schemes are mainly divided into two approaches, namely provable data possession (PDP) and proof of retrievability (POR). Ateneise et al. [6] pioneered PDP model which follows challenge–response protocol. PDP models work well for statically stored data which are never updated once outsourced. Juels et al. [7] introduced a “proof of retrievability” (POR) model which uses spot-checking and error-correcting codes for guaranteeing “possession” and

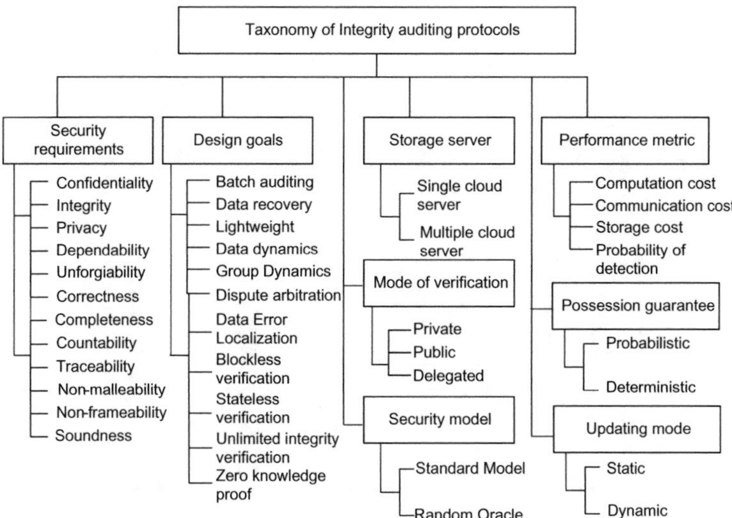

Fig. 2 Taxonomy of integrity auditing protocols

"retrievability" of outsourced data. POR checks the possession of remote data and also recovers it in case of failure/data corruption by making use of error-correcting codes.

On the other hand, PDP only assesses that data are truly possessed by CSP and does not guarantee data recovery if small portion of data has been corrupted/lost by the server. PDP scheme can be modified to POR by introducing error-correcting codes in PDP. Compact POR [8] scheme supports private verification using homomorphic authenticators and public verification using BLS signatures.

According to importance and sensitivity of data, data owner outsources the data on multiple cloud servers to ensure availability. This leads to need of checking the possession of data distributed on multiple clouds hosted by multiple cloud service providers. The issue of checking integrity of data stored on multiple clouds is handled in cooperative PDP (CPDP) scheme [9] using hash index hierarchy and homomorphic verifiable response. CPDP follows multiprover zero-knowledge proof system (MP-ZKPS). Similar to CPDP, the issue of distributed PDP has been handled in identity-based distributed PDP (ID-DPDP) by Wang [10]. This method involves client, CSP, combiner, and public key generator. Combiner is used to distribute data blocks on multiple cloud servers and also distributes challenge queries among them. The responses obtained by each server are combined, and aggregated responses are sent to client to complete the verification process. Public key generator is used to generate public key for private key of client.

Generally, PDP method follows public key infrastructure (PKI) protocol which involves heavy tasks like certificate generation, verification, revocation, renewals. Generally client has resource-constrained device, and so following all the procedure of PKI becomes overhead. Therefore, ID-DPDP scheme eliminates the certificate

Table 1 Comparative study of state-of-the-art approaches for integrity auditing

Paper	Technique	Updating mode	Mode of auditing	Security model	Possession guarantee	Batch auditing		Achieved security requirement	Security assumption
						Multi-owner	Multi-cloud		
PDP [6]	HVT	Static	Public	RO	Probabilistic	No	No	Unforgeability	RSA, KEA
POR [7] and Compact POR [8]	Sentinel-based approach, RS code, BLS signs, HLA	Static	Private [7, 8] and public [8]	ST	Probabilistic [7], Deterministic [8]	No	No	Data recovery, privacy, correctness	Symmetric encryption [7], MAC, RSA [8]
CPDP [9] and ID-DPDP [10]	HVT and hash index hierarchy, bilinear pairings	Dynamic	Public [9, 10], private, and delegated [10]	RO	Probabilistic	No	Yes	Completeness, soundness, zero-knowledge, unforgeability	Collision-resistant hash function [9], CDH [10]
DPDP [12] and FlexDPDP [13]	HVT, authenticated skip list	Dynamic	Private	ST	Probabilistic	Yes	Yes	Dependability	Collision-resistant hash function
IBDO [16]	Bilinear pairings	Dynamic	Public, delegated	RO	Probabilistic	Yes	No	Soundness	CDH
GUR [18] and NPP [19]	Victor commitment, bilinear groups, symmetric group key agreement, homomorphic group signs	Dynamic	Public	RO	Probabilistic	No	No	Confidentiality, countability, traceability, unforgeability, non-malleability	SDH, decision linear assumption [18], discrete logarithm [19]

generation task of client to improve the efficiency and allow clients to perform verification efficiently. To resolve potential disputes between misbehaving third-party auditor and cloud server, fair arbitration scheme is put forth by Jin et al. [11]. Authors used signature exchange method for designing arbitration protocols to settle potential disputes among client, auditor, and CSP. This method also supports dynamic updates on data by introducing an index switcher to map block indices with tag indices, thus removing overhead of block indices in tag computation.

Erway et al.'s scheme of dynamic PDP (DPDP) [12] focuses on two versions of DPDP in which DPDP-I provides optimized logarithmic computation and communication cost for PDP based on rank-based authenticated dictionary. Another scheme DPDP-II focuses on improved detection probability but higher computation overhead using rank-based authenticated dictionary with RSA tree. Esiner et al. [13] devised data structure—FlexList to support dynamic updates of variable sized blocks and put forth FlexDPDP scheme for multiple proofs and updates with improved scalability.

OPOR scheme [14] aims to decrease computation cost at client's side in cloud-based public variability context. In this, heavy computation of tag generation is outsourced to semi-honest cloud audit server, alleviating the burden on clients with resource-constrained devices. This framework of integrity auditing exerts too much trust on the cloud audit server, so there are chances of collusion by audit server. Another work on public integrity auditing scheme for clients with low powered end devices is put forth in Lightweight Privacy-Preserving Public Auditing Protocol (LPAP) [15] in which the online/offline signatures are used by clients to perform lightweight computations. LPAP supports batch auditing and dynamic updates.

Identity-based data outsourcing (IBDO) [16] works for multiuser setting scenario in which outsourcing task is handled by authorized proxies with identities. This eliminates the overhead of certificate management in distributed cloud servers. The beauty of IBDO scheme is that it allows conventional integrity auditing and auditing of origin, consistency, and type of data. Efficient dynamic PDP (EDPDP) [17] scheme works for multiple users to provide shared data access and revision control. To avoid collusion attack of revoked users and dishonest cloud servers, Jiang et al. [18] developed auditing scheme supporting secure group user revocation (GUR) with additional support for group-based data encryption and decryption. It uses victor commitment, asymmetric group key agreement, and verifier-local revocation group signature.

Most of public auditing approaches with multiuser settings rely on single data owner to manage the group activities. This leads to abuse of single authority power, and there are chances that single data owner may become over-powerful. To eliminate the single authority power of data owner, New Privacy-Aware Public Auditing Scheme (NPP) [19] shares data among multiple group users (managers) in which minimum "k" number of group managers are required to recover a key used for tracing. This method follows homomorphic verifiable group signature and achieves non-frameability, traceability, identity privacy requirement of auditing protocol with group user revocation. Dynamic POR scheme (DPOR) [20] works on the principle of Oblivious RAM (ORAM) for proof of retrievability. ORAM lets users outsource the memory to cloud server and still supports read and write operations to be performed randomly in private manner. Authenticated ORAMs guarantee that most recent data will be retrieved by client upon read request, and thus refrain cloud server to roll back the updates.

It can be noted that there is need of lightweight protocols so that clients with low powered devices perform auditing without depending on powerful server for auditing. Apart from this, use of high-performance computing for storing data at cloud side and fulfilling the demands of dynamic updates and user queries at real time is still untouched research area in database outsourcing.

4 Conclusion

The popularity of the computation outsourcing and data outsourcing brings the concept of "verifiability" in the light. The proliferation of big data has resulted into use of cloud services for faster deployment of business units. Therefore, auditing the integrity of outsourced data is the need of the hour. This paper targeted to assess the integrity auditing protocols both in single cloud and multi-cloud environments.

Integrity auditing is a method of offloading the computation of a function or data to third party service provider and letting the service provider calculate the results or show the proof that data are correctly stored. This is verified by the trusted entities like data owner (private verifiability) or third-party auditor (public verifiability). In this paper, a thorough assessment of integrity auditing protocols is done on the basis of thematic taxonomy. The thematic taxonomy of auditing protocols has been put forth in this paper. The state-of-the-art approaches are discussed to get clear notion of security requirements to be fulfilled by auditing protocols.

References

1. Digital universe, https://www.emc.com/leadership/digital-universe/2014iview/executive-summary.htm
2. Pathak, A.R., Padmavathi, B.: A secure threshold secret sharing framework for database outsourcing. In: IEEE International Conference on Advanced Communications, Control and Computing Technologies, pp. 1642–1649 (2014)
3. Miller, R.: Amazon addresses EC2 power outages. Data Center Knowledge 1 (2010)
4. Pathak, A.R., Padmavathi, B.: Survey of confidentiality and integrity in outsourced databases. Int. J. Sci. Eng. Technol. **2**(3), 122–128 (2013)
5. Pathak, A.R., Padmavathi, B.: Analysis of security techniques applied in database outsourcing. Int. J. Comput. Sci. IT **5**(1), 665–670 (2014)
6. Ateniese, G., Burns, R., Curtmola, R., Herring, J., Kissner, L., Peterson, Z., Song, D.: Provable data possession at untrusted stores. In: ACM CCS, pp. 598–609 (2007)
7. Juels, A., Kaliski Jr, B.S.: Pors: proofs of retrievability for large files. In: 14th ACM Conference on Computer and Communications Security, pp. 584–597 (2007)

8. Shacham, H., Waters,B.: Compact proofs of retrievability. In: 14th International Conference on Theory and Application of Cryptology and Information Security, pp. 90–107 (2008)
9. Zhu, Y., Hu, H., Ahn, G.J., Yu, M.: Cooperative provable data possession for integrity verification in multicloud storage. IEEE Trans. Parallel Distrib. Syst. **23**(12), 2231–2244 (2012)
10. Wang, H.: Identity-based distributed provable data possession in multicloud storage. IEEE Trans. Serv. Comput. **8**(2), 328–340 (2015)
11. Jin, H., Jiang, H., Zhou, K.: Dynamic and public auditing with fair arbitration for cloud data. IEEE Trans. Cloud Comput. **13**(9) (2014)
12. Erway, C., Küpçü, A., Papamanthou, C., Tamassia, R.: Dynamic provable data possession. ACM Trans. Inf. Syst. Secur. **17**(4) (2015)
13. Esiner, E., et al.: FlexDPDP: flexlist-based optimized dynamic provable data possession. ACM Trans. Storage **12**(4) (2016)
14. Li, J., Tan, X., Chen, X., Wong, D.S., Xhafa, F.: OPoR: enabling proof of retrievability in cloud computing with resource-constrained devices. IEEE Trans. Cloud Comput. **3**(2), 195–205 (2015)
15. Li, J., Zhang, L., Liu, J. K., Qian, H., Dong, Z.: Privacy-preserving public auditing protocol for low-performance end devices in cloud. IEEE Trans. Inf. Forensics Secur. **11**(11) 2572–2583 (2016)
16. Wang, Y., et al.: Identity-based data outsourcing with comprehensive auditing in clouds. IEEE Trans. Inf. Forensics Secur. **12**(4), 940–952 (2017)
17. Zhang, Y., Blanton, M.: Efficient dynamic provable possession of remote data via update trees. ACM Trans. Storage **12**(2) (2016) (Article 9)
18. Jiang, T., et al.: Public integrity auditing for shared dynamic cloud data with group user revocation. IEEE Trans. Comput. **65**(8), 2363–2373 (2016)
19. Fu, A., Yu, S., Zhang, Y., Wang, H., Huang, C.: NPP: a new privacy-aware public auditing scheme for cloud data sharing with group users. IEEE Trans. Big Data (2017)
20. Cash, D., Küpçü, A., Wichs, D.: Dynamic proofs of retrievability via oblivious RAM. J. Cryptol. **30**(1), 22–57 (2017)

A New Term Weight Measure for Gender Prediction in Author Profiling

Ch. Swathi, K. Karunakar, G. Archana and T. Raghunadha Reddy

Abstract Author profiling is used to predict the demographic characteristics such as gender, age, native language, location, and educational background of the authors by analyzing their writing styles. The researchers in author profiling proposed various features such as character-based, word-based, structural, syntactic, and semantic features to differentiate the writing styles of the authors. The existing approaches in author profiling used the frequency of a feature to represent the document vector. In this work, the experimented carried with various features with their frequency and observed that only frequency is not suitable to assign better discriminative power to the features. Later, a new supervised term weight measure is proposed to assign suitable weights to the terms and analyzed the accuracies with various machine learning algorithms. The experimentation carried out on review domain and the proposed supervised term weight measure obtained good accuracy for gender prediction when compared to existing approaches.

Keywords Term weight measure · Gender prediction · BOW approach Author profiling

Ch. Swathi (✉)
Department of CSE, G Narayanamma Institute of Technology and Science, Hyderabad, Telangana, India
e-mail: chswathi0508@gmail.com

K. Karunakar
Department of CSE, Swarnandhra Institute of Engineering and Technology, Narsapur, Andhra Pradesh, India
e-mail: karunakar.mtech@gmail.com

G. Archana
Department of CSE, Swarnandhra College of Engineering and Technology, Narsapur, Andhra Pradesh, India
e-mail: ksj.archana@gmail.com

T. Raghunadha Reddy
Department of IT, Vardhaman College of Engineering, Hyderabad, India
e-mail: raghu.sas@gmail.com

V. Bhateja et al. (eds.), *Intelligent Engineering Informatics*, Advances in Intelligent Systems and Computing 695, https://doi.org/10.1007/978-981-10-7566-7_2

1 Introduction

In recent times, the text in the Internet is growing exponentially through bogs, social media, twitter tweets, and reviews. Most of the text generated by the authors in the Internet is anonymous. The researchers are attracted to know the demographic characteristics of these texts by analyzing the writing styles of the authors. Author profiling is one such area to predict the demographic characteristics of the text.

Author profiling is used in various applications such as security, marketing, educational domain, forensic analysis, and psychology. In forensic analysis, the experts analyze the problematic text based on their writing styles. In marketing domain, most of the people are posting their comments on the product without specifying their correct details. The author profiling techniques are used to analyze the anonymous reviews, and based on the results, the companies take strategic decisions about their business. In educational domain, the author profiling techniques are used to find the extraordinary students by analyzing the writing styles of the students. In security aspect, the threatening mails are analyzed to detect the characteristics of the mail. In psychology domain also, the author profiling techniques are used to know the mood of the authors while writing the text.

This paper is planned in six sections. Section 2 explains the related work in author profiling done by the researchers. The traditional approach, dataset characteristics, and evaluation measures are described in Sect. 3. The proposed term weight measure is explained in Sect. 4. Section 5 discusses the results of set of features, POS n-grams, and proposed term weight measure. Finally, Sect. 6 concludes this work with future directions.

2 Related Work

Most of the existing approaches for author profiling were used by Bag of Words (BOW) approach. In this approach, each document is represented with vector of word frequencies. In author profiling, the main concentration of the researchers is to extract the features that are suitable for differentiating the writing styles of the authors. Various researchers proposed different types of features such as lexical features, structural features, content-based features, syntactic features, and semantic features to represent the documents [1].

Juan Soler Company et al. worked [2] on the corpus of New York Times opinion blogs. They tried with 83 features of character-based, word-based, dictionary-based, sentence-based, and syntactic features for gender prediction and achieved good accuracy of 82.83% using Bagging classifier with all combination of features. In their observation, the accuracy was reduced when 3000 words having most TFIDF values used as Bag of Words. Argamon et al. [3] experimented on blog posts of 19320 blog authors. They used content-based and style-based features. It was observed that the style-based features were most useful to discriminate the gender and Bayesian multinomial regression classifier obtained an accuracy of 76.1% for gender prediction.

Estival et al. experimented [4] on emails dataset and extracted 689 features of lexical character and structural features. Different machine learning algorithms such as IBK, JRip, Random Forest, J48, SMO, Bagging, libSVM, and AdaBoost were applied on the dataset. It was observed that SMO obtained best accuracy of 69.26% for gender prediction compared to other classifiers. In Koppel et al. [5], 566 documents are collected from the British National Corpus (BNC). They extracted 1081 features and achieved an accuracy of 80% for gender prediction using exponential gradient algorithm. In another work [6], they achieved better accuracy of 80.1% using multiclass real Winnow algorithm with 1502 features of content-based and stylistic features on 37478 blogs corpus.

In Dang Duc et al. work [7], they considered Greek blogs and extracted standard stylometric features and 300 most frequent character n-grams and word n-grams. In their observation, longer sequences of character n-grams and word n-grams are suitable to increase the accuracy of gender prediction. Sequential minimal optimization (SMO) classifier obtained a good accuracy of 82.6% for gender prediction. In another work [8], they extracted 298 features of word-based and character-based features from the Vietnamese blogs. It was observed that the character-based features contributed less to gender prediction than word-based features, and the classifier IBK obtained a good accuracy of 83.34% for gender prediction using combination of word- and character-based features.

3 Existing Approach

Most of the author profiling approaches used the Bag of Words approach to represent the document vector.

3.1 Bag of Words (BOW) Approach

In BOW approach, first the preprocessing techniques are applied on the collected dataset. Extract the most frequent terms or features that are important to discriminate the writing styles of the authors from the modified dataset. Consider these terms or features as bag of words. Every document in the dataset is represented with this bag of words. Each value in the document vector is the weights of the bag of words. The existing approaches for author profiling used frequency of bag of words to represent the document vector. We observed that only frequency is not effective to assign discriminative power to the bag of words. In this work, a new term weight measure is proposed to assign suitable weights to the bag of words. Finally, the document vectors are used to generate classification model.

3.2 Dataset Characteristics

The dataset was collected from www.tripadvisor.com, and it contains 4000 reviews about different hotels. The dataset is carefully tested, and to ensure its quality, the reviews written by the authors whose gender details were given in their user profile are only considered. The corpus is balanced in terms of gender profile for male and female profile groups containing 2000 reviews of each.

3.3 Evaluation Measures

The evaluation measures are used to evaluate the system performance. Various evaluation measures such as precision, recall, F1-measure, and accuracy are used by the researchers in author profiling to test the accuracy of gender prediction. In this paper, the accuracy measure is used to evaluate the performance of the classifier. Accuracy is the ratio of number of documents correctly predicted their gender and the number of documents considered.

4 Proposed Supervised Term Weight Measure

The existing term weight measures are Binary, TF, and TFIDF which are originated from information retrieval domain. Binary weight measure assigns 1 or 0 to the term based on the term presence or absence in a document. TF measure assigns weight to the terms based on the number of times the term occurred in a document. TF may assign large weights to the common terms (a, an, the, of, etc.,) which are weak in text discrimination. To overcome this shortcoming, inverse document frequency (IDF) is introduced in the TFIDF measure. IDF measure assigns more weight to the terms that occurred in less number of documents. Although the term frequency inverse document frequency (TFIDF) has been proved in information retrieval domain and several text mining tasks for quantifying the term weights, it is not most effective for text classification because TFIDF disregards the class label information of the training documents. Therefore, researchers are looking for alternative effective term weight measures in text categorization.

The term weight measures are categorized into two types such as unsupervised term weight measures and supervised term weight measures based on the usage of the class label information. An unsupervised term weight measure does not use any information regarding class label. The supervised term weight measure uses class label information in the process of assigning weights to the terms. In this work, a new supervised term weight measure is proposed.

The proposed term weight measure is represented in Eq. (1). In this measure, tf (t_i, d_k) is the frequency of the term t_i in document d_k, DF_k is the total number of terms in a document d_k, $DC(t_i, p_{male})$ is the number of documents in male profile, which contains term t_i, and $DC(t_i, p_{female})$ is the number of documents in female profile, which contains term t_i.

$$W(t_i, d_k \in p_{male}) = \frac{tf(t_i, d_k)}{DF_k} * \frac{\sum_{x=1, d_x \in p_{male}}^{m} tf(t_i, d_x)}{1 + \sum_{y=1, d_y \in p_{female}}^{n} tf(t_i, d_y)} * \frac{DC(t_i, p_{male})}{1 + DC(t_i, p_{female})} \quad (1)$$

$\sum_{x=1,\ d_x \in p_{male}}^{m} tf(t_i, d_x)$ gives the total count of the term t_i in all the documents of male profile.

$\sum_{y=1, d_y \in p_{female}}^{n} tf(t_i, d_y)$ gives the total count of the term t_i in all the documents of female profile.

The basic idea of this term weight measure is the terms which are having high frequency in a document, high frequency in positive category of documents, and the number of documents of positive category contains this term having more discriminative power to select positive samples than negative samples.

5 Empirical Evaluations

5.1 Results of Stylistic Features

The experiment started with a set of 21 stylistic features as a bag of words which include number of capital letter words, number of sentences, number of words, number of characters, number of capital letters, number of small letters, average sentence length, average word length, ratio of capital letters to small letters, number of words length greater than 6, number of contraction words, number of special characters, number of words followed by digits, the ratio of white space characters to non-white space characters, the ratio of number of words length greater than six to total number of words, the ratio of number of words length less than three to total number of words, the number of punctuation marks, the number of words with hyphens, ARI, LIX, RIX.

The 21 features of frequencies are used to represent a document vector. Naïve Bayes Multinomial and Random Forest classifiers are used to generate the classification model with these document vectors. Table 1 shows the accuracies of gender prediction on reviews dataset.

The Random Forest classifier achieved an accuracy of 54.53% for gender prediction. This accuracy is not a good accuracy. In the next section, experimentation is carried out with part-of-speech (POS) n-grams.

Table 1 Accuracy of gender prediction using 21 features

Number of features	Naive Bayes Multinomial	Random Forest
21 features	52.98%	54.53%

5.2 *Weighted Representation of Part-of-Speech (POS) N-Grams*

Part-of-speech (POS) n-grams are a type of syntactic features. Stanford POS tagger is used to assign tags to the text. The POS tags' information is represented as a contiguous POS tags, also known as POS n-grams. The most frequent POS unigrams, POS bigrams, POS trigrams are identified as bag of words. The frequencies of POS n-grams are exploited to represent a document vector. Table 2 represents the accuracies of gender prediction when POS n-grams are used.

The Naïve Bayes Multinomial classifier achieved 64.85% for gender prediction when most frequent 600 POS n-grams are used to represent a document. The POS n-grams also failed to increase the accuracy of gender prediction, but they achieved good accuracy than 21 features.

5.3 *BOW Model for Gender Prediction Using Most Frequent Terms as Features*

In this work, initially the experimentation is carried out with 1000 most frequent terms using various classifiers such as Naive Bayes Multinomial (Probabilistic), IBK (lazy), Logistic and Simple Logistic (functional), Random Forest (Decision Tree), and Bagging (ensemble/meta). The features were considered from 1000 to 8000 most frequent terms with an increment of 1000 frequent terms in each iteration.

The accuracies of gender prediction in BOW model using various classifiers are depicted in Fig. 1. The Simple Logistic classifier obtained best accuracy (69.25%) than all other classifiers when the number of features is 7000.

Table 2 Accuracies of gender prediction using POS n-grams

Number of features	Naïve Bayes Multinomial	Random Forest
200 POS n-grams	64.11	62.58
400 POS n-grams	64.32	62.79
600 POS n-grams	64.85	61.63
800 POS n-grams	64.48	60.62
1000 POS n-grams	64.42	60.89

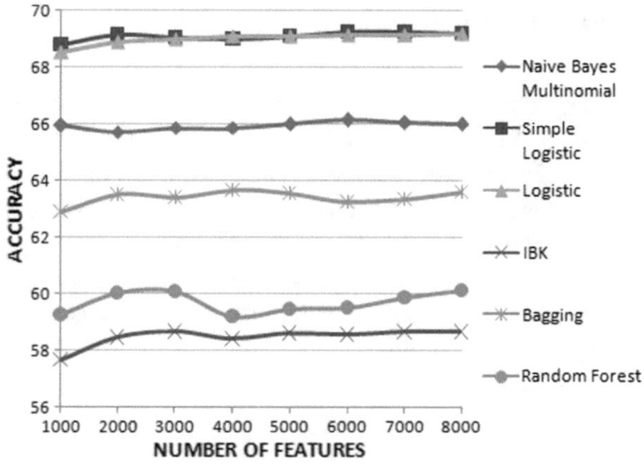

Fig. 1 Performance of BOW model for gender prediction

5.4 Proposed Supervised Term Weight Measure for Gender Prediction Using Most Frequent Terms as Features

The accuracies of gender prediction using proposed supervised term weight measure for various classifiers are displayed in Fig. 2. The Naïve Bayes Multinomial classifier obtained good accuracy than all other classifiers that are experimented in this work. All the classifiers' accuracy is increasing when the number of features is varying from 1000 to 8000. The proposed supervised term weight measure achieved an accuracy of 91.45% for gender prediction when the number of features is 8000.

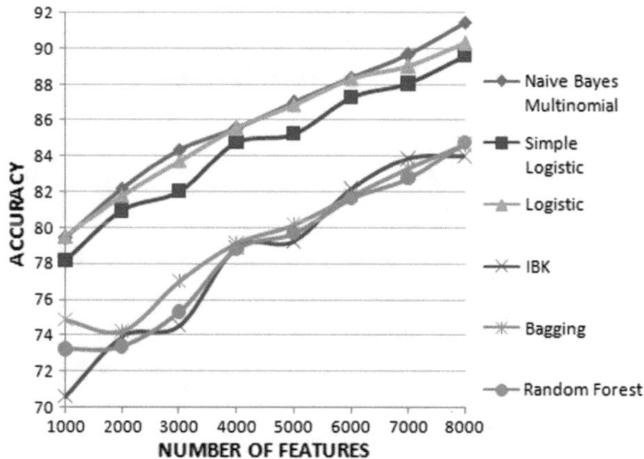

Fig. 2 Performance of proposed supervised term weight measure for gender

6 Conclusions and Future Scope

In this work, the experimentation is carried out with combination of stylistic features, POS n-grams, and proposed supervised term weight measure. The proposed supervised term weight measure achieved good accuracy of 91.45% for gender prediction than other features. The proposed term weight measure performs well than most of the existing approaches in author profiling.

In our future work, it is planned to predict other demographic profiles of the authors and also planned to propose a new term weight measure to increase the prediction accuracy of author profiles.

References

1. Reddy, T.R., Vardhan, B.V., Reddy, P.V.: A survey on authorship profiling techniques. Int. J. Appl. Eng. Res. **11**(5), 3092–3102 (2016)
2. Soler-Company, J., Wanner, L.: How to use less features and reach better performance in author gender identification. In: The 9th edition of the Language Resources and Evaluation Conference (LREC), pp. 1315–1319, May 2007
3. Argamon, S., Koppel, M., Pennebaker, J.W., Schler, J.: Automatically profiling the author of an anonymous text. Commun. ACM **52**(2), 119–123 (2009)
4. Estival, D., Gaustad, T., Pham, S.B., Radford, W., Hutchinson, B.: Author profiling for english emails. In: 10th Conference of the Pacific Association for Computational Linguistics (PACLING, 2007), pp. 263–272 (2007)
5. Argamon, KM.S., Shimoni, A.: Automatically categorizing written texts by author gender. In: Literary and Linguistic Computing, pp. 401–412 (2003)
6. Schler, J., Koppel, M., Argamon, S., Pennebaker, J.: Effects of age and gender on blogging. In: Proceedings of AAAI Spring Symposium on Computational Approaches for Analyzing Weblogs, March 2006
7. Dang Duc, P., Giang Binh, T., Son Bao, P.: Authorship attribution and gender identification in greek blogs. In: 8th International Conference on Quantitative Linguistics (QUALICO), pp. 21–32, April 26–29, 2012
8. Dang Duc, P., Giang Binh, T., Son Bao, P.: Author Profiling for vietnamese blogs. In: Asian Language Processing, 2009 (IALP '09), pp. 190–194 (2009)

Image Encryption Based on Arnold Cat Map and GA Operator

Subhajit Das, Satyendra Nath Mondal and Manas Kumar Sanyal

Abstract In this paper, three steps of encryption technique are used to encrypt image. At first, crossover and mutation operators of genetic algorithm are applied on each pixel of image to make modified image. The crossover points are decided from a random sequence generated by logistic map. Second, pixels of modified image are shuffled like transposition cipher based on Arnold cat map. Finally, the encryption image is obtained by performing logical operation between modified image and random sequence produced by logistic map. Some important tests have been performed on encrypted images to check security level of proposed algorithm. It has been observed that the proposed algorithm is able to resist various attacks and provide sufficient security, and it is giving better result among some of the published algorithms.

Keywords Logistic map · Arnold cat map · Crossover and mutation
Entropy analysis · Histogram analysis and correlation analysis

1 Introduction

In the age of globalization, data communication plays an important role and nowadays secure transmission of image data is becoming a real challenge to the researchers [1]. We have some efficient text encryption algorithm, but they are not suitable for image data because image data have some intrinsic features, such as

S. Das (✉)
Nayagram Bani Bidypith Nayagram, Nayagram 741235, India
e-mail: subhajit.batom@gmail.com

S. N. Mondal
Department of Information Technology, Kalyani Government Engineering College, Kalyani, Nadia, India
e-mail: satyen_kgec@rediffmail.com

M. K. Sanyal
Department of Business Administration, University of Kalyani, Kalyani 741235, Nadia, India

V. Bhateja et al. (eds.), *Intelligent Engineering Informatics*, Advances in Intelligent Systems and Computing 695, https://doi.org/10.1007/978-981-10-7566-7_3

huge amount of data, high redundancy and high correlation between its pixels in all directions. Many modern scientific theories and tools have been used by researchers to improve the efficiency of image encryption algorithms. In recent year, chaos theory-based image encryption has attracted a great deal of attention. Chaotic system has some important properties such as initial condition-based sensitivity, pseudorandomness and periodicity. These features are very much effective for image encryption. Most of the chaos-based encryption technique uses different types of chaotic maps to achieve a sufficient level of complexity, high security and expanded key space. Guan et al. [2] have used Arnold cat map and Chan map in their encryption algorithm. Fu et al. [3] proposed a novel shuffling algorithm which performs a bit-level permutation using two stages, Arnold cat map and chaotic sequence sorting. Soleymani et al. [4] proved that among many chaos map, Arnold cat map is most commonly used map that works with the main purpose of shuffling pixels of an image in a pseudorandom order. At present, most of the Arnold cat map-based algorithm uses one or more chaotic functions for pixel shuffling that increases the execution time of algorithm, but here apart from pixel shuffling, we proposed a bit shuffling phase based on genetic operator that shuffles the bits of pixel rather than pixel location.

2 Phases of Proposed Algorithm

The proposed encryption process consists of four phases, pseudorandom number generation, bit permutation, pixel permutation and pixel modification. In the first phase, logistic map has been used to generate two sets of random number that used in-bit permutations and in-pixel modification phase. Bit permutation phase is mainly based on two common genetic operators, i.e. crossover and mutation. Arnold cat map has been used to permute locations of each pixel. Both of these phases have been used to break the strong correlation-ship of pixels. A bitwise logical operation has been used in pixel modification phase to obtain a new cipher image.

2.1 Pseudorandom Number Generation

Logistic map is a general form of the chaotic map. It is a nonlinear polynomial of second degrees and can be expressed by using the following equation:

$$x_{n+1} = \mu x_n(1 - x_n) \tag{1}$$

where system parameter is defined by μ which lies between 0–4, x_n is map variable lies between 0–1, it is initial condition of logistic map, and n is the number of iteration used for generating iterative values. As the initial value of logistic map x_0 is taken and generate two series of m + n × n elements.

$$\mathbb{Z}_1 = \{floor(x_i \times 8) | 1 \le x_i \le m + n \times n\} \forall x_i \in [0, 8] \tag{2}$$

$$\mathbb{Z}_2 = \{floor(x_i \times 256) | m \le x_i \le m + n \times n\} \forall x_i \in [0, 255] \tag{3}$$

In order to avoid the harmful effect of the transition procedure, we choose mth element to (m + n × n)th element to construct two sets from the logistic chaotic equation.

2.2 *Bit Permutation Phase*

The greyscale image is a set of pixels represented by matrix $\mathbb{M}$

$$\mathbb{M} = \{a_{ij}\}_{n \times n}$$

Vertical folding of the image matrix exactly at the half line leads to the change of the elements of $\mathbb{M}$, such that

$$a_{ij} \leftrightarrow a_{ik}, \forall\, i = 1, 2, 3, \ldots, \text{n}; \, j = 1, 2, 3, \ldots, \frac{n}{2}; \, k = \frac{n}{2} + 1, \frac{n}{2} + 2, \ldots \text{n} - 2, \text{n} - 1, \text{n}$$

Greyscale value of every pixel is converted into 8-bit binary equivalent. Then, tools of genetic algorithm are used on the face-to-face greyscale value of the image to compute crossover and mutation operation, and a new image matrix is generated.

Eight-bit binary equivalent of each element of $\mathbb{M}$ can be represented as

$$a_{ij} = a_0^{ij} a_1^{ij} \ldots a_7^{ij}, \text{ and } a_{ik} = a_1^{ik} a_2^{ik} \ldots a_8^{ik} \; \forall\, i = 1, 2, 3, \ldots, \text{n}; \, j = 1, 2, 3, \ldots, \frac{n}{2};$$
$$k = \frac{n}{2} + 1, \frac{n}{2} + 2, \ldots \text{n} - 2, \text{n} - 1, \text{n where } a_1^{ij}, \ldots, a_1^{ik}, \ldots = 0 \text{ or } 1$$

The crossover operation may change the elements as shown below: *l*st position crossover results in

$$a'_{ij} = a_0^{ij} a_1^{ij} \ldots a_l^{ij} a_{l+1}^{ik} a_{l+2}^{ik} \ldots a_7^{ik}$$
$$a'_{ik} = a_0^{ik} a_1^{ik} \ldots a_l^{ik} a_{l+1}^{ij} a_{l+2}^{ij} \ldots a_7^{ij}$$

The index l may take any value between 0 and 7, and value of l has been taken from $\mathbb{Z}_1$ in sequential order.

The mutation operation (here last bit NOT operation) on the resulting elements after the crossover operation leads to

$$a''_{ij} = a_1^{ij} a_2^{ij} \ldots a_8^{ij} a_1^{ij} a_2^{ij} \ldots \sim a_8^{ij}$$
$$a''_{ik} = a_1^{ik} a_2^{ik} \ldots a_8^{ik} a_1^{ik} a_2^{ik} \ldots \sim a_8^{ik}$$

where $\sim a_8^{ij}$ = NOT anda_8^{ij} & $\sim a_8^{ik}$ = NOT a_8^{ik}.

New image matrix after vertical folding and subsequent crossover and mutation operation is $\mathbb{M}_1 = \left\{a''_{ij}\right\}_{m \times n}$.

Following the vertical folding in the next step, horizontal folding on the new image matrix again exactly at the half line leads to

$$b_{ij} \leftrightarrow b_{kj}, \forall\, i = 1, 2, 3, \ldots, \frac{m}{2}; j = 1, 2, 3, \ldots, \mathrm{n};\ k = \frac{m}{2} + 1, \frac{m}{2} + 2, \ldots \mathrm{m} - 2, \mathrm{m} - 1, \mathrm{m}$$

Then, the crossover and mutation result in new image matrix $\mathbb{M}_2 = \left\{b''_{ij}\right\}_{m \times n}$.

2.3 *Pixel Permutation Phase*

In the proposed method, permutation of pixel is based on Arnold cat map. Russian mathematician Vladimir I discovered Arnold cat map. It is a 2D transformation based on a matrix with a determinant of 1 that makes this transformation reversible. The transformation is defined as

$$\begin{bmatrix} x' \\ y' \end{bmatrix} = \begin{bmatrix} 1 & p \\ q & pq+1 \end{bmatrix} \begin{bmatrix} x \\ y \end{bmatrix} mod\, n \tag{4}$$

Here, (x, y) is the location of the pixel which is mapped to a new position (x', y'), and $p\, and\, q$ are integers.

Original order of pixels in plain image is shuffled by this transformation matrix. Reconstruction of original image from shuffled image is possible after a sufficient number of iterations. In decryption phase, we obtain original image from shuffled image using reverse mapping matrix presented in (5).

$$\begin{bmatrix} x \\ y \end{bmatrix} = \begin{bmatrix} pq+1 & -p \\ -q & 1 \end{bmatrix} \begin{bmatrix} x' \\ y' \end{bmatrix} mod\, n \tag{5}$$

In our algorithm, each pixel of diffused image matrix $\mathbb{M}_2$ is permuted for a specified number of r rounds with specified parameters p and q so each b''_{ij} is converted into $b''_{I,j}$ where

$$b''_{ij} = b''_{IJ} \forall i \neq I; j \neq J;$$

And we obtained a new image matrix $\mathbb{M}_3 = \{b''_{IJ}\}_{m \times n}$.

2.4 Pixel Modification Phase

At this point, logical XOR operation is performed between elements by elements of diffused image matrix $\mathbb{M}_3 = \{b''_{IJ}\}_{n \times n}$ and element by element with the elements of set $\mathbb{Z}_2$.

$$c_{ij} = \{b_{IJ} \oplus \mathrm{y}\colon \forall\, i = 1, 2, 3, \ldots, \mathrm{n}; j = 1, 2, 3, \ldots, n;\ y = \mathrm{m}, \mathrm{m}+1, \mathrm{m}+2, \ldots, \mathrm{m}+\mathrm{n}^*\mathrm{n}\}$$

In this way, the image is encoded in the form $\mathbb{M}_4 = \{c_{ij}\}_{m \times n}$.

3 Proposed Image Encryption and Decryption Algorithm

3.1 Algorithm: Key Generation

Input: Initial value of logistic map $(x_{0,}\ \mu)$ and a integer (m).
Output: Two sets of integer numbers.
Method
Step 1. To generate a set of random number using a logistic Eq. (1)
Step 2. To generate two series of m + n × n elements $(\mathbb{Z}_1\ \mathbb{Z}_2)$ as (2) and (3).

3.2 Bit Permutation Phase

Input: An greyscale image, a sets of random numbers. $(\mathbb{Z}_1)$
Output: Distorted image.
Method
Step 1. To fold the image vertically.
Step 2. To apply genetic algorithm.
Step 2.1. To select the grey value of pixels placed face to face after vertical folding as initial parent.
Step 2.2. To convert grey value of the pixels into 8-bit numbers.

Step 2.3: To perform crossover operation between two parents to generate two children using at crossover point taking from set $(\mathbb{Z}_1)$ sequentially.
Step 2.4. To take two new children as new parents.
Step 2.5. To perform step 2.3. and step 2.4. from l to LSB.
Step 2.6. To perform mutation operation between two new generated parents at LSB.
Step 2.7. To repeat the step 2.1 to 2.6 for all face-to-face pixels of folding image.
Step 2.8. To reconstruct the image by unfolding.
Step 3. To fold the image horizontally and apply the step 2.

3.3 Pixel Permutation Phase

Input: Distorted image, parameters of Arnold cat map (p, q), number of iterations(r).
Output: Extremely destroyed image.
Step 1. For each pixel of destroyed image do.
Step 1.1 Input its location in Arnold cat map and get a new location.
Step 1.2 Swap value of pixels with old and new locations.
End of for.
Step 2. Continue step 1 for r times.

3.4 Pixel Modification Phase

Input: Extremely destroyed image. A set of random numbers $(\mathbb{Z}_2)$
Output: Encrypted image.
Step 1. To perform bitwise XOR between elements of set $\mathbb{Z}_2$ taking in sequential order and grey value of pixels starting from left corner of the extremely destroyed image.

3.5 Algorithm Image Decryption

Input: Encrypted image, initial parameters of logistic map, Arnold cat map and number of iterations.
Output: Decrypted image.

As it is a symmetric key image encryption, we perform all the steps in reverse order to obtain original image.

4 Experimental Result

In this section, we analyse the performance of our proposed method. Three benchmark greyscales image-lena, babun and cameraman with size 256 × 256 have been tåken for our experiment. Here, key value is in from $K = \{x_0, \mu, m, p, q, r\}$ where x_0 is the initial condition, μ is system parameter, m is number of element from which set $\mathbb{Z}_2$ is constructed, $p\, and\, q$ are Arnold cat parameter, and r is number of iterations of cat map. For result analysis, our proposed key $K = \{0.86541, 3.9987, 254, 86, 85, 8\}$.

A stepwise outcome has been furnished in Table 1 that proves no similarity between cipher images.

4.1 Histogram Analysis

Histogram of an image states the statistical nature of pixels of plain image out of the encrypted image. Table 2 shows histograms of different cipher images which is very flat, and it proves statistical attack is not effective to proposed algorithm.

Table 1 Output of different stages

Input image	Output of different stages		
	Bit permutation	Pixel permutation	Pixel modification

	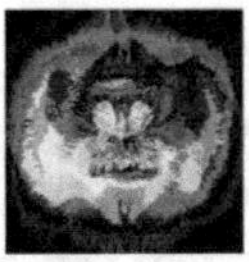		

Table 2 Histogram analysis of different input images

Input image	Histogram (plain image)	Histogram (cipher image)
	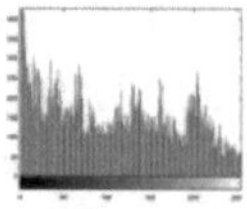	
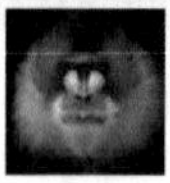	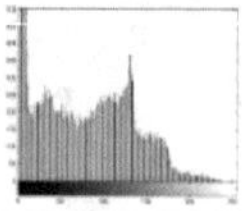	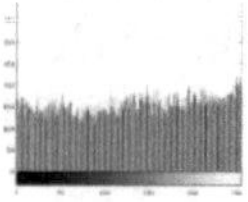
	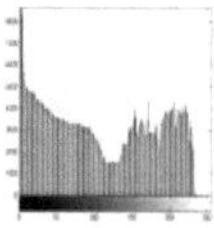	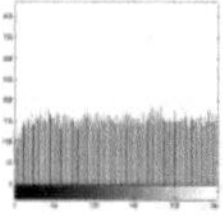

4.2 Information Entropy Analysis

In a true random system, 2^8 symbols will present with equal probability, i.e. $m = \{m_1, m_2, m_3, \ldots, m_{2^8}\}$, for bit depth 8. Information entropies of ciphered images obtained by the proposed algorithm are listed in Table 3, from the table; the ciphered images obtained by the proposed algorithm could hardly divulge information.

4.3 Correlation Analysis

In this paper, 1500 pairs of two adjacent (in horizontal, vertical, and diagonal directions) pixels from plain and encrypted images are tested. The correlation coefficient of plain and encrypted images has been calculated as nearly 0.0097 and 1. The correlation coefficients of some sample images have been furnished in Table 4.

Table 3 Entropy of plain images and encrypted images

Image	Plain image	Ciphered image
Lena	7.7460	7.9954
Babun	7.4069	7.9937
Camera man	7.7684	7.9898

Table 4 Correlation coefficient of two adjacent pixels in plain image and encrypted image

Image name	Plain image			Encrypted image		
	Vertical	Horizontal	Diagonal	Vertical	Horizontal	Diagonal
Lena	0.9374	0.9692	0.9107	0.0055	0.0012	0.0035
Babun	0.9833	0.9823	0.9109	0.0008	0.0032	0.0027
Cameraman	0.9809	0.9792	0.966	0.0005	0.0045	0.0017

Table 5 Comparisons with other algorithms

Image		Ref.[7]	Ref.[8]	Our Algorithm
	Correlation coefficient	0.0129	-0.0164	-0.0082
	Entropy	7.9929	7.9896	7.9938
	Entropy	7.9948	7.9882	7.9958
	Correlation coefficient	.0093	-0.00027	0.0048

5 Comparative Study

Performance of proposed algorithm has been compared between ref [5, 6]. The comparative study has been made between two images of size (128 × 128), and the result has furnished in Table 5.

6 Conclusion

Our proposed image encryption algorithm is based on simple logistic equation, genetic operator and Arnold cat map. Logistic map is used for random numbers. Arnold cat map forces for pixel shuffling and genetic operator focused on bit shuffling. The proposed method is a symmetric key image encryption algorithm. The efficiency of proposed algorithm has been tested by several statistical tests. Experimental results state that encrypted images have very low neighbouring pixel correlations at different levels, almost flat and uniform distribution of pixels at each level that can be considered as a nearly random image. A large key space proves that our proposed method is sensitive to the plain image and the encryption key. The algorithm is giving better result among some of the published algorithm.

References

1. Shujun, L., Xuanqin, M., Yuanlong, C.: Pseudo-random bit generator based on couple chaotic systems and its application sin stream-cipher cryptography. In: Progress in Cryptology—INDOCRYPT 2001, vol. 2247 of Lecture Notes in Computer Science, pp. 316–329 (2001)
2. Guan, Z., Huang, F., Guan, W.: Chaos-based image encryption algorithm. Phys. Lett. A: Gen. Atomic Solid State Phys. **346**(1–3), 153–157 (2005)
3. Fu, C., Lin, B., Miao, Y., Liu, X., Chen, J.: A novel chaos-based bit-level permutation scheme for digital image encryption. Opt. Commun. **284**(23), 5415–5423 (2011)
4. Soleymani, A., Nordin, M.J., Sundararajanl, E.: A chaotic cryptosystem for images based on henon and Arnold cat map. Sci. World J. **2014**, 21. (Article ID 536930)
5. Das, S., Mandal, S.N., Ghoshal, N.: Diffusion and encryption of digital image using genetic algorithm. In: Ficta 2014 Bhubaneswar. AISC Springer. vol 327 pp. 729–736
6. Sethi, N., Vijay, S.: Comparative image encryption method analysis using new transformed—mapped technique. In: Conference on Advances in Communication and Control Systems 2013 (CAC2S 2013), pp. 46–50

SVM Based Method for Identification and Recognition of Faces by Using Feature Distances

Jayati Ghosh Dastidar, Priyanka Basak, Siuli Hota and Ambreen Athar

Abstract In this paper, a scheme was presented to identify the locations of key features of a human face such as eyes, nose, chin known as the fiducial points and form a face graph. The relative distances between these features are calculated. These distance measures are considered to be unique identifying attributes of a person. The distance measures are used to train a Support Vector Machine (SVM). The identification takes place by matching the features of the presented person with the features that were used to train the SVM. The closest match results in identification. The Minimum Distance Classifier has been used to recognize a person uniquely using this SVM.

1 Introduction

The need for uniquely identifying a human being by using their biological traits is a long foregone conclusion. Hence the interest in the study of biometric authentication systems is obvious. Over the years, natural attributes such as fingerprints, retinal scans, iris imprints and acquired attributes such as voice, handwriting, gait have been studied and used widely for the purpose of authenticating human beings. The problem of automatic face recognition involves three key steps/subtasks: (1) detection of faces, (2) feature extraction, and (3) identification and/or verification. The first step deals with detecting a human face in an image or a video frame.

J. G. Dastidar (✉) · P. Basak · S. Hota · A. Athar
Department of Computer Science, St. Xavier's College, Kolkata, India
e-mail: jghoshdastidar@gmail.com

P. Basak
e-mail: mailpriyankabasak20@gmail.com

S. Hota
e-mail: siuli3335@gmail.com

A. Athar
e-mail: ambreenathar.0704@gmail.com

V. Bhateja et al. (eds.), *Intelligent Engineering Informatics*, Advances in Intelligent Systems and Computing 695, https://doi.org/10.1007/978-981-10-7566-7_4

Once a face has been detected, the second step deals with the extraction of features from the detected face. The final step involves using the extracted features to identify a person uniquely or verify whether the face belongs to a purported owner or not.

We have designed a system which is able to detect a face in an image or a video frame. The positions of key features (landmarks) such as eyes, nose tip, and chin in the face are then located. The positions of these extracted features are known as fiducial points. The distances between these points have been used by us to train a Support Vector Machine (SVM). Later, in order to identify a person uniquely, the Minimum Distance Classifier (MDC) has been used to look for a match in the training database of the SVM.

In this paper, Sect. 2 does a literature survey of related works, Sect. 3 discusses the detailed design methodology that has been followed, and Sect. 4 presents the results obtained from our implementation along with related discussion. Finally, Sect. 5 puts forth the concluding remarks.

2 Literature Survey

The different techniques that may be used for the facial recognition systems are eigenfaces, neural network, Fisher faces, elastic bunch graph matching, template matching, geometrical feature matching, and landmark localizing. In this paper, the landmark localization has been used. Landmark defines the geometry of the face, and location is represented by the coordinate of corresponding pixel. An example of this is the nose tip, which has a well-defined location. Landmark features are defined by the frequency information of the local regions that surround the landmark locations. Landmarks are parts of the face which are easily located and have similar structure across all faces. Some obvious examples of landmarks are the eyes, nose, and mouth. Each of these landmarks is well defined, common to all faces and has distinct representations in image space.

Typically, landmarks are defined such that their location has a very small error tolerance. Instead of defining a landmark as the "nose," the landmark is defined as the "nose tip." The nose tip can be located within a few pixels. The nose in its entirety is a large structure, and so difficult to select a single point to represent its location. There are other parts of the face that would serve as poor landmarks. Some examples of these would be the forehead or cheeks. Although these are common structures of the face, it is very difficult to determine an exact location for these structures. Even if one could define such a point, such as a cheekbone, it is very difficult to locate it to a high degree of accuracy because there may be no heuristic in the image that directs the algorithm to such a point.

The authors in [1] selected 26 landmark points on the facial region to facilitate the analysis of human facial expressions. They applied Multiple Differential Evolution-Markov Chain (DE-MC) particle filters for tracking these points through the video sequences. A kernel correlation analysis approach was proposed by the authors to find the detection likelihood by maximizing a similarity criterion between the target points and the candidate points [1].

Automatic feature extraction and face detection were done based upon local normalization, Gabor wavelets transform, and Adaboost algorithm in [2]. The authors incorporated normalization using local histogram with optimal adaptive correlation (OAC) technique to make the detection process invariant to fluctuating illumination levels [2].

The authors in [3] have presented a comprehensive survey on all of these methods and more. A model based on 3-D Point Distribution Model (PDM) that was fitted without relying on texture, pose, or orientation information was proposed in [4]. Face detection was performed by classifying the transformations between model points and candidate vertices based on the upper bound of the deviation of the parameters from the mean model [4].

The effectiveness of landmarks and their accompanying geometry was studied in depth in [5]. Different statistical distances such as the norm, Mahalanobis distance, Euclidean distance, eigenvalue-weighted cosine (EWC) distance, Procrustes distance were studied, and they were statistically correlated for analyzing and detecting facial features [5].

The authors in [6] have achieved an 84.5% recognition rate by using learning-based descriptors. Principal Component Analysis (PCA) was used to improve the discriminative ability of the descriptors [6]. A pose-adaptive matching method was proposed in [6] to make the system compatible with different poses.

The concept of landmarks has been ably explained and discussed in [7]. The authors have presented a comparative study of different techniques using which landmarks may be extracted and utilized in real time. Efficient extraction techniques are particularly important when the problem at hand deals with detection and recognition from a video clip at real time [7].

3 Design Methodology

Face recognition systems work by trying to find a match for a given face in a pre-created database. Thus, all face recognition systems consist of two distinct sub-systems, viz., the creation of the database, also known as the *Registration* phase and *Identification* phase of the given face. The registration module involves the process of registering a person's face into a database. The identification module takes the help of the database created during registration and tries to find a match for a test image in it.

During the registration phase, features are extracted from the image of a person's face. These features include a combination of landmark/fiducial points and some distance parameters, together forming a face graph. The extracted features are then stored in a database. During the identification phase, the image of an unknown face

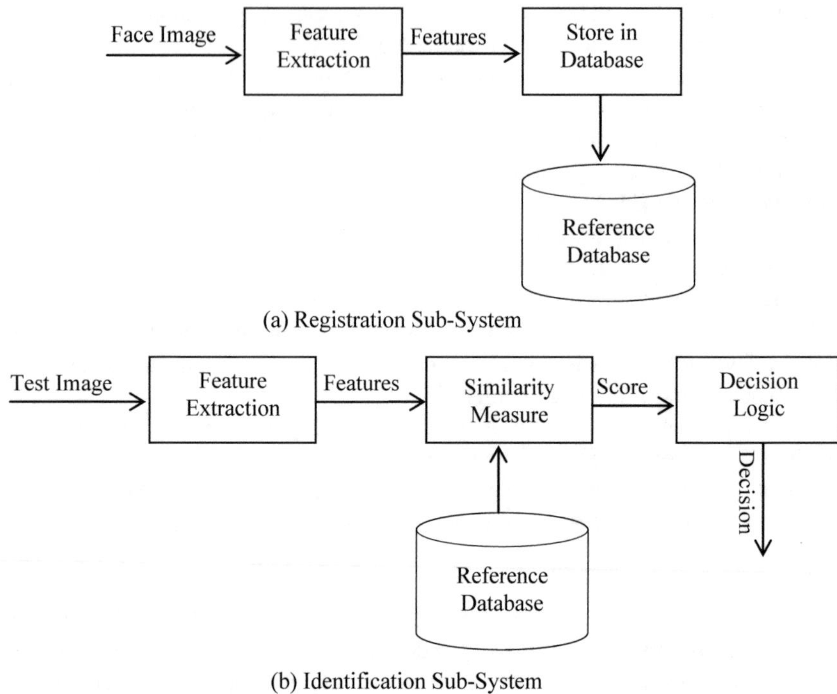

Fig. 1 Registration and identification sub-system

is presented to the system. The system then computes a face graph for the face. It finally tries to find a match for the presented face with the preregistered faces stored in the database. The registration and identification phases have been explained in Fig. 1.

3.1 Feature Extraction

As is obvious from Fig. 1, both the registration and identification modules rely upon an efficient *Feature Extraction* process. This process works by first detecting and localizing landmark points such as the iris of the two eyes, nose tip, and lip center. The process has been discussed below and the outcome is shown in Fig. 2.

- *Feature Detection*

Here, the input image is accepted and facial landmark features like left eye, right eye, nose, and lips are detected. Detected features are the highlighted using a green rectangular box for each. Figure 2a shows the highlighted landmarks.

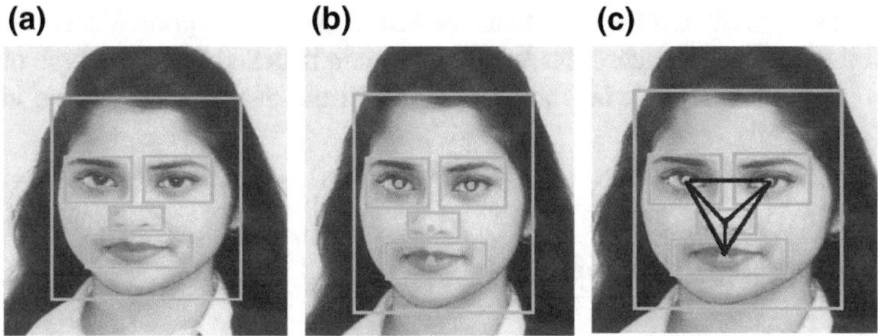

Fig. 2 **a** Extracting facial features; **b** localizing fiducial points; **c** making face graph (The subject of the image is the second author of the paper and has given her consent for her photo to be used.)

- *Localization of Fiducial Points of Each of the Features*

In this phase, midpoints of the rectangular boxes are calculated. These midpoints act as the fiducial points of the face. The midpoint of a rectangle is the bisection point of any of the two diagonals present in it, which can be easily obtained, given the coordinates of the four corners. Thus, if the two opposite ends of a rectangle have the coordinates (x_1, y_1) and (x_2, y_2), respectively, then the coordinate of the fiducial point is expected to be within a radius of one/two pixels of the coordinate, $\left(\frac{x_1+x_2}{2}, \frac{y_1+y_2}{2}\right)$. The fiducial points can thus be localized for all the landmarks. Figure 2b shows the localized fiducial points marked with green dots.

- *Making the Face Graph*

In this final step of feature extraction, the distance between each of the fiducial points is calculated to obtain a face graph. We consider 6 such distances and together these distances comprise the features obtained from a face, as illustrated in Fig. 3. Every face is thus represented by a set of distances. The face graph is considered to be unique for a person [5].

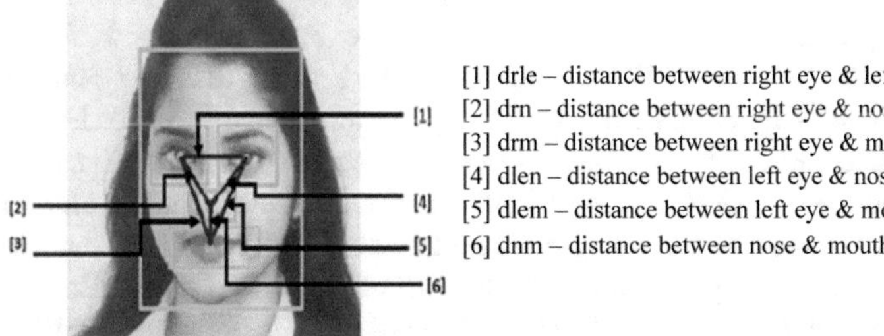

Fig. 3 Calculating distance between fiducial points (The subject of the image is the second author of the paper and has given her consent for her photo to be used.)

This "face graph" has been exploited and used to train a Support Vector Machine (SVM) during the registration phase and has again matched the face graph of a test person (to be identified) with the ones stored in the database during the identification phase.

3.2 Registration Sub-system

The process of registering a person involves storing the extracted features of his/her face into a database. Features obtained from multiple face images of a person are stored in the database. Storing multiple records for one person gives allowances for different poses, different illumination levels, and different distances from the camera. In our system, the features come down to the computed distances as shown in Fig. 3. Storing the distance parameters instead of the face images themselves also helps us save memory space. Later on during the identification phase, these distance parameters would again be required for the purpose of matching and they would be readily available from the database. Thus, bulky face matrices would not have to be dealt with repeatedly, saving on time and space. The distance parameters which have been considered are as follows:

a. drle—distance between right eye iris and left eye iris
b. drn—distance between right eye iris and nose tip
c. drm—distance between right eye iris and lip center
d. dlen—distance between left eye iris and nose tip
e. dlem—distance between left eye iris and lip center
f. dnm—distance between nose tip and lip center

3.3 Identification Sub-system

The identification module accepts a test image to be identified. It takes the help of the database produced in the previous module. As explained in Fig. 1, features are extracted for the test image. These features are the set of 6 distance parameters as explained above. The features extracted from the test image now need to be matched with the features of the registered faces. The success of the identification phase depends upon the choice of a precise *Similarity Measurement* routine along with a deterministic *Decision-making Logic*. The identification phase has been designed along the lines of a Support Vector Machine (SVM). Using a supervised learning model like SVM aids in the classification of an unknown pattern into an appropriate pattern class [6]. Similar methods have been successfully used in the field of audio processing [8]. In the proposed system, separate pattern classes have been defined for the different registered persons. Each class contains multiple vectors representing different face images of the same person. The attributes of the vectors are the 6 distance parameters *drle, drn, drm, dlen, dlem,* and *dnm*. Our SVM works as follows [9]:

a. All pattern vectors of each class are extracted from the database and stored in respective variables.
b. The mean vector for each class is calculated.
c. The decision-making function that has been used by us is a Minimum Distance Classifier (MDC). This function makes use of the Euclidean distance to arrive at the classification decision. The Euclidean distance is found for each pair: mean vector of ith class, m_i, and unknown pattern vector, x. It is given by $\|x - m_i\|$. Given two vectors A and B containing n attributes each, the Euclidean distance between A and B is calculated using (1).

$$\|A - B\| = \sqrt{\sum_{i=1}^{n} (A_i - B_i)^2}. \tag{1}$$

Here, A_i and B_i are the ith attributes of the vectors A and B, respectively. In our case, the attributes would be the 6 distance parameters for each face computed previously.

d. The unknown pattern belongs to the class for which the distance is the least.

This method of classification is known as the Minimum Distance Classification scheme. Lesser the value of the Euclidean Distance of the input test image from a class is, greater is the chance of the image belonging to the class; as the resemblance of the face with the class is more. Thus, the class for which the distance is the minimum may be the class to which the input test image belongs to. The input test image may be assumed to belong to the person the class is representing.

4 Results and Discussion

The system has been implemented using the NetBeans IDE for Java. For core image processing activities, OpenCV has been used. OpenCV can be easily integrated with the open-source environment provided by NetBeans. Multiple image samples from different persons were subjected to the processing. Table 1 shows the 6 distance parameters (rounded off to 2 decimal places) obtained for four different samples for one person. The mean vector representing the person X is shown in Table 2.

The Euclidean distances of a test image from 6 different persons have been shown in Table 3.

As seen from the table, the test image has the minimum value for the Euclidean distance with person X. Hence, it may be assumed that the test image is that of

Table 1 Pattern class for person X

Sample	drle	Drn	drm	dlen	dlem	dnm
1	13925.0	5070.25	10104.25	7897	301.25	5460.25
2	13693.0	4981.0	13581.25	8082.5	429.25	5554.25
3	13576.25	5162.0	11969.0	7837	432.5	4076.5
4	14529.25	4857.25	12121.25	5566.25	457.5	4872.5

Table 2 Mean vector representing person X

drle	drn	Drm	dlen	dlem	dnm
13930.88	5017.63	11943.94	7345.69	405.13	4990.88

Table 3 Euclidean distances

Person name	Euclidean distance
X	2719.15
Y	8804.17
Z	22450.43
W	9375.35
U	9982.96
V	22075.52

person X. The designed system has behaved relatively well with most samples baring a few exceptions. The false accept ratio (FAR) and the false reject ratio (FRR) were calculated with a population size of over 100. The figures obtained for FAR and FRR, respectively, are 0.25 and 0.15.

5 Conclusion

In this paper, a scheme has been presented for registration and identification of faces of different persons. The face graph has been used to obtain features of the face image samples. These features were then used to train an SVM. The final classification was done by using Minimum Distance Classifier. The results obtained have been analyzed by calculating the FAR and the FRR. The values show that the technique that has been adopted has shown promising results. The system has also worked for bespectacled persons. The instances of false accepts or false rejects are few. However, the few times that the system has misbehaved are because of several factors. To begin with, the Minimum Distance Classifier (MDC) has been solely used for the purpose of matching. The performance of the system is expected to improve by combining MDC with statistical parameters such as correlation. The conditions under which the face images are captured also play a vital role in all face recognition systems. Thus, illumination levels, angle of placement of the camera, distance of the subject from the camera lens, and face profile presented to the camera (side or frontal) determine the correctness of the extracted features. The deviations that have been observed are because of all of these factors.

The designed system may be used in any environment where authentication is based on face recognition. The system may be further enhanced by combining the matching process with body temperature sensing and pulse sensing devices.

Declaration
Requisite informed consent has been taken by authors prior to usage of the photograph in simulations.

References

1. Tie, Y., Guan, L.: Automatic landmark point detection and tracking for human facial expressions. J. Image Video Proc. **2013**, 8 (2013). https://doi.org/10.1186/1687-5281-2013-8
2. Yun, T., Guan, L.: Automatic face detection in video sequences using local normalization and optimal adaptive correlation techniques. Pattern Recogn. **42**(9), 1859–1868 (2009)
3. Zhao, W., Chellappa, R., Phillips, P.J., Rosenfeld, A.: Face recognition: a literature survey. J. Comput. Surv. (CSUR) **35**(4), 399–458 (2003). https://doi.org/10.1145/954339.954342
4. Nair, P., Cavallaro, A.: 3-D face detection, landmark localization, and registration using a point distribution model. IEEE Trans. Multimed. **11**(4), 611–623 (2009). https://doi.org/10.1109/tmm.2009.2017629
5. Shi, J., Samal, A., Marrx, D.: How effective are landmarks and their geometry for face recognition? Comput. Vis. Image Underst. Elseivier **102**(2), 117–133 (2006)
6. Cao, Z., Yin, Q., Tang, X., Sun, J.: Face recognition with learning-based descriptor. Comput. Vis. Pattern Recogn. (CVPR), IEEE. (2010) https://doi.org/10.1109/cvpr.2010.5539992
7. Celikutan, O., Ulukaya, S., Sankur, B.: A comparative study of face landmarking techniques. EURASIP J. Image Video Process. (2013). Springer. https://doi.org/10.1186/1687-5281-2013-13
8. O'Shaughnessy, D.: Automatic speech recognition: history, methods and challenges. Pattern Recogn. **41**(10), 2965–2979 (2008). Elseiveier
9. Gonzalez, R.C., Woods, R.E.: Digital Image Processing. 3rd. ed. Pearson Education (2009)

Content-Aware Reversible Data Hiding: URHS

Velpuru Muni Sekhar, P. Buddha Reddy and N. Sambasiva Rao

Abstract This paper demonstrates a Reversible Data Hiding (RDH) technique by considering content-aware (region) information. This technique uses uniform region of histogram shift (URHS) technique to embed secret data into cover-content. Proposed technique can embed more data compared to traditional Histogram Modification (HM)-based RDH techniques. The experimental results also proved that image visual quality in terms of Peak Signal-to-Noise Ratio (PSNR) is more compared to RDH techniques based on HM. However, hybrid HM techniques such as histogram shift (HS) and differential expansion (DE), HS and prediction-error expansion (PEE), DE, PEE have more embedding capacity (EC) and approximately same PSNR. Computational cost of proposed technique is less compared to other techniques. Moreover, the experimental results suggest that proposed technique is more significant to less number of edges.

1 Introduction

Data embedding is a process to hide data into cover-content such as image, video, audio, [1, 2]. Digitization of human society made data hiding as an essential tool for providing steganography and watermarking. A traditional data hiding contains the following components: (1) cover-content 'C,' (2) stego-image 'stego,' (3) secret message 'M,' (4) secret info 'K,' (5) embedding algorithm 'Em,' and (6) extraction algorithm 'Ex' [2]. Embedding is a process which combines two sets of data into one called as embedding process (1).

V. Muni Sekhar (✉) · P. B. Reddy · N. S. Rao
Vardhaman College of Engineering, Hyderabad, India
e-mail: munisek@gmail.com

P. B. Reddy
e-mail: buddhareddy.polepelli@gmail.com

N. S. Rao
e-mail: snandam@gmail.com

V. Bhateja et al. (eds.), *Intelligent Engineering Informatics*, Advances in Intelligent Systems and Computing 695, https://doi.org/10.1007/978-981-10-7566-7_5

$$Stego = Em(C, M) \tag{1}$$

A process that separates secret message from stego-object with secret info is called as extractions process (2).

$$[C', M'] = Ex(\text{stego, k}) \tag{2}$$

Depending on the embedding process input data and extraction process output data, data hiding techniques are classified into three categories: RDH [3–16], Near Reversible Data Hiding (NDH) [17], and Irreversible Data Hiding (IDH) [18].

Definition 1 ***RDH***: *Data hiding is said to be reversible, iff it extracts the concealed data or cover-content exactly after the extraction.*

Definition 2 ***NDH***: *Data hiding is said to be near-reversible, iff it extracts the concealed data or cover-content approximately after the extraction.*

Definition 3 ***IDH***: *Data hiding is said to be irreversible, iff it cannot be able to extract the concealed data or cover-content after the extraction.*

Most of existing data hiding techniques have the following challenges:

(1) Invert back the cover-content after extraction. Medical diagnosis [19], law order [3], etc., applications, it is a desirable feature [3, 5, 7, 13, 19, 20].
(2) Trade-off between the *EC*, *PSNR,* and *Complexity* (time complexity) is also a difficult task [2, 7].
(3) Considering the cover-content information while embedding secret message into cover-content. It is a desirable feature to limit the smoothing effect at edges and flipped pixel problems [2].

This paper demonstrates a RDH technique based on URHS. It considers the region information while embedding data and improves EC with less distortion in visual quality. This technique also proves that embedding capacity and visual quality are not always inversely proportional.

Structure of the paper is as follows: Sect. 2 discusses related work on HM-based RDH. Section 3 discusses proposed technique, embedding and extraction processes and algorithms. Section 4 discusses result analysis and comparison with existing techniques. Finally, Sect. 5 summarizes the paper outcomes.

2 Related Work

RDH is a new direction in data hiding research. RDH techniques are generally three types: (1) HM-based techniques, (2) differential expansion-based techniques, and (3) quantization/compression-based techniques. Among that, this paper focuses on HM-based techniques. Thus, HM data hiding techniques are described as follows:

2.1 Histogram Modification

Ni [3] proposed a new category of RDH based on HM. It can embed data into cover-content while maintaining good visual quality (PSNR). Here, the PSNR is calculated between cover-content and stego-image. HM is also famous for its ease of implementation and less computational complexity [3, 5]. In embedding, the histogram of cover-content is slight modified to hide the data. In extraction, with the reverse operations corresponding to HM, both the cover-content and embedded data can be extracted exactly.

2.2 Literature

Vleeschouwer et al. [19] introduced HM technique based on circular interpretation histogram bin shift-based watermarking. Then [20] improved [19] by using bijective transformations. In 2006, Ni [3] introduced a reversible watermarking technique using a set {*peak*, *zero*} points in histogram, to increase EC of HM watermarking techniques. Lin et al. [4] reported 'a multilevel reversible watermarking technique that uses the histograms of different images for data embedding.' Ni et al. [5] broadly investigated the [20] approach to optimize results (EC, PSNR). Hence, [5] concluded their investigation by tackling modulo-256 addition overflow and underflow problem in [20] which lead to smoothing effect and salt-and-pepper noise problems. In [5], Ni introduced a technique to limit the smoothing effect and salt-and-pepper noise problems caused due to modulo-256. Moreover, in [6] Gao et al., mentioned the failures of [5] and enhanced the [5] Ni et al.'s work by 'Reversibility improved lossless data hiding.' Present HM-based techniques hide data into selected fields of histogram block by modular addition/subtraction operation [3, 10, 11], and it is also prone to flipped pixel problem [13]. Recently, Li et al. [14–16] introduced a hybrid RDH technique based on HM and PEE. In all HM technique, EC is equal to peak point or pair of point.

Existing histogram-based techniques manifest the subsequent merits, i.e., (1) Content-adaptive data embedding [10, 11, 14–16] and (2) Reduce distortion in stego-object [3, 14]. Furthermore, HM-based techniques manifest the subsequent demerits, i.e., (1) Increased pixel modification in histogram shift [3, 5, 10, 11, 14–16] and (2) Low EC [3, 5]. Best of our knowledge on a content-aware RDH technique based on HM including all merits and limiting all demerits, no such techniques exist in literature.

2.3 Histogram Modification Embedding Process

Embedding process of HM technique has following steps:

(1) Generate the histogram

$$h(x) = \#\{1 \leq i \leq m \times n;\ x_i = x\} \forall x \in [0, 255] \tag{3}$$

where 'h(x)' is the histogram bins and x_i is the ith pixel intensity. The pixel intensity with maximum occurrences in histogram, denoted by $h(p)$, is labeled as 'max point,' while no occurrence in histogram denoted by $h(z)$ is labeled as 'zero point.' If $h(z) > h(p)$, then increment by 1 unit from left to right $(h(p),\ h(z))$. Otherwise, decrement by 1 unit from right to left $(h(z),\ h(p))$.

(2) Embed the data [3]: after HM embed secret data at $h(p)$ using Eq. (4)

$$\text{stego}(i,j) = \begin{cases} A(i,j) & \text{if } S(k) = 0 \,\&\, A(i,j) = h(p) \\ A(i,j) \pm 1 & \text{if } S(k) = 1 \,\&\, A(i,j) = h(p) \\ A(\mathrm{i},\mathrm{j}) & \text{otherwise} \end{cases} \tag{4}$$

Shown in Fig. 1d stego-image histogram disappeared peak point considered secret info. Here, number of image pixels modified is approximately greater than or equal to half of image pixels. Hence, HM techniques [3] can modify more than half of the image pixels.

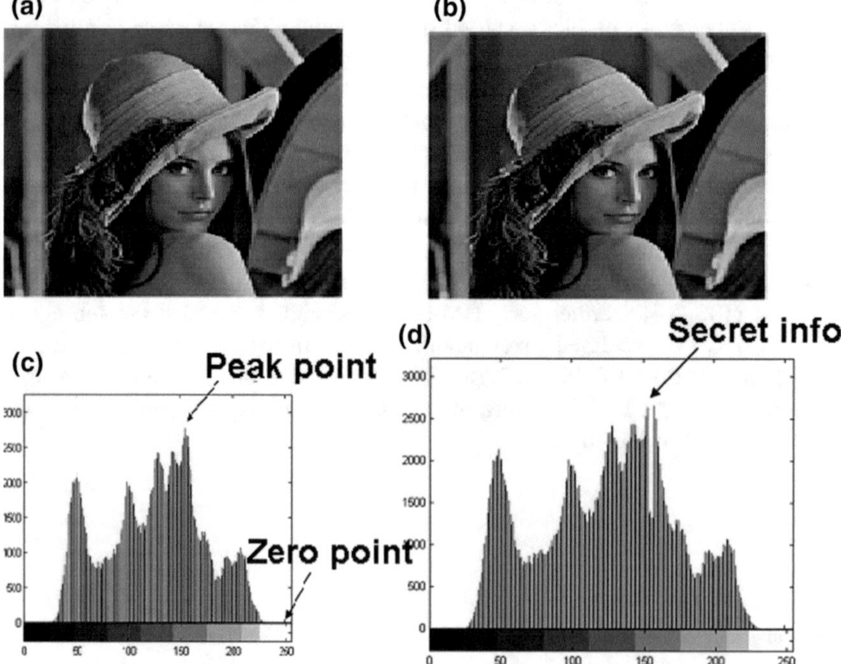

Fig. 1 Lena image **a** cover-content **b** stego-image **c** cover-content histogram **d** stego-image histogram

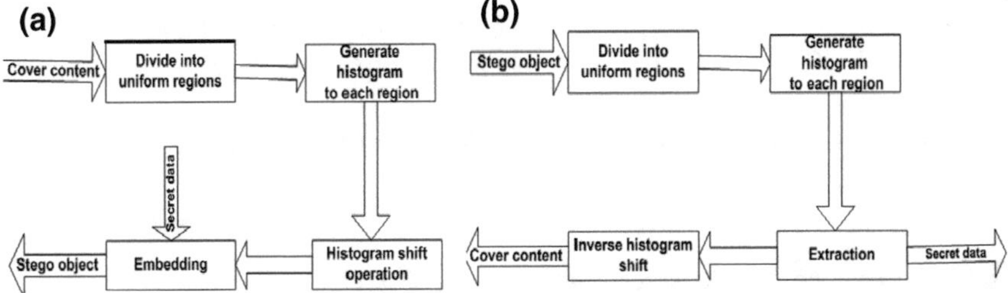

Fig. 2 URHS **a** embedding process **b** extraction process

3 Proposed Technique

The proposed URHS technique is the continuation to Ni [3] technique. This URHS RDH technique is designed to improve visual quality (PSNR) by increasing EC. This URHS technique considers cover-content/content-aware information (uniform region information) to embed the data into cover-content. URHS technique embedding and extraction processes state chart diagram have been defined in Fig. 2.

3.1 Divide into Uniform Regions

An image can be divided into uniform regions of sizes m by m (m = 32/64/128) using Eq. (5).

$$A_{ij} = \mathrm{A}((\mathrm{i}-1)\times m + 1 : \mathrm{i}\times m, (\mathrm{j}\ -1)\times m + 1 : \mathrm{j}\times m); 1 \le i, j \le n \quad (5)$$

where A_{ij} is ijth uniform region, m is block size, and n is number of blocks.

3.2 Generate Histogram Bins to Each Region

Generate histogram bins to each region A_{ij} using Eq. (6).

$$\mathrm{A}_{ij}(x) = \#\{1 \le i \le m \times n;\ x_i = x\} \forall x \in [0, 255] \quad (6)$$

where '#' is cardinal number and 'm' is block size

3.3 Histogram Shift Operation

For each region, we have to calculate peak point '$A_{ij}(p)$' and zero point '$A_{ij}(z)$' (let us consider $p > z$) as said in Sect. 2.3. Hence, histogram shift can be performed using Eq. (7).

$$A_{ij}(x, y) = \begin{cases} A_{ij}(x, y) & \text{if } z < A_{ij}(x, y) \&\& p >= A_{ij}(x, y) \\ A_{ij}(x, y) - 1 & \text{if } z > A_{ij}(x, y) \&\& p < A_{ij}(x, y) \end{cases} \quad (7)$$

where p is peak point and z is zero point indexes of region '$A_{ij.}$'

3.4 Proposed Technique Embedding Process

Proposed URHS technique embeds data on region peak points by histogram shift operation as shown in Eq. (8). In URHS embedding algorithm, if data bit is 0 $S(k) = 0$, then pixel intensity value of cover-content does not change. Otherwise, pixel intensity value is decremented by 1 as shown in Eq. (8).

$$A'_{ij}(x, y) = \begin{cases} A(x, y); & \text{if } S(k) = 0 \,\&\, A_{ij}(x, y) = p \\ A(x, y) - 1; & \text{if } S(k) = 1 \,\&\, A_{ij}(x, y) = P \end{cases} \quad (8)$$

where p is histogram peak point index value count and $S(k)$ is secret bit at k.

URHS embedding algorithm: URHS embedding algorithm takes input as 512×512 grayscale image. Then, divide a cover-content/input image into blocks (uniform regions), generate histogram to each block/ uniform region, carry out histogram shift, and embed data using URHS embedding algorithm as given in algorithm 1.

Time order analysis of Algorithm 1: Algorithm 1 is an iterative algorithm from step count method. URHS embedding algorithm's time complexity is $O(m \times n)$. There m and n are rows and column of cover-content. Furthermore, the computational time taken to compute URHS embedding using tic and toc function in MATLAB is approximately 0.067488 s.

```
Algorithm 1:  URHS Embedding Algorithm
//Input: Grayscale image of size 512×512 &S is secret data
//Output: A' is stego-object of size 512×512
  1. Mention the parameters [uniform_region_size, number
     of uniform_regions].
  2. Construct histogram using equation (5)
  3. Histogram generation to using equation (6)
  4. Apply histogram shift operation using equation (7)
  5. Embed secret data S into peak points of regions
  6. Embed secret data S in cover A using equation (8)
  7. Inverse histogram shift to generate stego-object.
```

3.5 Proposed Technique Extraction Process

Extraction processes of proposed URHS depend on peak point of each block ($A_{ij}(p)$). Here, each block's peak point is considered as secret info. Hence, once peak point to a block is identified, then extraction of embedded data is performed by Eq. (9). However, cover-content will be reconstructed as shown Eq. (10). Finally, reconstruct cover-content by inverting histogram shift operation as shown in Eq. (11).

$$S(m) = \begin{cases} 0; & \text{if } A'_{ij}(x,y) = A_{ij}(p) \\ 1; & \text{if } A'_{ij}(x,y) = A_{ij}(p) - 1 \\ \text{Do nothing}; & \text{otherwise} \end{cases} \tag{9}$$

$$A''(i,j) = \begin{cases} A'_{ij}(x,y) & \text{if } A'_{ij}(x,y) = A_{ij}(p) \\ A'_{ij}(x,y) + 1; & \text{if } A'(i,j) = A_{ij}(p) - 1 \end{cases} \tag{10}$$

$$A_{ij}(x,y) = \begin{cases} A'_{ij}(x,y) & \text{if } z < A'_{ij}(x,y) \&\& p >= A'_{ij}(x,y) \\ A'_{ij}(x,y) + 1 & \text{if } z > A'_{ij}(x,y) \&\& p < A'_{ij}(x,y) \end{cases} \tag{11}$$

```
Algorithm 2:    URHS Extraction Algorithm
//Input: A' is stego-object (grayscale) of size 512×512
//Output: A is extracted cover-content and secret data S
  1. Mention the parameters such as [Block_size, Number
     of Regions]
  2. Construct histogram using equation (5)
  3. Histogram generation using equation (6)
  4. Apply histogram shift operation using equation (7)
  5. Extract secret data and reconstruct cover content
     by applying equations (9)
  6. Reconstruct cover-content using equation (10)
  7. Inverse histogram shift of A'_ij(x,y)
```

URHS extraction algorithm: URHS extraction algorithm takes input as 512×512 stego-images. Then, divide each stego-image into block/uniform regions, generate histogram to each block/uniform region, identify peak point $A_{ij}(p)$. Furthermore, extract the embedded data by histogram shift operation (9) and recover cover-content by inverse histogram shift operation (10) and (11). The entire process is demonstrated on algorithm 2.

Time order analysis of algorithm 2: Algorithm 2 is an iterative algorithm from step count method. URHS extraction algorithm's time complexity is $O(m \times n)$.

Here, m and n are number of rows and column, respectively, in cover-content. Furthermore, the computational time taken to compute URHS extraction using tic and toc function in MATLAB is approximately 0.0624s.

4 Results Analysis

Experiments were conducted on MATLAB R2012b and input dataset as shown in Fig. 3, and embedding data is generated from MATLAB *rand()* function. This paper only focuses on following data hiding measures such as normalization cross-correlation coefficients (NCC) for reversibility, RSNR, and mean square error (MSE) for visual quality and EC.

$$NCC = \frac{\sum_{x,y \in m} I(x,y) I'(x,y)}{\sum_{x,y \in m} I^2(x,y)} \tag{11}$$

$$MSE = \frac{1}{m^*n} \sum_{x=1}^{M} \sum_{y=1}^{N} (\mathrm{A}(x,y) - B(x,y)) \tag{12}$$

$$PSNR = 10 \log_{10} \frac{255^2}{MSE} \tag{13}$$

Result analysis is demonstrated in Fig. 4, where 4a represents comparison of proposed technique PSNR with respect to different block sizes, 4b represents comparison of proposed technique EC with respect to different block sizes, 4c represents comparison of PSNR with literature techniques and proposed technique, 4d represents comparison of EC with literature techniques and proposed technique,

Fig. 3 Input grayscale cover-contents

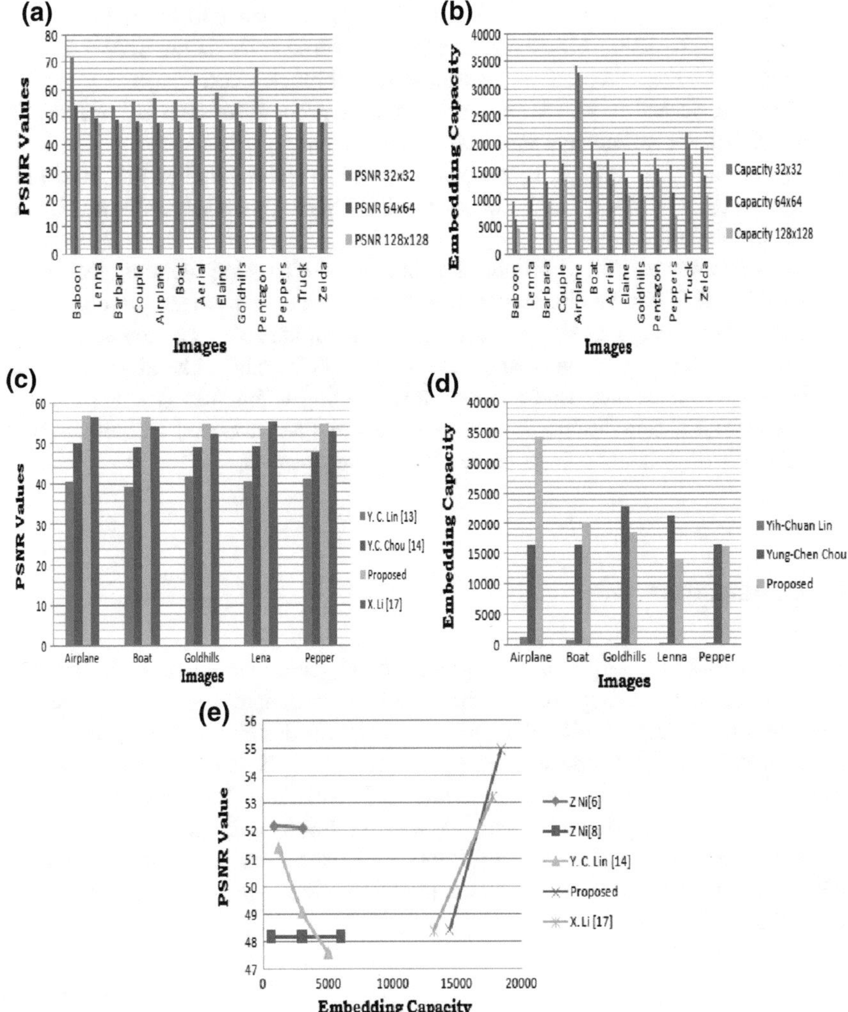

Fig. 4 Result analysis

and 4e represents change of PSNR with respect to EC for different literature techniques and proposed technique.

Observations:

- In Fig. 4a, PSNR is high in block size 32×32 compared to other block sizes ($64 \times 64, 128 \times 128$). Moreover, it can be observed that if an image has more number of edges, then PSNR is more in small block sizes rather than in bigger block sizes.

- In Fig. 4b, EC is high in block size 32×32 compared to other block sizes ($64 \times 64, 128 \times 128$). Moreover, it can be observed that if an image has less number of edges, then EC is maximum for all block sizes.
- From Fig. 4a and b, we can observe a unique property of data hiding, i.e., EC and PSNR are not always inversely propositional.
- Fig. 4c compares PSNRs of proposed technique with existing literature techniques [10, 11, 14]. It can be observed that PSNR values are more or same as existing techniques.
- Fig. 4d compares ECs of proposed technique with existing literature techniques [10, 11, 14]. It can be observed that ECs are more or less comparatively exist in proposed techniques. However, proposed technique has an advantage over existing techniques, if an image has less number of edges like airplane image.
- Fig. 4e compares proposed and literature existing techniques [10, 11, 14] with respect to EC and PSNR. It can be observed that proposed techniques have better PSNR and EC compared to literature techniques.

5 Conclusions

URHS is a RDH technique based on HM. URHS technique guarantees maximum EC approximately 0.1bpp and minimum EC approximately 0.05. Moreover, it guarantees maximum PSNR value approximately 72dB and minimum PSNR value approximately 48 dB. Proposed technique uses region pixels information to embed data, so it can be called as a content-aware RDH technique. Moreover, it maintains the trade-off between EC and PSNR and also it exhibits a unique property, that is, by increasing EC with increasing PSNR. It takes very less time to execute embedding and extraction functions compared to other data embedding algorithms like DE and PEE.

References

1. Cox, I., Miller, M., Bloom, J., Fridrich, J., Kalker, T.: Digital Watermarking and Steganography, vol. 2. MK Publication, USA (2007)
2. Muni Sekhar, V., Rao, K.V.G., Sambasive Rao, N., Gopichand, M.: Comparing the capacity, NCC and fidelity of various quantization intervals on DWT. In: Proceedings of the Third ICICSE, Springer, AISC, vol. 413, pp. 45–56 (2015)
3. Ni, Z., Shi, Y., Ansari, N., Su, W.: Reversible data hiding. IEEE Trans. Circuits Syst. Video Technol. **16**(3), 354–362 (2006)
4. Lin, C., Tai, W.L., Chang, C.C.: Multilevel reversible data hiding based on histogram modification of difference images. J. Pattern Recogn. **41**(12), 3582–3591 (2008)
5. Ni, Z., Shi, Y.Q., Ansari, N., Su, W., Sun, Q., Lin, X.: Robust lossless image data hiding designed for semi-fragile image authentication. IEEE Trans. Circuits Syst. **18**(4), 497–509 (2008)

6. Gao, X., An, L., Li, X., Tao, D.: Reversibility improved lossless data hiding. J. Signal Process. **89**(10), 2053–2065 (2009)
7. Tian, J.: Reversible data embedding using a difference expansion. IEEE Trans. Circuits Syst. Video Technol. **13**(8), 890–896 (2003)
8. Jung, S.W., Le, T., Ha, S., Ko, J.: A new histogram modification based reversible data hiding algorithm considering the human visual system. IEEE Signal Process. Lett. **18**, pp. 95–98 (2011)
9. Zhang, W., Hu, X, L, X, Yu, N: Recursive histogram modification: establishing equivalency between reversible data hiding and lossless data compression. IEEE Trans. Image Process. **22** (7), 2775–2785 (2013)
10. Lin, Y.C.: Reversible image data hiding using quad-tree segmentation and histogram shifting. J. Multimed. **6**(4), 349–358 (2011)
11. Chou, Y.C.: High payload reversible data hiding scheme using difference segmentation and histogram shifting. J. Electron. Sci. Technol. **11**(1), 9–14 (2013)
12. Zandi, Y., Mehran, M., Alireza, N., Nazanin, N., Mehran, Z.: Histogram shifting as a data hiding technique: an overview of recent developments. In: Proceedings of International Conference on Digital Information and Communication Technology and its Applications, Iran, pp. 770–786 (2011)
13. Fridrich, J.: Invertible authentication. In: Proceedings International Conference of SPIE, vol. 4314, pp. 197–208, NY (2001)
14. Li, X., Zhang, W., Gui, X., Yang, B.: Efficient reversible data hiding based on multiple histogram modification. IEEE Trans. Inf. Forensics Secur. **10**(9), 2016–2027 (2015)
15. Hu, X., Zhang, W., Li, X., Yu, N.: Minimum rate prediction and optimized histogram modification for reversible data hiding. IEEE Trans. Inf. Forensics Secur. **10**(3), 653–664 (2015)
16. Li, X., Zhang, W., Gui, X., Yang, B.: A novel reversible data hiding scheme based on two-dimensional difference-histogram modification. IEEE Trans. Inf. Forensics Secur. **8**(7), 1091–1100 (2013)
17. Barni, M., Bartolini, F., Piva, A.: An improved wavelet based watermarking through pixel wise masking. IEEE Trans. Image Process. **10**(5), 783–791 (2001)
18. Wang, W.J., Huang, C.T., Wang, S.J.: VQ applications in steganographic data hiding upon multimedia images. IEEE Syst. J. **5**(4), 528–537 (2011)
19. De Vleeschouwer, C., Delaigle, J.E., Macq, B.: Circular interpretation of histogram for reversible watermarking. In: IEEE: Proceedings of Fourth Workshop on Multimedia Signal Processing, France, pp. 345–350 (2001)
20. De Vleeschouwer, C., Delaigle, J.E., Macq, B.: Circular interpretation of bijective transformations in lossless watermarking for media asset management. IEEE Trans. Multimed. **5**(1), 97–105 (2003)

Application of TF-IDF Feature for Categorizing Documents of Online Bangla Web Text Corpus

Ankita Dhar, Niladri Sekhar Dash and Kaushik Roy

Abstract This paper explores the use of standard features as well as machine learning approaches for categorizing Bangla text documents of online Web corpus. The TF-IDF feature with dimensionality reduction technique (40% of TF) is used here for bringing in precision in the whole process of lexical matching for identification of domain category or class of a piece of text document. This approach stands on the generic observation that text categorization or text classification is a task of automatically sorting out a set of text documents into some predefined sets of text categories. Although an ample range of methods have been applied on English texts for categorization, limited studies are carried out on Indian language texts including that of Bangla. Hence, an attempt is made here to analyze the level of efficiency of the categorization method mentioned above for Bangla text documents. For verification and validation, Bangla text documents that are obtained from various online Web sources are normalized and used as inputs for the experiment. The experimental results show that the feature extraction method along with LIBLINEAR classification model can generate quite satisfactory performance by attaining good results in terms of high-dimensional feature sets and relatively noisy document feature vectors.

Keywords Bangla text classification · Term frequency · Inverse document frequency · LIBLINEAR · Corpus

A. Dhar (✉) · K. Roy
Department of Computer Science, West Bengal State University, Kolkata, India
e-mail: ankita.ankie@gmail.com

K. Roy
e-mail: kaushik.mrg@gmail.com

N. S. Dash
Linguistic Research Unit, Indian Statistical Institute, Kolkata, India
e-mail: ns_dash@yahoo.com

V. Bhateja et al. (eds.), *Intelligent Engineering Informatics*, Advances in Intelligent Systems and Computing 695, https://doi.org/10.1007/978-981-10-7566-7_6

1 Introduction

The expeditious developments in computer hardware have made it possible to collect and store large volume of text data from all possible domains. Nowadays, the number of text documents dispersed over online by different sectors is in the orders of millions. These online documents may contain valuable and useful information which can benefit the people coming from different fields and domains of knowledge. The process of analyzing and processing these documents manually by various domain experts is extremely hard and time-consuming with respect to the volume of text documents and number of text domains. Therefore, there is a pressing need for developing an intelligent technique that can automatically categorize these text documents into respective domains. Thus, automatic text categorization or classification (TC) has turned into one of the key domains for organizing and handling textual data. The TC is the task which categorizes text documents into predefined categories based on the contents available in texts. Currently, text classification techniques are applied in many domains of application like text filtering, document organization, classification and indexing of news stories into hierarchical category, searching for interesting information on the Web, etc. What confirms this observation is that classification of text documents either automatically or manually has been a regular practice in the domain of text analysis and information retrieval for the purpose of knowledge generation and machine learning.

The remaining part of the paper is organized in the following manner: Sect. 2 refers to some early works of text categorization on various languages; Sect. 3 casts lights on the overall view of the proposed work; Sect. 4 discusses the analysis of experimental setup and results; Sect. 5 presents a short conclusion with some remarks on future works.

2 Existing Work

A simple survey can reveal that considerable amount of research is done in text categorization for English. Many well established supervised techniques have been used often in English such as Chen et al. [1] used Naive Bayes (NB), Joachims [2] and Cortes and Vapnik [3] used SVM. Also, reasonable amount of comparative study has been done in English. In the work of Bijalwan et al. [4] KNN, NB and term-gram have been used for their experiment which shows that the accuracy of KNN is better than NB and term-gram. Besides this, authors in [5] provided a comparative study on DT, NB, KNN, Rocchio's algorithm, back propagation network and SVM. In their comparisons, it has been shown that SVM performed better than all other approaches they used in their experiment for 20 newsgroups dataset.

For Arabic text classification, Mohammad et al. [6] carried out the experiment on dataset obtained from several Arabic news sources using KNN, C4.5, and Rocchio algorithm and achieves accuracies of 82% for Rocchio, 65% for C4.5, and 82% for KNN when the value of K is 18 based on term selection and weighting.

In case of Urdu text documents, scholars [7] have compared statistical techniques for Urdu text classification using Naive Bayes and SVM. The experimental result shows that the classification using NB is effective but not as specific as SVM.

For Chinese text classification, Wei et al. [8] used Chinese corpus TanCorpV1.0 consists of more than 14,000 texts belonging to 12 classes. They have used TF-IDF weighting schemes and achieve accuracy in terms of macro-F1 and micro-F1 applying LIBSVM and Naive Bayes classifiers. Macro-F1 and micro-F1 values for experiment Ex02 and Ex12 are 0.83–0.86 and 0.89–0.91, for Ex01 and Ex11 the values are 0.82–0.85 and 0.88–0.90, for Ex03 and Ex13 values are 0.79–0.86 and 0.87–0.91, and for Ex04 and Ex14 values obtained are 0.79–0.85 and 0.87–0.90.

In case of some Indian languages, attempts are also made for categorizing text documents, although the volume of work is not as much as noted for English. For instance, automatically categorizing Marathi text documents based on the user's profile is presented in [9]. The system uses Label Induction Grouping (LINGO) algorithm based on VSM. Their dataset consists of 200 documents of 20 categories. The result shows that the algorithm is efficient for classifying Marathi documents.

In case of Hindi, researchers have proposed a rule-based, knowledge-based tool to classify Hindi verbs in syntactic perspective automatically [10]. They have claimed of developing the largest lexical resource for Hindi verbs providing information on the class based on valency and some syntactic diagnostic tests as well as their morphological type.

In case of Tamil, scholars [11] have proposed an efficient method for extracting C-feature for classifying Tamil text documents. They claimed that using C-feature extraction, one can easily classify the documents because C-feature contains a pair of terms to classify a document to a predefined category.

For Punjabi text documents, a team of scholars [12] has used a hybrid classification algorithm by combining Naive Bayes and ontology-based classification algorithm and also carried out the experiment using ontology-based classification, Naive Bayes, and centroid-based classification algorithm separately for Punjabi text classification. It is observed that hybrid classification as well as ontology-based classification gives better result in comparison to the centroid-based and the Naive Bayes classifiers.

For Telugu, scholars [13] proposed supervised classification using Naive Bayes for Telugu news articles in four major categories consisting of near about 800 documents. Here, TF-IDF was used as the feature values.

For Bangla, to the best of our knowledge, limited effort has been made in this direction. To mention a few, in [14], N-gram based technique has been used by the researchers to categorize Bangla newspaper text corpus. However, the text corpus is very small with regards to the number of text documents used for the experiment. The documents have been collected from a single newspaper (i.e., Pratham Alo) covering just one year data. Another study [15] compared four supervised learning techniques for labeled news Web documents into five categories. Classification has been made on 1000 documents with a total number of words being 22,218. Accuracies of 89.14% for SVM, 85.22% for NB, 80.65% for DT (C4.5), and 74.24% for KNN have been achieved. Other researchers [16] have used stochastic gradient descent (SGD) classifier to classify some Bangla text documents of nine different text categories along with other classifiers used for comparisons. Accuracies of 93.44% for ridge classifier, 91.42% for perceptron, 93.19% for passive-aggressive, 93.78% for SVM, 91.89% for NB, 93.05% for logistic-regression, and 93.85% for SGD classifier are achieved. Comparative study on distinct types of approaches to categorize Bangla text document has been performed in [17].

3 Proposed Model

Prior to the classification of the text documents, it is necessary to do tokenization which is the segmentation of sentences and preprocessing which includes removal of tokens that have no pivotal role in identifying and classifying the text documents into domains. At preprocessing stage, eviction work is performed for stop words, punctuations, postposition, conjunctions, English equivalent words, and English and Bangla numerals which are quite frequently used in a Bangla text. After that, extraction of proper feature sets are also required before training and constructing the model for successful text document categorization. The following block diagram (Fig. 1) illustrates the overall process of Bangla text classification system employed in this experiment. Detailed description of the model is provided in the following sections (Sect. 4).

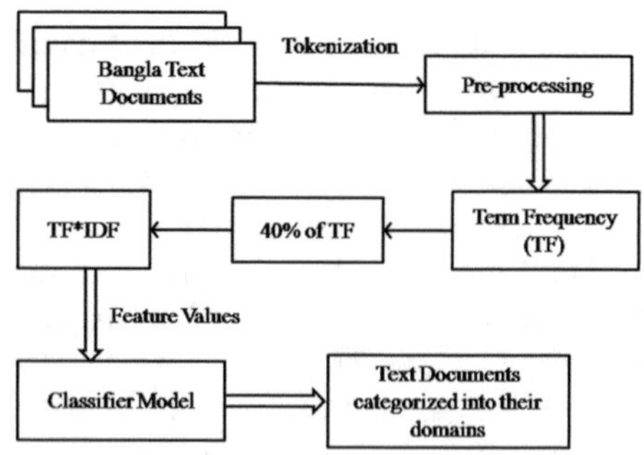

Fig. 1. Bird's eye view of the proposed model

4 Experimental Setup

4.1 Data Collection

The text documents used for the experiment have been collected from various online Bangla news text corpus produced by some Bangla daily newspapers:

(a) AnandabazarPatrika: http://www.anandabazar.com/
(b) Bartaman: http://allbanglanewspapers.com/bartaman/
(c) Ebelatabloid: http://www.ebela.in/

Moreover, some more data for medical domain are collected from some online magazines. From this corpus, total 1960 Bangla text documents are obtained from five categories. At the initial stage, these are considered as our 'primary dataset.' The number of text documents taken into consideration from each domain is presented in Table 1. After tokenization, the number of tokens generated is 8,12,540. After performing preprocessing task which includes removal of punctuation marks, postpositions, conjunctions, stop words, pronouns, English and Bangla numerals, and English equivalent words, the number of tokens finally obtained is 6,32,924.

4.2 Feature Extraction

For the experiment, standard feature like term frequency-inverse document frequency (TF-IDF) has been selected to extract the features from the text documents. To represent the weight for each token numerically, term frequency together with inverse document frequency is mostly used. Tokens received after preprocessing task are arranged in the ascending order of their number of occurrence in the text documents. The number of tokens results after removal of repetitive tokens is 3,68,738.

Prior to feeding the feature set to the classifier, it is subjected to the reduction technique. Top 40% tokens are taken into account as the features based on the dimensionality reduction. Forty percentage is chosen on the basis of the previous knowledge that guides us to infer that most frequently occurring terms carry no relevant information and are not domain specific. They are the terms that occur mostly in every text documents. Hence, they have been discarded in our

Table 1. Text documents of five domains

Domain	Number of text documents
Business	310
Medical	500
State	410
Sports	540
Technology	200

experiment. In total, 1,47,495 features out of 3,68,738 tokens are selected using TF-IDF. For normalizing TF-IDF, we considered the maximum value to be 1.0. Finally, top 35,060 features among 1,47,495 tokens are selected for the experiment.

TF-IDF method [2] is the combination of two terms: term frequency (TF) and inverse document frequency (IDF). TF here denotes the frequency (number of occurrence) of a term t in a document d.

IDF generally estimates the major role of a particular term t in a text document d. IDF of term t is measured based on the following (Eq. 1) where N is the total number of text documents and document frequency (DF) is the number of documents where term t occurs.

$$\mathrm{IDF_t} = \log \frac{\mathrm{N}}{\mathrm{DF_t}} \tag{1}$$

Now, TF-IDF weighting scheme for a term t in a document d is computed using the (Eq. 2) stated below.

$$\mathrm{TF} - \mathrm{IDF} = \mathrm{TF} * \mathrm{IDF_t} \tag{2}$$

4.3 *Experimental Result*

LIBLINEAR [18], a linear classification algorithm, is used for training the model with 5000 iterations and 0.001 epsilon value. For a given set of instance-label pairs (x_i, y_i) where i = 1, …, l, $x_i \in R^n$, $y_i \in \{-1, +1\}$, C > 0 is a penalty parameter. The following unconstrained optimization problem is solved with loss functions $\xi(\mathrm{w}; x_i, y_i)$ for vector w using (Eq. 3). The values are esteemed based on trial run. The detail of the result for the five domains is laid out in the following Table 2.

$$\min_{w} \frac{1}{2} w^T w + C \sum_{i=1}^{l} \xi(\mathrm{w}; x_i, y_i) \tag{3}$$

It can be observed from the table that in average, the number of text documents correctly classified is more for sports domain followed by that of medical science and science domain. The reason behind this is primarily due to the fact that the

Table 2. Number of documents classified correctly for every domain

Domains	Total	Correctly classified	Incorrectly classified
Business	310	280	30
Medical	500	481	19
State	410	375	35
Sports	540	528	12
Science	200	188	12

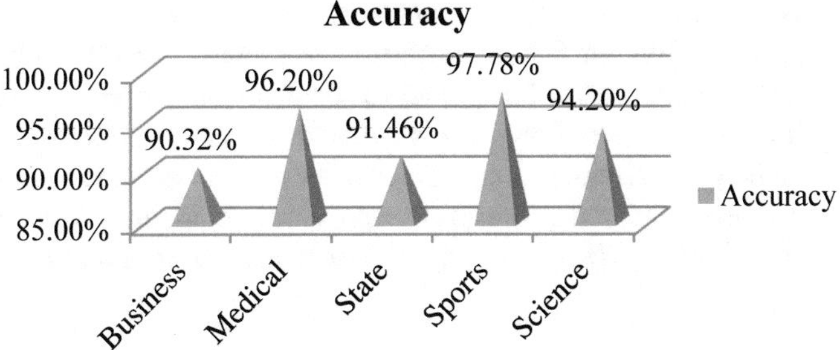

Fig. 2. Obtained accuracies for five domains

Table 3. Statistical analysis using Friedman test

Classifier	Dataset					Mean rank
	#1	#2	#3	#4	#5	
LIBLINEAR	97.81 (1)	97.38 (1)	98.33 (1)	98.49 (1)	96.70 (1)	$R_1 = 1.0$
Naive Bayes Multinomial	96.45 (2)	95.63 (2)	97.02 (2)	97.11 (2)	97.08 (2)	$R_2 = 2.0$
Naive Bayes	95.42 (3)	95.56 (3)	96.32 (3)	96.18 (3)	95.34 (3)	$R_3 = 3.0$
J48	93.19 (5)	93.26 (5)	94.07 (5)	93.99 (5)	93.05 (5)	$R_4 = 5.0$
Random forest	95.28 (4)	94.32 (4)	95.94 (4)	95.77 (4)	93.20 (4)	$R_5 = 4.0$

terms that occur in business and state domains are comparatively close to the terms that occur in normal/general text domains. On the contrary, in case of medical, sports, and science domain, the text documents contain more domain specific terms. The domain-wise accuracies are further enumerated in the following diagram (Fig. 2).

A few other commonly used classifiers like Naive Bayes Multinomial, Naive Bayes, decision tree (J48), and random forest are also tested on this dataset, and their statistical analysis using Friedman nonparametric test are presented below (Table 3).

5 Conclusion

This paper shows that the application of TF-IDF feature with dimensionality reduction technique can bring in precision in the process of lexical matching for identification of domain categories of a text document. The present experiment shows that higher level of accuracy is possible to achieve based on the reduction approach that may be adopted for classifying Bangla text documents of news text corpus. The system can also be tested on a larger database with increasing number of text categories. In future, other various standard reduction techniques can be applied along with different feature extraction technique as well. Different commonly used classifiers can also be applied for comparison purposes for testing the levels of performance of the systems. It is hoped that a close comparison as well as cross-interpolation of techniques may elicit higher level of accuracy in case of text domain identification and text categorization provided the input texts are adequately processed for extracting required data and information to train an automated system.

Acknowledgements One of the authors would like to thank Department of Science and Technology (DST) for support in the form of INSPIRE fellowship.

References

1. Chen, J., Huang, H., Tian, S., Qu, Y.: Feature selection for text classification with Naive Bayes. Expert Syst. Appl. **36**, 5432–5435 (2009)
2. Joachims, T.: Text categorization with support vector machines: learning with many relevant features. In: Proceedings of the 10th European Conference on Machine Learning, pp. 137–142 (1998)
3. Cortes, C., Vapnik, V.: Support-vector networks. Mach. Learn. **20**, 273–297 (1995)
4. Bijalwan, V., Kumar, V., Kumari, P., Pascual, J.: KNN based machine learning approach for text and document mining. Int. J. Database Theor. Appl. **7**, 61–70 (2014)
5. Pawar, P.Y., Gawande, S.H.: A comparative study on different types of approaches to text categorization. Int. J. Mach. Lear. Comput. **2** (2012)
6. Mohammad, A.H., Al-Momani, O., Alwada'n, T.: Arabic text categorization using k-nearest neighbour, Decision Trees (C4.5) and Rocchio classifier: a comparative study. Int. J. Curr. Eng. Technol. **6**, 477–482 (2016)
7. Ali, A.R., Ijaz, M.: Urdu text classification. In: Proceedings of the 7th International Conference on Frontiers of Information Technology, pp. 21–27 (2009)
8. Wei, Z., Miao, D., Chauchat, J.H., Zhao, R., Li, W.: N-grams based feature selection and text representation for Chinese text classification. Int. J. Comput. Intel. Syst. **2**, 365–372 (2009)
9. Patil, J.J., Bogiri, N.: Automatic text categorization marathi documents. Int. J. Adv. Res. Comput. Sci. Manage. Stud. 2321–7782 (2015)
10. Dixit, N., Choudhary, N.: Automatic classification of Hindi verbs in syntactic perspective. Int. J. Emerg. Technol. Adv. Eng. **4**, 2250–2459 (2014)
11. ArunaDevi, K., Saveetha, R.: A novel approach on tamil text classification using C-Feature. Int. J. Sci. Res. Dev. 2321–0613 (2014)
12. Gupta, N., Gupta, V.: Punjabi text classification using Naive Bayes, centroid and hybrid approach. In: Proceedings of the 3rd Workshop on South and South East Asian Natural Language Processing (SANLP), pp. 109–122 (2012)
13. Murthy, K.N.: Automatic Categorization of Telugu News Articles. Department of Computer and Information Sciences, University of Hyderabad (2003)
14. Mansur, M., UzZaman, N., Khan, M.: Analysis of N-gram based text categorization for Bangla in a newspaper corpus. In: Proceedings of International Conference on Computer and Information Technology (2006)
15. Mandal, A.K., Sen, R.: Supervised learning methods for Bangla web document categorization. Int. J. Artif. Intell. Appl. (IJAIA) **5**, 93–105 (2014)

16. Kabir, F., Siddique, S., Kotwal, M.R.A., Huda, M.N.: Bangla text document categorization using stochastic gradient descent (SGD) classifier. In: Proceedings of International Conference on Cognitive Computing and Information Processing, pp. 1–4 (2015)
17. Islam, Md.S., Jubayer, F.E. Md., Ahmed, S.I.: A comparative study on different types of approaches to Bengali document categorization. In: Proceedings of International Conference on Engineering Research, Innovation and Education (ICERIE), 6 pp (2017)
18. Fan, R.-E., Chang, K.-W., Hsieh, C.-J., Wang, X.-R., Lin, C.-J.: LIBLINEAR: a library for large linear classification. J. Mach. Learn. Res. **9**, 1871–1874 (2008)

An Ensemble Learning-Based Bangla Phoneme Recognition System Using LPCC-2 Features

Himadri Mukherjee, Santanu Phadikar and Kaushik Roy

Abstract An array of devices have emerged lately for easing our daily life but one concern has always been towards designing simple user interface (UI) for such devices. A speech-based UI can be a solution to this, considering the fact that it is one of the most spontaneous and natural modes of interaction for most people. The process of identification of words and phrases from voice signals is known as Speech Recognition. Every language encompasses a unique set of atomic sounds termed as Phonemes. It is these sounds which constitute the vocabulary of that language. Speech Recognition in Bangla is a bit complicated task mostly due to the presence of compound characters. In this paper, a Bangla Phoneme Recognition system is proposed to help in the development of a Bangla Speech Recognizer using a new Linear Predictive Cepstral Coefficient-based feature, namely LPCC-2. The system has been tested on a data set of 3710 Bangla Swarabarna (Vowel) Phonemes, and an accuracy of 99.06% has been obtained using Ensemble Learning.

Keywords Phoneme · LPCC-2 · Ensemble Learning · Mean · Standard deviation

1 Introduction

Technology has had a large impact on our life in the form of countless devices for assisting us in our day-to-day activities. There has always been a demand of devices which are easy to interact with leading to the development of graphical user interface (GUI) to succeed the command user interface (CUI). There is still need for

H. Mukherjee (✉) · K. Roy
Department of Computer Science, West Bengal State University, Kolkata, India
e-mail: himadrim027@gmail.com

K. Roy
e-mail: kaushik.mrg@gmail.com

S. Phadikar
Department of Computer Science & Engineering,
Maulana Abul Kalam Azad University of Technology, Kolkata, India
e-mail: sphadikar@yahoo.com

V. Bhateja et al. (eds.), *Intelligent Engineering Informatics*, Advances in Intelligent Systems and Computing 695, https://doi.org/10.1007/978-981-10-7566-7_7

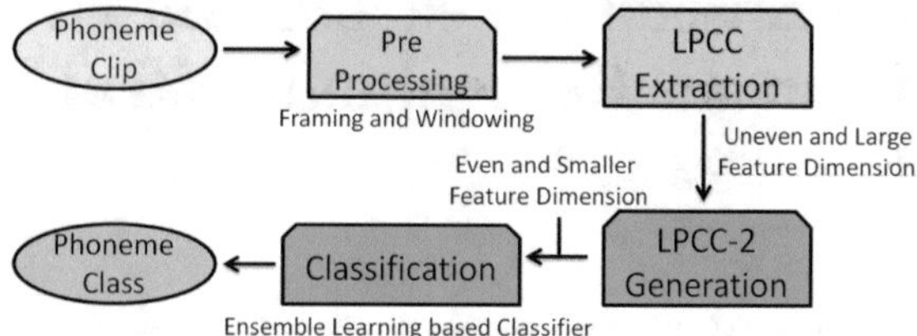

Fig. 1 Block diagram of the proposed system

simplifying the means of communication between us and the devices, and a speech-based UI can be a solution to this since we are accustomed to verbal communication from birth. Speech Recognition refers to the task of identification of words and phrases from voice signals. Though accurate Speech Recognizers are available in various languages [10] but such is not the case for Bangla which is the 6th most spoken language in the world, spoken by approximately 874 Million people [1] thereby establishing the importance and need for the development of a Bangla Speech Recognizer. Such a system can help those people to use IT who are not very much proficient in English as well as simplify the task of feeding Bangla data into the devices which is a tedious task when it comes to typing the same due to the presence of compound characters. Every language has a set of unique sounds called Phonemes which constitute the entire vocabulary of the language. Among them, vowel (Swarabarna in Bangla) Phonemes are extremely important and only a few meaningful words can be found for a language which do not have at least one vowel Phoneme. In this paper, a newly proposed Linear Predictive Cepstral Coefficient (LPCC) based feature, namely LPCC-2, has been used to distinguish Swarabarna Phonemes with the help of an Ensemble Learning based classifier. The proposed system is illustrated in a nutshell in Fig. 1.

In the rest of the paper, Sect. 2 discusses the Related Works followed by the details of the Data set in Sect. 3 and that of the Proposed System in Sect. 4. The obtained results are analysed in Sect. 5, and finally the conclusion is presented in Sect. 6.

2 Related Work

English Speech Recognition has come a long way since the attempt of Forgie et al. [2] to recognize English vowels in 1959. Speech Recognition in English and few other languages has developed remarkably and has been commercialized as well but the case has not been so for Bangla Speech Recognition which started way back in 1975 [3] and databases like SHRUTI [4] were also developed. Eity et al. [5] presented a Bangla Phoneme Recognition System with a two-stage multilayer neural network and hidden Markov model. They converted Mel Frequency Cepstral Coefficient (MFCC) values into Phoneme probabilities in the 1st stage and inserted the deltas and double deltas of the same in the second stage for improvement. They used 3000 sentences from 30 volunteers to train the system and another 1000 sentences from ten volunteers for testing and obtained a highest Phoneme Correct Rate of 68.78%. Ahmed et al. [6] designed a Speech Recognition system with the aid of Deep Belief Network and MFCC features. Their database consisted of 840 utterances of words by 42 speakers, and an accuracy of 94.05% was obtained using generative pretraining and enhanced gradient for adjustment of model parameters. Debnath et al. [7] used 3rd differential MFCC coefficients along with HMM for recognizing Bangla speech. They experimented in various phases by varying the database size with the highest of 1600 sentences by 16 volunteers. They obtained a highest accuracy of 98.94% for their proposed method which was slightly better than the obtained accuracy of 98.39% for 39 dimensional features. Hasnat et al. [8] used a 39 dimensional MFCC feature along with Dynamic Time Warping to characterize real-time inputs. In order to improve the accuracy of the system, K-nearest neighbours classifier was applied as well which produced an accuracy of 90%. Hasan et al. [9] designed a Speech Recognition system using a 39 dimensional MFCC feature along with HMM classifier. They used a database composed with the aid of 40 volunteers and obtained a word correct rate of 90.33%. Mukherjee et al. [10] differentiated Bangla Swarabarna Phonemes with the help of MFCC based features on a database of 1400 Phonemes uttered by 20 speakers and obtained an accuracy of 98.35% using an artificial neural network based classifier. Hassan et al. [11] constructed a Phonetic feature table in their 1st part of their experiment. In the 2nd part, they extracted MFCC features, embedded phone feature extraction with the help of a neural network and finally fed the 22 dimensional feature to a HMM based classifier. They used 3000 sentences for the purpose of training and another 100 sentences uttered by ten volunteers for testing. They obtained a word correct rate of 92.25%, word accuracy of 91.64% and a sentence correct rate of 91.60%. Kabir et al. [12] proposed a three stage method for Bangla Speech Recognition involving mapping of local features into Phonetic features followed by enhancement and inhibition of the dynamic patterns. Finally the features were normalized using Gram–Schmidt algorithm and fed to a HMM based classifier. The system was trained using 3000 sentences and tested with another 1000 sentences which produced highest word correct rate, word accuracy and sentence correct rate of 98.57%, 98.39% and 97.90%, respectively, on the test set.

3 Data set Development

The quality of data has a significant impact on the outcome of an experiment. Care needs to be taken during the data collection phase so that inconsistencies do not cast a shadow on the data. A data set of Bangla Swarabarna Phonemes was put together with the help of 32 male and 21 female volunteers (aged between 20 to 30) due to the fact that there is no such standard database to the best of our knowledge. The volunteers pronounced the seven Swarabarna Phonemes as presented in Table 1 in a single take which was repeated ten times for each of the volunteers. This led to a database of 3710 (53 * 7 * 10) Phonemes which were separated using a semi-

Table 1 IPA symbol of the Phonemes along with their Bangla alphabetic representation as well as Bangla and equivalent English pronunciation

IPA Symbol	ɔ	a	i	u	e	o	æ
Bangla Representation	অ	আ	ই, ঈ	উ, ঊ	এ	ও	অ্যা
Bangla Pronunciation as in	অমর	আধার	বিলাস	তুল্য	ছেলে	ভোজ	টেরা
Equivalent English Pronunciation	Lot	Beta	Bee	Fool	Said	Coal	Map

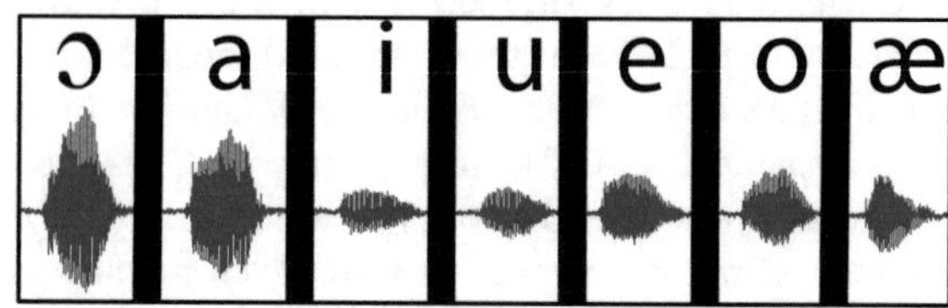

Fig. 2 Amplitude-based representation of the seven Phonemes whose boundaries are shown in black

supervised amplitude based approach. The data was recorded using various locally made microphones including Frontech JIL-3442 with Audacity [13] in stereo mode. The data was stored in .wav format at 1411 kbps bitrate. The microphones were kept at various distances and angles each time to incorporate real-life scenario, and the recordings were also done in various rooms to make our data set robust. The amplitude-based representation of a single channel of one of the tracks is presented in Fig. 2.

4 Proposed Method

4.1 Preprocessing—Framing and Windowing

In order to make the clips spectrally quasi-stationary, they were framed into smaller parts which were individually analysed. Framing was done in overlapping mode (to maintain inter frame continuity) with a frame size of 256 sample points and an overlap of 100 points and then multiplied with Hamming Window as presented in [10, 14] to minimize spectral leakage during spectrum based analysis. A clip consisting of R sample points can be divided into N frames of size X with an overlap of O points as presented in Eq. (1).

$$N = \left\lceil \frac{R - X}{O} + 1 \right\rceil \quad (1)$$

4.2 Feature Extraction

Linear Predictive Cepstral Coefficients (LPCC) were obtained for every frame of the clips with the help of Linear Predictive Analysis which predicts a signal sample as a linear combination of previous signal samples. The nth sample of a signal can be estimated with the help of previous T samples as presented in Eq. (2) where a_1, a_2, a_3, a_4,..,a_T are termed as the linear predictive coefficients or predictors.

$$s(n) \approx a_1 s(n-1) + a_2 s(n-2) + a_3 s(n-3) + a_4 s(n-4) + ... + a_T s(n-T) \quad (2)$$

The error of prediction e(n) between the actual (s(n)) and predicted ($\hat{s}$(n)) samples is presented in Eq. (3).

$$e(n) = s(n) - \hat{s}(n) = s(n) - \sum_{k=1}^{T} a_k s(n-k) \quad (3)$$

The error of the sum of squared differences is presented in Eq. (4), which is minimized to engender a unique set of predictor coefficients involving m samples in a frame.

$$E_n = \sum_{m} \left[s_n(m) - \sum_{k=1}^{T} a_k s_n(m-k) \right]^2 \quad (4)$$

The Cepstral Coefficients (C) are calculated with a procedure which is recursive in nature as shown in Eq. (5)

$$\left.\begin{aligned} C_0 &= log_e T \\ C_m &= a_m + \sum_{k=1}^{m-1} \frac{k}{m} C_k a_{m-k}, for\ 1 < m < T\ and \\ C_m &= a_m + \sum_{k=m-p}^{m-1} \frac{k}{m} C_k a_{m-k}, for\ m > T \end{aligned}\right\} \quad (5)$$

4.3 LPCC-2 Generation

Since clips of different lengths produced variable number of frames, features of disparate dimensions were obtained. A clip of 1 s sampled at 44100 Hz produces 440 frames (in accordance with Eq. (1)) each of which have 256 sample points with an inter frame overlap of 100 sample points. If only ten LPCC features are extracted for every frame, then a total of 4400 (10×440) feature values are obtained. Features of larger dimensions are obtained for clips of larger length which burdens the system with computational load. In order to deal with these two issues of large and uneven

Table 2 Dimensions of F_1–F_5 pre- and post reduction along with the percentages of reduction

Feature set	F_1	F_2	F_3	F_4	F_5
Original dimension	210	315	420	525	630
Dimension post reduction	94	127	160	189	216
Dimension reduction (%)	55.24	59.68	61.90	64.00	67.54

dimensionality, LPCC-2 feature is proposed whose dimension does not change with the length of a clip, thereby solving the problem of uneven dimensionality. Moreover, the lower dimension reduces the load of computation on the system.

The energy values of the LPCC coefficients were analysed to find the global maximum and minimum which were used to define N equally spaced classes which was set to 18 after trial runs. The class-wise percentage of occurrence of energy values out of the total number of energy values for all the bands was calculated. Along with this, the bands were also graded in descending order based on the total energy content. The band-wise mean and standard deviation were also added to the feature set. Thus, when 10 features per frame were extracted for the aforementioned clip, then a final feature of 210 ($10 \times 18 + 10 + 2 \times 10$) dimension was obtained which is quite less as compared to 4400-dimensional features. The final feature set pre classification was reduced to remove such features which had the same value for all the instances which reduced the feature dimension even more. In our experiment, 10 (F_1), 15 (F_2), 20 (F_3), 25 (F_4) and 30 (F_5) dimensional LPCCs were used for the generation of LPCC-2s. The size of the feature sets pre and post reduction is presented in Table 2.

4.4 *Ensemble Learning-Based Classification*

The technique of combining multiple learners at the time of training and thereafter testing is termed as Ensemble Learning [15]. A set of hypothesis is constructed in the training phase, and a decision is taken by their combination in contrary to various standard machine learning techniques. A set of learners designated as base learners is contained in an ensemble. These learners are sometimes termed as weak learners because the generalization capability of an ensemble is greater than that of a base learner. Most of the ensemble based methods use a single base learner to generate a homogeneous set of learners but heterogeneous sets are also produced in some cases. In the current experiment, a Random Forest [16] based classifier which embodies a set of decision trees was used to incorporate Ensemble Learning. Random Forests have the capability of estimating missing data efficiently, and the accuracy is also maintained when a large chunk of data is missing as often observed in real-world cases. Random Forests can be parallelized completely, thereby making them suitable for parallel processing systems. Random Forest consists of an array of decision trees ($T_{1...n}$), and a random vector ($\Theta_{1...n}$) is generated for every tree having similar distribution but Θ_k is independent of Θ_1 to Θ_{k-1}. The training set as well as Θ_k helps to grow the kth tree and produce the classifier h(x, Θ_k) where x is the input vector. The margin function for an ensemble consisting of k classifiers with a training set which is randomly drawn from the random vector (Y) distribution is shown in Eq. (6),

$$mg(X, Y) = av_k I(h_k(X) = Y) - max_{j \neq Y} av_k I(h_k(X) = j) \tag{6}$$

where $h_k(X)=h(X,\Theta_k)$ and I is the Indicator function. Higher value of this margin portrays more confidence in the task of classification. The generalization error is shown in Eqs. (7) and (8) which present its convergence.

$$PE^* = P_{X,Y}(mg(X, Y) < 0) \tag{7}$$

$$P_{X,Y}(P_\Theta(h(X, \Theta) = Y) - max_{j \neq Y} P_\Theta(h(X, \Theta) = j) < 0) \tag{8}$$

5 Result and Discussion

Each of the five feature sets were tested using a five fold cross validation technique at 100, 200, 300, 400 and 500 iterations. The obtained accuracies for the sets for each of the iterations are presented in Table 3a. It can be observed from the Table that the highest accuracy was obtained for F_3 at both 300 and 400 iterations (highlighted in Blue) out of which the result for the lower number of iterations was considered. The individual Phoneme accuracies for the same are presented in Fig. 3a with the highest individual accuracy highlighted in green. The percentages of confusions among all the Phoneme pairs for the same are presented in Table 3b which shows that Phoneme pairs ‘i’ and ‘e’ as well as ‘ɔ ’ and ‘a’ constituted the most confused pairs (12 confusions for each pair). It was observed during data collection that the pronunciations of the aforesaid Phonemes were pair-wise very close for many volunteers which is one of the reasons for such confusions. A few other machine learning algorithms were applied to F_3 whose results are presented in Fig. 3b.

Table 3 (a) Obtained accuracies for F_1–F_5 at various training iterations. (b) Confusions (%) among the phoneme pairs fog F_3

Feature	Iterations				
Set	100	200	300	400	500
F1	96.23	96.31	96.31	96.17	96.23
F2	97.49	97.68	97.79	97.68	97.74
F3	98.87	98.98	**99.06**	**99.06**	99.03
F4	98.30	98.28	98.25	98.27	98.27
F5	98.71	98.84	98.89	98.89	98.81

(a)

	ɔ	a	i	u	e	o	æ
ɔ	-	0.94	0	0	0	0.38	0
a	1.32	-	0	0	0	0.57	0
i	0	0	-	0.19	1.13	0	0
u	0	0	0.38	-	0.19	0	0.19
e	0	0	1.13	0	-	0	0
o	0	0	0	0.19	0	-	0
æ	0	0	0	0	0	0	-

(b)

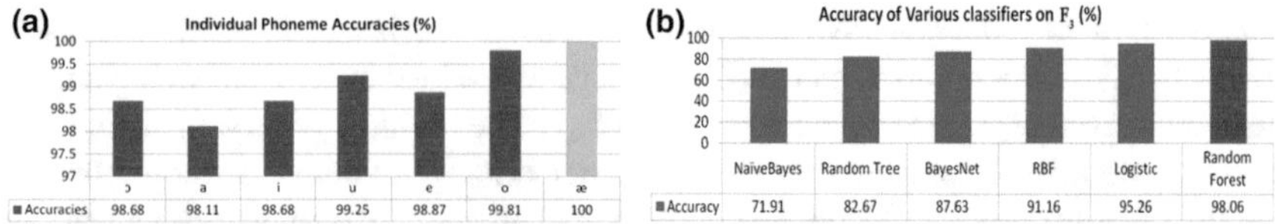

Fig. 3 **a** Individual phoneme accuracies for F_3 at 300 iterations. **b** Performance of various classifiers on F_3

6 Conclusion

A Phoneme Recognition system for Bangla Swarabarnas has been presented using a newly proposed LPCC-2 feature and Ensemble Learning based classification to help in the development of a Bangla Speech Recognizer. The system has been tested on a database of 53 volunteers, and an encouraging accuracy has been obtained with a precision of 0.994. We plan to test our system on a larger data set of consonant Phonemes and also use various preprocessing and pre-emphasis techniques to reduce the influence of noise. The silent sections at the header and trailer of the clips will also be identified and removed for further improvement of performance. We also plan to use other features as well as machine learning techniques in the future.

Acknowledgements The authors would like to thank the students of West Bengal State University for voluntarily providing the voice samples during data collection.

References

1. Ethnologue. http://www.ethnologue.com (2017). Accessed 1 May 2017
2. Forgie, J.W., Forgie, C.D.: Results obtained from a vowel recognition computer program. J. Acoust. Soc. Am. **31**, 1480–1489 (1959)
3. Pramanik, M., Kido, K.: Bengali speech: formant structures of single vowels and initial vowels of words. Proc. ICASSP **1**, 178–181 (1976)
4. Das, B., Mandal, S., Mitra, P.: Bengali speech corpus for continuous automatic speech recognition system. In: 2011 International Conference on Speech Database and Assessments (Oriental COCOSDA), pp. 51–55. IEEE (2011)
5. Eity, Q.N., Banik, M., Lisa, N.J., Hassan, F., Hossain, M.S., Huda, M.N.: Bangla speech recognition using two stage multilayer neural networks. In: Proceeding of the International Conference on Signal and Image Processing (ICSIP), pp. 222–226 (2010)
6. Ahmed, M., Shill, P.C., Islam, K., Mollah, M.A.S., Akhand, M.A.H., Acoustic modeling using deep belief network for Bangla speech recognition. In: Proceeding of the International Conference on Computer and Information Technology (ICCIT), pp. 306–311 (2015)
7. Debnath, S., Saha, S., Aziz, M.T., Sajol, R.H., Rahimi, M.J.: Performance comparison of MFCC based bangla ASR system in presence and absence of third differential coefficients. In: Proceeding of the International Conference on Electrical Engineering and Information Communication Technology (ICEEICT), pp. 1–6 (2016)
8. Sayem, Asm: Speech analysis for alphabets in Bangla language: automatic speech recognition. Int. J. Eng. Res. **3**(2), 88–93 (2014)

9. Hasan, M.M., Hassan, F., Islam, G.M.M., Banik, M., Kotwal, M.R.A., Rahman, S.M.M., Muhammad, G., Mohammad, N.H.: Bangla triphone hmm based word recognition. In: Proceeding of Asia Pacific Conference on Circuits and Systems (APCCAS), pp. 883–886 (2010)
10. Mukherjee, H., Halder, C., Phadikar, S., Roy, K.: READA Bangla phoneme recognition system. In: Proceeding of 5th International Conference on Frontiers in Intelligent Computing: Theory and Applications (FICTA), pp. 599–607 (2017)
11. Hassan, F., Kotwal, M.R.A., Huda, M.N., Bangla phonetic feature table construction for automatic speech recognition. In Proceedings of 16th International Conference on Computer and Information Technology (ICCIT), pp. 51–55 (2013)
12. Kabir, S.M.R., Hassan, F., Ahamed, F., Mamun, K., Huda, M.N., Nusrat, F.: Phonetic features enhancement for Bangla automatic speech recognition. In: Proceeding of the 1st International Conference on Computer and Information Engineering (ICCIE), pp. 25–28 (2015)
13. Audacity, http://www.audacityteam.org/. Accessed 25 April 2017
14. Mukherjee, H., Phadikar, S., Rakshit, P., Roy, K.: REARC—A Bangla Phoneme Recognizer. In: Proceeding of the International Conference on Accessibility to Digital World (ICADW), pp. 177–180 (2016)
15. Rokach, L.: Ensemble-based classifiers. Artif. Intell. Rev. **33**, 139 (2010)
16. Breiman, L.: Random forests. Mach. Learn. **45**(1), 5–32 (2001)

Solving Uncapacitated Facility Location Problem Using Monkey Algorithm

Soumen Atta, Priya Ranjan Sinha Mahapatra and Anirban Mukhopadhyay

Abstract The Uncapacitated Facility Location Problem (UFLP) is considered in this paper. Given a set of customers and a set of potential facility locations, the objective of UFLP is to open a subset of facilities to satisfy the demands of all the customers such that the sum of the opening cost for the opened facilities and the service cost is minimized. UFLP is a well-known combinatorial optimization problem which is also NP-hard. So, a metaheuristic algorithm for solving this problem is natural choice. In this paper, a relatively new swarm intelligence-based algorithm known as the Monkey Algorithm (MA) is applied to solve UFLP. To validate the efficiency of the proposed binary MA-based algorithm, experiments are carried out with various data instances of UFLP taken from the OR-Library and the results are compared with those of the Firefly Algorithm (FA) and the Artificial Bee Colony (ABC) algorithm.

Keywords Uncapacitated Facility Location Problem (UFLP)
Simple Plant Location Problem (SPLP) · Warehouse Location Problem (WLP)
Monkey Algorithm

1 Introduction

The *Uncapacitated Facility Location Problem* (UFLP) deals with finding a subset of facilities from a given set of potential facility locations to meet the demands of all the customers such that the sum of the *opening cost* for each of the opened facilities and the *service cost* (or *connection cost*) is minimized [1–3]. In UFLP, unrestricted capacity of each facility is considered and each facility has some nonnegative

S. Atta (✉) · P. R. S. Mahapatra · A. Mukhopadhyay
Department of Computer Science and Engineering, University of Kalyani, Nadia, W.B., India
e-mail: soumen.atta@klyuniv.ac.in

P. R. S. Mahapatra
e-mail: priya@klyuniv.ac.in

A. Mukhopadhyay
e-mail: anirban@klyuniv.ac.in

V. Bhateja et al. (eds.), *Intelligent Engineering Informatics*, Advances in Intelligent Systems and Computing 695, https://doi.org/10.1007/978-981-10-7566-7_8

opening cost associated with it. UFLP is also known as the *Simple Plant Location Problem* (SPLP) [4, 5] and the *Warehouse Location Problem* (WLP) [1]. Since UFLP is an NP-hard problem [6, 7], different approaches have been applied to solve it. Some of the significant methods used to solve UFLP are mentioned here. Different deterministic methods such as branch-and-bound [1, 8], linear programming [9], constant factor approximation algorithm [10], semi-Lagrangian relaxation [11], surrogate semi-Lagrangian dual [12], dual-based procedure [13], primal-dual-based approach [14] have been developed for solving UFLP. Many metaheuristic methods such as tabu search [2, 15], unconscious search [3], discrete particle swarm optimization [16], artificial bee colony optimization [17, 18], genetic algorithm [5], simulated annealing [19], ant colony optimization [20], firefly algorithm [21] have been developed for solving UFLP.

The *Monkey Algorithm* (MA) is a relatively new *swarm intelligence*-based algorithm. MA was proposed by Zhao and Tang in 2008 [22]. MA simulates the mountain-climbing processes of monkeys [22]. It consists of mainly three processes. The *climb process* is used to gradually improving the objective value by searching local optimal solutions. The *watch-jump process* is employed to speed up the convergence rate of the algorithm. The *somersault process* avoids from falling into local optima. Due to its simple structure, strong robustness, and the capability of avoiding to fall into local optima, MA has been successfully applied to the intrusion detection problem [23], the sensor placement problem [24], the project scheduling problem [25], the cluster analysis problem [26], the integer knapsack problem [27], to name a few.

In this paper, the binary version of MA is applied to solve UFLP. We have compared the performance of the proposed method with that of the Firefly Algorithm (FA) [21], FA with local search (FA-LS) [21], and the Artificial Bee Colony (ABC) algorithm [18]. The organization of the paper is as follows: The problem is formally defined in Sect. 2. In Sect. 3, the MA-based method is described in detail. Section 4 provides the experimental results, and comparisons are also done in this section with FA-, FA-LS-, and ABC-based solution. Finally, Sect. 5 concludes the paper.

2 Problem Definition

Let $C = \{c_1, c_2, \dots, c_m\}$ be the set of m customers and $\mathcal{F} = \{f_1, f_2, \dots, f_n\}$ be the set of n potential facilities that can be opened, where $c_i, f_j \in \mathbb{N}$ and $1 \le i \le m$, $1 \le j \le n$, $\mathbb{N} = \{1, 2, \dots\}$. Note that for simplicity here we assume that the customers and potential facilities are represented by natural numbers. Let $\mathcal{D} = [d_{ij}]_{m \times n}$ be the *service matrix* (or *distance matrix*) where d_{ij} represents the service cost of customer i if it gets service from facility j, and $\mathcal{G} = \{g_1, g_2, \dots, g_n\}$ be the set of *opening cost* where g_j denotes the opening cost for facility j. Now, we define two binary decision variables s_{ij} and y_j which are defined as follows:

$$s_{ij} = \begin{cases} 1 \text{ if customer } i \text{ gets service from facility } j \\ 0 \text{ otherwise;} \end{cases}$$

$$f_j = \begin{cases} 1 \text{ if facility } j \text{ is opened} \\ 0 \text{ otherwise.} \end{cases}$$

Using the above descriptions of parameters and decision variables, the mathematical formulation of the Uncapacitated Facility Location Problem (UFLP) [2] is given as follows:

$$minimize \sum_{i=1}^{M} \sum_{j=1}^{N} d_{ij} s_{ij} + \sum_{j=1}^{N} g_j y_j$$

subject to the constraints

$$\sum_{i=1}^{M} s_{ij} = 1, \quad \forall j \in \mathcal{F},$$

$$s_{ij} \leq y_j, \quad \forall i \in \mathcal{C}, \forall j \in \mathcal{F},$$

$$s_{ij},\ y_i = \{0, 1\}, \quad \forall i \in \mathcal{C}, \forall j \in \mathcal{F}.$$

The first term in the objective function denotes the total service cost, and the second term denotes the total opening cost of the opened facilities.

3 Proposed MA-based Method

In this section, binary MA-based algorithm (BMA) is proposed. The following subsections give the details of the proposed algorithm.

3.1 Encoding

The position of the monkey i is denoted by a vector $X_i = \{x_{i1}, x_{i2}, \dots, x_{in}\}$, where n is the number of potential facilities and $x_{ik} \in \{0, 1\}$, $1 \leq k \leq n$. If facility k is open, then $x_{ik} = 1$; otherwise, it is 0.

3.2 Creation of Initial Population

For the monkey i, its position vector X_i encodes a candidate solution and each element of X_i is assigned by either 1 or 0 randomly. In this way, the initial population of size *nPop* is created.

3.3 Climb Process

The climb process is used to gradually improve the objective value. The climb process [27] for the monkey i is as follows: At first, we create two vectors ΔX_1 and ΔX_2 each of length n such that each element of these vectors is either 1 or -1 created randomly. Here, the *climb step* is 1. Now, we determine $X_i' = \mid X_i - \Delta X_1 \mid$ and $X_i'' = \mid X_i - \Delta X_2 \mid$. Note that this may create vectors X_i' and X_i'' which do represent feasible solutions. So, x_{ik}' and x_{ik}'' are changed to 0, if they were previously more than 0.5; otherwise, they are changed to 1. If the objective value corresponding to X_i' is better than the objective values for both X_i'' and X_i, then X_i is replaced by X_i', and if the objective value corresponding to X_i'' is better than the objective values for both X_i' and X_i, then X_i is replaced by X_i''. In all other cases, X_i is unchanged. This step is repeated for all the monkeys for a predetermined number of steps, known as the *climb number*. In the proposed algorithm, it is set to 2 to reduce the computational time.

3.4 Watch-Jump Process

The watch-jump process [22, 27] is used to find a better solution locally for faster convergence. The watch-jump process for the monkey i is as follows: A new vector Y_i of length n is created where y_{ik} is chosen randomly in the range $[y_{ik} - 1, y_{ik} + 1]$. Note that this may create a vector Y_i which does not represent a feasible solution. So, y_{ik} is changed to 0, if it was previously more than 0.5; otherwise, it is changed to 1. Now, if the objective value corresponding to Y_i is less than X_i, then X_i is replaced by Y_i. The same process is repeated for all the monkeys for a predetermined number of times, known as the *watch-jump number*. In the proposed algorithm, we have experimentally set it to 2.

3.5 Cooperation Process

The *cooperation process* introduced by Zhou et al. [27] is used here. The cooperation method for the ith monkey is as follows: A new vector Y_i of length n is created where y_{ik} is set to x_{ij} with the probability 0.5 and y_{ik} is set to x_{bk} with the probability 0.5, where X_b is the best monkey in terms of objective value. If the objective value of Y_i is less than that of X_i, then X_i is replaced with Y_i.

3.6 Somersault Process

The somersault process prevents monkeys from falling into local optima. We have observed that this process is useful for initial few iterations. So, we perform this process only for the initial ten iterations of the proposed algorithm. The somersault process for the ith monkey is described here. At first, a random number $\theta \in [-1, 1]$ is generated and we select a monkey p from the population randomly. This randomly selected monkey is known as the *somersault pivot* (p). Then, a vector Y_i of length n is created where $y_{ik} = x_{pk} + \theta(x_{pk} - x_{ik})$. Again, this may create a monkey Y_i which corresponds to an invalid solution. So, y_{ik} is changed to 0, if it was previously more than 0.5; otherwise, it is changed to 1. Now, if the objective value corresponding to Y_i is less than X_i, then X_i is replaced by Y_i. The same process is repeated for other monkeys.

3.7 Termination Condition

The loop of the climb process, the watch-jump process, the cooperation process, and the somersault process will stop when either a predetermined number of iterations are reached or there is no improvement of the best monkey in terms of the objective value for five consecutive iterations. The objective value corresponding to the best monkey is returned as the output of the algorithm.

4 Experiments and Results

The proposed algorithm BMA is coded with MATLAB R2013a, and all the experiments are performed in a machine with Intel Core i5 3.10 GHz processor having Windows 10 operating system with 8 GB of RAM. The twelve instances of UFLP used in this paper are taken from the Beasley's OR-Library [28]. Here, the performance of the proposed BMA is compared with that of the FA [21], FA-LS [21], and ABC-based [18] algorithms. The performance of BMA is compared with respect to the *average relative percent error* (ARPE) and the *hit to optimum rate* (HR). ARPE is defined as follows [17, 21]:

$$ARPE = \sum_{i=1}^{R} \left(\frac{H_i - U}{U} \right) \times \frac{100}{R},$$

where H_i denotes the objective value corresponding to the monkey i; R is the number of solutions in population, and U is the optimal value given in [28]. Note that R is nothing but *nPop*. HR is defined as the number of times that the algorithm finds the optimal solutions over all repetitions [21]. In our experiments, we have repeated each experiment for 20 times and then the average value of ARPE and HR is reported in Table 1. The performance of the proposed BMA with that of the other algorithms is compared in Table 1 and depicted in Fig. 1 using box plots in terms of ARPE and HR.

It is clear form Table 1 that the proposed BMA is far better than all the other algorithms in terms of average ARPE. In terms of average HR, the proposed BMA is better than FA and it is comparable with FA-LS and ABC. Note that ARPE is much more robust parameter for performance comparison than HR.

Table 1 Comparison with other algorithms

Instance	FA [21]		FA-LS [21]		ABC [18]		BMA	
	ARPE	HR	ARPE	HR	ARPE	HR	ARPE	HR
cap71	0.000	1.00	0.000	1.00	0.00	1.00	0.0000	1.00
cap72	0.000	1.00	0.000	1.00	0.00	1.00	0.0000	1.00
cap73	0.012	0.71	0.000	1.00	0.00	1.00	0.0000	1.00
cap74	0.000	1.00	0.000	1.00	0.00	1.00	0.0000	1.00
cap101	0.097	0.49	0.000	1.00	0.00	1.00	0.0057	0.65
cap102	0.094	0.34	0.000	1.00	0.00	1.00	0.0003	0.95
cap103	0.073	0.22	0.025	0.71	0.00	1.00	0.0057	0.65
cap104	0.246	0.62	0.000	1.00	0.00	1.00	0.0000	1.00
cap131	0.993	0.00	0.073	0.46	0.143	0.24	0.0297	0.10
cap132	1.063	0.00	0.003	0.72	0.039	0.41	0.0121	0.05
cap133	1.121	0.00	0.044	0.27	0.040	0.36	0.0011	0.15
cap134	1.516	0.04	0.039	0.81	0.000	1.00	0.0056	0.50
Average	0.4346	0.4517	0.0153	0.8308	0.0185	0.8342	0.0050	0.6708

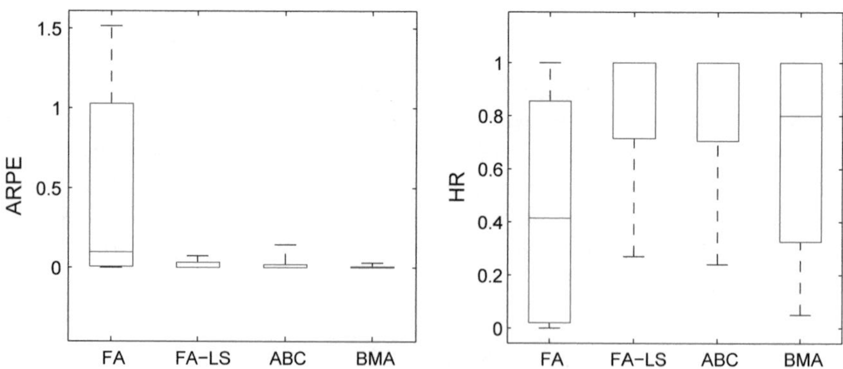

Fig. 1 Performance comparison among FA, FA-LS, ABC, and BMA in terms of ARPE and HR

5 Conclusion

In this paper, a binary-coded Monkey Algorithm (MA)-based solution is presented for solving the Uncapacitated Facility Location Problem (UFLP). The performance of the prosed algorithm is compared with that of the Firefly Algorithm (FA) and the Artificial Bee Colony (ABC) algorithm. It has been observed that the proposed MA-based solution technique is comparable with both FA- and ABC-based solutions. For future work, we will improve the convergence rate of the proposed solution technique by incorporating local search method(s) and perform more experiments on other UFLP instances available in the literature.

References

1. Akinc, U., Khumawala, B.M.: An efficient branch and bound algorithm for the capacitated warehouse location problem. Manage. Sci. **23**(6), 585–594 (1977)
2. Al-Sultan, K., Al-Fawzan, M.: A tabu search approach to the uncapacitated facility location problem. Ann. Oper. Res. **86**, 91–103 (1999)
3. Ardjmand, E., Park, N., Weckman, G., Amin-Naseri, M.R.: The discrete Unconscious search and its application to uncapacitated facility location problem. Computers & Industrial Engineering **73**, 32–40 (2014)
4. Krarup, J., Pruzan, P.M.: The simple plant location problem: survey and synthesis. Eur. J. Oper. Res. **12**(1), 36–81 (1983)
5. Kratica, J., Tošic, D., Filipović, V., Ljubić, I.: Solving the simple plant location problem by genetic algorithm. RAIRO-Operations Research **35**(01), 127–142 (2001)
6. Garey, M.R., Johnson, D.S.: Computers and Intractability: A Guide to NP-Completeness (1979)
7. Lenstra, J., Kan, A.R.: Complexity of Packing, Covering and Partitioning Problems. Econometric Institute (1979)
8. Bilde, O., Krarup, J.: Sharp lower bounds and efficient algorithms for the simple plant location problem. Ann. Discret. Math. **1**, 79–97 (1977)
9. Van Roy, T.J.: A cross decomposition algorithm for capacitated facility location. Oper. Res. **34**(1), 145–163 (1986)
10. Shmoys, D.B., Tardos, É., Aardal, K.: Approximation algorithms for facility location problems. In: Proceedings of the twenty-ninth annual ACM symposium on Theory of computing, pp. 265–274. ACM (1997)
11. Beltran-Royo, C., Vial, J.P., Alonso-Ayuso, A.: Solving the uncapacitated facility location problem with semi-Lagrangian relaxation. Statistics and Operations Research, Rey Juan Carlos University, Mostoles, Madrid, España (2007)
12. Monabbati, E.: An application of a Lagrangian-type relaxation for the uncapacitated facility location problem. Jpn J. Ind. Appl. Math. **31**(3), 483–499 (2014)
13. Erlenkotter, D.: A dual-based procedure for uncapacitated facility location. Oper. Res. **26**(6), 992–1009 (1978)
14. Körkel, M.: On the exact solution of large-scale simple plant location problems. Eur. J. Oper. Res. **39**(2), 157–173 (1989)
15. Sun, M.: Solving the uncapacitated facility location problem using tabu search. Comput. Oper. Res. **33**(9), 2563–2589 (2006)
16. Guner, A.R., Sevkli, M.: A discrete particle swarm optimization algorithm for uncapacitated facility location problem. J. Artif. Evol. Appl. **2008** (2008)

17. Tuncbilek, N., Tasgetiren, F., Esnaf, S.: Artificial bee colony optimization algorithm for uncapacitated facility location problems. J. Econ. Soc. Res. **14**(1), 1 (2012)
18. Watanabe, Y., Takaya, M., Yamamura, A.: Fitness function in ABC algorithm for uncapacitated facility location problem. In: ICT-EurAsia/CONFENIS, pp. 129–138 (2015)
19. Aydin, M.E., Fogarty, T.C.: A distributed evolutionary simulated annealing algorithm for combinatorial optimisation problems. J. Heuristics **10**(3), 269–292 (2004)
20. Kole, A., Chakrabarti, P., Bhattacharyya, S.: An ant colony optimization algorithm for uncapacitated facility location problem. Artif. Intell. **1**(1) (2014)
21. Tsuya, K., Takaya, M., Yamamura, A.: Application of the firefly algorithm to the uncapacitated facility location problem. J. Intell. Fuzzy Syst. **32**(4), 3201–3208 (2017)
22. Zhao, R.Q., Tang, W.S.: Monkey algorithm for global numerical optimization. J. Uncertain Syst. **2**(3), 165–176 (2008)
23. Zhang, J., Zhang, Y., Sun, J.: Intrusion detection technology based on monkey algorithm. Comput. Eng. **37**(14), 131–133 (2011)
24. Yi, T.H., Li, H.N., Zhang, X.D.: A modified monkey algorithm for optimal sensor placement in structural health monitoring. Smart Mater. Struct. **21**(10), 105033 (2012)
25. Mali, Z.T.X.Y.Z.: Optimization of gas filling station project scheduling problem based on monkey algorithm. Value Eng. **8**, 058 (2010)
26. Chen, X., Zhou, Y., Luo, Q.: A hybrid monkey search algorithm for clustering analysis. Sci. World J. **2014** (2014)
27. Zhou, Y., Chen, X., Zhou, G.: An improved monkey algorithm for a 0–1 knapsack problem. Appl. Soft Comput. **38**, 817–830 (2016)
28. Beasley, J.E.: OR-Library: distributing test problems by electronic mail. J. Oper. Res. Soc. 1069–1072 (1990)

AISLDr: Artificial Intelligent Self-learning Doctor

Sumit Das, Manas Sanyal, Debamoy Datta and Aniruddha Biswas

Abstract In recent decades, observation is that there are numerous corruptions in medical diagnosis; instead of proper diagnosis, some corrupted practitioners follow the money-earning-diagnosis-path by trapping the patient at critical stage in some countries. The common people are suffering from lack of diagnosis due to high diagnostic cost and lack of certified practitioners. This paper analyzes this shortcoming and 'design and implement' an intelligent system (AISLDr) which can perform the same without any corruption in a cost effective manner as like honest human doctor. In this work, the disease Tuberculosis has taken as a prototype because many people of our country are suffering from this disease and they know it at critical stage as India is highest Tuberculosis (TB) burden country. Here, our AISLDr performs diagnosis as well as draws awareness in the society to serve the nation in sustainable manner using fuzzy logic, probabilistic reasoning, and artificial intelligence (AI).

Keywords AISLDr · Fuzzy logic (FL) · Artificial intelligence (AI) Knowledge base (KB) · Tuberculosis (TB)

S. Das (✉) · A. Biswas
Information Technology, JIS College of Engineering, Kalyani 741235, India
e-mail: sumit.it81@gmail.com

A. Biswas
e-mail: biswas.aniruddha@gmail.com

M. Sanyal
Department of Business Administration, University of Kalyani, Kalyani 741235, India
e-mail: manas_sanyal@rediffmail.com

D. Datta
Electrical Engineering, JIS College of Engineering, Kalyani 741235, India
e-mail: datta.debamoy@gmail.com

V. Bhateja et al. (eds.), *Intelligent Engineering Informatics*, Advances in Intelligent Systems and Computing 695, https://doi.org/10.1007/978-981-10-7566-7_9

1 Introduction

Artificial intelligence (AI) has found numerous applications in the field of medicine. However, as the time goes a single type of disease evolved into numerous varieties. Nowadays, diagnoses of them are really costing for the life of common peoples. Doctors may not be available always and the complexities of symptoms of the disease are increasing in order to combat these doctors should be dynamic. Like these days, we have superbugs that are resistant to the antibiotics presently available. Artificial neural networks (ANNs) used in various studies but they too have their limitations. What is required is an alternative way of thinking. Our intelligent system does not use ANN for self-learning. It uses an independent set of equations that gives the result of whether to update the database of the current person or not. The user is required to enter his/her feedback and our intelligent system studies this file and tells whether to update the knowledge base or not. In the paper, the key idea is that doctors are not available at emergency or doctors may do some mistake by coincidence [1]. In this world, diseases have become very common problem. Diagnosing the patients and detecting the disease is a great challenge, especially Tuberculosis in our country.

2 Background Study

In this study, various types of Tuberculosis (TB) that has spread over the world, naturally huge database is needed to perform various operations on it. This database, called as knowledge base, is the heart of this study. In making this, following types of data classification is done: symptoms and disease name. A patient may not feel the same symptoms that the other patient may feel exactly likely; suppose one feels very tired and also lost weight and other feels the same but his weight loss is not significant. That is some symptoms may show while others may not be prominent. In this study, taking the example of renal TB, if a person has pulmonary TB and at the same time a renal failure he definitely has a renal TB, but this may not happen in case of all other persons that is where the Boolean logic comes into action. A.B + C.D = conclusion [where A, B, C, and D are the symptoms]. Judgment in medical field is not only a composite process but it is also time consuming. It can lead to errors in the diagnosis of a particular medical problem, which in turn will lead to erroneous medication [2]. In this paper, we not only dealt with the diagnosis of different TB based on various infectious agents but also shown the acuteness of the disease based on computations.

3 Methodology

A first-order predicate logic amalgamates with probability and fuzzy logic in this work. A fifth-generation language, supporting querying and extending the knowledge base from a command-line interface using a format oriented on natural language, is used for programming. The input format defines a Definite Clause Grammar (DCG), which is the generalization of Context-Free Grammar (CFG), description language built into Prolog. It supports parsing which is one of the important applications of Prolog and Logic Programming. The knowledge represents in text files containing Prolog predicates with defined operators to make it a domain-specific language, which is not too far from the actual input format, making it maintainable by the domain expert without requiring Prolog or general programming skills. The architecture of the proposed AISLDr is depicted in Fig. 1.

Our method has used fuzzy logic to implement the concept of probability, where membership function is represented as,

$$\mu = (\text{Actual Input} - \text{Minimum input})/(\text{Maximum input} - \text{Minimum input})$$

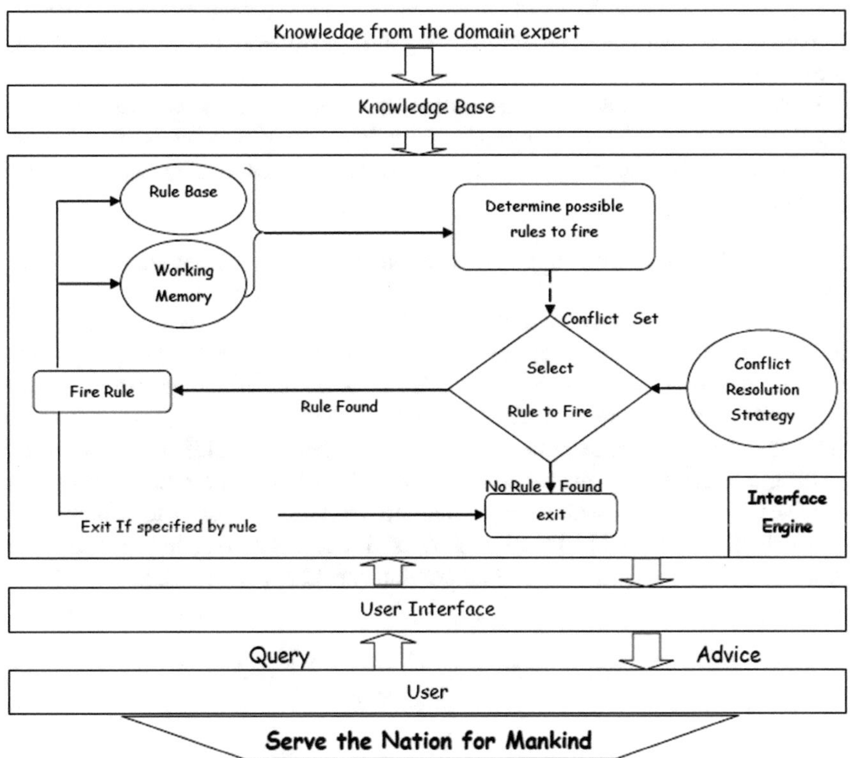

Fig. 1 Architecture of AISLDr

Table 1 Linguistic variables as per symptoms and assigned values

Symptoms	Linguistic variable	Assign values (x)
intense_cough_lasting_more_than_3_weeks	Extreme	5
intense_cough_lasting_more_than_3_weeks	Very much	4
intense_cough_lasting_more_than_3_weeks	Moderate	3
intense_cough_lasting_more_than_3_weeks	Somewhat	2
intense_cough_lasting_more_than_3_weeks	A little bit	1
intense_cough_lasting_more_than_3_weeks	Not at all	0

For our case,

$$\mu_{\text{linear}} = \text{x}/5 \tag{1}$$

Following logics are designed based on the symptoms of TB with the help of linguistic variables as elaborated in Table 1. How much does a patient feel?

Enter the degree, to which you feel, the user enters the value 4 then the percentage of uncertainty is calculated as (4 * 100)/5, which turns out to be 80%. It not only takes the average of all symptoms and finally returns the final uncertainty. Now it also takes the feedback of the patient, and just as a doctor stores the name of the patient along with his symptoms in a logbook, it stores all the symptoms in a file in directory.

In this work, the following assumptions are considered: (a) the patient speaks absolutely truth and (b) if a symptom occurs in a person then he has the disease.

Consider F be the fuzzy set that contains the membership value (μ) and symptoms:

$$\text{F} = \{(\mu_1, \text{symptom}_1), (\mu_2, \text{Symptom}_2), \ldots, (\mu_n, \text{ Symptom}_n)\} \tag{2}$$

We assume that probability of having Tuberculosis increases with the increase in the percentage of feeling. Suppose someone who feels extreme symptom is more likely to have disease than one who is feeling somewhat of the symptoms. Hence, probability of occurrence of the disease (P) α membership value (μ) that is, P = k.μ. Now when a person has extreme of the symptoms then probability of occurrence of the disease is 100%, then mathematically μ = 1, P = 1 and k becomes 1. Therefore, it implies P = μ, that is membership values arrive at a sound estimate of probability in which the membership function μ(x) is linear. Moreover, for getting better estimate, the following nonlinear membership function can be utilized.

$$\mu_{exponential} = \left(\exp\left(-\frac{x^2}{(x-5)^2}\right) - 1\right)^2, 0 \le x \le 5 \tag{3}$$

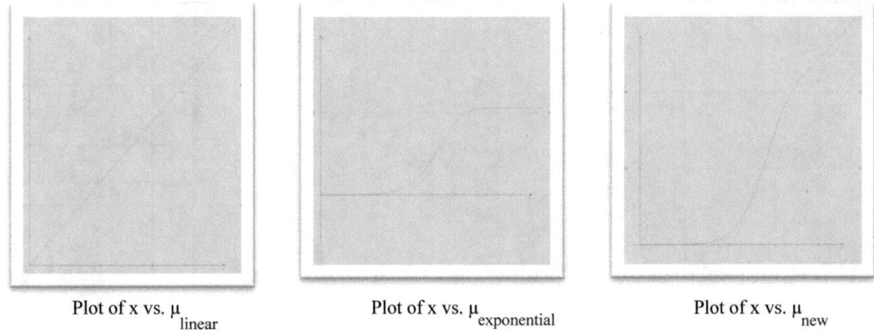

Plot of x vs. μ linear
Plot of x vs. μ exponential
Plot of x vs. μ new

Fig. 2 Characteristic of different membership functions

where x is the linguistic variable let us see how does it work when $x = 0, \mu_2 = 0$ when $x = 5, \mu_2 = 1$. Now one can arrive at similar membership functions and therefore a generalized theory of our probability measure is,

$$P \propto \mu_1 {}^* \mu_2 {}^* \mu_3 {}^* \ldots {}^* \mu_n$$

$$P = k * \mu_1 * \mu_2 * \mu_3 * \ldots * \mu_n \tag{4}$$

Now when $\mu_1 = \mu_2 = \mu_3 = \ldots = \mu_n = 1$, then we are sure that P = 1 or 100% and therefore k = 1, with the same reasoning as we had done previously.

$$P = \mu_1 * \mu_2 * \mu_3 * \cdots * \mu_n \tag{5}$$

The more we use fuzzy logic functions, the more accurate our probability estimate becomes. So, in our case the probability estimate becomes $P = \mu_{\text{linear}} * \mu_{\text{exponential}}$.

Now new membership function becomes $\mu_{\text{new}} = \mu_{\text{linear}} * \mu_{\text{exponential}}$.

In Fig. 2, the new membership function combines the nature of both types of membership function linear as well as our complicated exponential function, which is essential in the proposed system because it ensures the accurate estimate.

3.1 The Mathematics Behind Self-learning of Knowledge Base

Now consider the mathematics used in the project.

Figure 3 depicts the logic behind the self-learning approach. The variables described here are, P it gives the probability (Not in percent, we previously calculated this in percent). It is stored in knowledge base and retrieved for a particular patient.

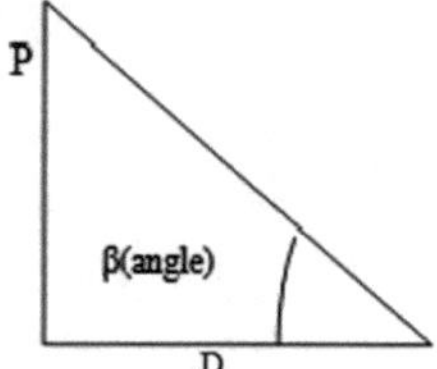

Fig. 3 Self-learning triangle

D is the dissatisfaction level of the patient. D is calculated as the total number of words the patient writes in his or her file. Now it is a common sense that the more the patient writes more is she dissatisfied, now consider the unitary method, D P Or 1 → P/D, i.e., for every unit word, the patient writes, we have P/D amount of probability of occurrence of the disease. Assuming whatever she writes in the file is not in our knowledge base, it gives the relative value of the patient's words as compared to others. If it exceeds a particular value, then the database of the patient needs to be updated, we have visualized this in the form of a triangle, β in the triangle represents the criticality factor $\tan(\beta) = P/D$. For small β, it can be approximated as $\beta = P/D$. We now want to determine the complicatedness of the patient's condition from intuition, we can assume that

$$\text{Complicatedness}\,\alpha\,\text{P(probability of occurrence)} \tag{6}$$

$$\text{Complicatedness}\,\alpha\,\text{D(dissatisfaction level)} \tag{7}$$

So from (6) and (7) we get,

$$\text{Complicatedness}\,\alpha\,\text{(P).(D)} \tag{8}$$

But we have a more firm basis of analyzing the situation by considering the area of the triangle, $A = (1/2) * (base) * (Height)$, under this situation, the area of our triangle becomes,

$$A = 1/2 * (D) * (P) \tag{9}$$

Now comparing (3) and (4), we get complicatedness is same as the area of triangle with constant of proportionality as half. Now consider the hypotenuse of the triangle and let us see what it gives,

$$H = \left(P^2 + D^2\right)^{\frac{1}{2}} \quad \text{or,} \; H = P * \left(1 + \left(\frac{D}{P}\right)^2\right)^{1/2} \quad \text{or,} \; H = P * \left(1 + \left(\frac{1}{\beta}\right)^2\right)^{\frac{1}{2}}$$

Let, $\gamma = 1/\beta$, γ physically represents the simplicity factor, i.e., more this value more can we be assured that the patient's probability of occurrence of the disease from the unknown symptoms is less (i.e., the degree of valuelessness of the words he/she has written in the file). $H = P * \left(1 + (\gamma)^2\right)^{1/2}$

Now $\left(1 + (\gamma)^2\right)^{1/2}$ can be considered like a control variable in our process, i.e., more its value more can we be assured of patient's wellness and P can be considered as the main variable, i.e., more the P increases more the patient is helpless. Hence, we conclude that H represents the foreign Improbability factor, i.e., what is the ultimate condition of the patient, when the value of foreign Improbability factor is high we are assured that there is no link of the foreign symptom with our knowledge base. We will describe what foreign symptom is shortly $(\mathrm{dH}/\mathrm{d}\gamma)$ rate of change of foreign Improbability factor with simplicity factor; we use a bit of differential calculus. It is given as, $dH/d\gamma = P * \gamma * (1 + \gamma^2)^{-\frac{1}{2}}$ or, $\mathrm{dH}/\mathrm{d}\gamma = \mathrm{P} * \left(\left(\frac{1}{\gamma}\right)^2 + 1\right)^{-\frac{1}{2}}$.

Lim $\gamma \rightarrow 0$ (dH/dγ) approaches P. This is the extreme case that can ever happen provided P as constant(i.e., we already determined P from program and this calculation is done after user enters feedback in the prolog just ordinary text, i.e., we use file input–output concept in prolog then store the data entered and count the number of words).

We note that P and D are asserted, i.e., stored in the knowledge base from which we retrieve the values for the current patient and we use crisp set to determine whether the person has particular TB or not then we use fuzzy logic to get the percentage of uncertainty then the above calculations follow. When it is said to update, someone has to manually see the file of the patient and update the knowledge base.

We see that considering P as a parameter when P increases, H rises more rapidly with γ. As we have previously mentioned, when P increases the rate of rise of foreign Improbability factor is also increased, this can be justified on the grounds that the patient's symptoms are well known to us, that is why we are sure that the disease of patient had occurred is within our knowledge base. That is higher the values of P more is the foreign improbability factor so more are we assured that there is no link of foreign symptom with our knowledge base. Foreign symptoms are the symptoms that need to be discovered from the feedback file of the database.

3.2 Description of the Work

Consider when a patient interacts with the system (in the console user just types 'run.' command) Run module is called that uses the Boolean LOGIC that was described in Background study section to determine whether the person has the particular type of Tuberculosis or not. Now consider we have various types of Tuberculosis, our focus in this study was to determine the various types of

Tuberculosis and to diagnose them properly. Therefore, we collected various types of symptoms related to various types of Tuberculosis like lung TB [3], bone TB [4], articular TB [5], genital TB [6], renal TB [7], spinal TB [8], miliary TB [9], abdominal TB [10], ocular TB [11], Tuberculosis of Tonsils [12], hepatic TB [13], multidrug-resistant TB [14]. Based on this information, we organized our knowledge base that used these data. Now that after the identification is done, percent module is called (in console user just enters 'percent.' command) in that we implement the fuzzy logic described in methodology section and it gives the probability in percentage. Not only that we store this value in P variable. Then as we know that, the bacteria that causes the disease is constantly evolving, we now have the so-called superbugs [15] so the symptoms exhibited by the disease may change owing to evolution [16]. Hence, we prompt the user to enter any other symptoms that he/she is feeling by calling More module (user just enters 'more.' in the console). Then the user is prompted to run Display module. It takes the Name of the patient and creates a file with that name that contains his feedback just as a physician maintains his LOGBOOK, he maintains it because to study if the patient is showing any other new symptoms that are not in his knowledge base, i.e., he studies and learns and becomes more experienced each time. Our system stores the word count in the variable D. Now, the physician is intelligent he knows which symptom to consider and which not to consider. We do this intelligently using mathematics described in the Sect. 3.1, the mathematics behind updating of Knowledge Base (KB). This updating is done finally by calling the update module. This module tells us whether to update the knowledge base of the current patient or not. It is the responsibility of a human to update manually the database from his file. We thought it was the best for a human to do the task. It also gives web support.

3.3 Accuracy of the Estimate of Membership Function

We will try to derive a formula for this. First let us see what is the physical significance of the area under the μ versus x, i.e., $A = \int_0^5 \mu * dx$. Now before discussing about the significance let us do a bit of mathematics. Consider Heaviside's function or unit step function. In crisp set, the μ versus x graph there would be a Heaviside's function with its peak value at 5, i.e., at boundary. (We use Heaviside instead of Dirac's delta function as delta function gives infinite value at x = 5) $\mu = H(x - (5 + \varepsilon))$ where $\varepsilon \to 0$. Such that $A_{\mu = H(x - (5 + \varepsilon))} = 1$, but in taking crisp set, ignore other values and hence it is a kind of deviation. To have a very sharp rising peak somewhere and nearly flat regions somewhere else in the μ versus X curve, we want an evenly distributed function not necessarily linear. So greater the value of A differs from this value better is our estimate of the membership function. Hence, if we denote:

$$M = \left| \left| A_{H(x-(5+\varepsilon))} \right| - \left| A_{\mu} \right| \right|$$

Then M can be a measure of our criterion, if M $\rightarrow$ 0 the choice of our membership function is really bad else good. But there can be a function that has a unit area but does not behave like H(x) in particular we note that if our function rises very rapidly then it cannot give us a very good estimate of probability. Hence, modify the definition of M slightly:

$$M = \frac{\left| \left| A_{H(x-(5+\varepsilon))} \right| - \left| A_{\mu} \right| \right|}{r} + 1.1$$

where $r = (d\mu/dx)_{x=5}$ at x = 5. Now when M $\rightarrow$ 1, made a bad estimate, let us analyze this equation. When the graph rises very rapidly, r is very high and when r $\rightarrow$ ∞ M $\rightarrow$ 1. So, as M $\rightarrow$ ∞, implies a better estimate. Hence M is ultimate: $M = \frac{\left|1 - \left|A_{\mu}\right|\right|}{r} + 1.$

4 Result and Analysis

It is to be noted that after designing the Prolog program, the respective files where the program is written are consulted in the SWI-Prolog environment. After that, various queries are typed to get the desired result as shown in Fig. 4. Now, detection of a variety of diseases can be performed by typing various queries. Let us see where our intelligent system stands in this real-life scenario: Suppose this person approaches our AISLDr.

Here, modified the program to incorporate the $\mu_{new} = \mu 1 * \mu 2$. However, to investigate the works execute correctly, we need to improve the precision and define deviation $\delta = \mu_{linear} - \mu_{new}$. So we need a minimum of deviation, as linear was a good estimate but our modified membership function gives a better estimate, mathematically, $\frac{d\delta}{dx} = 0$, implies $x \approx 1.766351032859$. The deviations as per linguistic variable are depicted in Table 2.

So, maximum deviation of 34.85% for x = 2 is acceptable for this work.

$$\delta(\text{average}) = \frac{1}{5}\left(\int_{0}^{5} \delta . dx\right)$$

Now this part of the integral as shown below cannot be performed by ordinary methods so use a technique of numerical integration using Simpson's rule and the result comes out to be 0.1448 or 14.48%. Upon running our system, the probability of occurrence and the better estimate of the disease are shown in Fig. 5.

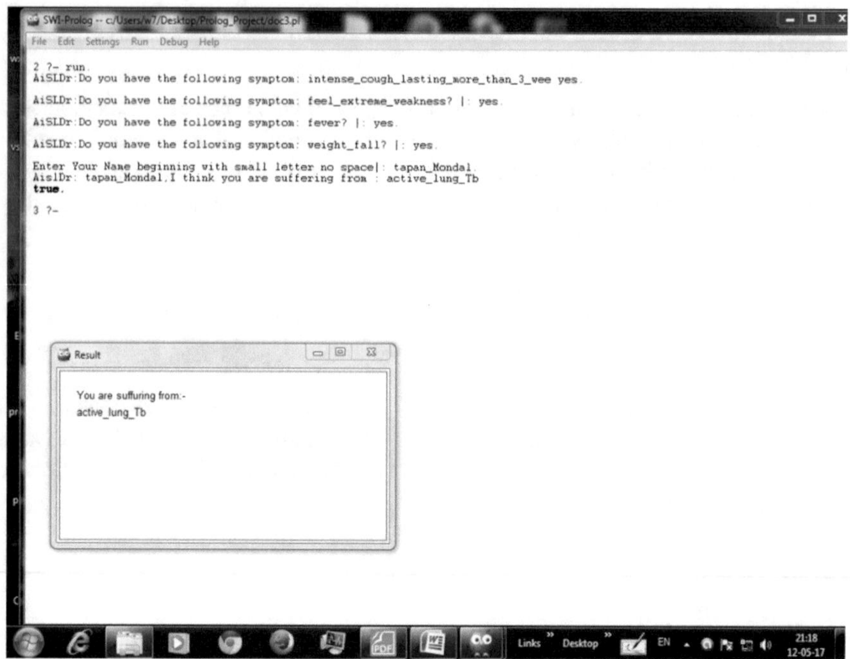

Fig. 4 Intelligent system detected the lung Tuberculosis

Table 2 Deviations as per linguistic variable

x	μ_{linear}	μ_{new}	δ
0	0	0	0
1	0.2	$7.43 * 10^{-4}$	0.199
2	0.4	0.0515	0.3485
3	0.6	0.4801	0.1199
4	0.8	0.79	0.01
5	1	1	0

5 Conclusion and Future Scope

Use of AI in medical field has proved to be an effective tool in providing proper diagnosis and hence in providing proper medication. In this paper, different types of disease based on various infectious agents have been implemented using SWI-Prolog which show better diagnosis compare to others [17]. It can also be used for diagnosing other diseases. The method [18] for implementing various diseases using AI have been studied but our method is much more intelligent and user friendly as it is simple and lucid. This method is still in its developing stage. With further study and assessment, medical professionals would be able to use it in fields

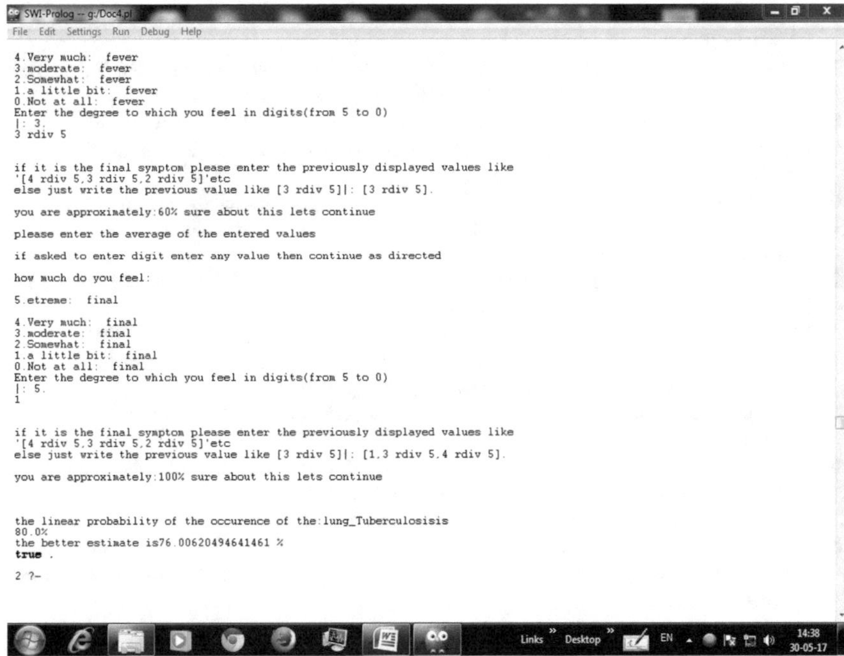

Fig. 5 Probability of occurrence of the lung Tuberculosis

like diagnosis, treatment of diseases, providing proper medication, development of auto-prescription generating machine and many more [2].

With further study and research assessment, enhance the system in such a way that the medical professionals would be able to use it in fields like diagnosis, treatment of diseases, providing proper medication, development of auto-prescription generating machine and many more in sustainable manner. A Web page could be designed in such a way that the results of programs are stored in a statistical manner and by giving access to real doctors; the mentioned hazards could be minimized to some extent. Recommender system could be recommended for such enhancement.

References

1. Amanda Page, Health centre: 9 ways artificial intelligence affecting medical field, 9 April 2012
2. Lallemand, N.C.: Health policy brief: reducing waste in health care. Health Aff. 13 Dec 2012
3. Tuberculosis, Lung. Springer Reference. https://doi.org/10.1007/springerreference_137647
4. Sankaran, B.: Tuberculosis of bones & joints. Indian J. Tuberc. (1993)
5. Reuter, H.: Overview of extrapulmonary tuberculosis in adults and children. Tuberculosis 377–390. https://doi.org/10.1016/b978-1-4160-3988-4.00034-2 (2009)

6. Jassawalla, M.J.: Genital tuberculosis—a diagnostic dilemma. J. Obstet. Gynecol. India **56**, 203–204 (2006)
7. Reuter, H.: Overview of extrapulmonary tuberculosis in adults and children. Tuberculosis 377–390. https://doi.org/10.1016/b978-1-4160-3988-4.00034-2 (2009)
8. Jain, S.: Spinal Tuberculosis Diagnosis & Management. Retrieved from www.aiimsnets.org
9. Harris, H. W., Menitove, S.: Miliary tuberculosis. Tuberculosis, 233–245. https://doi.org/10.1007/978-1-4613-8321-5_21 (1994)
10. Rathi, P., Gambhire, P.: Abdominal tuberculosis. J. Assoc. Physicians India **64** (2016)
11. Mehta, S.: Ocular tuberculosis in adults and children. Tuberculosis, 476–483. https://doi.org/10.1016/b978-1-4160-3988-4.00046-9 (2009)
12. Cowan, D., Jones, G.: Tuberculosis of the tonsil: case report and review. Tubercle **53**(4), 255–258 (1972). https://doi.org/10.1016/0041-3879(72)90003-7
13. Purl, A.S., Nayyar, A.K., Vij, J.C.: Hepatic tuberculosis. Indian J. Tuberc. (1994)
14. Clinical Diagnosis of Drug Resistant Tuberculosis. Drug Resistant Tuberculosis, 18–21. https://doi.org/10.2174/9781681080666115010006 (2015)
15. Antimicrobial resistance. (11 April 2017). In Wikipedia, The Free Encyclopedia. Retrieved 15:49, 15 April 2017
16. Denamur, E., Matic, I.: Evolution of mutation rates in bacteria. Mol. Microbiol. **60**(4), 820–827. https://doi.org/10.1111/j.1365-2958.2006.05150.x (2006)
17. http://www.differencebetween.com/ difference between prolog and lisp
18. E. P Ephzibah School of Information Technology and Engineering, VTT University, Vellore, Tamil Nadu, India, A Hybrid Genetic Fuzzy Expert System For Effective Heart Disease Diagnosis

Kernel-Based Naive Bayes Classifier for Medical Predictions

Dishant Khanna and Arunima Sharma

Abstract Researchers and clinical practitioners in medicine are working on predictive data analysis at an alarming rate. Classification methods developed using different modeling methodologies is an active area of research. In this paper, live dataset in clinical medicine is used to implement recent work on predictive data analysis, implementing a kernel-based Naïve Bayes classifier in order to validate some learned lessons for predicting the possible disease. With the medical diagnosis prediction, the aim is to enable the physician to report the disease, which might be true. The input training dataset for the classifier was taken from a government hospital.

Keywords Classifiers · Naïve Bayes · Kernel density · Modeling Live dataset · Prediction · Precision · Kernel Naïve Bayes

1 Introduction

In the recent times, 'data mining' has become a quite frequent term to be employed in the medicinal literature along with computational sciences. Altogether, the term 'data mining' does not elucidate any meticulous characterization but a common apprehension of its illustration: the practice of novel approaches as well as tools for analyzing voluminous data and datasets. Data mining can be defined as the progression of opting, reconnoitering, and modeling voluminous data for the sole purpose of discovering unfamiliar/unidentified patterns or associations between the data provided as the input for the evaluating system, which provides a comprehensible and beneficial result to the data analyst [1]. Devised during the period of

D. Khanna (✉)
Bharati Vidyapeeth's College of Engineering, New Delhi, India
e-mail: dishant.khanna1807@gmail.com

A. Sharma
Columbia University, New York, NY 10027, USA
e-mail: s.arunima@columbia.edu

V. Bhateja et al. (eds.), *Intelligent Engineering Informatics*, Advances in Intelligent Systems and Computing 695, https://doi.org/10.1007/978-981-10-7566-7_10

mid-1990s, today, the terminology 'data mining' has become an alternative expression for 'Knowledge Discovery in Databases,' also known as KDD process, which, as anticipated by Fayyad et al. [2], accentuated on the enactment of the process of data analysis instead of using other explicit analysis approaches on the data.

The process for 'Knowledge Discovery of Databases' includes many steps for gaining knowledge from the given dataset. The dataset which is provided is usually in the raw format and needs to be preprocessed, and various selection algorithms are applied on the dataset to select useful features from the large voluminous dataset. In the second step, preprocessing and transformation of data is done to convert the dataset into something useful and obtain knowledge from the dataset. Next, different data mining algorithms are applied on the processed and transformed dataset to get the knowledge out of the raw data that we were provided with. The complete process is explained by Fayyad et al. [2], in the overview of the 'Knowledge Discovery of Databases.'

The process by which continuous variables are supervised is processed by discretization in case of Bayesian network-based classifiers, or it is presumed from the beginning that they trail a Gaussian distribution. This paper presents the kernel-based Naïve Bayesian net model which is used in case of a problem which uses supervised classification method, for predicting probable tuberculosis disease in a person. This model is a Naïve Bayesian network by means of which the true density of a continuous variable is evaluated by means of different kernels that can be implemented in the Naïve Bayesian network, in this case for the prediction of tuberculosis for a person. Above and beyond, the tree-augmented Naïve Bayesian network, complete graph classifier, and k-dependence Bayesian classification model are confirmed to the innovative kernel-based Naïve Bayes net paradigm. Additionally, the robust steadiness assets of the offered Naïve Bayesian classifiers are demonstrated and an estimator, for tuberculosis disease, of the reciprocated data constructed on kernels is offered by the means of this paper.

Supervised classification [3, 4] is a formidable exercise in pattern recognition and data analysis. This type of classification entails the creation of a classification algorithm, which is defined by a function which provides a requisite tag or an identification label of a particular class to the instances that are described by a group of variables described in a set, provided in the dataset. Among a large number of classifier paradigms available for use, the Bayesian network (BN) [5, 6] majorly incorporates PGMs which are defined as probabilistic graphical models [7, 8] and are well acknowledged for effectiveness as well as efficiency in the areas that constitute ambiguity. Probabilistic graph models of similar nature make an assumption which is for every arbitrary variable trail, a process which is known as the conditional probability distribution, with a provided explicit value belonging to the variable's super or parent variables. Among the domain variables, for the process of encoding the joint distribution's factorization, Bayesian networks are used. They are generally designed by means of using the property of conditional independencies which are generally characterized by means of a directed graph structure. This property, when it is used collectively along with the Bayes rule, is utilized for the process for performing classification on the provided data. For the purpose of instigating the classification algorithm on the data provided at the input of the classifier, only two kinds of variables are taken under consideration: first, those variables or set of discrete predictors apart from the aforementioned variable types have the class variable of class C. The project that is being depicted in the paper illustrates a collection of generative classification algorithms comprising of a novel genus of Bayesian networks which are named as kernel-based Naïve Bayesian Classifier networks.

One of the most simpleton classification algorithms which is based on Bayesian networks is known Naive Bayesian (NB) networks [9–11]. Naïve Bayes classification algorithm is founded on the hypothesis of conditionally independent predictors of the provided classes. Despite the strong assumption, the performance shown by these classifiers is astonishingly high, even with the databases that do not hold the conditional independence assumption [12]. A Bayesian network-based classifier is determined by the composition of the graph, density functions, and the independence relations between the variables modeled into distributions. The following mechanisms are to be considered for the purpose of modeling a continuous variable's density function:

(1) Direct estimation of the density function performed in a parametric way using variables like Gaussian densities.
(2) Estimating the probability distribution of the discretized variable by using the multinomial probability distribution.
(3) Nonparametric estimation of the density function using variables like the kernel density functions [13].
(4) Direct estimation of the density performed in a semi-parametric way using variables like finite mixture models [14].

Supervised classification is considered the most popular approach for classification according to the research literature, which can be described as the estimation of discretized values using multinomial distribution. Bayesian networks tend to incorporate an understanding that the variables within the system can be constituted in multinomial probability distribution that in turn is called the Bayesian multinomial network [6, 15] (BMN). The aforementioned paradigm is specifically used to manage the discrete variables, and hence in case of a continuous variable imbibed in the data, the process of discretization is followed in turn there is consequent loss of information and knowledge [16]. Despite the loss, the process of using discretization combined with multinomial operation ensures adequately accurate estimation of the true density required for classification purposes [15]. However, negative repercussions of the process occur when a graph is supposed to be modeled along a complicated construction along with iteratively discretized variables for the reason that the quantity of the constraints required to be predictable can be excessive. Also, the quantity of constraints essential to be predictable is directly related to the chance of overfitting, thereby increasing numeral of constraints increases the menace of occurrence overfitting. Furthermore, the amount of applicable cases necessary for computing each and every constraint can be extremely little and, consequently, the strength and durability of statistic evaluated might be affected as explained above [17]. A Bayesian multinomial network-based classification induction algorithms of numerous structural convolution-based battery are offered in the works: Bayesian network-augmented Naive Bayes [15], general Bayesian network [15], tree-augmented Naive Bayes (TAN) [18], Naive Bayes (NB) [4, 10, 11], semi-Naive Bayes [19, 20], and k-dependence Bayesian (kDB) classifier [21].

2 Density Estimator for Kernel

D-dimensional estimator for kernel [8]:

$$f(x:H) = n^{-1} \sum_{i=1}^{n} K_H\left(x - x^i\right) \tag{1}$$

where K (•) is the kernel—a function with nonnegative values that integrate to one and comprises of mean value that equates to zero; whereas, **H** is a smoothing parameter, which is positive, never zero, and is described as the bandwidth. A scaled kernel is described as a kernel with subscript **H**, which is mathematically described as follows:

$$\mathrm{KH(x)} = 1/\mathrm{h}\ \mathrm{K(x/h)} \tag{2}$$

H is spontaneously chosen whose value is reserved to be as small as allowed by the dataset which is provided as the input of the system; nevertheless, there always exists a trade-off among the variance of the estimator and the bias. With an assumption that K is d-dimensional density function, it can be characterized by:

(1) The bandwidth matrix H.
(2) The kernel density K selected (Fig. 1).

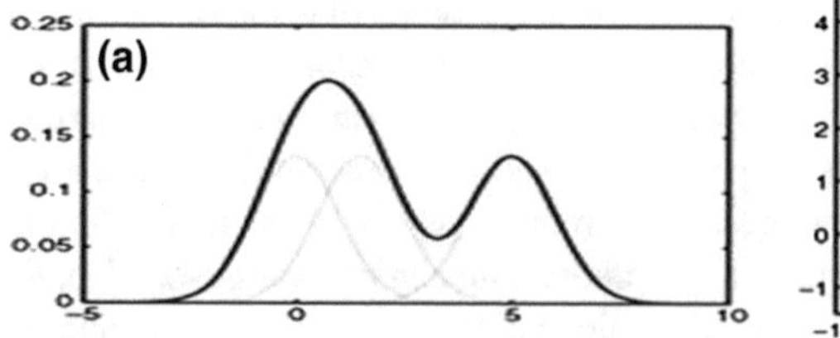

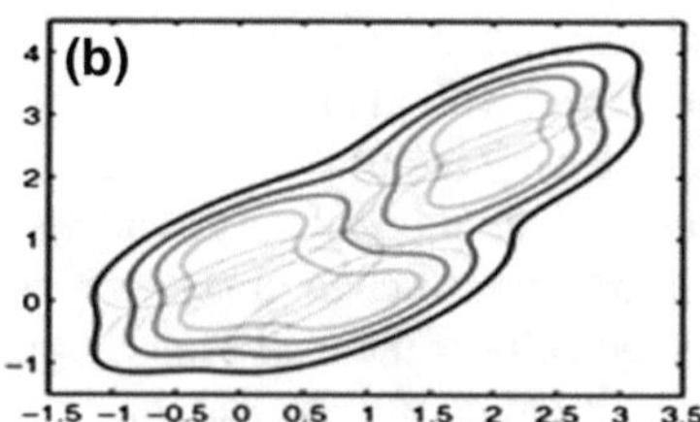

Fig. 1 Univariate (**a**) and bivariate Gaussian kernel density estimators (**b**). Straight bold lines: contribution of each kernel to the density estimation based on kernel. Crooked likes: contribution of each kernel to density estimation

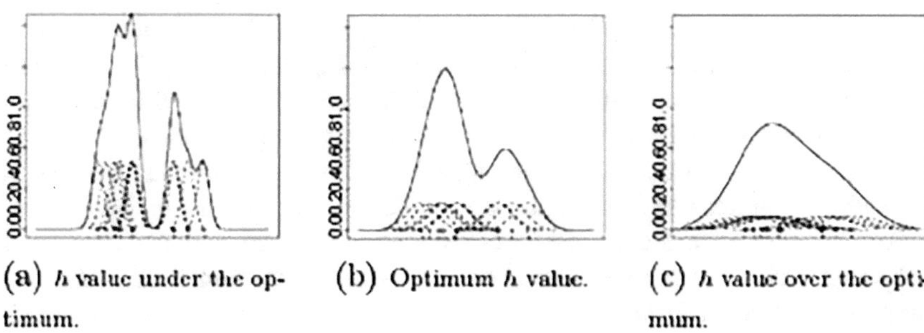

(a) h value under the optimum. (b) Optimum h value. (c) h value over the optimum.

Fig. 2 Effect in the density estimation with variable degree of smoothing, controlled by the parameter 'H'

'H' depicts a factor describing the scale used to determine the extent of kernel at apiece of the coordinate axes' direction. The process of estimating the kernel density is constituted along with a scaled kernel centered on individual observations which are provided in the dataset as input for the system. So, the summation of bumps that are placed at the individual observations is described as the kernel density estimator; the shape of those bumps is determined by the kernel function K. At any point or value x, the value of the kernel estimate is mathematically described as the mean of the n kernels which are described at that position. The kernel could be viewed as diffusing a 'probability mass' of magnitude n = 1 in relation with every individual data point existing within the neighborhood of that point described. On merging the contributions from every individual data point would entail explaining for areas having numerous observations, the true density is anticipated to have relatively larger value. It is not an essentially crucial part for choosing the shape of the kernel function. Nevertheless, for the process of estimating the density, it is considered a crucial process for selecting the value which describes the bandwidth matrix [22]. Examples for the bivariate and univariate estimation of density at points are shown in the Fig. 2a, b.

3 Design and Implementation

A Naive Bayes classifier is described as an elementary probabilistic classifier which is constructed on the implementation of Bayes' theorem along a dedicated set of assumptions for conditional independence. 'Independent feature mode' would be considered as a more vivid terminology which explains the original probability model. In straightforward terms, the algorithm behind the working of a Naive Bayesian classification process presumes that the occurrence (or non-occurrence) of any distinct property of the provided class also known as attributes remains unaffected by the occurrence (or non-occurrence) of any additional feature existing in the data.

The key benefit of using the Naive Bayesian classification algorithm is only a small amount of data is required as the training data for training the system to estimate the variances as well as means of all the variables provided in the dataset, which is an essential condition for the process of performing classification. Only the variance of the variables for every individual label is needed to be evaluated and not the entire covariance matrix for the data which is provided as an input for the system for the reason that all the independent variables are grouped into classes. The kernel Naïve Bayes net can also be implemented on the numerical attributes in contrast to the Naive Bayesian classifier.

A weighting function which is utilized in nonparametric estimation procedures is known as a kernel. Different kernels are implemented for the process of kernel density estimation technique for estimation of the density functions of the random variables, or in kernel regression mechanism for the estimation of the conditional expectation of a random variable in the dataset. These kernel density estimators belong to a special class of estimating functions which are known as nonparametric density estimators. In contrast to parametric estimators where the density estimator has a fixed functional form and the parameters of this function are the only information that is needed to be stored, there is no fixed structure for nonparametric density estimators and they rely on the data points provided in the data to reach an estimate.

The implemented process for medical predictions includes the following steps:

1. Dataset retrieval.
2. Data preprocessing.
   a. Data cleaning.
   b. Replacing missing values.
   c. Outliers identification and noise removal.
   d. Data transformation (scaling, conversion, and normalization).
3. Machine-learning algorithm (kernel-based Naïve Bayes classifier).
4. Prediction and performance calculation.

4 Datasets

The associated data is a real-time data taken from a software called NIKSHAY by Government of India and comes under a national program named RNTCP. It is monitored and supervised by CTD and WHO. NIKSHAY is a solution that is Web-based which is used for keeping track of patients suffering from TB or either came for TB test, for the purpose of supervising Revised National Tuberculosis Program (RNTCP) in an effective manner, for monitoring the records of the patients registered, is advanced by the National Informatics Centre (NIC). RNTCP is an initiative for controlling the spread of tuberculosis (TB) and is state run by the Government of India which aims for an undertaking for accomplishing 'TB-free

Table 1 Preprocessing done on the dataset

Category	Old values	New values
Sex	M/F	1/0
Sputum tests	Neg, SCN1 to SCN8, +1 to +3	0, 1 to 8, 100 to 300
Category	Category 1/Category 2	½
Classification	Pulmonary/Extra pulmonary	½
Type	New, Retreatment others, Treatment after default, Relapse, New others, Failure, Transfer in	1 to 7, respectively

India' as mentioned in the strategic plan of 2012–2017. A formal permission has been taken officially from the concerned authorities. The data concerns a general population during 2013 and is spread over all age groups, genders, and living conditions in district Meerut, Uttar Pradesh, India.

Because the data is on ground real-time data, it is spread across myriad divisions of age, gender, and living conditions. The most enriching factor about the dataset is minimum bias. The live dataset is of tuberculosis disease with classes, ID of the patient, name of patient, sex of patient, address of patient, age of patient, mobile number of the patient, TB treatment (whether the patient had treatment before), TB year, DOT operator name, sputum results of the patient, category of TB, type of TB, HIV status, treatment outcome.

The obtained dataset was in raw format, which needed to be processed. The preprocessing of data was done. The values needed to be changed from polynomial to numerical. The tables show the preprocessing done on the dataset (Table 1):

5 Results

After preprocessing of the data is completed, the data is induced to the network for predicting the tuberculosis disease. This processed dataset was provided as an input to the kernel Naïve Bayes classifier Java code, and the output is shown as in Table 2.

Table 2 Confusion matrix

	True extra pulmonary	True pulmonary	Class precision (%)
pred. Extra pulmonary	364	95	79.30
pred. Pulmonary	37	903	96.06
Class recall	90.77%	90.48%	

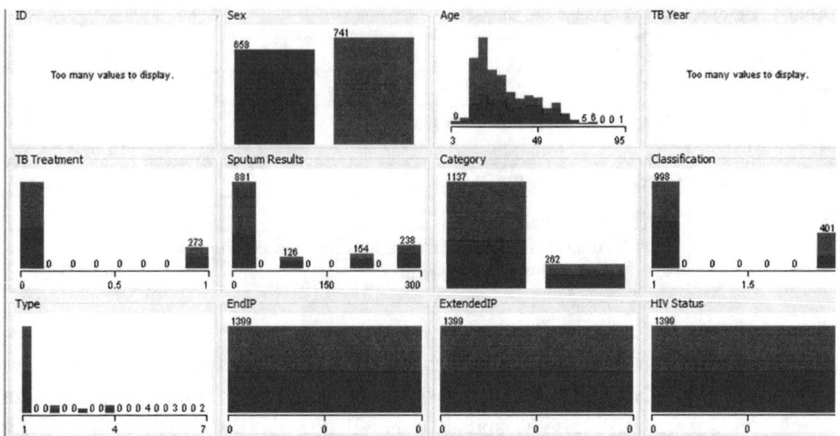

Fig. 3 Graphical representation of dataset (Class: Sex(M/F))

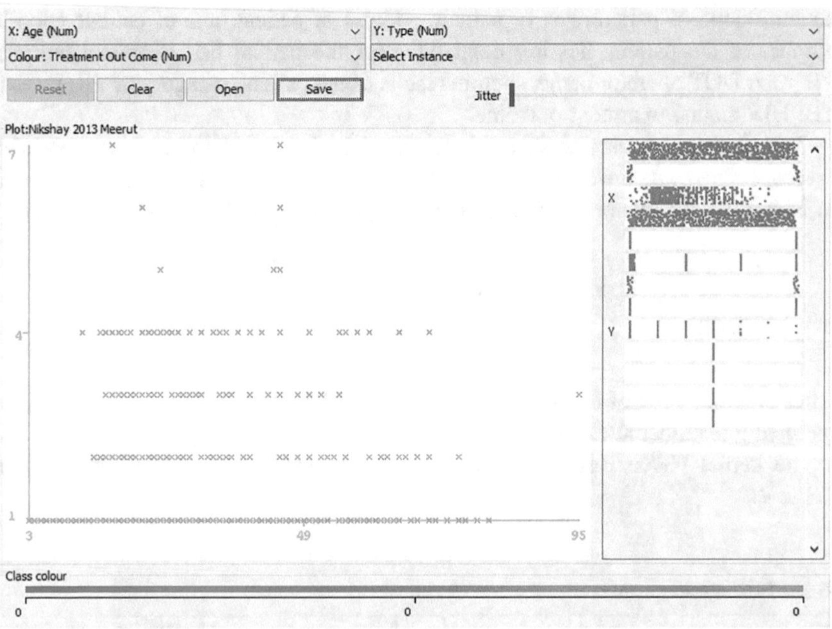

Fig. 4 Graphical representation (Age vs. TB Type)

6 Conclusion

The paper offers a kernel-based Naïve Bayesian classification net (KBN) model for the purpose of illustrating the process of supervised classification on the given live dataset for tuberculosis. The aforementioned network is defined as a Bayesian net that is used for the process for determining of the density of continuous variables by means of estimators which uses kernels. The implemented method for kernel-based Naïve Bayes network results in a performance of 90.56%. While the performance measured for the classical Naïve Bayesian network measure is 77.01%. This implemented Naïve Bayesian network based on kernel (KBN) models for the process of supervised classification results in higher accuracy than the classical Naïve Bayesian network (Figs. 3, 4, and 5).

The proposed technique implemented was tested with live dataset for tuberculosis, which was taken from the Government of India, from the RNTCP program. This implementation can be used for medical predictions for different diseases (Tables 3 and 4).

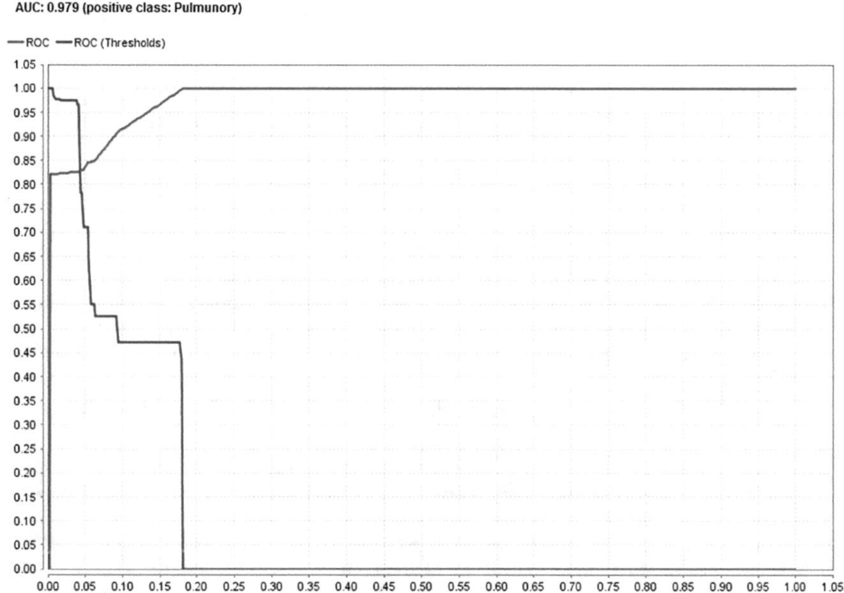

Fig. 5 AUC plot

Table 3 Detailed accuracy by class

	TP Rate	FP Rate	Precision	Recall	F-Measure	ROC Area	Class
Weighted Avg	0.983	0.05	0.793	0.9077	0.933	0.976	EP
	0.959	0.017	0.961	0.9048	0.971	0.976	P
	0.9606	0.027	0.905	0.9048	0.96	0.976	

Table 4 Evaluation on test set

Property	Value	Percentage (%)
Instances correctly classified	1273	90.56
Instances incorrectly classified	126	9.44
Kappa statistics	0.9035	
Mean absolute error	0.072	
Root mean squared error	0.1907	
Relative absolute error	17.5936%	
Root relative squared error	42.1831%	
Total number of instances	1399	

Acknowledgements This work has been possible thanks to Ms. Narina Thakur, Head of Department, Department of Computer Science Engineering, BVCOE, New Delhi.

References

1. Aladjem, M.: Projection pursuit fitting Gaussian mixture models. In: Proceedings of Joint IAPR, vol. 2396. Lecture Notes in Computer Science, pp. 396–404 (2002)
2. Fayyad, U. et al.: Knowledge discovery in databased: an overview. In: Releational Data Mining, pp. 28–47
3. Aladjem, M.: Projection pursuit mixture density estimation. IEEE Trans. Signal Process. **53** (11), 4376–4383 (2005)
4. Bilmes, J.: A gentle tutorial on the EM algorithm and its application to parameter estimation for gaussian mixture models. Technical Report No. ICSI-TR-97021, University of Berkeley (1997)
5. Bishop, C.M.: Neural Networks for Pattern Recognition. Oxford University Press (1995)
6. Bishop, C.M.: Latent variable models. In: Learning in Graphical Models, pp. 371–403 (1999)
7. Bishop, C.M.: Pattern Recognition and Machine Learning. Information Science and Statistics, Springer, Heidelberg (2006)
8. Bottcher, S.G.: Learning Bayesian Networks with mixed variables, Ph.D. Thesis, Aalborg University (2004)
9. Castillo, E., Gutierrez, J.M., Hadi, A.S.: Expert Systems and Probabilistic Network Models. Springer (1997)
10. Chow, C., Liu, C.: Approximating discrete probability distributions with dependence trees. IEEE Trans. Inf. Theory **14**, 462–467 (1968)

11. Cormen, T.H., Charles, L.E., Ronald, R.L., Clifford, S.: Introductions to Algorithms. MIT Press (2003)
12. Domingos, P.: A unified bias-variance decomposition and its applications. In: Proceedings of the 17th International Conference on Machine Learning, Morgan Kaufman, pp. 231–238 (2000)
13. Domingos, P., Pazzani, M.: On the optimality of the simple Bayesian classifier under zero-one loss. Mach. Learn. **29**, 103–130 (1997)
14. Fayyad, U., Irani, K.: Multi-interval discretization of continuous-valued attributes for classification learning. In: Proceedings of the 13th International Conference on Artificial Intelligence, pp. 1022–1027 (1993)
15. Casella, G., Berger, R.L.: Statistical Inference. Wadsworth and Brooks (1990)
16. Friedman, J.H.: On bias, variance, 0/1—loss, and the curse-of-dimensionality. Data Min. Knowl. Disc. **1**, 55–77 (1997)
17. Friedman, N., Geiger, D., Goldszmidt, M.: Bayesian network classifiers. Mach. Learn. **29**, 131–163 (1997)
18. Cheng, J., Greiner, R.: Comparing Bayesian network classifiers. In: Proceedings of the 15th Conference on Uncertainty in Artificial Intelligence, pp. 101–107 (1999)
19. Chickering, D.M.: Learning equivalence classes of Bayesian-network structures. J. Mach. Learn. Res. **2**, 445–498 (2002)
20. Cover, T.T., Hart, P.E.: Nearest neighbour pattern classification. IEEE Trans. Inf. Theory **13**, 21–27 (1967)
21. DeGroot, M.: Optimal Statistical Decisions. McGraw-Hill, New York (1970)
22. Fukunaga, K.: Statistical Pattern Recognition. Academic Press Inc. (1972)
23. Bouckaert, R.: Naive Bayes classifiers that perform well with continuous variables. In: Proceedings of the 17th Australian Conference on Artificial Intelligence, pp. 1089–1094 (2004)
24. Gurwicz, Y., Lerner, B.: Rapid spline-based kernel density estimation for Bayesian networks. In: Proceedings of the 17th International Conference on Pattern Recognition, vol. 3, pp. 700–703 (2004)

AKM—Augmentation of K-Means Clustering Algorithm for Big Data

Puja Shrivastava, Laxman Sahoo, Manjusha Pandey and Sandeep Agrawal

Abstract Clustering for big data analytics is a growing subject due to the large size of variety data sets needed to be analyzed in distributed and parallel environment. An augmentation of K-Means clustering algorithm is projected and evaluated here for MapReduce framework by using the concepts of genetic algorithm steps. Chromosome formation, fitness calculation, optimization, and crossover logics are used to overcome the problem of suboptimal solutions of K-Means clustering algorithm and reduction of time complexity of genetic K-Means algorithm for big data. Proposed algorithm is not dealing with the selection of parents to be sent to mating pool and mutation steps, so the performance time is improved.

Keywords Big data analytics · K-Means · Genetic clustering Chromosome · Optimized cluster

1 Introduction

Big data clustering is a challenging task due to the variety of large size data collected every day at various locations and needed to be analyzed in collective manner to visualize the messages/patterns/customer behavior and future scope hidden in it. This paper presents enhancement of MapReduce-based K-Means clustering algorithm by utilizing the logic of genetic algorithm. In the Map step,

P. Shrivastava (✉) · L. Sahoo · M. Pandey · S. Agrawal
School of Computer Engineering, KIIT University, Bhubaneswar, Odisha, India
e-mail: pujashri@gmail.com

L. Sahoo
e-mail: laxmansahoo@yahoo.com

M. Pandey
e-mail: manjushapandey82@gmail.com

S. Agrawal
e-mail: sandygarg65@gmail.com

V. Bhateja et al. (eds.), *Intelligent Engineering Informatics*, Advances in Intelligent Systems and Computing 695, https://doi.org/10.1007/978-981-10-7566-7_11

chromosome formation concept gives facility to fix the no. of clusters in each data node; fitness function optimizes the chromosomes after every iteration of K-Means. In Reduce step, chromosomes from all data nodes are collected and crossover logic is performed by combining the similar chromosomes and taking the average of chromosomes in which encoded centroids having very near values. Once again run K-Means on finally obtained chromosomes and the result will be global cluster, since it has seen whole data processed in every node.

K-Means is one of the most used algorithms from traditional data mining clustering techniques because of its simplicity, ability to deal with large no. of attributes, and providing good quality clusters with the **n*k*d** computational complexity where **n** is the no. of elements in data space, **k** is count of clusters to be identified, and **d** is the no. of attributes/dimensions [1]. The logic of K-Means algorithm is to compute Euclidean distance among objects and make clusters in the way that intra-cluster distance should be minimum and interclusters distance should be maximum.

Results of K-Means clustering are dependent on cluster center initialization; the center initialization will decide the resultant clusters. The major drawback of K-Means clustering algorithm is the local convergence which means it is not able to provide globally optimum results. For different data sets, diverse versions of K-Means clustering are chosen. The enhancement of K-Means clustering algorithm is presented, implemented, and compared in this paper through the concepts of genetic algorithm.

Genetic algorithms are search and optimization techniques based on the logic of natural genetics and part of computational intelligence domain. Dimensions of data set (search space) are encoded in chromosome (string), and collection of such chromosomes is called a population. The first step of algorithm is initialization of random population (selection of data points from data set). Each chromosome is having one objective and fitness function to represent the degree of its goodness. Fittest strings are selected as parents to breed new generation with the help of crossover and mutation steps. The selection, crossover, and mutation steps are repeated till the no. of iterations is fixed [2].

The approach of implementing genetic algorithm logic in the MapReduce framework for the improvement of K-Means clustering algorithm is the core contribution in the domain of big data analytics.

Presented paper is organized as introduction, state of the art, enhanced K-Means clustering algorithm, results and discussions, plus conclusion parts.

2 State of the Art

Implementation of data mining techniques for big data analytics needs several modifications and enhancements in algorithms. Right now not a single algorithm is suitable for the big data analytics. Every algorithm suffers from the problem of high computing time and stability. Combining of clustering algorithms and including distributive environment concept can provide better algorithms [3]. MapReduce-based parallel K-Means clustering algorithm, with three functions map (key, value), combine (key, value), and reduce (key, value) due to the framework of MapReduce, can process huge data set efficiently on commodity hardware [4]. MapReduce based K-Means algorithm does not provides globally optimized clusters in the comparison of MapReduce based K-Medoid algorithm; but the time-complexity of K-Means is less than K-Medoid [5].

Selection of primary centroids and improving the scalability of algorithm to process high volume data are yet to be achieved [6]. Initial selection of centroids from dense area of data optimizes center selection and reduces no. of iterations for the convergence, but the distance calculation is a time-consuming process [7].

Since conversion of traditional K-Means clustering algorithm in MapReduce-based parallel and distributed environment to process massive and variety data set is not faultless and suffers problem of local optimum convergence, there are two ways to select initial centroids: First is random selection, and second is choosing certain k test data points, and both suffer the problem of local optimum convergence. To overcome this problem and obtain global optimum clusters, idea of including evolutionary algorithm with K-Means algorithm is arisen and a comparison of K-Means clustering and Genetic algorithm-based clustering is discussed in [8]. Genetic algorithm-based clustering techniques perform better, but the increased time complexity reduces its importance. Particle swarm optimization (PSO) with parallel K-Means clustering algorithm in Hadoop MapReduce is proposed in [9] with three combinations—first K-Means + PSO, second PSO + K-Means, and third is K-Means + PSO + K-Means—to improve the scalability of algorithm. Hadoop MapReduce-based implementation of genetic algorithm needs to be parallelized in two ways: One is coarse-grained parallel genetic algorithm, and another is fine-grained parallel genetic algorithm. The coarse-grained technique distributes the partitions of data set to data nodes for the processing purpose (map phase), after which data points transfer from one data node to another to optimize the obtained clusters. In fine-grained parallelization, each data point is allocated to separate data node and to compute fitness value of it then adjacent nodes communicate to each other for the rest of the steps of genetic algorithm [10]. Generally, genetic algorithms are search and optimization techniques which are unable to partition data set, and K-Means is well-known technique for partitioning the data set. Combination of both genetic and K-Means algorithm is implemented in MapReduce framework to obtain globally optimized clusters successfully with increased time complexity [11]. The available work is not sufficient, and still typical problems of K-Means algorithm such as selection of initial centroids, fixing the no. of k are present, and implementation of K-Means in MapReduce framework faces the problem of local optimum convergence, plus inclusion of evolutionary techniques with K-Means clustering mechanism in MapReduce framework improves the cluster quality but increases the time complexity.

3 Augmented K-Means Clustering for MapReduce

A combination of genetic and K-Means algorithms in MapReduce is implemented in [11] which shows improvement in the obtained clusters by providing globally optimized clusters but increases the time complexity. To overcome the problem of time complexity and obtain globally optimized clusters, author has projected an enhancement of genetic K-Means algorithm for big data clustering in MapReduce framework, by opting only the logic of chromosome, fitness, optimization, and crossover from genetic algorithm.

3.1 *Basic Theory*

Authors' idea is to form chromosomes as per the number of data nodes is available. If the number of data nodes available is **p,** then **p** numbers of chromosomes are formed from the data set.

Initial centroids are selected from the data set and grouped into the chromosome. If **k** clusters are required in the **d**-dimensional data set then the length of each chromosome will be **k*d** where first **d** values in chromosome will be the attributes of first centroid [12]. **k*p** centroids are selected from the whole data sets to form chromosome population of size **p**, and it is the job of same node to distribute **p** chromosomes to **p** data nodes with the segmented data sets.

The fitness computation step of genetic algorithm is similar to distance computation and cluster formation step of K-Means algorithm where initial clusters are formed according to initial cluster centroids and then recomputed centroids by taking mean of all data points in one cluster.

3.2 *Augmented K-Means Algorithm (AKM) for MapReduce*

Step1 Form chromosome—a string of length **k*d** where **k** = no. of clusters expected and **d** = no. of attributes.

Step2 Initialize chromosome by randomly selecting data points from the data set.

Step3 Create **p** no. of chromosomes = population.

Step4 Divide data set in **p** parts; each part with one chromosome allocated by name node to **p** no. of data nodes.

Step5 Shape clusters as per the centroids encoded in chromosome at every data node by allocating data points x_i, i = 1,2,…….n, to clusters C_j with center z_j where
$\|x_i - z_j\| < \|x_i - z_p\|, p = 1, 2, \ldots k, \; and\, p \neq j$

Step6 Here $\|x_i - z_j\|$ is Euclidean distance. Euclidean distance between two points x and y is calculated as $d = \sqrt{\sum_{i=1}^{n} (x_i - y_i)^2}$

Step7 Update chromosome by substituting new centroid z_i^* which is the mean point of the respective cluster, computed as
$z_i^* = \frac{1}{n_i}\sum_{x_j \in c_i} x_i$, $i = 1, 2 \ldots\ldots k$.

Step8 Iterate for the till convergence or fix no. of iteration.

Step9 Reduce step of MapReduce framework collects chromosomes from all data nodes. Compare all chromosomes and merge similar chromosomes in one. If the centroids encoded in chromosomes are in a near range, take average of those chromosomes and then compute clusters according to the centroids encoded in chromosomes through equation 1.

Step10 Obtain clusters as per newly combined chromosomes which will provide globally optimized clusters.

Selection of chromosomes on the basis of fitness, sending them to mating pool, crossover, and mutation steps of genetic algorithms are altered in the proposed clustering algorithm, thus it reduces the computation time.

4 Results and Discussions

Presented algorithm implemented with R on Hadoop2.7.0 single node installed on Ubuntu 14.0 with the configuration of core i5 processor, 8 GB RAM, and 500 GB hard disk on the data set telecom sector to get the clusters of similar behavior customers and performed as per the expectations. Problem of local optimal clusters of big data K-Means clustering algorithm is solved in the proposed algorithm by providing the concept of chromosome which finally provides optimal clusters from all data nodes. The problem of more time complexity of big data genetic K-Means algorithm is reduced by removing crossover, mutation, and intermediate selection steps of genetic algorithm. Proposed algorithm applied on telecom data to categorize customer behavior and results is shown in Figs. 1 and 2.

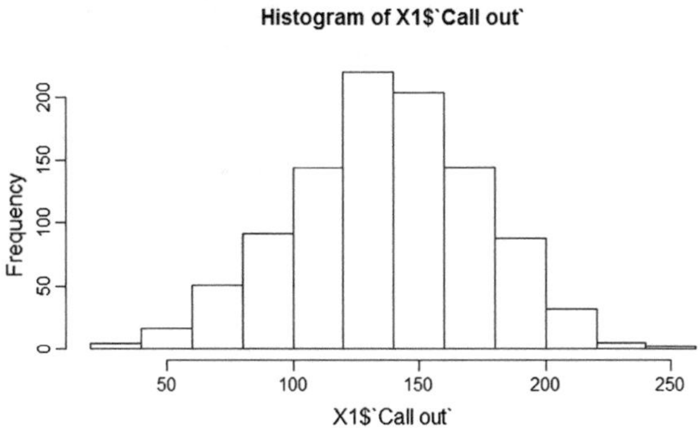

Fig. 1 Call out versus frequency

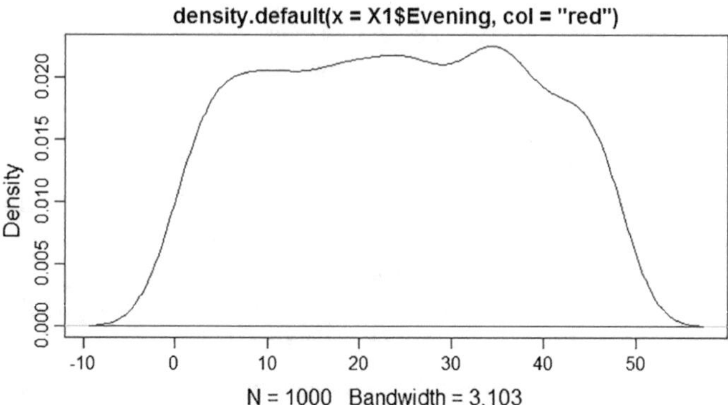

Fig. 2 Call density in evening time

Figure 1 shows frequency of calls made by customers in histogram and depicts that no. of calls made by customers is more in range of 100 to 150, and Fig. 2 depicts the density of calls made in the evening.

5 Conclusion and Future Scope

Big data clustering techniques are trending due to the inclusion of evolutionary technique in it. Genetic K-Means clustering algorithm for big data analytics provides better clusters, but it faces the problem of increased time complexity. Presented AKM clustering algorithm is augmentation of genetic K-Means clustering algorithm, expected to provide globally optimized clusters from the data set in less computing time.

AKM clustering algorithm is framed for Hadoop/MapReduce, and it works smoothly as per the expectations by providing globally optimized clusters within less computing in the comparison of others.

Technical contribution of this paper is the novel idea of utilizing the logic of genetic algorithm, which is an algorithm of computational intelligence, in MapReduce framework for the analysis purpose. The K-Means clustering algorithm is just a part of proposed algorithm for the computation of distance among data points.

Implementation of proposed algorithm in multiple machine and comparison of it with existing MapReduce-based K-Means clustering algorithm is the future work.

References

1. Ayed, A.B., Halima, M.B., Alimi, A.M.: Survey on clustering methods: towards fyzzy clustering for big data. In: International Conference of Soft Computing and Pattern Recognition, pp. 331–336. IEEE (2014)
2. Rajashree, D., Rasmita, D.: Comparative analysis of K-Means and genetic algorithm based data clustering. Int. J. Adv. Comput. Math. **3**(2), 257–265 (2012)
3. Adil, F., Najlaa, A., Zahir, T., Abdullah, A., Ibrahim, K., Albert, Y.Z., Sebti, F., Abdelaziz, B.: A survey of clustering algorithms for big data: taxonomy and empirical analysis. IEEE Trans. Emerg. Top. Comput. **2**(3), 267–279 (2014)
4. Weizhong, Z., Huifang, M., Qing, H.: Parallel K-Means clustering based on MapReduce. In: CloudCom 2009, LNCS 5931, pp. 674–679. Springer (2009)
5. Preeti, A., Deepali, Shipra, V.: Analysis of K-Means and K-Medoids algorithm for big data. Procedia Comput. Sci. **78**, 507–512 (2016) (Elsevier)
6. Prajesh, P.A., Anjan, K.K., Srinath, N.K.: MapReduce Design of K-Means Clustering Algorithm. IEEE (2013)
7. Nadeem, A., Mohd, V.A., Shahbaaz, A.: MapReduce model of improved K-Means clustering algorithm using hadoop MapReduce. In: 2nd International Conference on Computational Intelligence and Communication Technology, pp. 192–198. IEEE (2016)
8. Rajshree, D., Rasmita, D.: Comparative analysis of K-Means and genetic algorithm based data clustering. Int. J. Adv. Comput. Math. Sci. **3**(2), 257–265 (2012)
9. Kaustubh, C., Gauri, C.: Improved K-means clustering on Hadoop. Int. J. Recent Innov. Trends Comput. Commun. **4**(4), 601–604
10. Nivranshu, H., Sana, M., Omkar, S.N.: Big data clustering using genetic algorithm on hadoop Mapreduce. Int. J. Sci. Technol. Res. **4**(4), 58–63 (2015)
11. Pooja, B., Kulvinder, S.: Big data mining: analysis of genetic K-Means algorithm for big data clustering. Int. J. Adv. Res. Comput. Sci. Softw. Eng. **6**(7), 223–228 (2016)
12. Ujjwal, M., Sanghamitra, B.: Genetic algorithm-based clustering technique. Pattern Recognit. **33**, 1455–1465 (2000) (Elsevier)

Face Recognition Using Local Binary Pattern-Blockwise Method

Nikhil Kumar and Sunny Behal

Abstract Face recognition is the most challenging facets of image processing. Human brain uses the same process to recognize a face. Brain extracts essential features from the face and stores in his database. Next time, brain tries to recognize the same face by extracting the features and compares it with already stored features. After comparing both faces, human brain signifies the result. But for a computer system, this process is very complex because of the image variations in terms of location, size, expression and pose. In this paper, we have used a novel LBP-blockwise face recognition algorithm based on the features of local binary pattern and uniform local binary pattern. We have validated the proposed method on a set of classifiers computed on a benchmarked ORL image database. It has been observed that the recognition rate of the novel LBP-blockwise method is better than the existing LBP method.

Keywords Face recognition · Digital image processing
Local binary pattern [LBP] · Uniform local binary pattern [ULBP]

1 Introduction

Face Recognition has been one of the most eye-catching subjects in the recent years. Automatic recognition of face is a daunting task, and it is in high-identification due to its rapid popularity worldwide. This becomes most commonly used technology in various fields such as commercial stores and malls, high-security official offices, banks, private companies. Face recognition is facing many hurdles when it is about implementation of accurate scanned. Many applications used in sectors are unable to provide a solid solution for monstrous problems. Face recognition technology is the

N. Kumar · S. Behal (✉)
SBS State Technical Campus, Ferozepur, Punjab, India
e-mail: sunnybehal@sbsstc.ac.in

N. Kumar
e-mail: Nikhilfdk1639@gmail.com

V. Bhateja et al. (eds.), *Intelligent Engineering Informatics*, Advances in Intelligent Systems and Computing 695, https://doi.org/10.1007/978-981-10-7566-7_12

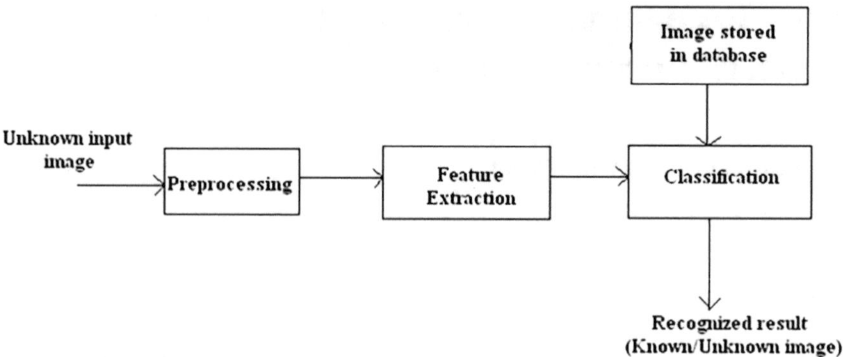

Fig. 1 A generalized framework of face detection method

most emerging technology in digital security arena. It can be seen in high-level confidential research laboratories established by world powers.

The term of face recognition takes place when scanned result occurs a recheck of many faces in a database using the various features. The environment conditions like lighting, prospective view of a face, facial expressions and ageing. The similarity in faces makes it hard to recognize the exact face from database. Face recognition technology is used for identity authentication, access control and other highly featured applications. In the present time, still there is more detailed research which is conducting for improvements and reducing flaws of biometric technology. Basically, face recognition simply works with face scanned images, face viewpoint and other preset expressions. Therefore, the functionality of the face recognition system is always a matter of doubt.

The face tracking motion depends on the prominent face features such as face eyebrows, eyes, nose, mouth and chin in a whole scene. In next step, all the possible angles and set of expressions are used with device to provide facial authority codes. Face tracking or recognition has always upper hand when compared to the other biometric technologies present in tech market. Fingerprint and iris scan are high-frequency used modalities. Although these technologies are still present in wide common devices like surveillance cameras, camera in mobile phones, the emergence of face recognition as future technology remains unquestionable [1–4] (Fig. 1).

1.1 Feature Extraction Methods

There are two types of feature extraction methods, namely global feature extraction and local feature extraction method. Both of these techniques play a major role in face recognition. Local information depicts comprehensive deviations of facial appearances, while global information deals with antique essential patterns of face expressions. So, from the point of view of frequency deviation analysis, the global features correspond to lower frequencies, whereas local features correspond to high frequency and are reliant on location and alignment in an image [5].

1.2 Challenges of Face Recognition

There are numbers of hurdles in face recognition task as mentioned below:

- Facial Appearance: The face appearance is directly affected by a person's facial expression. For example, a smiling face, a crying face, a face with closed eyes, a face with glasses and even a small nuance in the face expression can affect recognition system's accuracy.
- Varying Illumination: The face exposures under lights have an effect on the process of face scanning. If any portion of face is more exposed to light as compared to other, then this makes face recognition a tough task [6].
- Pose: The images of a face from a different angle as front or side view and some facial features such as nose or eyes may turn into partially or wholly covered.
- Ageing: Age is the prime factor that affects the facial features the most. Faces are supposed to change in every 2–5 year's period. This is a monstrous problem and reduces the reliability of the face recognition technology [7].
- Rotation: Rotation may be clockwise or anti-clockwise can affect the recital of the face recognition system.
- Size of the Image: A test image of size 64×64 is hard to classify if the training image size is 128×128. In other words, the dimensions of all prototypes, as well as query face, must match to achieve the job of face recognition.
- Occlusion: The faces may be partially covered with other objects such as glasses, moustaches, breads and/or scarves, which affect the system performance [8].
- Identical Twins: Identical twins cannot be easily identified by the automatic face recognition systems [9–11].

2 Literature Review

Feng et al. [12] proposed an innovative method to distinguish facial lexes from static images. They used LBP method to characterize facial images and then applied an LP procedure to categorize several facial lexes such as fear, anger, sadness, disgust, happiness, surprise and neutral. They used JAFFE database for the validation of their proposed approach. Their proposed method outperforms all of the existing methods in the same database and achieved an average recognition accuracy of 93.8%.

Sukhija et al. [9] proposed a novel technique for face recognition using genetic algorithm (GA). Their proposed scheme compared unknown images with the known training images already stored in the database. They compared the results of proposed scheme with PCA and LDA algorithms. The reported results indicate increase in the recognition rate. Hassaballah et al. [10] proposed a novel method to identify eye and nose areas from greyscale facial images. Their proposed scheme used ICA algorithm to learn external facial features. They selected those regions having high response to ICA basis vectors. They validated their proposed approach different benchmarked databases. Geng et al. [7] proposed an automatic age estimation process called AGES. They compared their proposed approach with the existing age judgment approaches and human discernment ability. Gawande and Agrawal [13] used PCA with different distance classifiers for the face recognition. They validated their proposed approach using ORL database and computed recognition rate.

3 Proposed Work

In this work, the performance at different levels of implementation for the face recognition has been studied using ORL database. A basic face recognition method is composed of **two phases.** System is trained using feature vectors taken from sample images in the first phase, whereas classifiers are applied on the selected features to recognize the new image in the second phase.

Generally, the face recognition process can be subdivided into following stages:

1. Feature extraction using local binary pattern and uniform local binary pattern.
2. Training using simple approach and block approach.
3. Classification using different distance matrices algorithms.

3.1 Local Binary Pattern

Local Binary Pattern (**LBP**) was first defined in 1994. It is a type of feature used for the purpose of classification in computer vision. It has been observed that LBP produces the best classifier of humans when it is combined with histogram of oriented gradient (HOG) classifier [12, 14–16] (Fig. 2).

We used the notation $LBP_{P,R}$ for the LBP operator. Here, subscript represents the operator in (P, R) neighbourhood. The histogram of a labelled image $f(x, y)$ can be defined as

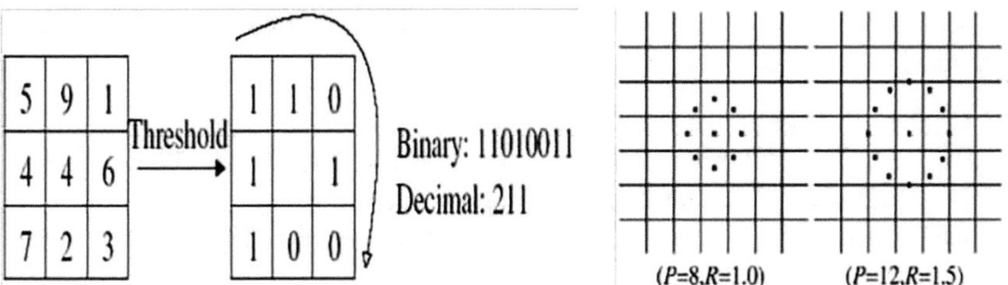

Fig. 2 Left: the basic LBP operator; Right: two examples of the extended LBP: a circular (8; 1) neighbourhood, and a circular (12; 1:5) neighbourhood

$$H_{i,j} \sum_{x,y} I\{f(x,y)=i\}, i=0, \ldots, n-1,$$

where n is the number of different labels produced by LBP operator and

$$I\{A\} = \begin{cases} 1, & A \text{ is true} \\ 0, & A \text{ is false} \end{cases}$$

The above-mentioned histogram specifies the distribution of local micropatterns, such as edges, flat areas and spots over the whole image. For representing face images efficiently, we should also retain spatial information. For this purpose, the image is divided into regions R_0, R_1,… R_{m-1}. The spatially enhanced histogram is then defined as

$$H_i = \sum_{x,y} I\{f_1(x,y)=i\} I\{(x,y) \in R_j\},\ i=0, \ldots, n-1; j=0, \ldots, m-1$$

3.2 Uniform Local Binary Pattern

Local binary pattern is said to be uniform if upon traversing the bit pattern in a circular way, the binary pattern contains at most two bitwise transitions from 0 to 1 or vice versa. These patterns are used in the computation of LBP labels so that there is a separate label for each uniform pattern and but a single label for all the non-uniform patterns [14].

Two variants (each) of both the techniques are used in this thesis work. One variant is applied to whole image as it is and in other variant, first image is divided into 16 blocks of size 16 × 16 and then the techniques are applied to each block and in the end, the features of all the blocks are appended together to form a combined feature vector for the whole image. So in total, we have four methods to compare. Let us name them as:

1. LBP_SIMPLE (Number of features per image = 256)
2. LBP_BLOCKS (Number of features per image = 16 * 256 = 4096)
3. ULBP_SIMPLE (Number of features per image = 58)
4. ULBP_BLOCKS (Number of features per image = 16 * 58 = 928)

3.3 Classification

Classification means matching or comparing of test/query image with training/sample images stored in the database. The used algorithms are explained below [13]:

Sr. No.	Distance measure	Mathematical formula
1	Euclidean distance	$d_E(x,y) = \sqrt{\sum_{i=1}^{n} (x_i - y_i)^2}$
2	Squared chord measure	$d_{sc}(x,y) = \sum_{i=1}^{n} \left(\sqrt{x_i} - \sqrt{y_i}\right)^2$
3	Chi-square distance	$d_{chi}(x,y) = \sum_{i=1}^{n} \frac{(x_i - y_i)^2}{x_i + y_i}$
4	Histogram intersection measure	$d_{HI}(x,y) = 1 - \frac{\sum_{i=1}^{n} \min(x_i, y_i)}{\sum_{i=1}^{n} \max(x_i, y_i)}$
5	Cosine measure	$d_{\cos}(x,y) = \cos^{-1}\left(\frac{\sum_{i=1}^{n} (x_i y_i)}{\sqrt{\sum_{i=1}^{n} x_i^2}\sqrt{\sum_{i=1}^{n} y_i^2}}\right)$
6	Extended Canberra distance	$d_{EC}(x,y) = \sum_{i=1}^{n} \frac{\lvert x_i - y_i\rvert}{\lvert x_i + u_x\rvert + \lvert y_i + u_y\rvert}$ *where* $u_x = \sum_{i=1}^{n} x_i/n \quad u_y = \sum_{i=1}^{n} y_i/n$

4 LBP-Blockwise Algorithm

In this work, we extract the features from the image by using LBP and ULBP in two ways. In the first step, LBP and ULBP are applied on whole image and we get LBP (256) and ULBP (58) features. In the second step, the image is divided into 16 × 16 size blocks and LBP and ULBP are applied on each block individually. We implemented these algorithms in MATLAB.

5 Results and Discussion

The designed methodology was tested against ORL face database in which images are in tiff format. Images of every person differ in lexes such as expressions, details, pose and scale. Figure 2 represents five sample images of a person. Each image is scaled down to the size of 64 × 64 pixels.

Sr. No.	Similarity measure	LBP_ SIMPLE	LBP_ BLOCKWISE	ULBP_ SIMPLE	ULBP_ BLOCKWISE
1	Euclidean distance	85	95	86	95
2	Chi-square distance	86.5	97	90	96.5
3	Cosine measure	85.5	95.5	85	95.5
4	Extended Canberra distance	84	97.5	89.5	97
5	Square chord measure	84.5	97	90	96.5
6	Histogram intersection measure	88.5	97	88	97

We have compared the recognition rate of two techniques in a simple way (in which image is used as it is) and blockwise way (in which an image is divided into 16 blocks of 16 × 16). The recognition rate is best archived by using large number of features. We can see the analysis of results, as we increase the number of features our recognition rate is also increasing.

6 Conclusion and Future Scope

This work used dimensionality reduction techniques to analyse their effectiveness in face recognition. In the first phase, local binary pattern (LBP) and uniform local binary pattern (ULBP) are used blockwise approaches and are applied on training data. The purpose is to extract the proper features for face recognition process. After applying these algorithms on benchmarked ORL database images, various coefficients are selected to construct a feature vector. In the second phase, we used different classification methods for matching the test image with training images already stored in ORL database. It has been observed that the results of blockwise methods are better as compared to simple methods. As part of the future work, we shall use other facial variation features like ageing, accessories, highly varying expressions for face recognition.

References

1. Singh, S., Sharma, M., Suresh Rao, N.: Accurate face recognition using PCA and LDA. In: International Conference on Emerging Trends in Computer and Image Processing, pp. 62–68 (2011)
2. Li, S.Z., Lu, J.: Face recognition using the nearest feature line method. IEEE Trans. Neural Netw. **10**(2), 439–443 (1999)
3. Roy, S., Podder, S.: Face detection and its applications. Int. J. Res. Eng. Adv. Technol. **1**(2), 1–10 (2013)
4. Dutta, S., Baru, V.B.: Review of facial expression recognition system and used datasets. Int. J. Res. Eng. Technol. **2**(12) (2013)
5. Mishra, S., Dubey, A.: Face recognition approaches: a survey. Int. J. Comput. Bus. Res. (IJCBR) **6**(1) (2015)
6. Prakash, N., Singh, Y.P., Rai, D.: Emerging trends of face recognition: a review. World Appl. Programm. **2**(4), 242–250 (2012)
7. Geng, X., Zhou, Z.-H., Smith-Miles, K.: Automatic age estimation based on facial aging patterns. IEEE Trans. Pattern Anal. Mach. Intell. **29**(12), 2234–2240 (2007)
8. Goel, S., Kaushik, A., Goel, K.: A review paper on biometrics: facial recognition. Int. J. Sci. Res. Eng. Technol. **1**(5) (2012)
9. Sukhija, P., Behal, S., Singh, P.: Face recognition system using genetic algorithm. Procedia Comput. Sci. **85**, 410–417 (2016)
10. Hassaballah, M., Murakami, K., Ido, S.: Eye and nose fields detection from gray scale facial images. In: IAPR Conference on Machine Vision Applications, pp. 406–409 (2011)
11. Khade, B.S., Gaikwad, H.M., Aher, A.S., Patil, K.K.: Face recognition techniques: a survey. Int. J. Comput. Sci. Mob. Comput. **5**(11), 65–72 (2016)
12. Feng, X., Pietikäinen, M., Hadid, A.: Facial expression recognition with local binary patterns and linear programming. Pattern Recognit. Image Anal. **15**(2), 546–548 (2005)
13. Gawande, M.P., Agrawal, D.G.: Face recognition using PCA and different distance classifiers. IOSR J. Electron. Commun. Eng. **9**(1) (2014)
14. Lee, K.-D., Kalia, R., Kwan-Je, S., Oh, W.G.: Face recognition using LBP for personal image management system and its analysis. IITA (2010)
15. Jiang, X., Guo, C., Zhang, H., Li, C.: An adaptive threshold LBP algorithm for face recognition. IJSSST 51.1–51.6
16. Singh, S., Amritpal Kaur, T.: A face recognition technique using local binary pattern method. Int. J. Adv. Res. Comput. Commun. Eng. **4**(3), 165–168 (2015)

Social Group Optimization (SGO) for Clustering in Wireless Sensor Networks

Neeharika Kalla and Pritee Parwekar

Abstract Wireless Sensor Network (WSN) is a domain which has its application in the variety of fields like military, disaster management, environment monitoring. Energy consumption is one of the key challenges in the field of WSN where researchers are strongly exploring and discovering new techniques or methods. Direct or hop-by-hop transmission of data from the node to the BS leads to more number of transmissions. Clustering is applied to reduce the number of transmissions. Nodes can consume less energy if the distance between node to node or from node to BS is less. An optimization technique is used to minimize the transmission distance and to dynamically select the number of cluster heads. Social Group Optimization (SGO) is implemented, and the results are compared with Genetic Algorithm (GA) and Particle Swarm Optimization (PSO).

Keywords Wireless Sensor Network · Social Group Optimization Clustering · Sensor nodes · Base station · Optimization

1 Introduction

Wireless Sensor Network (WSN) is a dynamic (ad hoc) network consisting of huge number of sensors which are controlled in distributed and cooperative manner [1]. Wireless Sensor Networks are one of the important technologies which are considered as the powerful application of Micro-Electro-Mechanical Systems (MEMS), thus attracting many researchers to work on the challenging issues in the domain of WSN [2]. The sensor nodes consist of limited power, limited memory, limited range of transmission, and limited data rate. The key functionality of WSN is to sense the physical environment and broadcast the data to the base station (BS). As the sensor

N. Kalla (✉) · P. Parwekar
Anil Neerukonda Institute of Technology and Sciences, Visakhapatnam, India
e-mail: neeharika.kalla@gmail.com

P. Parwekar
e-mail: pritee.cse@anits.edu.in

V. Bhateja et al. (eds.), *Intelligent Engineering Informatics*, Advances in Intelligent Systems and Computing 695, https://doi.org/10.1007/978-981-10-7566-7_13

nodes are positioned in the vast area, it must incorporate the feature of the long sensing capability [2]. Thus, continuous monitoring of the environment requires more energy. As the sensors are battery operated devices and cannot be recharged directly or replaced, the energy consumed by the sensors should be minimized. Thus, the energy consumption is one of the major issues in WSN [2, 3, 4]. One approach to consume the energy could be of making the sensor active only at some point of time as described in [5].

Sensors cannot directly transmit the data to the BS because of limited transmission range. The data is transmitted from one node to the next till it reaches the BS hop-by-hop [2]. In order to reduce the total count of transmissions and the processing overhead of the sensor nodes, clustering technique is used. Sensors are grouped in clusters with every cluster having a leader called as cluster head (CH). In this, CH is responsible for gathering the information from the sensors and broadcasting it to the BS [1, 4]. As the clustering technique is classically a NP-Hard problem, nature-inspired evolutionary algorithms also called as optimization techniques like Genetic Algorithm (GA), Particle Swarm Optimization (PSO), Artificial Bee Colony Optimization (ABC), etc., are applied to the clustering problems in order to minimize the energy consumption and to prolong the lifetime of the network [1]. A detailed study of different clustering techniques is addressed in [4]. Applications like military and civil require the location of nodes to be known that is described in [6]. Advancement in sensor technology gave rise to the new application of human detection, tracking, and recognition explained in [7]. Distributed applications of WSN require the feature of clock consistency described in [8].

The remaining part of the paper is organized as follows. Section 2 provides the literature survey of the three optimization techniques using different approaches. Section 3 reviews the proposed methodology of the SGO technique along with the problem statement and the fitness function. Section 4 illustrates the simulation results. Finally, our work is concluded in Sect. 5.

2 Literature Survey

Paper on PSO-based clustering [9] focuses on the residual nodes. The residual nodes either directly transmit the data to the BS or search for the next hop to transmit the data. The main idea is to perform clustering such that no node is left out without being a member of a cluster. Particle Swarm Optimization (PSO) algorithm is applied to update the positions and velocities of the sensor. Leapfrog algorithm is used to select the criteria for clustering the network. To identify the best and the least count of the cluster heads is termed as nondeterministic polynomial (NP-Hard) problem in [10]. Clustering performance greatly depends on the selection of the cluster heads. These CHs are responsible for generating the clusters, to transmit the data to the BS and to control the member nodes. This paper focuses on the selection of the CHs to improve the lifetime of the network. The CHs are selected in such a way that the sensor with higher remaining energy acts as a CH. Genetic Algorithm

(GA), Particle Swarm Optimization (PSO), and Differential Evolution (DE) algorithms are used to address this issue, and their performances are compared.

Generally, Wireless Sensor Network (WSN) has limited supply of energy. In [11], energy-aware clustering with Particle Swarm Optimization (PSO) technique is presented. The method is implemented at the base station (BS). Initially, the nodes send their energy information and their location to the sink or BS. Depending on this data, the BS calculates the levels of energy of the nodes. The sensor with the level of energy above the average is elected as the CH by the BS. Now, the BS runs the Particle Swarm Optimization (PSO) algorithm to find the optimal count of CHs from the list of selected CHs. After the selection of optimal CHs and their node members, the CHs adopt a TDMA schedule to transmit the data.

In [3], Genetic Algorithm (GA) is used to find the optimum count of CHs. In the setup stage, BS finds the number of CHs and assigns the nodes to the CHs. In the second phase, the nodes transmit the data to the CH and the CHs transmit to the BS. Initially, the chromosomes are formed by the BS and then GA is applied to obtain the optimal number of CH. In addition, an agent cluster head (CH) is used to reduce the distance among the BS and their respective CHs. The agent CH is the cluster head with maximum residual energy and is selected from the set of CHs. Clustering Hierarchy Protocol in [12] contains two stages: clustering setup stage and data transmission stage. Here, relay nodes are used to consume less energy by the nodes. Initially, the BS classifies the nodes into relay nodes, CHs, and the cluster members depending on the energy level of the node. In the clustering setup stage, CHs and relay nodes are determined and clusters are formed. Improved PSO algorithm is used for clustering. After the formation of clusters, in the second stage, cluster members transmit the data to the CHs, the CHs to the relay nodes, and the relay nodes to the BS.

Social Group Optimization (SGO) in [13] is a population-based evolutionary technique which mimics the concept of group of individuals trying to solve a complex problem based on their level of knowledge. In improving stage, each individual's knowledge is augmented with the knowledge of the best individual in the group who is having more knowledge than the remaining persons. In acquiring phase, each person in the group interacts randomly with the other person along with the best person among the group to acquire problem-solving skills if they are having more knowledge than him/her and vice versa. Energy consumption and conservation is the primary challenge for WSNs. Several techniques are proposed to decrease or optimize the energy consumed. One of the approaches to save on energy consumption could be that some nodes will be kept in an active mode and some will be kept in sleep mode. Another approach as described in [14] is where solar energy is used to provide power to the sensors through solar panels. This approach is an eco-friendly approach and in consonance with the changing times.

3 Proposed Methodology

3.1 Problem Statement

In WSN, the sensors are randomly deployed in the region of interest (ROI) and are responsible for sensing the data and transmitting it to the BS. If each sensor directly transfers the data to the BS, then the number of transmissions increases leading to increase in the energy consumption of the nodes. If there are "n" nodes, then it takes $n(n-1)/2$ transmissions to transmit the data from the node to the BS. Thus, clustering is performed to decrease the total number of transmissions [15]. In this, the sensors are grouped into clusters. Every cluster consists of one CH which is having high residual energy when compared to all the sensors. The cluster members relay the data to the CH which in turn transmits to the BS. After performing clustering, optimization technique is applied for the dynamic selection of the cluster heads depending on the remaining (leftover) energy of the node.

Figure 1i represents the direct transmission of data from the sensor nodes to the BS or sink. Figure 1ii represents the clustering technique to minimize the energy consumed by the nodes.

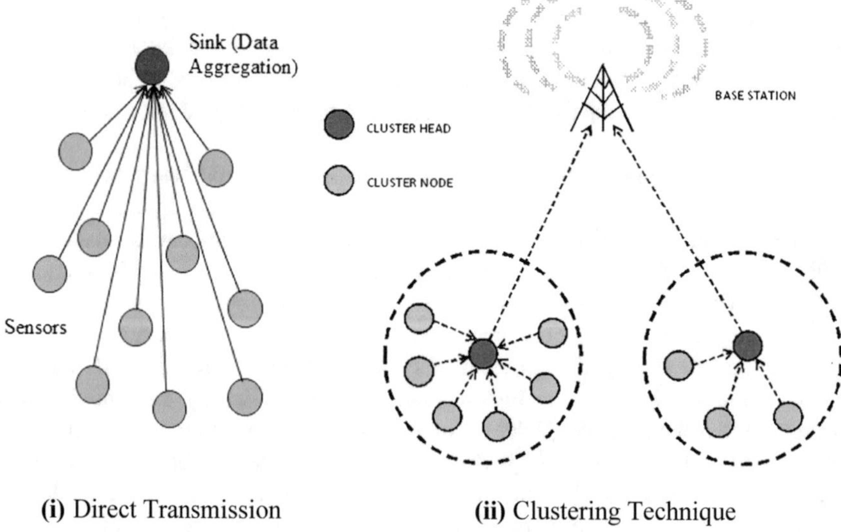

Fig. 1 i. Direct transmission, ii. clustering technique

3.2 Social Group Optimization (SGO)

For a problem to be solved, group approach is more effective than individual approach. As known, many behavioral traits like honesty, dishonesty, caring, fear, respectfulness, tolerance, courage are seen in humans, which are necessary to tackle situations in life. Solving the problems may vary from person to person and also among groups. But group solving capability is more effective and efficient way in solving problems. So by using the same concept, SGO, a novel technique in optimization is developed [13]. As SGO comes under population-based algorithm, here each person has some knowledge and is capable of solving a particular problem. Here population refers to known as group of persons. This can correspond to "fitness." The person who gains more knowledge has more chance/capability of solving a problem, thereby will be the best person and so the best solution [13].

SGO procedure can be of two stages. The initial stage is of improving stage and the next is the acquiring stage. In the first stage, the knowledge level of each person in the group is enhanced, so the best individual can assess the knowledge levels among all the other people and he/she can be able to resolve the problem. In the second stage, the individuals in the group will interact among themselves and also with the best individual among the group to enhance their knowledge for solving the problem [13]. The arithmetic explanation of the method of SGO is represented as follows.

Let X (i) where i = 1, 2, 3…n be the number of persons in the group. The group consists of "n" individuals, and every individual can be described as X (i) = (x (i, 1), x (i, 2), x (i, 3)….x (i, d)) where "d" refers to the behavioral traits of each person and this varies from person to person. The corresponding fitness values of the person will be f (i), where i = 1, 2, 3….n.

The person with the maximum fitness value will be the best individual among the group, i.e., gbest. Here gbestg = max (f (i), i = 1, 2, 3…….n) where g refers to the generation. This is for solving maximization problem. The best individual among the group (gbest) tries to enhance the knowledge of each person in the group through his/her knowledge. Here, increase in the knowledge refers to the change in the behavioral traits of the person [13].

Improving phase

In this phase, each individual in the group gets the knowledge from the individual who is best (gbest). The updating of the knowledge of each person is calculated as follows [13]:

```
For i=1: n
    For j=1: d
            Xnext(i, j) = c*Xprev(i, j) + rand(0,1)* (gbest-Xprev(i, j))
    End for
End for
```

where c is self-analysis parameter. Its value is $0 < c < 1$.

Xnext is accepted only if its fitness is better when compared with the old fitness Xprev.

Acquiring phase

In this phase, each individual interacts randomly with the other individual of the group and also with the best person of the group. The knowledge of an individual will be enhanced only if the other individual is having more problem-solving skills than him/her and vice versa. The knowledge of the best individual (gbest) has great influence on the other persons to learn from him/her [13].

Here, gbest = max (f (X (i), i = 1, 2, 3….n)) where X (i) is the updated value after the improving stage.

```
For i= 1: n
    Select one individual randomly Xr, where i ≠r
    If f (Xi) < f (Xr)
    For j= 1: d
            Xnext(i, j) = Xprev(i, j) + rand(0,1) * (X(i, j) − X(r, j) +
rand(0,1) * (gbest − X(i, j)))
End for
Else
    For j= 1: d
            Xnext(i, j) = Xprev(i, j) + rand(0,1)*(X(r, : )-X(i, : ) +
rand(0,1)*(gbest-X(i, j)))
    End for
    End If
End for
```

Xnext is accepted only if its fitness is better when compared with previous.

Algorithm

```
Step 1: Initialize the population
Step 2: Calculate the fitness of each individual
Step 3: Identify global best (gbest) solution
Step 4: Improving stage
        For i=1: n
                For j=1: d
                    Xnext(i, j) = c*Xprev(i, j) + rand(0,1)* (gbest-Xprev(i, j))
                End for
        End for
Step 5: If better solution, then accept and find gbest. Otherwise reject
Step 6: Acquiring stage
        For i= 1: n
        Select one individual randomly Xr, where i≠r
        If f (Xi) < f (Xr)
                For j= 1: d
                Xnext(i, j) = Xprev(i, j) + rand(0,1) * (X(i, j) − X(r, j) +
rand(0,1) * (gbest − X(i, j)))
                End for
                Else
                For j= 1: d
                Xnext(i, j) = Xprev(i, j) + rand(0,1)*(X(r, : )-X(i, : ) +
rand(0,1)*(gbest-X(i, j)))
                End for
        End If
        End for
Step 7: If better solution, then accept. Otherwise reject
Step 8: Repeat steps 3-7 till an optimal solution is obtained
```

3.3 *Fitness Function*

The objective function that is used to reduce the transmission distance and also to reduce the cluster heads count is as follows [15]:

$$f = w^*(D - d_i) + (1 - w)^*(N - CH_i)$$

where D is the total distance of all the sensors from the base station (BS) or sink; d_i is the summation of the distances from the sensors to their CH and distances from their CHs to BS; w is the predefined weight factor; N refers to count of sensors; CH_i refers to total count of CHs. The value of w lies between 0 and 1 ($0 < w < 1$).

4 Simulation, Discussion, and Results

The implementation and the execution of the proposed technique using MATLAB with the simulation parameters as shown in Table 1. The algorithm is performed in the sensor area consisting of 200 × 200 m^2. The BS is placed at the center of the ROI so that nodes that are near to the BS can transmit the data easily and requires less energy. Genetic Algorithm (GA) is inspired by the biological evolution of the human beings. The basic genetic operators that are used for reproduction are selection, crossover, and mutation. In this, each member of the population is represented by a set of bits called as chromosomes. GA mimics Darwin's theory of biological evolution. Particle Swarm Optimization (PSO) mimics the flocking behavior of birds. All the particles move in a search space for food with some position and velocity. After each generation, the particle tends to move toward the best position by comparing itself with the neighboring particles position. The process continues till the optimal solution is obtained. Social Group Optimization (SGO) technique is implemented and compared with the Genetic Algorithm (GA) and Particle Swarm Optimization (PSO) techniques, and the results are described below (Table 2).

The performance of SGO is tested with different population size by varying the node size and compared with GA and PSO. The results are depicted below.

Figure 2 shows that Social Group Optimization (SGO) performs better than Particle Swarm Optimization (PSO) and Genetic Algorithm (GA). SGO has the better fitness value when compared with GA and PSO.

Table 1 Simulation parameters

Sensor area	200 × 200
Population size	80
Node size	100
Inertia factor (w)	0.3
MaxIterations	1000

Table 2 Best fitness values using GA, PSO, and SGO

Population size	Sensor size	GA	PSO	SGO
10	30	375	706	720
20	40	548	902	973
30	50	740	1159	1177
40	60	927	1296	1433
50	70	1109	1533	1680
60	80	1266	1821	1873
70	90	1494	1891	2179
80	100	1631	2084	2330

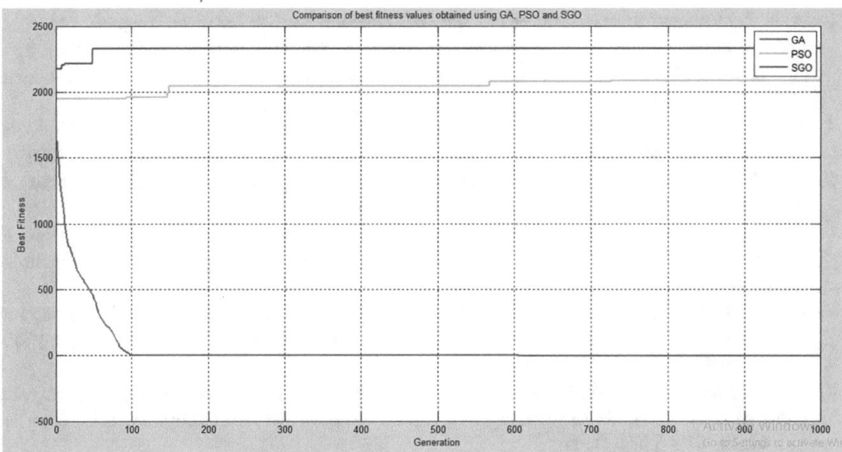

Fig. 2 Comparison of best fitness value using optimization techniques

5 Conclusion

The energy consumed by the nodes can be minimized by minimizing the transmission distance. In this approach, the energy can be utilized effectively by reducing the transmission distance. Hence, new optimization technique SGO is used to reduce the distance of transmission and it is compared with the most popular techniques, GA and PSO. The performance of Social Group Optimization (SGO) is compared with GA and PSO, and the results show that SGO performs better than GA and PSO in terms of convergence; i.e., SGO gets the best fitness value when compared with the other two techniques. The illustrated technique can be treated as a milestone in the domain, a unique solution or a significant improvement over an existing method. By this review, we tend to provide to the researchers a beneficial resource to explore in the field of WSN. The future work can be of applying SGO technique to the new fitness function that consists of two additional parameters and also calculating energy consumption of the nodes.

References

1. Abba Ari, A., Gueroui, A., Yenke, B.O., Labraoui. N.: Energy efficient clustering algorithm for Wireless Sensor Networks using the ABC metaheuristic. In: 2016 International Conference on Computer Communication and Informatics (ICCCI), Coimbatore, pp. 1–6 (2016)
2. Lalwani, P., Das, S.: Bacterial foraging optimization algorithm for CH selection and routing in wireless sensor networks. In: 2016 3rd International Conference on Recent Advances in Information Technology (RAIT), Dhanbad, pp. 95–100 (2016)

3. Aziz, L., Raghay, S., Aznaoui, H., Jamali, A.: A new approach based on a genetic algorithm and an agent cluster head to optimize energy in Wireless Sensor Networks. In: 2016 International Conference on Information Technology for Organizations Development (IT4OD), Fez, pp. 1–5 (2016)
4. Kalla, N., Parwekar, P.: A study of clustering techniques for Wireless Sensor Networks (WSN). In: 1st International Conference on Smart Computing & Informatics (SCI) (2017)
5. Al-kahtani, M.S.: Efficient cluster-based sleep scheduling for M2M communication network. Res. Artic. Comput. Eng. Comput. Sci. **40**(8), 2361–2373 (2015)
6. Alaybeyoglu, A.: An efficient monte carlo-based localization algorithm for mobile Wireless Sensor Networks. Arab. J. Sci. Eng. 1375–1384 (2015) (Springer Science & Business Media B.V.)
7. Kamal, S., Jalal, A.: A hybrid feature extraction approach for human detection, tracking and activity recognition using depth sensors. Res. Artic. Comput. Eng. Comput. Sci. **41**(3), 1043–1051 (2016)
8. Shi, X., Fan, L., Ling, Y., He, J., Xiong, D.: Dynamic and Quantitative Method of Analyzing Clock Inconsistency Factors among Distributed Nodes. Res. Artic. Comput. Eng. Comput. Sci. **40**(2), 519–530 (2015)
9. Akila, I.S., Venkatesan, R., Abinaya, R.: A PSO based energy efficient clustering approach for Wireless Sensor Networks. In: 2016 International Conference on Computation of Power, Energy Information and Commuincation (ICCPEIC), Chennai, pp. 259–264 (2016)
10. Elhabyan, R.S., Yagoub, M.C.E.: Evolutionary algorithms for cluster heads election in wireless sensor networks: performance comparison. In: 2015 Science and Information Conference (SAI), London, pp. 1070–1076 (2015)
11. Latiff, N.M.A., Tsimenidis, C.C., Sharif, B. S.: Energy-aware clustering for wireless sensor networks using particle swarm optimization. In: 007 IEEE 18th International Symposium on Personal, Indoor and Mobile Radio Communications, Athens, pp. 1–5 (20070
12. Zhou, Y., Wang, N., Xiang. W.: Clustering hierarchy protocol in wireless sensor networks using an improved PSO algorithm. In: IEEE Access, pp. 2241–2253 (2017)
13. Satapathy, S., Naik, A.: Social group optimization (SGO): a new population evolutionary optimization technique. Complex Intell. Syst. 173–203 (2016)
14. Parwekar, P., Rodda, S., Satapathy Dr. S.C.: Solar energy harvesting: an answer for energy hungry wireless sensor networks. In: CSI Communications, pp. 23–26 (2016)
15. Jin, S., Zhou, M., Wu. A.S.: Sensor network optimization using a genetic algorithm. In: Proceedings of the 7th World Multiconference on Systemics, Cybernetics, and Informatics, pp. 109–116 (2003)

Implementation of High-Performance Floating Point Divider Using Pipeline Architecture on FPGA

C. R. S. Hanuman and J. Kamala

Abstract Data intensive DSP algorithms mostly depend on double precision (DP) floating point arithmetic operations. With advanced FPGA devices, applications need more floating point arithmetic operations to accelerate reconfigurable logic. Important and complex applications heavily depend on the floating point divider (FPD) blocks. This paper illustrates implementation of low-latency dividers based on pipelining, which is operated at 30% faster than existing divider. DPFPD implemented using field programmable gate array (FPGA) outperforms other ULP dividers. Pipelining architecture of FPD increases the throughput and reduces the power considerably. The architecture is validated for standalone as well as integrated application levels.

Keywords ULP · Reconfigurable logic · DPFPD · FPGA

1 Introduction

The computation and RT complexity of defense and radar applications increasing extensively and FPD blocks are more important for data intensive operations. Most recent processors in IBM, Intel, and AMD use FPD hardware. Compared to other arithmetic operations, division operation is more complex and latency is also too high. The application which uses DPFPD algorithms has been modified so that number of division operations should be minimized. To meet the above requirements, design high-speed floating point dividers which use double precision formats [1]. The division algorithms are classified into two categories, namely subtractive and multiplicative. The advantage of subtractive (digit recurrence)

C. R. S. Hanuman · J. Kamala (✉)
Department of Electronics and Communication Engineering, College of Engineering, Guindy, Anna University, Chennai, India
e-mail: jkamalaa06@gmail.com

C. R. S. Hanuman
e-mail: crshanuman@sasi.ac.in

V. Bhateja et al. (eds.), *Intelligent Engineering Informatics*, Advances in Intelligent Systems and Computing 695, https://doi.org/10.1007/978-981-10-7566-7_14

algorithms is less hardware overhead and needs less area for implementation. In the other hand, it is slow such that acquiring each quotient bit requires unique recurrence cycle (linear convergence) [2].

1.1 IEEE 754 Standard

The technical standard 754 represents floating point numbers in both single (binary32) and double (binary64) precision formats, where number of digits are 24 and 53, respectively [3].

The format for single precision floating point is shown in Fig. 1.

The format for double precision floating point is shown in Fig. 2.

The IEEE 754 standard makes sure that all machines will provide similar length outputs for the same program, but this format is complex and hard to implement effectively for large data values [4, 5].

1.2 SRT Algorithm

In digit recurrence algorithms, SRT division algorithm is widely used. This algorithm uses iterations of recurrence, which generates one quotient digit per iteration [6]. The input vectors are represented by normalized double precision floating point notation with 'm' bit significant in sign-magnitude representation. The number of digits in the quotient is m radix r [7].

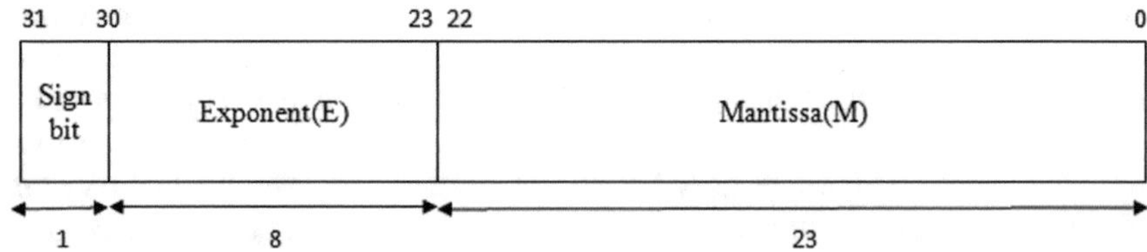

Fig. 1 Single precision floating point format

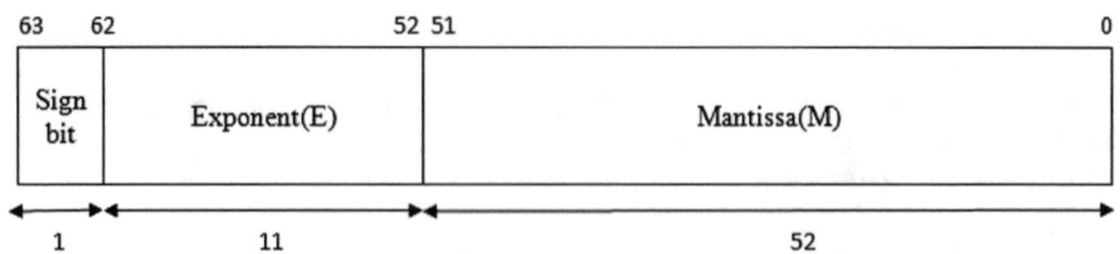

Fig. 2 Double precision floating point format

$$r = 2^b. \tag{1}$$

$$m = \frac{n}{b}. \tag{2}$$

where n is 24 bits for single precision and 53 bits for double floating point

m is no. of iterations taken to compute the result.
b is number of bits discarded for every iteration.

In each iteration, the radix-r algorithm leaves b bits of quotient, which needs m iterations to compute the output, hence the latency of m cycles.

The basic recurrence equations used for each iteration are

$$rP_0 = divd. \tag{3}$$

$$P_{i+1} = rP_i - Q_{i+1}.divr. \tag{4}$$

where P_i and P_{i+1} are the partial remainder at iteration i and i + 1.

The QDS function calculates one quotient digit for each iteration.

$$Q_{i+1} = SEL(rP_i; divr). \tag{5}$$

After 'm' iterations, the quotient will be

$$Q = \sum_{i=1}^{m} Q_i r^{-i}. \tag{6}$$

The remainder depends on the value of P_n.

$$remainder = \begin{cases} P_n \\ P_n + divr \end{cases}. \tag{7}$$

This paper is organized with Sect. 1 introducing the concept of floating point algorithm. Section 2 discusses about the existing methods. Section 3 explains the proposed method. Section 4 shows the simulation of proposed and optimization chart of double precision floating point division for proposed and existing methods. Future work and conclusion part are summarized in Sect. 5.

2 Existing Method

The structure of existing floating point divider is shown in Fig. 3. The stage U_1 performs subtraction operation on exponents E_0 and E_1 and generates flags for IEEE special cases. The priority encoder gives the amount of shift needed for mantissa in denormalized cases. In stage U_2, add bias to the exponent.

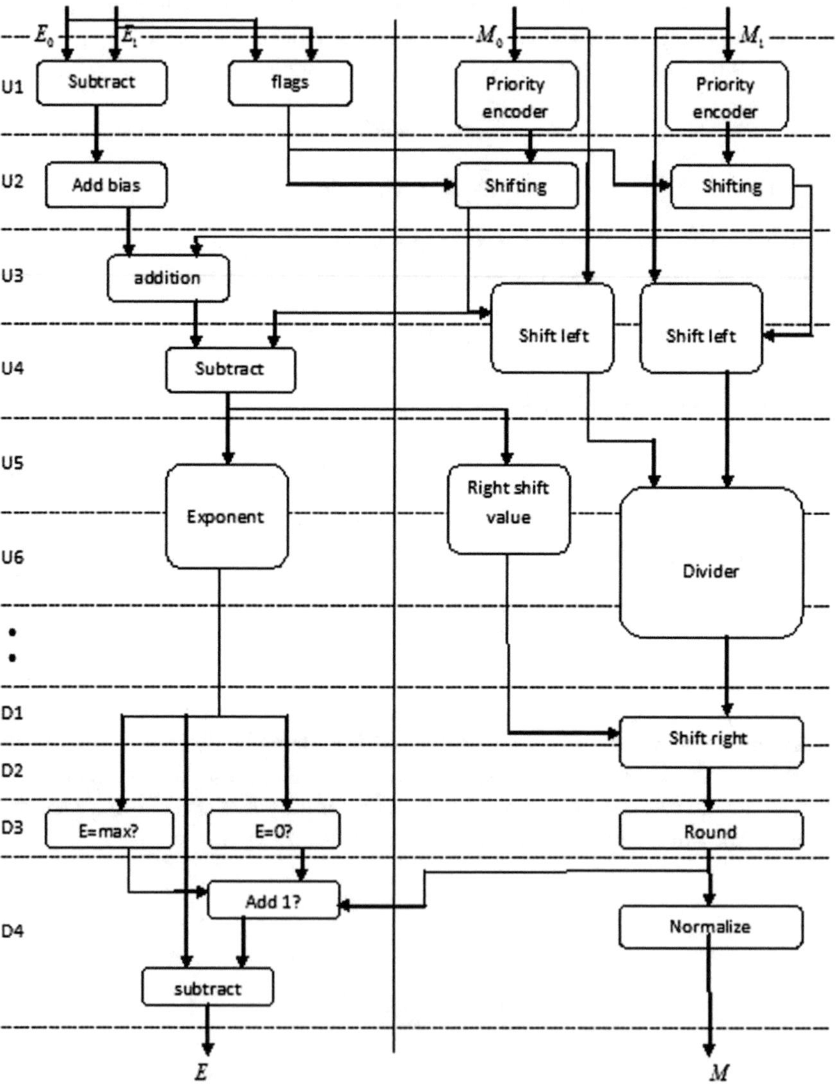

Fig. 3 Structure of floating point divider

At stages U_3 and U_4, the shift values, SV_0 and SV_1, are added and subtracted, respectively, by the exponent to get the final value. For better alignment, each mantissa passes through two shift blocks [8, 9]. Both U_5 and U_6 perform with divider core to generate the exponent and right shift value needed for later stages. The divider core uses iterative or pipelining technique which is performed up to 52 cycles [10].

On D_1 and D_2 stages, the mantissa is right shifted and finds the sticky bit. The rounding will induce the carry or IEEE special flags required to modify the mantissa, which is performed in stage D_4. The minimum and maximum values of exponents are calculated in stage U_3. By introducing high-speed processors day by day makes the instructions to be executed faster and faster, so algorithm for the efficient implementation of division is necessary.

Although various algorithms like functional iterative, subtractive method were developed for faster computation of division, they consume large power and occupy larger area to accommodate more multipliers, which is not desirable. The proposed pipelined digit recurrence algorithm [11] which uses sequential double precision floating point divider that consumes very less area when compared with other algorithms.

3 Proposed Method

3.1 Division Protocol

The standard protocol used in this paper is divided into three stages. In first stage, mantissa and exponent are generated during cycle 1 to cycle 11, followed by mantissa division up to cycle 53. In last stage, postnormalization is performed.

3.2 Implementation

The proposed double floating point precision divider is shown in Fig. 4. For each clock cycle, one quotient is calculated by comparing divr_reg and divd_reg. The higher mantissa value is stored in dividend, and lower mantissa value is stored in divisor. Subtract both mantissas, if the result is $\geq$ zero, the MSB is '1,' otherwise it is '0.' The partial result shifted to left by one bit, and it becomes the dividend for the next clock cycle. The hardware required for doing the mantissa division are carry save adder, registers to hold the inputs, and shift register. Total operation takes 55 clock cycles to produce the 55-bit value which is later normalized and rounded to get 53-bit quotient. The extra two bits in quotient are useful for rounding.

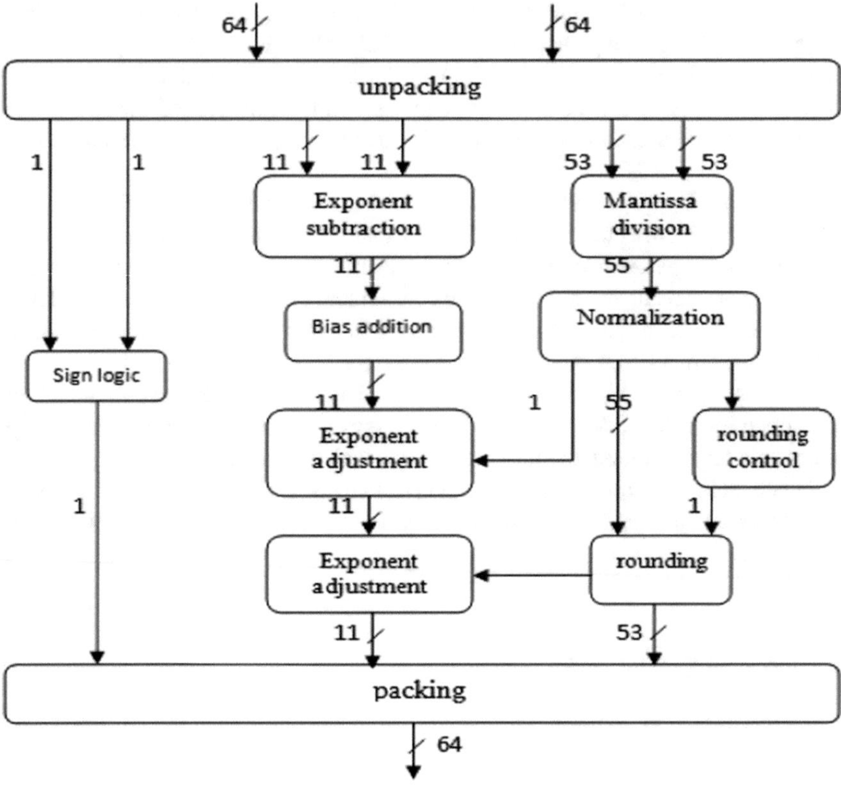

Fig. 4 Architecture of precision divider

The exponents of dividend and divisor are subtracted to find the exponent of the output. A value of 1023 is added to the exponent of dividend, and this sum is subtracted by the divisor exponent. The resultant exponent is output of the exponent subtraction. In 56th cycle, hold bit(h) is calculated by ORing of final remainder bits. No additional shift registers are required, since the remainder is stored in dividend register. In cycle 57, the exponent incremented by 1 and the value is transferred to the divr_reg. In next cycle, 55-bit remainder is rounded to 53 bits.

The inputs to the rounding operation taken from division module are 1-bit sign, 56-bit mantissa, and 12-bit quotient. The 56 bits in mantissa comprises MSB '0' bit, 52-bit mantissa with leading '1,' and two extra bits at LSB. The exponent has an extra '0' bit at MSB to avoid overflow. If the register has only 11 bits, exponent maintains '0' value and it would be incorrect. Rounding is operated in four modes, starts with code C0 for round_nearest, code C1 for round_zero, code C2 for round_infinity followed by round_negative infinity for code C3. In C0 mode, both LSB and extra remainder bits are high. No rounding is preformed in C1 mode. In last two modes, check the sign of the output infinity.

3.3 Pipelining Architecture

Figure 5 shows two-level pipelining of radix-4 single stage. A couple of registers are inserted between processing stages of the array divider. The main advantage of pipelining is high throughput, but the entire architecture occupies more area. To minimize area overhead, the circuit is partial unrolled to $2^n(n = 1, 2,..5)$ stages and in each stage, add the pipeline registers. This technique increases the throughput

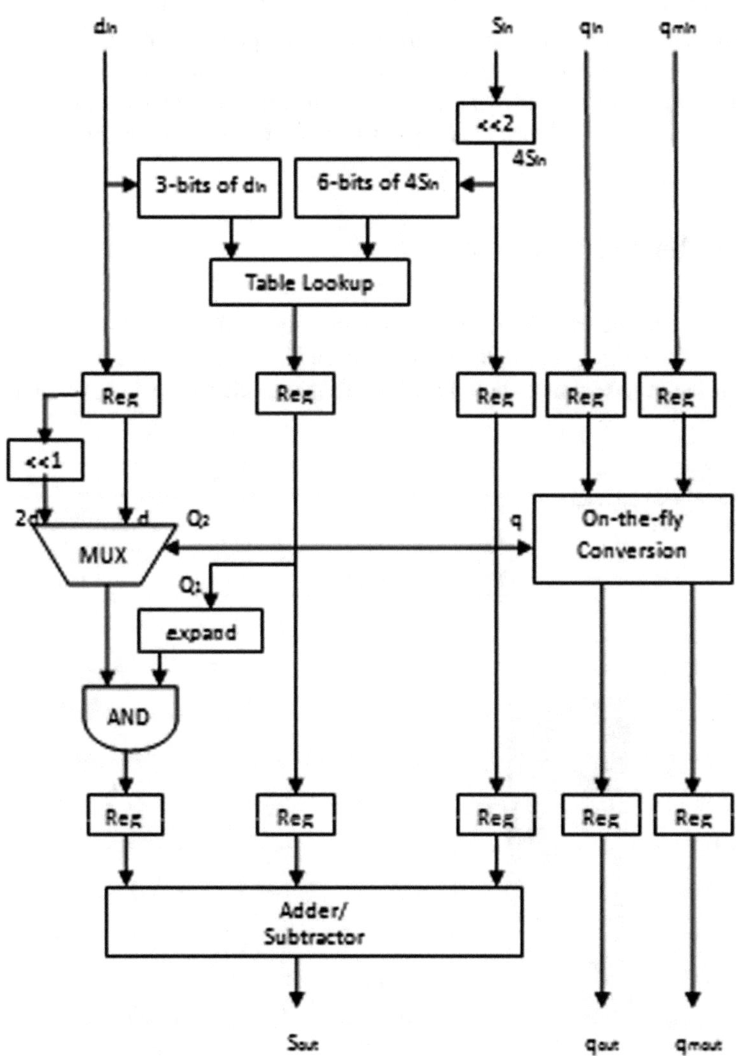

Fig. 5 Two-level subpipelining of radix-4 for single stage

with small area overhead. The area overhead is directly proportional to the latency, i.e., the higher the area higher the latency.

The throughput of pipeline is the number of instructions executed per second.

$$H_p = \frac{1}{kd}. \tag{8}$$

where k—number of iterations per stage

d—single stage execution delay

Compared to parallelism, pipelining technique improves throughput with expense of design size. The pipelining of two-stage implementation is twice shorter than previous method. The major limitations for pipelining are clock skew, overhead in pipelining, and hazards. In pipelining, maintaining proper clock skew is more important to maintain latency less compared to previous methods. We cannot avoid overhead in pipelining, but high data rates in execution make it simple.

4 Results and Simulation

The digit recurrence divider algorithm was implemented in Virtex-5. The algorithm was coded in Verilog, simulated using ISim and synthesized in Xilinx 14.2i. From Table 1, the proposed method improves in all design characteristics. The number of

Table 1 Optimization parameter of existing and proposed methods

Parameter	Existing	Proposed
BELS	3641	3628
Delay (ns)	4.720	4.21
Frequency (MHz)	211.87	356.55

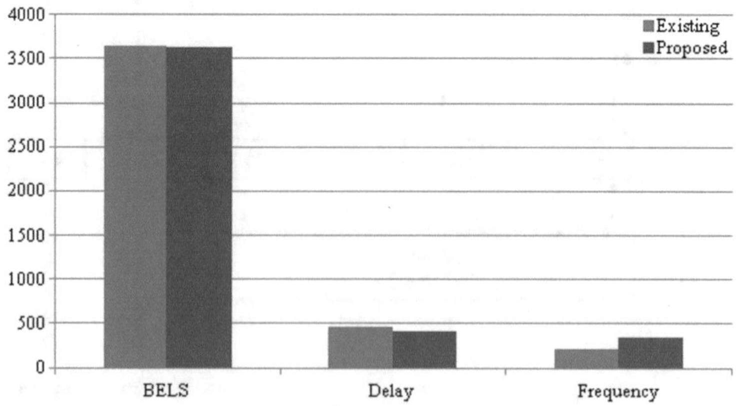

Fig. 6 Optimization parameter comparison chart

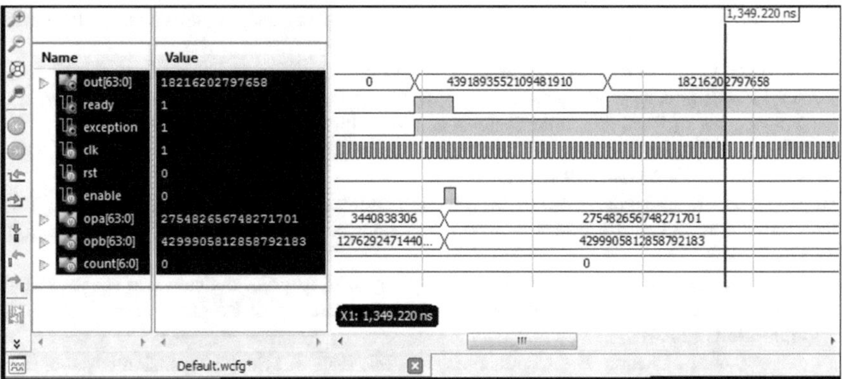

Fig. 7 Simulation results for two-stage pipelined floating point divider

BELS is decreased by 12%, and delay is reduced by 11%. The frequency is increased by a factor of 68%.

Figure 6 shows the comparison chart, in which delay is taken in μs and frequency is taken in MHz along y-axis. Figure 7 shows the snapshot of simulation results for two-stage pipelined floating point divider. The two input operands opa and opb are 64-bit numbers. The output is denoted as **out** (64-bit) produced after 58 clock cycles.

5 Conclusion and Future Work

This paper presents the design considerations of the pipelined designs of double precision floating point divider unit. The proposed design performed better than conventional iterative-based dividers, but it occupies more area. Partial unrolling makes the DPFPD more faster without affecting the performance. Further, we can implement the DPFPD with full unrolling which makes the design more faster, but the pipelining becomes more tedious with clock skew. We can also use secondary clock for mantissa execution which will further decrease the latency. The use of cache memory to store the intermediate values also minimizes the latency.

References

1. Wang, L.-K., Schulte, M.J.: A decimal floating point divider using newton-raphson iteration. The J. VLSI Signal Process. **49**(1), 3–18 (2007) (Springer, USA)
2. Amanollahi, S., Jaberipur, G.: Energy efficient VLSI realization of binary 64 division with redundant number systems. In: IEEE Transactions on Very Large Scale Integration systems, pp. 1–12 (June 2016)

3. Muller, Jean-Michel, Functions, Elementary: Algorithms and Implementation. Birkhauser Boston Publishers, Springer science (2016)
4. Govindu, G., Scrofano, R., Prasannna, V.K.: A library of parameterizable floating point cores for FPGAs and their application to scientific computing. In: International Conference on Engineering of Reconfigurable Systems and Algorithms (2005)
5. Wei, L., Nannarelli, Alberto: Power efficient division and square root unit. IEEE Trans. Comput. **61**(8), 1059–1070 (2012)
6. Hemmert, K.S., Underwood, K.D.: Floating point divider design for FPGAs. In: IEEE Transaction on very large scale integration systems, vol. 15, No. 1, pp. 115–118, (Jan 2007) (K. Elissa)
7. Wang, X., Nelson, B.E.: Tradeoffs of designing floating point division and square root on virtex FPGAs. In: International Conference on Engineering of Reconfigurable Systems and Algorithms (2004)
8. Jaiswal, M.K., Cheung, R.C.C., Balakrishnan, M., Paul, K.: Series expansion based efficient architectures for double precision floating point division. J. Circuits, Syst. Signal Process. **33** (11), 3499–3526 (2014) (Springer)
9. Soderquist, Peter, Leeser, Miriam: Division and square root: choosing the right implementation. IEEE Micro **17**(4), 56–66 (1997)
10. Paschalakis, S., Lee, P.: Double precision floating point arithmetic on FPGAs. In: IEEE Conference on Field Programmable Technology (2003)
11. Oberman, S.F., Flynn, M.J.: Division algorithms and implementations. IEEE Trans. Comput. **46**(8), 833–854 (1997)

GA_DTNB: A Hybrid Classifier for Medical Data Diagnosis

Amit Kumar and Bikash Kanti Sarkar

Abstract Recent trends in medical data prediction have become one of the most challenging tasks for the researchers due to its domain specificity, voluminous, and class imbalanced nature. This paper proposed a genetic algorithms (GA)-based hybrid approach by combining decision table (DT) and Naïve Bayes (NB) learners. The proposed approach is divided into two phases. In the first phase, feature selection is done by applying GA search. In the second phase, the newly obtained feature subsets are used as input to combined DTNB to enhance the classification performances of medical data sets. In total, 14 real-world medical domain data sets are selected from University of California, Irvine (UCI) machine learning repository, for conducting the experiment. The experimental results demonstrate that GA-based DTNB is an effective hybrid model in undertaking medical data prediction.

Keywords Prediction · GA · NB · DTNB · Medical

1 Introduction

Medical data mining is one of the most important tasks for finding useful patterns, meaningful information, or extraction of knowledge from data sets. Classification is a *supervised* learning technique of data mining used for extracting useful knowledge over a large *volumetric* data by predicting the *class values* based on the *relevant* attribute values. Therefore, designing an accuracy-based classification model has become a challenging task for the researchers.

A. Kumar (✉) · B. K. Sarkar
Department of Computer Science and Engineering, Birla Institute of Technology(DU), Mesra, Ranchi, India
e-mail: amit1022@gmail.com

B. K. Sarkar
e-mail: bk_sarkarbit@hotmail.com

V. Bhateja et al. (eds.), *Intelligent Engineering Informatics*, Advances in Intelligent Systems and Computing 695, https://doi.org/10.1007/978-981-10-7566-7_15

Recently, medical data mining greatly contributes in the discovery of disease diagnosis that provides the domain users (i.e., medical practitioners) with valuable and previously unknown knowledge to enhance diagnosis and treatment procedures for various diseases. A number of tools have been made to assist medical practitioners in making their clinical decisions. Few medical domain works have been cited here. Seera and Lim [1] proposed an efficient hybrid classifier for medical data sets using fuzzy min–max neural network. Selvakuberan et al. [2] presented a feature selection-based classification system using machine learning approaches for health care domain. Applying both artificial and fuzzy neural network, Kahramanli and Allahverdi [3] also developed a hybridized system that was very much helpful in the diagnosis of diabetes as well as heart diseases. Lee and Wang [4] built a fuzzy-based expert system for diabetes prediction. Kalaiselvi and Nasira [5] built a new prediction system for diabetes as well as cancer diseases, using adaptive neuro-fuzzy inference system (ANFIS). Chen and Tan [6] developed a predictive model for the diagnosis of type-2 diabetes based on distinct elements levels in blood and chemometrics. Garg et al. [7] published a systematic review on the significance of computerized clinical decision support systems over the practitioner performances and patient outcomes. Similarly, Kawamoto et al. [8] published a systematic review on identifying critical features for success in enhancing the clinical practices using decision support systems. Narasingarao et al. [9] designed a decision support system for diabetes prediction using multilayer neural networks. In 2003, Huang and Zhang [10] applied rough set theory in ECG recognition. Similarly, rough set approach is also applied for generating optimal rule sets for medical data by Srimani and Koti [11]. Ye et al. [12] investigated on fuzzy sets to derive rules for the prediction of degree of malignancy in brain glioma. Syeda-Mahmood [13] has given a plenary talk on the role of machine learning algorithms on clinical decision support systems. Wagholikar et al. [14] published a survey on modeling paradigms for medical diagnostic decision systems. Recently, Martis et al. [15] published a research paper on the development of intelligent applications for medical imaging processing and computer vision. Rajinikanth et al. [16] proposed a study on entropy-based segmentation of tumor from brain MR images using TLBO approach. Applying back propagation neural networks, Gautam et al. [17] proposed an improved mammogram classification technique. Dey et al. [18] published study on some algorithmic- and computer-based approaches to medical imaging in clinical applications.

The trend says that these systems are mostly being used in clinical data diagnosis, prediction, and risk forecasting for different diseases. However, it is found that medical data sets may contain hundreds of irrelevant features, but no standard feature selection *criterion* is properly set up. Obviously, a relevant feature subset selection is an effective way to improve the performance of a classifier because the small subset of informative features may contain enough relevant information to construct *diagnostic models* for distinct diseases.

The present study first adopts a GA over each data set to select an optimal set of features, and then, combined classifier DTNB is applied to train over the data set. Finally, the learned knowledge is applied over test data for evaluation.

2 Background

2.1 Feature Selection

Feature selection is a process of *identifying* a subset of original attributes by discarding the *irrelevant* and *redundant* attributes as possible. A raw data set may have a number of features or attributes, but feature selection technique identifies and removes the irrelevant features to *maximize* the classification accuracy. For better understanding the process, Fig. 1 is depicted below.

One may note that the stopping criterion is either *fixed number* of iterations or a particular *threshold value*. Anyway, the job can be successfully satisfied by GA search.

2.2 Genetic Algorithm

Genetic algorithm is a *probabilistic search algorithm* and also an *optimization technique*. It is based on the principles of evolution and *natural genetic systems*. During the last few decades, the *GAs* are being applied to a wide range of *optimization* and *learning* problems. The simplicity of algorithm and fairly accurate results are the main advantages of *GA* formulation. In particular, it is a method of finding an optimal solution to a problem, based on the feedback received from its repeated attempts at a set of feasible solutions.

The GA-based problem solving strategy was first introduced by Prof. John Holland in 1975 [19], but it gained its popularity as solution spaces in different

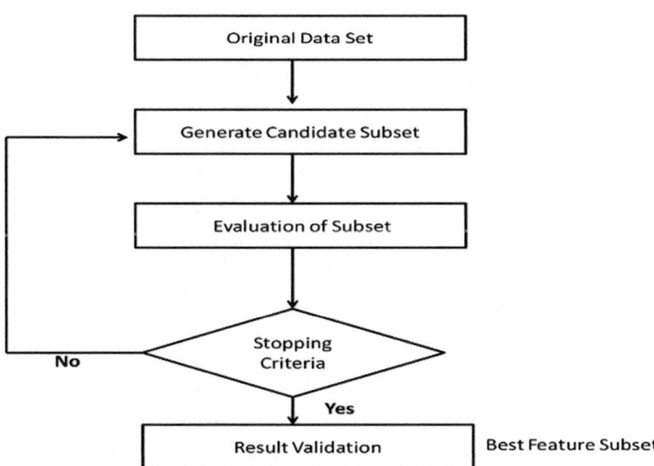

Fig. 1 Steps for feature selection

search and optimization problems. It is true that GA associates nature-inspired phenomenon like *genetic inheritance* and *survival of the fittest*. Any *GA*-based search technique starts with a population set of potential solutions, whereas all other methods proceed with a single point of the search space at a time. Interestingly, the evolution process of GA starts with *encoding* of population set of search space represented by *chromosomes*. It mimics the natural evolution process by repeatedly altering a population set of feasible solutions until an *optimal* solution is not found. Some basic steps of a simple GA are pointed out below.

Step 1. Initialize with a population set of *n chromosomes*.
Step 2. Determine the fitness function f(x) for each of the chromosomes x.
Step 3. If termination condition reaches, then stop the *process*.
Step 4. *Select* the encoded strings to evaluate new mating pool.
Step 5. Generate new population set by using *crossover* and *mutation* probability.
Step 6. Go to Step 2.

Important components of GA

Genetic algorithm consists of the following important components.

Encoding Techniques

Genetic algorithms initialize with the *chromosomal* representation of a population set encoded as a finite size string so that computer can understand. The important encoding schemes are, respectively, binary *encoding*, *permutation encoding*, *value encoding, tree encoding*, etc.

Evaluation technique

In genetic algorithms, feasible solution set follows the concept of '*survival of the fittest*' to obtain an optimal solution. Therefore, they use a fitness function as an objective function in order to select the *fittest* string that will generate new and probably better population set of strings.

Selection

According to *selection* or *reproduction* operator, the chromosomes are selected as parents to perform *crossover* and produce *offspring* usually from the population set. In this process, individual strings are copied into a provisional new population known as *mating pool*, for genetic operations. The number of copies that an individual receives for the next generation is usually based on Darwin's evolution theory '*Survival of the fittest*', i.e., the best one should survive and generate new offspring, thereby mimicking the natural evaluation. The most frequently used selection procedures are, respectively, *roulette wheel selection*, *rank selection, steady-state selection*, and *elitism*.

Crossover

The main objective of crossover operation is to combine (mate) two parent chromosomes to produce two new chromosomes (also known as offspring). It is true that new chromosomes may be better than both the parents if they inherit the best characteristics from the parents. Crossover occurs during evolution according to a user-defined probability. The most commonly used techniques are single point crossover, two point crossover, uniform crossover, arithmetic crossover, and tree crossover.

Mutation

Mutation takes place after a crossover operation. It is, indeed, used as a genetic operator that maintains *genetic diversity* from one generation of a population to the next generation. It occurs during evolution according to a user-defined *mutation probability*. Actually, mutation *alters* one or more *gene* values in a chromosome from its initial state. Hence, it results *entirely* new gene values being added to the gene pool. With the new gene values, the genetic algorithm may be able to produce a *better* solution than previous one.

2.3 Decision Table and Naïve Bayes (DTNB): A Hybrid Model

DTNB is a rule-based hybrid model by combining both DT and NB classifiers. The model is implemented in WEKA by Hall and Frank [20]. Conceptually, DTNB model starts with splitting the set of attributes into two groups. One group assigns class probabilities based on decision table (DT), whereas the other group assigns class probabilities based on Naïve Bayes (NB). Initially, all attributes are modeled by DT. At every step, the method considers dropping an attribute entirely from the classifier. Finally, the class probabilities are computed as follows.

$P(y/X) = k \times PDT(y/X>) \times PNB(y/X^{\perp})/P(y)$, where $PDT(y/X>)$ and PNB $(y/X^{\perp})$ are the class probability estimates derived from DT and NB, respectively. P (y) is the prior probability, and k is the normalization constant of the class.

The model is evaluated over 50 iterations on the basis of 66% of the data for training and rest for testing at each iteration. The authors have reported the improved performance over the base learners. However, computational complexity of the model increases by an order of magnitude.

3 Proposed Hybrid Model

The construction of *GA*-based hybrid model is described in this section. Certainly, the approach claims to build separate learning models by applying *GA* and the combined capabilities of Decision Table/Naïve Bayes (DTNB)-based rule inducer.

The model first receives an original data set with given *features*. GA process begins with the selection of individual population set of specified features. Next, it follows two operators, namely *crossover* and *mutation*, to produce a new generation. After each generation, the features are evaluated and tested for the termination of the algorithm. If the termination condition does not reach, then the process

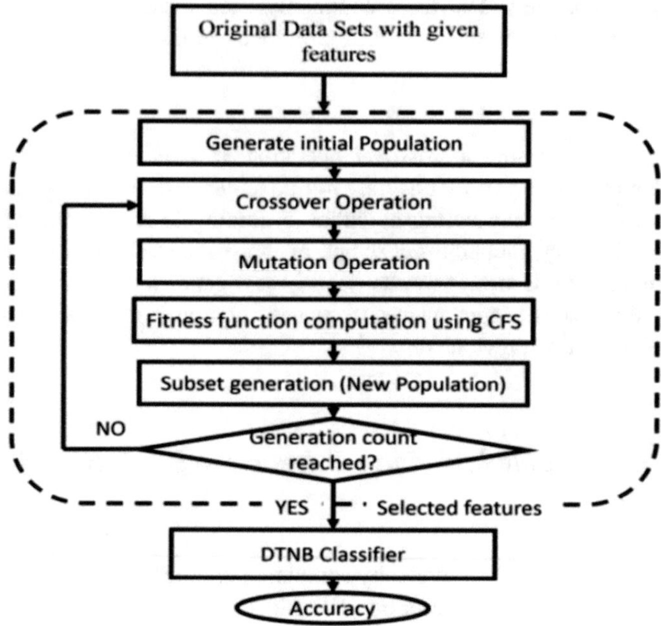

Fig. 2 Hybrid model

continues. Note that in the present study, the algorithm cycle proceeds until the termination criterion (generation count) reaches. Finally, a reduced feature set of data set is obtained.

In the next phase, the 10-fold cross-validation scheme is applied to evaluate the overall classification performance of the proposed model. For better understanding, the sketch of the algorithm is shown in Fig. 2.

4 Experimental Design and Analysis

This section discusses about conducted experimental design and their analysis.

Experiments

For showing the performance of the suggested model, two experiments are conducted. These are named as, respectively, e_1 and e_2, where e_1 is performed for selection of relevant features using GA, but e_2 is conducted over the combined model DTNB based on selected features.

The package WEKA-3.7.2 is adopted for the implementations rule-based classifier DTNB. It may be noted that the GA-based feature selection approach is also implemented by this package too. The proposed model is experimentally evaluated over the 14 medical data sets (selected from UCI repository) of real-world problems and compared based on *correctly* classified instances (measured in terms of accuracy).

Results and Discussions

It may be noted that 10-fold cross-validation scheme is used in the present study to measure the performances (in terms of mean accuracy) of the classifier. The in-built implementation of the scheme is taken from WEKA package. For each fold, a classifier is initially trained on the other folds and then it gets tested. Further, the average trained accuracies are calculated over all 10 outcomes. Finally, a standard deviation (s.d.) along with each mean result is recorded from 10 runs of the experiment. Standard deviation is an important measure, since it generalizes the overall performance of the classifier. The accuracy may be computed as follows.

Accuracy: The classification *accuracy* of a classifier is measured as follows.Accuracy (acc.) = $\frac{\mathrm{m}}{\mathrm{n}} \times 100$, where *m* denotes the number of correctly classified test examples and *n* is the total number of test examples. It is, indeed, the percentage of unseen data correctly classified.

In other way, it is equivalent to $\frac{TP+TN}{P+N} \times 100$ for binary class valued problem, where P and N represent, respectively, the number of *positive* and the *negative* examples, and TP, TN, FP, and FN refer, respectively, the number of *true positive*, *true negative*, *false positive*, and *false negative*. However, if the data set is *multi-class* classification data set, then its each class is considered as the positive class, whereas the examples of the rest classes fall under negative class.

Analysis of experiments

A deep comparison study on the basis of performance Table 1 reveals that the relevant feature selection approach through *GA* from original data sets improves classification performances of DTNB classifier. Obviously, it speeds up the computation time to model the classifier, as the number of features is reduced. This says that the relevant feature selection contributes an effective role to design a good model.

Overall findings of the model

- It can be observed from Table 1 that the hybrid model GA_DTNB performs better prediction results in 12 cases (namely, *breast cancer*, *dermatology*, *E coli*, *heart(swiss)*, *heart* (*cleveland*), *hepatitis*, *liver disorder*, *lymphography, new-thyroid*, *Pima-Indians*, *primary tumor*, *sick* data sets) in comparison to DTNB.
- The most noticeable point is that GA-based DTNB model results in higher classification accuracy with low standard deviation (*s.d.*) over the data sets with relevant features (extracted by GA approach). This claims the reliability of the hybrid model over medical data sets.
- Further, Fig. 3 (drawn from mean accuracy percentage values) concludes that the *accuracy rate* of the hybrid model (*GA_DTNB*) is higher than *DTNB* hybrid learner in most cases.

Table 1 Performance of DTNB and GA_DTNB classifiers on the UCI data sets using 10-fold cross-validation over 10 runs

Problem name	Total number of attributes	Number of reduced attributes	Number of classes	Number of examples	Accuracy percentage	
					DTNB *acc.* ± *s.d.*	GA_DTNB *acc.* ± *s.d.*
Breast cancer	10	6	2	699	71.67 ± 2.32	**75.17 ± 1.72**
Dermatology	34	21	6	366	97.54 ± 1.37	**97.91 ± 1.43**
Pima-Indians	8	4	2	768	73.82 ± 2.31	**76.91 ± 1.91**
E coli	8	6	8	336	81.25 ± 2.21	**81.25 ± 2.21**
Heart (Hungarian)	13	6	5	294	**82.65 ± 1.34**	81.97 ± 1.40
Heart (swiss)	13	1	5	123	38.21 ± 2.32	**39.52 ± 1.92**
Heart (cleveland)	13	8	5	303	79.86 ± 2.19	**82.83 ± 2.02**
Hepatitis	19	9	2	155	81.29 ± 3.41	**81.93 ± 2.31**
Liver disorder	6	1	2	345	57.68 ± 2.26	**57.68 ± 2.26**
Lung cancer	56	17	3	32	**81.25 ± 4.37**	78.75 ± 6.12
Lymphography	18	10	4	148	73.64 ± 4.03	**79.22 ± 1.75**
New-thyroid	5	5	3	215	93.28 ± 1.41	**93.89 ± 1.09**
Primary tumor	17	12	22	339	41.59 ± 1.73	**42.07 ± 1.67**
Sick	29	8	2	3772	97.26 ± 0.18	**97.53 ± 0.24**

Note: The value appearing just before '±' at each column indicates the *mean* accuracy (*acc.*), whereas the value appearing after '±' represents standard deviation value (s.d.)

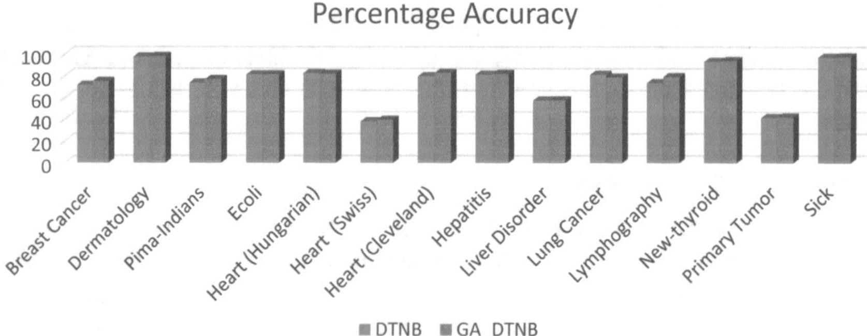

Fig. 3 Comparative mean accuracy percentage chart for GA_DTNB and DTNB hybrid models with given data sets

One may note that all the *experiments* discussed in this section are performed on an *ACER ASPIRE* notebook computer with *P6200* @ 2.13 *GHZ CPU* running Microsoft Windows 7 and 2 *GB RAM.*

5 Conclusion and Future Work

Many classification approaches for medical data mining have been proposed in the past decades, but they have drawbacks like disease *specificity* of model and *vagueness* of patient's data. In this study, the proposed hybrid model outperforms over most of the given data sets. On the basis of the analysis of the performed results, it may be concluded that GA-based hybrid model (GA_DTNB) constructed in this article claims to improve accuracy in *diagnosis*, *prognostication*, and *treatment* strategies of distinct diseases. It improves the reliability over medical domain data sets.

However, the hybrid approach with GA could not improve on *heart (Hungarian)* and *lung cancer* data sets in the present study. The reason is that these data sets are very much complex in nature, and our future research is to be extended on it.

References

1. Seera, M., Lim, C.P.: A hybrid intelligent system for medical data classification. Expert Syst. Appl. **41**(5), 2239–2249 (2014)
2. Selvakuberan, K., Kayathiri, D., Harini, B., Devi, M.I.: An efficient feature selection method for classification in health care systems using machine learning techniques. In: 3rd International Conference on Electronics Computer Technology (ICECT), vol. 4, pp. 223–226. IEEE (2011)
3. Kahramanli, H., Allahverdi, N.: Design of a hybrid system for the diabetes and heart diseases. Expert Syst. Appl. **35**(1), 82–89 (2008)
4. Lee, C.S., Wang, M.H.: A fuzzy expert system for diabetes decision support application. IEEE Trans. Syst. Man Cybern. Part B (Cybern.) **41**(1), 139–153 (2011)
5. Kalaiselvi, C., Nasira, G.M.: A new approach for diagnosis of diabetes and prediction of cancer using ANFIS. In: World Congress Computing and Communication Technologies (WCCCT), pp. 188–190. IEEE (2014)
6. Chen, H., Tan, C.: Prediction of type-2 diabetes based on several element levels in blood and chemometrics. Biol. Trace Elem. Res. **147**(1–3), 67–74 (2012)
7. Garg, A.X., Adhikari, N.K., McDonald, H., Rosas-Arellano, M.P., Devereaux, P.J., Beyene, J., Sam, J., Haynes, R.B.: Effects of computerized clinical decision support systems on practitioner performance and patient outcomes: a systematic review. JAMA **293**(10), 1223–1238 (2005)
8. Kawamoto, K., Houlihan, C.A., Balas, E.A., Lobach, D.F.: Improving clinical practice using clinical decision support systems: a systematic review of trials to identify features critical to success. BMJ **330**(7494), 765 (2005)
9. Narasingarao, M.R., Manda, R., Sridhar, G.R., Madhu, K., Rao, A.A.: A clinical decision support system using multilayer perceptron neural network to assess well being in diabetes. pp. 127–133. (2009)
10. Huang, X.M., Zhang, Y.H.: A new application of rough set to ECG recognition. In: International Conference on Machine Learning and Cybernetics, vol. 3, pp. 1729—1734. IEEE (2003)
11. Srimani, P.K., Koti, M.S.: Rough set (RS) approach for optimal rule generation in medical datawork. **2**(2), 9–13 (2014)

12. Ye, C.Z., Yang, J., Geng, D.Y., Zhou, Y., Chen, N.Y.: Fuzzy rules to predict degree of malignancy in brain glioma. Med. Biol. Eng. Compu. **40**(2), 145–152 (2002)
13. Syeda-Mahmood, T.F.: Role of machine learning in clinical decision support (Presentation Recording). In: SPIE Medical Imaging. International Society for Optics and Photonics 94140U–94140U (2015)
14. Wagholikar, K.B., Sundararajan, V., Deshpande, A.W.: Modeling paradigms for medical diagnostic decision support: a survey and future directions. J. Med. Syst. **36**(5), 3029–3049 (2012)
15. Martis, R.J., Lin, H., Gurupur, V.P., Fernandes, S.L.: Frontiers in development of intelligent applications for medical imaging processing and computer vision (2017)
16. Rajinikanth, V., Satapathy, S.C., Fernandes, S.L., Nachiappan, S.: Entropy based segmentation of tumor from brain MR images–a study with teaching learning based optimization. Pattern Recognit. Lett. (2017)
17. Gautam, A., Bhateja, V., Tiwari, A., Satapathy, A.C.: An improved mammogram classification approach using back propagation neural network. In: Data Engineering and Intelligent Computing, pp. 369–376. Springer, Singapore (2018)
18. Dey, N., Bhateja, V., Hassanien, A.E. (eds.): Medical Imaging in Clinical Applications: Algorithmic and Computer-Based Approaches, vol. 651. Springer (2016)
19. Holland, J.H.: Adaptation in Natural and Artificial Systems: An Introductory Analysis with Applications to Biology, Control, and Artificial Intelligence. MIT press (1992)
20. Hall, M.A., Frank, E.: Combining Naive Bayes and Decision Tables. In: FLAIRS Conference, vol. 2118, pp. 318–319. (2008)

Applications of Spiking Neural Network to Predict Software Reliability

Ramakanta Mohanty, Aishwarya Priyadarshini, Vishnu Sai Desai and G. Sirisha

Abstract In the period of software improvement, programming dependability expectation turned out to be exceptionally critical for creating nature of programming in the product business. Time to time, numerous product dependability models have been introduced for evaluating unwavering quality of programming in programming forecast models. However, building precise forecast model is hard because of intermittent changes in information in the space of programming designing. As needs be, we propose a novel procedure, i.e. spiking neural system to anticipate programming unwavering quality. The key goal of this paper is to exhibit another approach which upgrades the exactness of programming unwavering quality prescient models when utilized with the product disappointment dataset. The viability of quality of a product is exhibited on dataset taken from the literature, where execution is measured by utilizing normalized root mean square error (NRMSE) obtained in the test dataset.

Keywords Software reliability · Neural network · Spiking neural network Normalized root mean square error

R. Mohanty (✉) · V. S. Desai · G. Sirisha
Department of Computer Science & Engineering, Keshav Memorial Institute of Technology, Narayanaguda, Hyderabad 500029, Telengana, India
e-mail: ramakanta5a@gmail.com

V. S. Desai
e-mail: vishnudesai196@gmail.com

A. Priyadarshini
Department of Computer Science and Engineering, International Institute of Information Technology, Bhubaneswar, India
e-mail: aishwarya8skvs@gmail.com

V. Bhateja et al. (eds.), *Intelligent Engineering Informatics*, Advances in Intelligent Systems and Computing 695, https://doi.org/10.1007/978-981-10-7566-7_16

1 Introduction

Nowadays, numerous administration associations and corporates are digitized to expand their efficiency. As the product disappointments are expanding quickly, programming experts likewise focussed to enhance unwavering quality of programming. In this procedure, many models have been joined to gauge the product disappointments to meet the requests of telemarketing and digitization. Notwithstanding programming improvement, there are numerous other prerequisites to be satisfied with a specific end goal to handle the present situation of digitization, viz. the phases of dependability of the product, security, convenience, upkeep and viable cost. The productivity of any product ventures relies on its quality, sturdiness and refreshing of programming with the changing requests of the clients [1]. The key motivation behind programming unwavering quality designing is utilized to help Engineer, Manager or client of programming figure out how to bring out exceptionally precise items. A moment point is to create programming exactly mindful of programming unwavering quality by focussing on it. A correct choice can keep time and cash on a venture and amid the lifetime of the product in various ways [2].

Programming dependability is the most vital elements that are identified with shortcomings and deformities. An essential segment of programming dependability is programming quality, usefulness, execution, capacity, ease of use, openness, serviceability and support. As of late, industry had put developing significance on programming constancy and has embarked to utilize and apply the hypotheses created in the research. The use of dependable figures is a very much characterized way to take since the product business tries to give clients trust in its stock.

Programming experts can anticipate or measure programming unwavering quality [3] utilizing models of programming dependability. However, the cost and time are constrained in the genuine programming improvement handle. The essential drivers that impact programming unwavering quality are (i) error introduction, (ii) the environment and (iii) fault deletion. The error introduction depends for the most part on the elements of the product and the improvement procedure.

Programming building advances contain the trademark in the development procedure and the devices that are used in the unpredictability of necessities, the stratum of experience of the workforce and other trademark. The failure location and programming is not sufficient to find the reliability of a software. Uniformly, pretty much of the above components are probabilistic in nature and control after sometime, and the product dependability models have a rule being given decision as far as the irregular procedures in execution time. In a current past, top to bottom review has been completed in programming determining and unwavering quality, and it has been found that no single model can catch programming elements and dependability [4, 5].

Whatever is left, the paper is sorted out in the accompanying strategy. In the segment 2, literature study is introduced. In area 3, outline of spiking neural networks is displayed. Sect. 4 examines the test plan. Area 5 displayed an entire discussion about the resolutions and talks. Finally, segment 6 finishes up the paper.

2 Literature Survey

Because of technological advances, these days numerous new programming created at a steadily expanding rate, size and dimensionality of programming keep on growing by step. Therefore, it is important to develop efficient and effective machine learning methods that can be used to analyse the software reliability and extract useful knowledge and insights from this wealth of information to maintain the high standard of quality of software. Many authors have given numerous solutions with respect to software reliability from time to time. We will talk about one by one. Cai et al. [6] utilized backpropagation trained neural network on software failure dataset to predict the time between software failures with respect to time between software project developments. Tian and Noore [7] utilized evolutionary neural network on programming aggregate failure time. From their trial results, they performed better to anticipate programming software quality.

Su et al. [8] built up a combinational element weighted in the light of neural network to look at the outcomes in three angles, i.e. integrity of fit, capacity of short- and long-haul expectation. They discovered better expectation precision contrasted with different models. Ho et al. [9] proposed neural network and demonstrated for software quality of programming by reusing the past project from past undertakings. From their outcome, they watched better foreseeing precision in early phases of testing. Su and Huang [10] utilized dynamic weighted combinational model to anticipate programming unwavering quality. The outcome obtained by them is better than others. Huang et al. [11] used unified theory on reliability growth model by employing non-homogenous poison process. Costa et al. [12] utilized genetic programming to anticipate programming software quality, and their re-enacted results are high exactness contrasted with neural network approaches.

Pai and Hong [13] employed simulated algorithm of SVM on software failure dataset to predict the software reliability. They explored that tuning of parameters plays a crucial role to determine the high accuracies. Rajkiran and Ravi [14] proposed an ensemble model comprises of MLR, MARS and dynamic evolving neuro-fuzzy inference system derivation framework to anticipate software quality. Their ensemble model created by them came about high exactness as far as normalized root means square error. Promote, Ravi et al. [15] worked on software failure dataset by use of wavelet neural network and morlet neural network to predict the software quality. Their experimental outcome showed a high precision in terms of normalized root mean square error which is concerned. Mohanty et al. [16–18] proposed group method of data handling (GMDH), genetic programming and distinctive machine-learning strategies to foresee software reliability by utilizing software failure datasets. From their exploratory outcome, they found that outcomes obtained by them beat different strategies in the literature. Most recently, Mohanty et al. [19–21] developed an ensemble model consists of recurrent and hybrid approach of genetic programming and group method of data handling to predict software reliability and also they discussed about ant colony optimization techniques to arrive better prediction than earlier approach. More recently, Gautam et al. [22] presented mammogram classification to define texture anomaly by using greylevel co-occurrence matrix feature and later employed backpropagation trained neural network for separating the mammogram into abnormal and normal.

3 Methodology

3.1 Spiking Neural Network

Spiking neural networks [23, 24] are the most rising field of calculation in the third era of neural system. It is truly intense and developing contribution of scientists in the field of PC vision and natural associations. The spiking neural systems are depicted in Fig. 1.

The design of spiking neural system comprises of a feedforward neural system having spikes of neurons and synaptic terminals with different deferred [23, 24].

The relationship between input spikes and output spikes are determined by synapse at the point where two neurons interact. An input, i.e. presynaptic spike arrives at the synapse, which produces the neurotransmitter which then influences the state or membrane potential of the output neuron i.e. post synaptic neuron.

At the point when the estimation of this state achieves an edge level ϑ, the objective neuron delivers a spike and the estimation of the state resets in unmanageable reaction.

In SNN, the neuron j contains an arrangement of Γj presynaptic neurons. The presynaptic neuron gets an arrangement of spikes with terminating time $t_{i,}$ where $i \in \Gamma j$. When information example is encouraged to SNN organize, the neuron

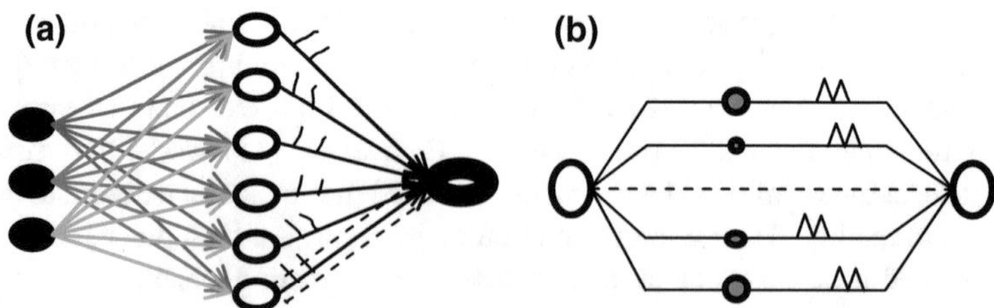

Fig. 1 **a** Spiking neural network architecture, **b** multiple synapses transmitting multiple spikes

creates no less than one spike and flames just when it achieves threshold level ϑ. The idea of inner state variable $x_j(t)$ by and large relies on impinging spikes, whose reaction can be discovered by spike response function ε(t). The adequacy of synaptic weight (w_{ij}) can be composed of

$$x_j(t) = \sum_{i \in \Gamma j} w_{ij} \in (t - t_i) \quad (1)$$

where $\in (t - t_i)$ is the spike reaction and is alluding to postsynaptic potential (PSP). The stature of the PSP can be transmitted by the synaptic weight w_{ij} to get post-synaptic potential at neuron because of neuron i.

The above equation can be composed of

$$x_j(t) = \sum\nolimits_{i \in \Gamma j} \sum\nolimits_{k=1}^{m} w_{ij}^{k} \varepsilon\left(t - t_i - d^k\right) \quad (2)$$

Here, m be the number of synaptic terminals and d^k be the postponement of the synaptic terminal k. The postponement is the contrast between the terminating or firing time of the synaptic neuron and postsynaptic potential. The unweighted commitment of a synaptic terminal can be communicated as

$$y_i^k(t) = \varepsilon\left(t - t_i - d^k\right) \quad (3)$$

where $\varepsilon(t)$be the spike response . The esteem $\varepsilon(t) = 0$ for $t < 0$

t_i be the terminating or firing time of presynaptic neuron and d^k defer identified with synaptic terminal k.

The standard PSP can be

$$\varepsilon(t) = \frac{t}{\tau} e^{1 - \frac{t}{\tau}} \quad (4)$$

where τ decides the membrane potential decay time constant which comes the rise and decay time of postsynaptic potential.

If we add multiple connection of synapses and adding the state variable x_j of neuron j from neuron i, then the sum of weighted presynaptic can be reproduced as

$$x_j(t) = \sum\nolimits_{i \in \Gamma j} \sum\nolimits_{k=1}^{m} w_{ij}^{k} y_i^k(t) \quad (5)$$

where w_{ij}^k refers to weight related to synaptic terminal k.

The firing of neuron takes place when it crosses the threshold value $\vartheta i.e. x_j(t) > \vartheta$.

Therefore, firing time becomes a nonlinear function of the state variable x_j. Accordingly, $t_j = t_j\left(x_j(t)\right)$, the threshold value becomes constant for all the neurons in the network.

3.2 Algorithm

Step 1: *Initialize the parameters.*
Step 2: *Initialize for j = 1 to n*
Step 3: *Apply the derivative to output network j.*
Step 4: *Next, Compute gradient of potential.*
Step 5: *If the gradient is less than 0.1, then the value is 0.1 and end of the loop.*
Step 6: *Loop*
Step 7: *Weight should be initialized for random values*
Step 7: *For i = 1 to m*
Step 8: *The first step: All weights of the connections from neuron i to neuron j.*
Step 9: *Then, calculate partial derivative networks all weights of input and output neuron and also gradient*
Step 10: *If there are any weight-changes then calculate delta weight of input neuron to output neuron with delay time k.*
Step 11: *End for*
Step 12: *For all weights input to output neurons with delay time k*
Step 13: *Calculate weights of input neuron to output neuron and changes of input to output neurons*
Step 14: *End*

4 Experimental Design

In our experimentation, a time series dataset is used to predict software reliability [17]. The dataset consists of one dependent variable and no explanatory variables. The time series can be explained as

$$P_t = f(P') \tag{5}$$

Here P′ is the vector of lagging variables.

It can be represented as $\{p_{t-1}, p_{t-2}, \ldots p_{t-m}\}$. Here the key is used to find out the solution for the problem and adjust of weight can be carried out iteratively. Figure 2 shows the design model of our training dataset. In Fig. 2, *m* is the number of variables in the dataset and *(t–m)* is the total number training samples in the dataset. From our model, *P* is a set of *(t–m)* vectors of dimension *P* and *Q* is a vector of dimension *(t–m)*. Here, we represent P as explanatory variable and *Q* as the dependent variable in the model.

We used Musa [1] software failure dataset for our simulation, and the dataset contains 101 samples of the pair (t, Q_t). We divided the dataset in the ratio of 80:20 for simulation. Here, Q_t is the time of failure after the *t*th modification carried out. Accordingly, we developed 5 lags, i.e. lag# 1, 2, 3, 4 and 5. To measure the accuracy, we used the normalized root mean square error (NRMSE) as follows:

Fig. 2 Design arrangement of training dataset

P				Q
p_1	p_2	$\cdots$	p_m	p_{m+1}
p_2	p_3	$\cdots$	p_{m+1}	p_{m+2}
p_3	p	$\cdots$	p_{m+2}	p_{m+3}
.	.	…	.	.
.	.	…	.	.
.	.	.		.
p_{t-m}	p_{t-m+1}	$\cdots$	p_{t-1}	p_t

$$\text{NRMSE} = \sqrt{\frac{\sum_{i=1}^{n}\left(d_i - \widehat{d_i}\right)^2}{\sum_{i=1}^{n} d_i^2}} \tag{6}$$

where n is the number of observations and d_i is the actual value of the period i and $\widehat{d_i}$ is predicted value of the software reliability at the period i.

5 Results and Discussions

We employed spiking neural network in MATLAB environment and simulated the experiment. For each lag, we created 25 sample files and 25 target files of each lag. Each sample file contains 50 observations. In each sample file, we apply the standard deviation method to calculate the target value of each sample files. Each lag folder consists of 25 sample files which is taken and loaded into the SNN cube. An SNN cube is a tool to perform the operations by loading the lag files and performing the operations. The steps that are taken to perform the experiment are data encoding, cube initialization, training of the SNN cube, training regressor and verify regressor.

A lag folder is chosen with the input samples and the operations mentioned above are being performed, and a file is obtained with data values which are again calculated using the cross-validation and parameter optimization. Then, an output value is obtained along with the actual value and predicted values, then we calculated the accuracy based on normalized root mean square error (NRMSE) as an overall final result.

From Table 1, we calculated the NRMSE values in the order as Lag1 of values 0.0252, Lag2 of value 0.0374, Lag3 of value 0.0426, Lag4 of value 0.0531 and Lag5 of 0.0382, respectively. We found that Lag1 provides very high accuracy of 0.0252. The advantages of using NRMSE value are that the lower the value of NRMSE, higher is the accuracy. However, this survey shows that SNN outperformed even that with a respectable leeway. Additionally, we noticed that Mohanty et al. [16, 17] proposed the same dataset to examine their effectiveness with number

Table 1 NRMSE values on test data [10]

Techniques	Lag1	Lag2	Lag3	Lag4	Lag5
BPNN	0.1713	0.1660	0.1514	0.1449	0.1455
TANN	0.1793	0.1837	0.1584	0.1520	0.1503
PSN	0.1868	0.1767	0.1659	0.1648	0.1579
MARS	0.1705	0.1709	0.1613	0.1548	0.1526
GRNN	0.2102	0.2114	0.1767	0.1798	0.1668
MLR	0.1714	0.1677	0.1565	0.1511	0.1478
TreeNet	0.1682	0.1678	0.1681	0.1569	0.1611
DENFIS	0.1709	0.1673	0.1542	0.1483	0.1476
Morlet-based WNN	0.1194	0.1187	0.1228	0.1157	0.1162
Gaussian-based WNN	0.1241	0.1206	0.1189	0.1186	0.1115
GP	0.1267	0.1271	0.1189	0.1152	0.1096
ACOT	0.1106	0.1371	0.1341	0.1030	0.1396
SNN	**0.0252**	**0.0374**	**0.0426**	**0.0531**	**0.0382**

of techniques, but SNN outperformed all the techniques discussed by them. This is a significant result obtained by us by using SNN.

The advantages of SNN is that it not only provides the best results as far as NRMSE value is concerned but also takes least time to process the information.

6 Conclusions

The paper presents the comparison of SNN with neural network and its variant. The paper also explores the ability of SNN to produce global best solution for complex problems like software reliability in the software industry. A comparative study of SNN with respect to BPNN, TANN, PSN, MARS, GRNN, MLR, TreeNet, DENFIS, Morlet-based WNN, Gaussian-based WNN and ACO techniques performs the best of all tested solutions of tunes parameters. The experimental analysis indicates that SNN outperforms the other machine-learning methods too.

References

1. Musa, J.D.: Software Reliability Engineering: More Reliable Software Faster and Cheaper, 2nd edn. Tata McGraw-Hill Edition
2. Lyu, M.R.: Handbook of Software Reliability Engineering. IEEE Computer Society Press and McGraw Hill (ed.) (1996)
3. Khoshgoftaar, T.M., Allen, E.B., Jones, W.D., Hudepohi, J.P.: Classification—Tree models of software quality over multiple releases. IEEE Trans. Reliab. **49**(1), 4–11 (2000)

4. Khohgoaftaar, T.M., Allen, E.B., Hudepohl, J.P., Aid, S.J.: Applications of neural networks to software quality modelling of a very large telecommunications system. IEEE Trans. Neural Netw. **8**(4), 902–909 (1997)
5. Cai, K., Yuan, C., Zhang, M.L.: A critical review on software reliability modelling. Reliab. Eng. Syst. Saf. **32**, 357–371 (1991)
6. Cai, K.Y., Cai, L., Wang, W.D., Yu, Z.Y., Zhang, D.: On the neural network approach in software reliability modelling. J. Syst. Softw. **58**, 47–62 (2001)
7. Tian, L., Noore, A.: Evolutionary neural network modelling for software cumulative failure time prediction. Reliab. Eng. Syst. Saf. **87**, 45–51 (2005)
8. Su, A., Chin, Y., Huang, U.: Neural-network based approaches for software reliability estimation using dynamic weighted combinational models. J. Syst. Softw. **80**, 606–615 (2007)
9. Ho, S.L., Xie, M., Goh, T.N.: A study of the connectionist models for software reliability predictions. Comput. Math Appl. **46**, 1037–1045 (2003)
10. Su, Y.S., Huang, C.Y.: Neural-network based approaches for software reliability estimation using dynamic weighted combinational models. J. Syst. Softw. (2006). https://doi.org/10.1016/j.jss.2006.06.017
11. Huang, C.Y., Lyle, M.R., Kuo, S.Y.: A unified scheme of some non-homogenous poisons process model for software reliability estimation. IEEE Trans. Softw. Eng. **29**(3), 261–269 (2003)
12. Costa, E.O.,Vergili, O.S.R., Souz, G.: Modeling software reliability growth with genetic algorithm. In: Proceedings of 16th IEEE International Symposium on Software Reliability Engineering, pp. 170–180 (2005)
13. Pai, P.F., Hong, W.C.: Software reliability forecasting by support vector machine with simulated annealing algorithms. J. Syst. Softw. **79**(6), 747–755 (2006). https://doi.org/10.1016/j.jss.2005.02.025
14. Rajkiran, N., Ravi, V.: Software Reliability prediction by soft computing technique. J. Syst. Softw. **81**(4), 576–583 (2007). https://doi.org/10.1016/j.jss.2007.05.005
15. Ravi, V., Chauhan, N.J., Rajkiran, N.: Software reliability prediction using intelligent techniques: application to operational risk prediction in Firms. Int. J. Comput. Intell. Appl. **8** (2), 181–194 (2009). https://doi.org/10.1142/S1469026809002588
16. Mohanty R.K., Ravi, V.: Machine learning techniques to predict software defects. In: Encyclopedia of Business Analytics and Optimization, vol. 5, pp. 1422–1434. Elsevier (2014)
17. Mohanty R.K., Ravi, V., Patra, M.R.: Hybrid intelligent systems for predicting software reliability. Appl. Soft Comput. **13**, 180–200 (2013) (Elsevier)
18. Mohanty, R.K., Ravi, V., Patra, M.R.: Machine learning and intelligent technique to predict software reliability. Int. J. Appl. Evol. Comput. **1**(3), 70–86 (2010)
19. Mohanty R.K., Ravi, V., Patra, M.R.: Software Reliability prediction using genetic programming. In: The IEEE International Conferences of the Biologically Inspired Computing and Applications (BICA-2009), Bhubaneswar, India, pp. 331–336 (2009)
20. Mohanty R.K., Ravi, V., Patra, M.R.: Software reliability prediction using group method of data handling. In: International Conferences on RSFDGrC' 2009', LNAI 5908, IIT-New Delhi, pp. 344–351. Springer (2009)
21. Mohanty R.K., Naik, V., Mubeen, A.: Application of ant colony optimization techniques to predict software reliability. In: IEEE International Conference on Communication Systems and Network Technologies (CSNT-2014), Bhopal, India, pp. 494–500, IEEE (2014)
22. Gautam, A., Bhateja V., Tiwary, A., Sathpathy S.C.: An Improved Memmogram Classification Approach Using Back propagation Neural network, Data Engineering and Intelligent Computing, pp. 369–376. Springer, Singapore (2018)
23. NatschlNager, T., Ruf, B.: Spatial and temporal pattern analysis via spiking neurons. Netw. Comput. Neural Syst. **9**(3), 319–332 (1998)
24. Meftah, B., Le'zoray, O., Chaturvedi, S., Khurshid, A., Benyettou, A.: Image Processing with Spiking Neuron Networks, Artificial Intelligence, Evolutionary Computing and Metaheuristic, SCI 427, pp. 525–544. Springer, Berlin (2013)

An Optimized Rendering Solution for Ranking Heterogeneous VM Instances

S. Phani Praveen and K. Thirupathi Rao

Abstract Upholding quality of service (QoS) parameters while ranking cloud-based Virtual Machines (VMs) that deliver the same service is a challenging task which has been addressed by prior approaches like VM resource deep analytics (RDA). But these approaches fail to consider the heterogeneous aspect of the VMs where higher resource-centric VMs tend to offer sublime performance and lower resource-centric VMs offer nominal throughput. This can also influence the VM RDA ranking algorithms where the former tends to be at the top of the ranks while the latter at margins. To counter this effect and to create an equal footing to most VMs and optimize the rankings despite the VMs varying resource centricity, we propose a VM packaging algorithm that addresses the heterogeneous aspect. We considered a maximization problem of Virtual Machine where each machine is assigned P pages of memory, a set of m servers, a group of V virtual machines, such that a version of the problem consists of one server which is developed by using the dynamic programming solution to deploy all VM instances simultaneously and consider their ranking despite their heterogeneous aspect. Aided with this new algorithm, we intend to reduce the delays and overheads experienced with the usage of heterogeneous complexity of VMs and tend to deliver an efficient ranking solution.

Keywords RDA · Virtual machines · SRS · VMPA

S. Phani Praveen (✉)
Department of Computer Science, Bharathiar University, Coimbatore, India
e-mail: phanisurapaneni.pvp@gmail.com

K. Thirupathi Rao
Department of Computer Science & Engineering, KL University, Guntur, India
e-mail: profktrao@gmail.com

V. Bhateja et al. (eds.), *Intelligent Engineering Informatics*, Advances in Intelligent Systems and Computing 695, https://doi.org/10.1007/978-981-10-7566-7_17

1 Introduction

In Internet, the most used keyword for the latest developments on storage and security of data is cloud computing. Services like storage of data, security for data, and for universal services are provided by Cloud computing. Every cloud service providers such as Amazon [1], Google [2] or Microsoft [3] provide the services based on the quality of the infrastructure. The cloud services are distributed and are also heterogeneous to execute in any environment. In cloud the resources are not restricted to hardware such as hard disk, processor, ram, and software-related applications. Cloud services are available to the users based on demand and are dynamic to provide the hardware flexibility to allow the resources using virtualization. Cloud computing provides all the software and hardware services irrespective of system configuration.

In the heterogeneous aspect, there are some failures to provide the services using VM resource deep analysis. In the VMs where higher resource-centric VMs tend to offer sublime performance and lower resource-centric VMs tend to offer nominal throughput. This can also influence the VM RDA ranking algorithms where the former tends to be at the top of the ranks while the latter at a margins. To counter this effect we create an equal footing to most VM's and optimize the rankings despite the VM's varying resource centricity. In this paper, we propose a VM packaging algorithm that addresses the heterogeneous aspect.

Heterogeneous cloud computing is the mixture of various cloud networks at one place. It is very important to know that all the cloud service providers have to provide their network within the heterogeneous environment.

For this environment, the present infrastructure is not suitable to provide the services. It is mandatory to improve the efficiency of the infrastructure in order to maximize the performance and power efficiency. In this study, a profit maximization model for sharing Virtual Machines is identified and to overcome this, we adopted VM RDA to improve the performance of the VMs.

Quality and quantity of services are calculated by cloud service providers by taking into account the quality of service (QoS). Queuing model provides best service in terms of resource provisioning in grid environment.

In the case of scientific workflow with assignment replication in the single cloud environment, leads to an issue of unsatisfied user on demands of QoS. Since on account of single cloud condition, the cloud supplier will look for proper VM inside it for fulfilling the clients QoS requests.

On the off-chance that none of the VM fulfills the clients QoS requests, still it will keep on processing that work due to the truant of multi-cloud condition. On the off-chance that the multi-cloud condition is existing at that point; additionally, there is a possibility of narrow-minded conduct where the cloud suppliers expect to expand their income. A novel approach of asset designation will enhance the quality of work process application. Assets are meant for solitary clients to make selective utilization in a given time bound.

The remaining paper is as follows: in Sect. 2, literature survey on ranking algorithms. In Sect. 3, review of various ranking approaches. In Sect. 4, comparison approaches, and finally, Sect. 5 conclusion and future work.

2 Related Work

Hai Zhong, Kun Tao, Xuejie Zhang researched the likelihood to dispense the Virtual Machines (VMs) adaptable to allow the greatest use of physical assets. The utilization of an improved genetic algorithm (IGA) for the mechanized planning strategy is done [4]. Gunho Leey, Byung-Gon Chunz, Randy H. Katz recommends a system architecture to apportion assets to an information examination bunch in the cloud, and propose a metric of offer in a heterogeneous group to understand a plan that accomplishes elite and reasonableness [5]. Dr. M. Dakshayini, Dr. H. S. Guruprasad proposed cloud architecture that has achieved high service completion rate with guaranteed Qos over traditional over the traditional scheduling policy which does not consider the priority and admission control techniques. Two undertaking planning calculation are proposed mulling over their computational many-sided quality and registering limit of handling component [7]. There is one more work that exhibits the execution of a productive quality of service (QoS)-based meta-scheduler and backfill methodology-based lightweight Virtual Machine scheduler for dispatching employments [2].

3 Ranking

Ranking is one of the sorting techniques used in all the cloud virtual aspects. In some cases, the ranking is implemented in general [8].

Implementing the ranking in cloud services and cloud-related services is new which shows some alerts in recent years. It is an assignment in cloud environment that ranking will provide for all the cloud infrastructures such as load balancing in virtual machines, servers, and other systems [9].

4 Quality of Service

Quality of administration speaks to utilitarian properties of the given administration in light of amount or quality. It additionally speaks to the capacity of a system or framework for introducing better administrations [10]. Administrations in distributed computing have subjective traits, like responsibility, deftness, cost, execution, affirmation, ease of use, security and protection [11, 12]. These qualities are connected for contrasting diverse cloud administrations. Their definition is assessed in the following subsection.

In this approach proposed by [13], subjective values are considered for ranking. Subjective values are to be measured beforehand. However, calculation of subjective values is difficult in cloud environment which involves high cost and time.

5 SRS Approach

In [14], an approach of ranking the system services has been proposed. Ranking is given to the systems that exist in static and dynamic states. If a cloud service does not consider the requirements of user, then it is treated as static state. If it considers requirements of users, then it is termed as dynamic state.

6 Existing System

1. The service level agreement (SLA) is related to every application in the cloud environment based on some cases such as response time and throughput requirements.
2. During runtime, it is often impossible to track workload measures to their core considering their volatility because of reasons such as CPU and memory usage of the virtual server where it is deployed to work as an application.
3. In the existing system, the issue has discussed the tracking and managing the performance of application in a virtual grid server environment.

7 VM Maximization Problem

After a detailed study, maximization problem of Virtual Machines is identified with an aim to maintain a subset of Virtual Machines on a single server by maximizing the profit. This problem has been shown as a resource deep analytics algorithm [6]. A greedy optimization algorithm is considered with an aim of maximizing the profit and effective utilization of page sharing in Virtual Machine allocation process. A ratio of approximation is calculated between the proposed sharing model of Virtual machine approximation and non-hierarchial, non-structured model of sharing Virtual Machines.

8 Proposed System

1. The heterogeneous nature of different virtual machines considering their dependence on the host happens to affect the VM loading performance.
2. So we propose to use a greedy VM packing algorithm to boost up the process along with deep analytics to ensure upholding QoS parameters.
3. First, we considered a maximization problem of Virtual Machine where each machine is assigned P pages of memory, a set of m servers, a group of V virtual machines, such that a version of the problem consists of one server which is developed by using the dynamic programming solution.
4. A cluster-tree T with V nodes and E edges T = (V,E) is constructed where each leaf node is treated as a virtual machine with single node V. Non-leaf node consists of k nodes which are termed as a supernodes. Nodes of non-leaf are uniformly distributed a zero profit value and this illustrates that profit can be gained only through leaf nodes where entire Virtual Machines are packed.
5. Aided with this new algorithm, we intend to reduce the delays and overheads experienced with usage of heterogeneous complexity of VMs.

9 Algorithms

Our packing pack tree T is as follows

(1) To calculate size and count of v for each node of a tree T execute the steps of load balancing.

(2) The tree is packed as one server by load-balancing process if the count of the root of the tree is equal to 1, i.e., count (root (T)) = 1, as such the entire tree can be packed into a single server by applying this algorithm.

(3) Or else if the count of the root of the tree is greater than 1, i.e., count (root (T)) > 1, then follow these steps.

   (a) Consider the value of k as the lowest value of the count of the root of the tree T such that k > 1. Consider a supernode V having count k and count of all its children with value 1. Children of v are considered as a set of virtual machines, i.e., as a single item. Using best bin-packing algorithm, all the items are packed into servers of capacity P.

   (b) All the children of root node v are removed to form a new tree T. Apply greedy technique on T.

10 Performance Measures

The performance of VM models is compared with the metric [15]; first capture the performance overhead of VMs, the first type of metric is the efficient degradation of various applications running on cloud, compared with the speed of application obtained in the VMs running in an abstract environment. Precisely, the performance degradation of VMs P_d can be measured as

$$P_d = |x_{iaas} - x_{isolation}|/x_{isolation} \tag{1}$$

Performance of the second metric that captures the performance overhead of VMs is the variation of VM performance obtained in the VM of cloud over a given time period. Precisely, the VM performance variation P_v

$$P_v = \frac{1}{x}\sqrt{\frac{1}{(nx-1)}\sum_{i=1}^{nx}(xi - x)2} \tag{2}$$

The third metric, resource consumption in a VM is calculated with I

$$I = \alpha_0 + \sum_{i-1}^{p} \alpha_i X_i \tag{3}$$

The fourth metric, CPU and I/O bandwidth consumption of resource

$$I = \alpha_0 + \left(\sum_{I=1}^{4} \alpha_I X_I\right)^2 \tag{4}$$

The fifth metric, CPU and I/O bandwidth consumption of VM

$$I = \alpha_0 + \alpha_1 \exp\left(\sum_i \Upsilon_I X_i^{cpu}\right) + \alpha_2 \exp\left(\sum_i \omega_I X_i^{io}\right) \tag{5}$$

Estimating the performance of two VM load with CPU, I/O load, and network

$$I_{xy} = P\sum X_i Y_i - \sum X_i \sum y_i / \left(\sqrt{p\sum xi^2 - (\sum Xi)^2}\sqrt{p\sum yi^2 - (\sum Yi)^2}\right) \tag{6}$$

Dynamic prediction of online VM with respect to physical machine

$$P_i = \Phi(P_{i=1}, P_{i-n}, u_i \ldots\ldots, u_{i-m}) \tag{7}$$

Overheads caused for linear migration VM

$$T_{mig} = \frac{\beta}{U_{cpu}^k} \tag{8}$$

Overheads caused for nonlinear migration of VM

$$T_{mig} = \sum_{i=1}^{n} t\,t_I = \begin{pmatrix} \frac{t_i - i, D}{r} & i \geq 2 \\ \frac{Vmem}{r} & i = 1 \end{pmatrix} \tag{9}$$

Predication for traffic migration, traffic down, energy consumption, and performance migration.

$$P_{mig} = \alpha, T_{mig} + b.T_{down} + c.N_{mig} + d.E_{mig} \tag{10}$$

11 Results and Discussions

This project is implemented in Java programming language with Java Server Pages (JSPs), and it is Web application. Mysql is backend and Netbeans is the IDE to implement the project very easy.

The implementation is done using all the java libraries and packages. So we can say that Java is the better option for implementation of performance metrics. Table 1 shows the VM prediction errors. Our approach VMPA shows the prediction errors less than three from all the resource consumption, network disk, and number of VM.

Table 1 Comparative results of various approaches used to predict the performance of VM and its prediction errors based on performance metric

Approaches	Pre-run of workload	Scale of VMs	Kind/ Type of model	Resource consumption	Prediction errors
Zhu et al.	✓	Two	Linear	Cpu, memory, network, disk	Less than 8%
Koh et al.	✓	Two	Linear	Cpu, memory, network, disk, cache, number of VM switches, etc.	Less the 5%
TRACON	✓	Two	Nonlinear	Cpu, network	19%
Bu et al.		Multiple	Nonlinear	Cpu, network, disk	12%
Cuanta	✓	Multiple	n/A	Cache, memory	Less than 4%
pSciMapper	✓	Two	Nonlinear	Cpu, memory, network, disk	n/a
RCRP	✓	Multiple	Dynamic	Cpu, memory, network, disk, number of VM	Less than 4
RDA	✓	Multiple	Dynamic	Cpu, memory, network, disk, number of VM	Less than 4
VMPA	✓	Multiple	Dynamic	Cpu, memory, network, disk, number of VM	Less than 3

Table 2 Loading time for number of virtual machines

System	No of VMs	Loading time (Sec)
RDA (existing system)	1	12.93
VMPA (proposed system)	1	8.02
RDA	2	15.8
VMPA	2	12.6
RDA	3	17.9
VMPA	3	14.2
RDA	4	19.98
VMPA	4	15.78

By applying all the above metric for checking, the performance of VM model is shown in Table 1.

Table 1 provides the comparative results of various approaches used to predict the performance of VM, and its prediction errors are based on the performance metric.

The loading time for the no of VM is as follows in Table 2.

12 Conclusion

In this paper, the proposed system focuses on the VM packaging algorithm that addresses the heterogeneous aspect. We considered a maximization problem of Virtual Machine where each machine is assigned P pages of memory, a set of m servers, a group of V virtual machines, such that a version of the problem consists of one server which is developed by using the dynamic programming solution to deploy all VM instances simultaneously and consider their ranking despite their heterogeneous aspect. Aided with this new algorithm, we intend to reduce the delays and overheads experienced with usage of heterogeneous complexity of VMs and tend to deliver an efficient ranking solution.

References

1. Amazon, Aws.amazon.com
2. Google App Engine, Appengine.google.com
3. Microsoft, www.microsoft.com
4. Zhong, H., Tao, K., Zhang, X.: An approach to optimized resource scheduling algorithm for opensource. In: The Fifth Annual ChinaGrid Conference Cloud Systems. IEEE (2012)
5. Lee, G., Chun, B.G., Katz, R.H.: Heterogeneity-Aware Resource Allocation and Scheduling in the Cloud. University of California, Berkeley (Yahoo! Research)

6. Dakshayini, M., Guruprasad, H.S.: An optimal model for priority based service scheduling policy for cloud computing environment. IJCA. **32**(9) (2011)
7. Mittal, S., Katal, A.: An optimized task scheduling algorithm in cloud computing. IACC. (2016)
8. Sridevi, K.: A novel and hybrid ontology ranking framework using semantic closeness measure. IJCA. **87**(4) (2014)
9. Vaidya, O.S., Kumar, S.: Analytic hierarchy process: an overview of applications. EJOR. **169** (1) (2006)
10. Raisanen, V.: Service quality support—an overview. CC. **27**(15) (2004)
11. Garg, S.K., Versteeg, S., Buyya, R.: A framework for ranking of cloud computing services. FGCS. **29**(4) (2013)
12. Yau, S., Yin, Y.: QoS-based service ranking and selection for service-based systems. IEEE ICSC, pp. 56–63. (2011)
13. Choudhury, P., Sharma, M., Vikas, K., Pranshu, T., Satyanarayana, V.: Service ranking systems for cloud vendors. Adv. Mater. Res. **433–440**, 3949–3953 (2012)
14. Preeti Gulia, S.: Automatic selection and ranking of cloud providers using service level agreements. IJCA. **72**(11) 2013
15. Phani Praveen, S., Tirupathi Rao, K.: An algorithm for rank computing resource provisioning in cloud computing. IJCSIS. **14**(9), 800–805 (2016)

Gender Prediction in Author Profiling Using ReliefF Feature Selection Algorithm

T. Raghunadha Reddy, B. Vishnu Vardhan, M. GopiChand and K. Karunakar

Abstract Author Profiling is used to predict the demographic profiles like gender, age, location, native language, and educational background of the authors by analyzing their writing styles. The researchers in Author Profiling proposed various set of stylistic features such as character-based, word-based, content-specific, topic-specific, structural, syntactic, and readability features to differentiate the writing styles of the authors. Feature selection is an important step in the Author Profiling approaches to increase the accuracy of profiles of the authors. Feature selection finds the most relevant features for describing the dataset better than the original set of features. This is achieved by removing redundant and irrelevant features according to important criteria of features using feature selection algorithms. In this work, we experimented with a ReliefF feature selection algorithm to identify the important features in the feature set. The experimentation carried on reviews domain for predicting gender by using various combinations of stylistic features. The experimental results show that the set of features identified by the ReliefF feature selection algorithm obtained good accuracy for gender prediction than the original set of features.

T. Raghunadha Reddy (✉) · M. GopiChand
Department of IT, Vardhaman College of Engineering, Hyderabad, India
e-mail: raghu.sas@gmail.com

M. GopiChand
e-mail: gopi_merugu@yahoo.com

B. Vishnu Vardhan
Department of CSE, JNTUH Jagtiyal, Karimnagar, India
e-mail: mailvishnu@jntuh.ac.in

K. Karunakar
Department of CSE, Swarnandhra Institute of Engineering and Technology, Narsapur, Andhra Pradesh, India
e-mail: karunakar.mtech@gmail.com

V. Bhateja et al. (eds.), *Intelligent Engineering Informatics*, Advances in Intelligent Systems and Computing 695, https://doi.org/10.1007/978-981-10-7566-7_18

1 Introduction

The World Wide Web has become unmanageably big day by day, and the text is increasing exponentially mainly through Twitter, social media, blogs, and reviews. Most of this text is written by various authors in different contexts. The availability of text put a challenge to researchers and information analysts to develop automated tools for analyzing the text. In this regard, Author Profiling is a popular technique to extract as much information as possible from the texts by analyzing author's writing styles. Author Profiling is a popular technique in the present information era which has applications in forensic analysis, marketing, and security.

An Author Profiling system majorly built up from the three components such as text representation, feature selection, and training a classifier. Text representation converts documents into a vector representation with a set of features that can be handled by classification algorithms. Feature selection identifies the effective features to make classification algorithms efficient which can be implemented by methods including dimension reduction. Training a classifier builds up an autonomous classifier by using supervised learning algorithms, and the trained classifier is used to predict class labels for new documents.

The vector space model ("bag-of-words") is the most common text representation technique adopted by most machine learning algorithms. In this representation method, each document is represented by a list of extracted features (phrases, words, terms, tokens, or attributes). Their associated weights are computed from the frequency of each feature in a given document.

The features with highly informative or high relevancy are considered to be important for text classification. In general, the dataset is found to be a mixture of incomplete, irrelevant, outlier, and noisy features in nature. In this situation, feature selection plays a vital role to speed up the processing time and getting appropriate and accurate result. For selecting the appropriate and feasible feature from the dataset, there are lot of algorithms and statistical approaches available. In general, information gain, gain ratio, Gini index, and distance measures are used by many researchers to estimate the quality of attributes. They assume that the attributes are independent and there are poor chances of revealing quality attributes when there is a strong conditional dependency between attributes. In this work, the experimentation is carried by using ReliefF attribute selection algorithm.

This paper is structured as follows. The related work for gender prediction is described in Sect. 2. The Sect. 3 explains the various stylistic features used in our work for differentiating the writing styles of the authors. In Sect. 4, the ReliefF feature selection algorithm is illustrated. Section 5 discusses the experimental results. The conclusion of this work is discussed in Sect. 6.

2 Literature Review

The researchers proposed various stylistic features such as character-based, word-based, structural, topic-specific, content-based, syntactic, and readability features to recognize the writing style differences in the text for predicting the demographic profiles of the authors [1].

Upendra Sapkota et al. used [2] 208 features such as syntactic, stylistic, and semantic features. It is observed that their system gave a poor performance for gender prediction in English language when n-grams of POS tags were used. Lucie Flekova et al. [3] used five classes of features including content-based features, surface features, syntactic features like n-grams of POS tags, readability measures, and semantic features. They observed that the POS unigrams for English language and POS trigrams and quadrigrams for Spanish language play an important role to increase the prediction accuracy of gender and age and also find that the content-based features perform overall better than the stylistic features.

Yasen Kiprov et al. extracted [4] four types of features such as lexicon features, Twitter-specific features, orthographic features, and term-level features like word n-grams and POS n-grams. In their observation, word 3-grams and word 4-grams were not improve the performance, but the POS tags, word unigrams, and bigrams perform most sustainable performance on test datasets. It is also observed that the orthographic features were useful to improve the accuracy of gender and age, but these are not useful to improve the accuracy of big five personality traits.

Michał Meina et al. experimented [5] with 311 features and 476 features including part of speech n-grams, structural features, exploration of sequences of part of speech, test difficulty, dictionary-based features, errors, topical features, topical centroids and structural centroids for English language and Spanish language, respectively. It is observed that the exploration of sequences of part of speech n-gram features alone gave a good result for gender and age prediction.

S. Francisca Rosario experimented [6] with a dataset of cotton disease provided by cotton research station. They used ReliefF attribute selection algorithm to reduce the number of features used for classification. They conducted experiment with J48 and Naïve Bayes classifiers. It was observed that the precision value is increased from 0.444 to 0.917 after reducing features through ReliefF algorithm. Similarly, the recall measure is increased from 0.667 to 0.889 after reducing the number of features.

3 Stylistic Features

3.1 Character-Specific Features

Number of characters, number of capital letters, number of small letters, ratio of capital letters to small letters, the ratio of white space characters to non white space characters, the number of punctuation marks, ratio of numeric data in the text, ratio of tab spaces to total number of characters, ratio of capital letters to total number of characters and ratio of white space to total number of characters.

3.2 Word-Specific Features

Number of negative words, number of positive words, capital letters words, number of words, average word length, contraction words, the ratio of number of words length greater than six to total number of words, the ratio of number of words length less than three to total number of words, the number of words with hyphens, words length greater than 6, words followed by digits, unique terms, ratio of number of words which contain more than 3 syllables to total number of words, number of words that occur twice (hapax dislegomena), number of acronyms and number of foreign words.

3.3 Structural Features

Number of paragraphs, number of sentences, number of words per paragraph, number of sentences per paragraph, average sentence length in terms of words, average sentence length in terms of characters.

3.4 Readability Features

Flesch Kinkaid Grade Level, Coleman Liau Index, RIX Readability Index, Flesch Reading Ease, LIX, SMOG index, Automated Readability Index(ARI) and Gunning Fog Index,.

3.5 Topic-Specific Features

Topic models provide a easiest way to classify large volumes of text. A “topic” is a set of related words that frequently occurred together. Fifteen topics are identified as features which are extracted by using popular MALLET tool.

Assign all attribute weights to 0, $W[A] = 0.0$
for i = 1 to *m (number of instances)* do
begin

Select an instance R_i Randomly from a set of instances
Find k nearest hits H_j from positive class;
for each class $C \neq class(R_i)$ do
Find k nearest misses $M_j(C)$ from negative classes
for $A = 1$ to a (number of attributes) do

$$W[A] = W[A] - \sum_{j=1}^{k} diff(A, R_i, H_j)/(m\Box k) +$$

$$\sum_{C \neq class(R_i)} \left[\frac{P(C)}{1 - P(class(R_i))} \right] \sum_{j=1}^{k} diff(A, R_i, M_j(C))/(m\Box k)$$

end

Fig. 1 ReliefF algorithm

4 ReliefF Feature Selection Algorithm

Among the various feature selection algorithms, Relief algorithm is considered to be a simple, efficient, and successful attribute estimator. The original Relief algorithm deals with the numerical and nominal attributes only. Its function is limited to two class issues only. It is not dealing with incomplete, noisy, and duplicate attributes in the dataset. Many researchers have worked on the Relief feature selection algorithm and have proposed new variants of Relief algorithm based on its functionality of the concept. Kononenko et al. [7] have proposed the new algorithm of Relief. It is the extended version of Relief algorithm called as ReliefF for handling multi-class classifications. Figure 1 explains the algorithm of ReliefF feature selection algorithm.

5 Experimental Results

The dataset used in this work was collected from hotel reviews Web site www.TripAdvisor.com, and it contains 4000 reviews about various hotels in which 2000 male reviews and 2000 female reviews. In this work, accuracy measure is used to measure the efficiency of the classifier. Accuracy is the ratio of number of documents correctly predicted their gender to the total number of documents considered for evaluating the classifier. Random forest classifier is used to test the classification accuracy. Gender is the class attribute.

Table 1 Accuracies of gender prediction

Features	Random forest classifier accuracy (%)
55 features (CHAR + WORD + STR + TOPIC + READ)	91.17
55 features–12 negatives (CHAR + WORD + STR + TOPIC + READ)	93.35

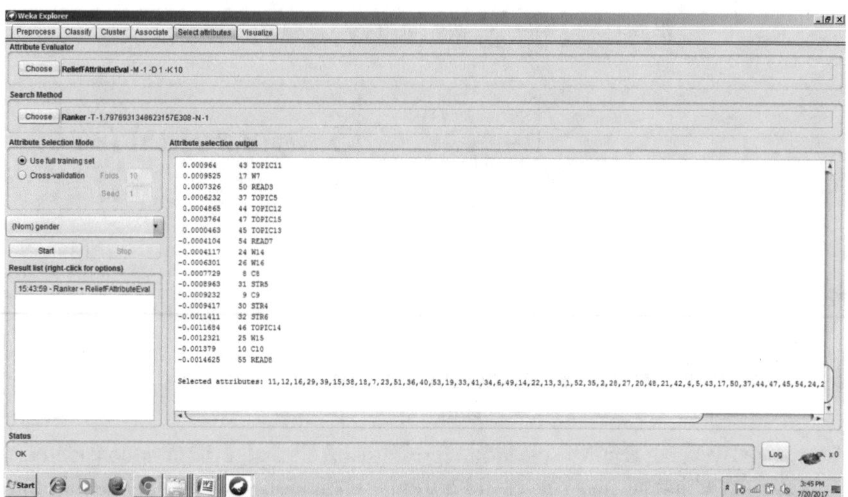

Fig. 2 Representation of ranking of attributes in WEKA tool using ReliefF algorithm

The experimentation starts with 55 features of various combinations of stylistic features such as character (CHAR), word (WORD), structural (STR), topic (TOPIC), and readability (READ) features, and 91.17% accuracy is achieved for gender prediction by using random forest classifier. Table 1 shows the accuracies of gender prediction for various combinations of features.

The experimentation with ReliefF algorithm is performed in a powerful java-based open source machine learning tool WEKA. The ReliefF algorithm gives ranks to the features based on the weights of the features. Figure 2 shows the ranking of features and their weight values using ReliefF algorithm in WEKA tool.

The positive weight values in the output of ReliefF algorithm indicate the feature is more relevant for classification, and negative weight indicates the feature is irrelevant for classification. The experiment continued with 43 features by removing negative weighted features, and 93.35% accuracy is obtained for gender prediction. Figure 3 depicts the accuracy of gender prediction by using 43 features in WEKA tool.

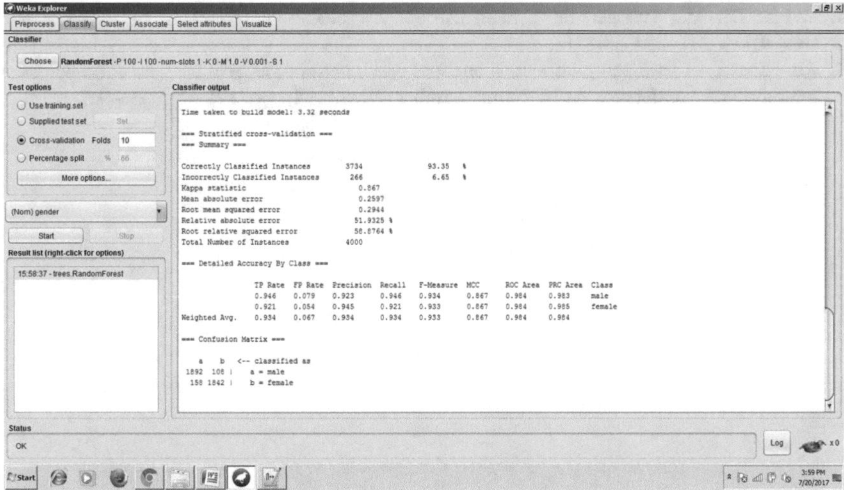

Fig. 3 Representation of classification of gender after removing negative ranking attributes in WEKA tool using random forest classifier

6 Conclusion

The set of features are used to differentiate the writing style of the authors in Author Profiling. Finding suitable set of features for predicting gender is a difficult task in Author Profiling. In this work, ReliefF feature selection algorithm is used to improve the accuracy of gender prediction. The accuracy 93.35% is obtained for gender prediction after removing less weighted features from the original set of features.

References

1. Raghunadha Reddy, T., VishnuVardhan, B., Vijaypal Reddy, P.: A survey on authorship profiling techniques. Int. J. Appl. Eng. Res. **11**(5), 3092–3102 (2016)
2. Sapkota, U., Solorio, T. Montes-y-Gómez, M., Ramírez-de-la-Rosa, G.: Author profiling for english and spanish text. In: Proceedings of CLEF 2013 Evaluation Labs (2013)
3. Flekova, L., Gurevych, I.: Can we hide in the web? large scale simultaneous age and gender author profiling in social media. In: Proceedings of CLEF 2013 Evaluation Labs (2013)
4. Kiprov, Y., Hardalov, M., Nakov, P., Koychev, I.: SU@PAN'2015: experiments in author profiling. In: Proceedings of CLEF 2015 Evaluation Labs (2015)
5. Meina, M., Brodzi´nska, K., Celmer, B., Czoków, M., Patera, M., Pezacki, J., Wilk, M.: Ensemble-based classification for Author Profiling using various features. In: Proceedings of CLEF 2013 Evaluation Labs (2013)

6. Rosario Dr, S.F., Thangadurai, K.: RELIEF: feature selection approach. In: Int. J. Innov. Res. Dev. **4**(11), 218–224 (2015)
7. Kononenko, I.: Estimating attributes: analysis and extensions of RELIEF. In: Proceedings of the European Conference on Machine Learning (ECML'94), pp. 171–182 (1994)

Optimal Coordination of Overcurrent Relays Using Simulated Annealing and Brute Force Algorithms

Himanshu Verma, Seebani Mahaptara, S. P. Anwaya, Abhirup Das and O. V. Gnana Swathika

Abstract With the advancement of technology, the electrical world has grown colossally through invention and optimization, hence adopting to the changing environment. With the arrival of power grids, smart grids, micro-grids, etc., there has always been a baggage of responsibility on the protection engineers to come up with a novel, operative, and optimized way to clean faults in the distribution on transmission lines, which focuses on maintaining a maximum tripping time for the point located at the farthest position from the actual location of the fault. Radial networks are susceptible to overcurrent fault conditions. So, for the protection against the faults, the overcurrent relays are used. For a power-protection engineer, it is imperative to maintain and optimize the coordination time interval (CTI) between relays in the course of overcurrent relay coordination. For the clearance of the fault, optimization algorithms are often fetched. In this paper, brute force algorithm and simulated annealing algorithm are implemented on a radial network to optimize the time dial setting (TDS) of relays. Thus, helping to maintain coordination among OC relays in the network.

1 Introduction

In the electrical power distribution system, the power is carried through the transmission system and distributed among the end users. The power distributed to the end users is carried out a low voltage level. The fault occurs chiefly due to impulsive current built-up in the network, thence, bringing overcurrent relays into limelight. Short-circuit protection for the network is executed with help of overcurrent relays. During the fault conditions, it is obligatory to isolate the faulty network section from the healthy part. The relay system is coordinated to protect the adjacent equipment. Relays are of two sorts: the primary relay which is near the faulted section and the backup relay [1–4]. Appropriate relay coordination is crucial

H. Verma · S. Mahaptara · S. P. Anwaya · A. Das · O. V. Gnana Swathika (✉)
VIT University Chennai, Chennai, India
e-mail: gnanaswathika.ov@vit.ac.in

V. Bhateja et al. (eds.), *Intelligent Engineering Informatics*, Advances in Intelligent Systems and Computing 695, https://doi.org/10.1007/978-981-10-7566-7_19

for the fault identification and its isolation from the healthy network. If the backup relay fails, mal-operation may occur, thence, causing a serious damage. The number of power equipment used in a network depends upon the size of the power system. Therefore, it is the sole duty of the protection engineers to work on new ways to protect these equipments.

This paper proposes on identifying the optimized time dial settings (TDSs) and operating time (T_{op}) by using brute force algorithm and simulated annealing algorithm method. The optimized TDS leads to quick fault clearance which is a necessary for any protection scheme.

2 Two Bus Relay Time Synchronization System

Overcurrent relays are of two types: directional and non-directional. Usually, directional is favored over non-directional since it does not require any coordination between the relays. As per Fig. 1, a radial system consisting of two sections and feeders is shown explaining that for a fault at point F, relay R2 will be the first to operate. The relay R1 will start operating after a delay time of 0.1 s in addition to the operation time of circuit breaker at primary bus, i.e., bus 2 and the overshoot time for R1.

2.1 Theoretical Verification

To achieve overcurrent protection in minimum time, the coordination between relays and their operating time need to be optimized [5–8]. Thus, proceeding to solve it as an optimization problem, we can minimize the value of top (operating time) as in (1).

$$\underset{j=1}{\overset{n}{\text{Min}}}\ z = \sum t_{opj} \tag{1}$$

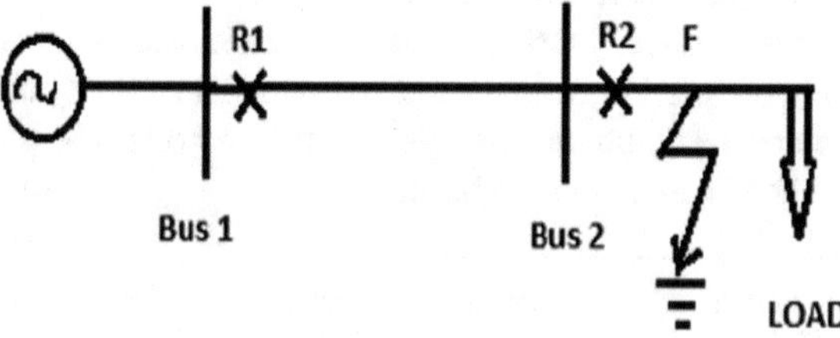

Fig. 1 A radial two bus system

here,

- n number of relays/bus;
- t_{op} time for operation for primary relay j, for a fault at j under the following constraints

Criteria for time coordination
Coordination time is the least time required between operation of the two relays. CTI (coordination time criteria) must be lesser than or equal to the difference between the operation times of the two relay.

$$t_{opjb} - t_{opj} \geq \Delta t \tag{2}$$

- t_{opjb} the operation time of backup relay j, for a fault at j
- t_{opj} primary relay operation time
- Δt the coordination time difference

Relay Characteristics
The relays follow inverse definite minimum time (IDMT) features. The relay operating time maybe hence represented as (3).

$$t_{opj} = \alpha(TDS) \tag{3}$$

The required function achieved in (4):

$$\mathrm{Min}\, z = \sum \alpha_i (TDS)_{j=1}{}^{n} \tag{4}$$

Thereby, determining the value of TDS, the relay operation time maybe minimized and eventually the fault is cleared quickly from the network.

Operating time
Relay response should be rapid for the quick identification of the fault. Thus, a time constraint is incorporated.

3 Algorithm Overview

3.1 Simulated Annealing

Simulated annealing (SA) is a technique to obtain the approximate global optimum of a given function using the idea of probability. Precisely, it is an investigative approach to approximate global optimization in a large search space. This method incorporates a probability condition to determine the winning solution.

Steps

- First set some initial value and come up with a random function
- Calculate the time using the generated function
- Make small changes and select neighboring values to generate a new random function
- Calculate the time for the newly generated function
- Check whether the new solution has smaller values as compared to the previous solution. In this case, consider the new value as base value and start searching for neighboring values to get the optimized solution.
  Continue the iteration until we get the globally optimized solution

3.2 Brute Force Algorithm

Brute force algorithm, also called as exhaustive search algorithm, is a systematic search of all solution for a constrained equation satisfying function. A brute force algorithm is mostly used to find the divisor of a natural number and would enumerate all numbers from 1 to n till answer is achieved.

Steps

- A sequential search is incorporated to find the desired value from an array of elements A(0, n) and the key k which is to be achieved.
- Take the array sequence k as (n + 1). Thus, now the array has now n + 1 elements instead of n elements for ease of achieving the optimum value.
- Start taking i as – > 0 for performing the test operation
- While A(i) = I K
  move to the next element in the array to check for the desired value i.e. i- > i+1
- This loop is run till A(i) = K. this is the exit condition.
- If i < n then return the i index of i or the result is achieved and is successful
- If the above condition is not true, i.e., i = n+1 then none of the element from the array is equal, then our result is a failure and return –1

4 Application of Algorithm

4.1 2—Bus Radial Network

Two bus radial systems are taken into consideration as shown in the given Fig. 1. The network as considered is specified at 220 kV, 10^5 KVA. The coordination time of the relay is designed to be 0.57 s. The maximum discrepancies (faults) beyond R1 are considered as 2108 A and beyond R2 as 1703 A. By considering the four

Table 1 PSM and α values

Fault location	Relay	
	Secondary	Primary
(1) Just beyond secondary relay PSM α	2.63	—
(2) Just beyond primary relay PSM α	2.97	2.00

mentioned data, α and PMS are calculated from (2) and (3) resp. and mentioned in Table 1.

Fault is considered just after first-line defense, i.e., R2 and just after second line of defense, i.e., R1. Conferring to the fault location section, the relays are selected. The first action, after the identification of primary fault, is acknowledged by primary relay. If the primary relay fails to act within the suitable time, then the other relay is activated.

As R2 is considered the primary relay, it acts first if a fault is detected. Let R2 operate in 0.2 s after the fault occurred. The operation of secondary relay should take place after CTI (sum of circuit breaker t_{op} at bus 2, overshoot time of secondary relay, and 0.2 s.)

Let X1 and X2 be the TDS values of secondary and primary, respectively. The problem statement as given in (5) and the constraints are given in (6).

$$\text{Min}\, z = 2.63x_1 + 2x_2 \tag{5}$$

Subject to

$$\begin{aligned} &2.97x_1 - 2x_2 \geq 0.57 \\ &2.63x_1 \geq 0.2 \\ &2x_2 \geq 0.2 \end{aligned} \tag{6}$$

4.2 Inference

The aforementioned Table 1 derives the values of PMS and α. Tables 2 and 3 give the values received from simulated annealing method for a series of iterations giving us the optimized values of TMS that is x1 and x2, respectively. Table 4

Table 2 Simulated annealing application on test system (1)

Sl.No	X1	X2	f
1	0.076	0.076	0.352
2	0.2	0.076	0.678
3	0.2	0.2	0.926
4	0.076	0.2	0.352

Table 3 Simulated annealing application on test system (2)

Iterations	Del x1	Del x2	X1	X2	Del E
1	0	0	0.138	0.138	–
2	0.006	–0.022	0.144	0.116	–0.0283
3	–0.025	–0.017	0.119	0.099	–0.0997
4	–0.035	–0.021	0.084	0.078	–0.194

Table 4 Brute force application on test system

Iteration	Del x1	Del x2	X1	X2	f
1	0	0	0.071	0.1	0.39988
2	0.05	0	0.081	0.1	0.41303
3	0.05	0	0.1381	0.1	0.5632
4	0.05	0	0.1881	0.1	0.694203
5	0.2	0	0.2381	0.1	0.8262
6	02	0	0.2581	0.1	0.8788
7	0.001	0	0.2591	0.1	0.881433
8	0.0005	0.0005	0.2596	0.10056	0.882748

Table 5 Comparison of results obtained from simulated annealing and brute force algorithms

	X1	X2
Simulated annealing algorithm	0.076	0.2
Brute force algorithm	0.2596	0.10056

denotes the TMS values: x1 and x2 received after a series of iterations using the brute force method. Table 5 indicates that while applying the simulated annealing and brute force algorithms on the two bus radial networks, the results of simulated annealing arrive at the best-optimized values of TMS x1 and x2.

5 Conclusion

The optimized value of the time multiplier settings of two bus radial networks is calculated using simulated annealing and brute force algorithms. Simulated annealing algorithm arrives at better optimized TMS results, than the brute force algorithm. Simulated annealing algorithm is applicable for conditions when it is allowed to arrive at approximate global optimum. Brute force algorithm involves very simple implementation but speed is not appreciable.

References

1. Gnana Swathika, O.V., Hemamalini, S.: Prims-Aided Dijkstra Algorithm for adaptive protection in microgrids. IEEE J. Emerg. Sel. Top. Power Electron. **4**(4), 1279–1286 (2016)
2. Madhumitha, R., Sharma, P., Mewara, D., Swathika, O.G., Hemamalini, S.: Optimum coordination of overcurrent relays using dual simplex and genetic algorithms. In: International Conference on Computational Intelligence and Communication Networks, pp. 1544–1547 (2015)
3. Swathika, O.G., Das, A., Gupta, Y., Mukhopadhyay, S., Hemamalini, S.: Optimization of overcurrent relays in microgrid using interior point method and active set method. In: Proceedings of the 5th International Conference on Frontiers in Intelligent Computing: Theory and Applications, pp. 89–97. Springer (2017)
4. Gupta, A., Swathika, O.G., Hemamalini, S.: Optimum coordination of overcurrent relays in distribution systems using Big-M and dual simplex methods. In: International Conference on Computational Intelligence and Communication Networks, pp. 1540–1543 (December 2015)
5. Swathika, O.G., Mukhopadhyay, S., Gupta, Y., Das, A., Hemamalini, S.: Modified Cuckoo search algorithm for fittest relay identification in microgrid. In: Proceedings of the 5th International Conference on Frontiers in Intelligent Computing: Theory and Applications, pp. 81–87. Springer (2017)
6. Swathika, O.G., Hemamalini, S.: Review on microgrid and its protection strategies. Int. J. Renew. Energy Res. **6**(4), 1574–1587 (2016)
7. Gnana Swathika, O.V., Bose, I., Roy, B. Kodgule, S., Hemamalini, S.: Optimization techniques based adaptive overcurrent protection in microgrids. J. Electr. Syst. **3**(10) (2016) (Special Issue)
8. Gnana Swathika, O.V., Hemamalini, S.: Adaptive and intelligent controller for protection in radial distribution system. In: Advanced Computer and Communication Engineering Technology, vol. 36, pp. 195–209. Springer (2016)

The Close-Loop Antenna Array Control System for Automatic Optimization in LTE and 3G Network

Archiman Lahiry

Abstract This paper will introduce some methods for the automatic close-loop optimization and troubleshooting by using the operation support subsystem (OSS)-controlled automated parameter detection and control for the mitigation of untoward radio and inapt antenna parameter settings in the 4G long-term evolution (LTE) and 3G wideband code division multiple access (W-CDMA) network. This article will propose the methods to eliminate the drudging and tedious physical cell site optimization and the extravagant radio frequency drive tests. The upgraded close-loop system will monitor and mitigate the unseemly cell site parameters for the complete automatic optimization and troubleshooting of the mobile radio network. The proposed close-loop automated control system can reduce the operational expenditures (OPEX) and capital expenditures (CAPEX) of the service provider. The novel features of the upgraded close-loop system make it a suitable candidate for the fully automated self-organizing network (SON) and energy-efficient network. The multiple case studies and the numerical results reveal the advantages of the close-loop control system.

Keywords Intelligent self-organizing network · Automation
Energy-efficient networks · Self-healing · Antenna array systems
Sensors · Close-loop control system · Troubleshooting
Inter-radio access technology · Operational expenditure
Capital expenditure

1 Introduction

The objective of self-organizing network (SON) is to fully automate the operational task in the mobile radio network. Traditionally, the coverage holes and other problems are detected through the drive tests. The physical parameter error detection and rectification of the sector antenna parameters such as detecting and mitigating the unseemly antenna azimuths, inapt mechanical tilts, and measurements of antenna

A. Lahiry (✉)
Cuttack 753003, Odisha, India

V. Bhateja et al. (eds.), *Intelligent Engineering Informatics*, Advances in Intelligent Systems and Computing 695, https://doi.org/10.1007/978-981-10-7566-7_20

heights above the ground level is an expensive and tedious process. The contribution of this paper is to introduce an upgraded close-loop system for the complete automation of the network troubleshooting and operations including the physical parameter detection and control. This article will propose an upgraded version of a close-loop system which is controlled from the operation support subsystem (OSS) with the ability to reconfigure the antenna tilt, antenna azimuth, and transmission radio frequency power after detecting the feedbacks from the sensors installed at eNodeB and NodeB ends.

In a related prior work [1], the method to eliminate the overshooting cells in W-CDMA radio network was proposed. This paper will propose an upgraded and more pragmatic approach, in order to detect the flaws in LTE and W-CDMA network [2] for a high-performance inter-radio access technology (RAT). In the prior work [1], it was stated that the laser meter may be implemented for antenna height above ground level measurement, but it may not be a very pragmatic approach for measuring the antenna height above ground level, because although the laser meters are more accurate, but the laser rays may be obstructed by the objects, therefore using MEMS-based barometric altimeters [3, 4] can be a better option for some occasions. The objective of the paper is to completely automate the optimization task in the mobile radio network by reducing the operational expenditure (OPEX) and capital expenditure (CAPEX). In a related prior work [5], a method to detect the coverage hole in LTE network was proposed, but this paper will introduce an enhanced version of a close-loop system specially to monitor the cell site parameters, for root cause analysis, automated troubleshooting, and coverage hole mitigation. The objective is to implement an advanced, smart root cause analysis and troubleshooting. The system is controlled from the operation support subsystem (OSS) after detecting the feedbacks, and therefore antenna footprints can be measured and controlled with highest possible accuracy to enable a paradigm shift in the field of fully automated self-organizing networks and energy-efficient networks.

The system and the process can minimize or completely eliminate the requirements of the drive tests [6, 7], due to the feedback and control system. Minimizing the drive tests is one of the major criteria of the self-organizing networks [6, 7], in order to reduce operational and capital expenditures of the service provider. The physical optimization of the cell site, such as altering antenna mechanical tilts and altering antenna azimuths, requires highly skilled manpower, still the effective cell site optimization cannot be guaranteed, and therefore it increases the probability of untoward parameter settings due to the human errors. The cell site physical optimization tasks are dangerous due to the life risks involved in carrying out the various site activities, and therefore the paper will try to eliminate all these challenges by proposing the control system.

This paper will try to propose the methods to detect the cell site coverage and automatically troubleshoot [8] the related problems associated with the unseemly physical cell site optimizations which may lead to the generation of the coverage holes, overshooting cells [1], or untoward cell site footprint overlaps due to inept physical optimization of sites, which is usually detected through the radio frequency drive tests. The paper will discuss some cases, and the final objective of the paper will be to increase the degree of intelligence of the self-organizing network. Furthermore, the close-loop automatic physical optimization will enhance the key performance indicators (KPI) by constantly monitoring and controlling the coverage and capacity of the mobile radio network.

2 Close-Loop Control Systems for LTE and W-CDMA Network

Figure. 1 illustrates the architecture [5] of LTE and 3G (W-CDMA) inter-radio access technology. Our paper adds a close-loop control system in the architecture for a better root cause analysis and automatic control. The objective of the paper is to introduce an intelligent system for enhanced network performance. The feedbacks are monitored, and the antenna parameters are controlled from the operation support subsystem (OSS). The close-loop control system for real-time monitoring, control, and troubleshooting in the LTE radio network is illustrated in detail in Fig. 2. The close-loop control system for real-time monitoring, control, and troubleshooting in the 3G radio network is illustrated in detail in Fig. 3. The proposed hardware shown in Fig. 2 and Fig. 3 is installed at the eNodeB and NodeB ends with a multiplexer for the transmission of the data to the operations supports

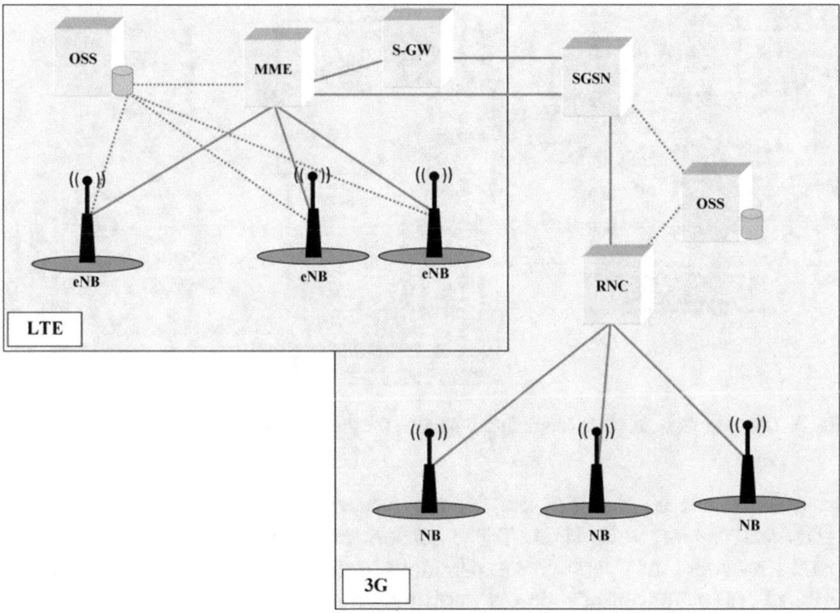

Fig. 1 Operation support subsystem (OSS) monitoring and controlling the radio parameters to ameliorate the key performance indicator in the inter-RAT technology

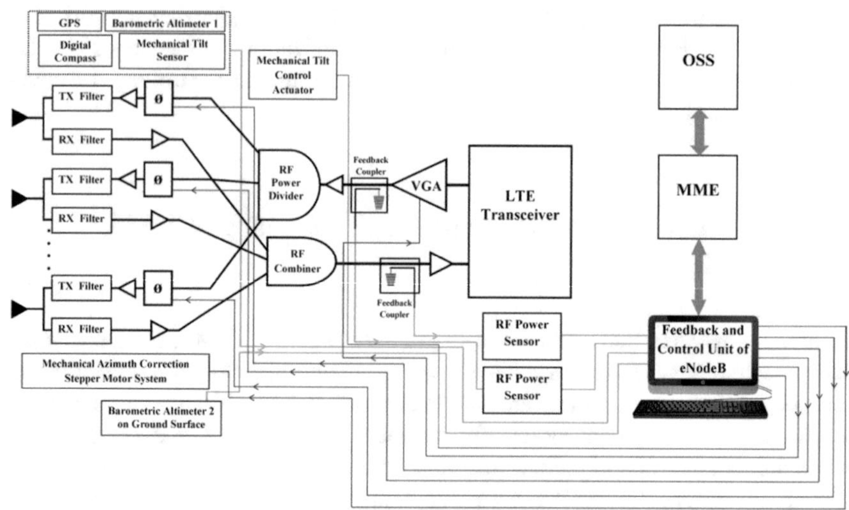

Fig. 2 Detailed view of close-loop control system for the LTE network

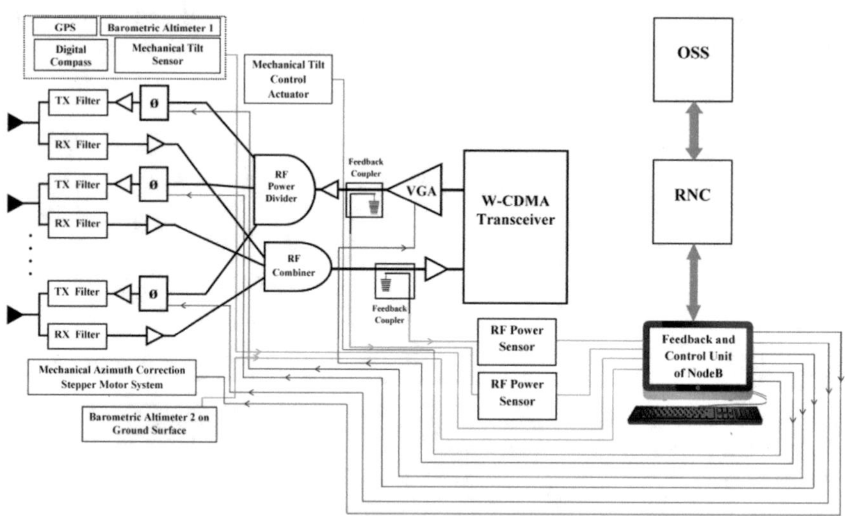

Fig. 3 Detailed view of close-loop control system for the W-CDMA network

subsystems and a demultiplexer for the reception of control signals from the operations support subsystem. These sensors can reduce the operational tasks, thereby reducing the operational expenditure of the service provider. Traditionally, physical parameter optimization is performed by radio frequency drive test engineers and technicians which is a drudging, expensive, and tedious task.

The GPS sensors and the compass [9, 10], sensors will monitor geographic position and azimuth of the antenna. The MEMS-based capacitive tilt sensors [11, 12] will monitor the antenna mechanical tilts. Two MEMS-based barometric pressure sensor-based altimeters [3, 4] will monitor the antenna heights above the ground level by monitoring the relative pressure difference of the two altitudes, one on the ground surface and another sensor behind the central elements of the antenna array. The traffic load of a sector is directly proportional to the total received power, and therefore coupler feedback at the receiver end is for monitoring the traffic load of the sector, and the power amplifier after the coupler is to compensate the insertion loss of the coupler. The coupler feedback at the transmitter end is for monitoring the transmission radio frequency power. Secondly, all the physical parameters will be monitored, and then the control signals will be generated such as the control signals for altering the inapt antenna azimuths, antenna electrical tilts, antenna mechanical tilts, and increasing or decreasing the radio frequency transmission power from the operation support subsystem (OSS). The antenna azimuth can be reconfigured by a stepper motor, and the digital compass [9, 10] will measure the changes of the antenna azimuths in the clockwise direction with respect to the true north in degrees. The antenna array mechanical tilts will be controlled by a tilt control actuator, and the tilt changes will be monitored by the MEMS-based tilt [11, 12], sensors.

Nomenclature given in Figs. 4 and 5

- AT = antenna array tilts in degrees
- VB = antenna array vertical half-power beam width in degrees
- AH = antenna height above the ground level
- C = total cell site coverage
- AC = azimuth coverage
- HB = horizontal half-power beam width in degrees.

The relative traffic loads of the sectors can be detected by monitoring the total received radio frequency power of each sector antenna. The traffic load is directly proportional to the total received power of a sector; therefore, the congested sector

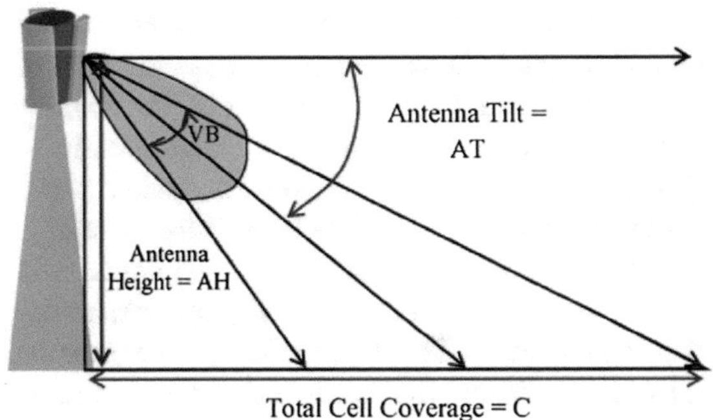

Fig. 4 Detailed view of cell site coverage

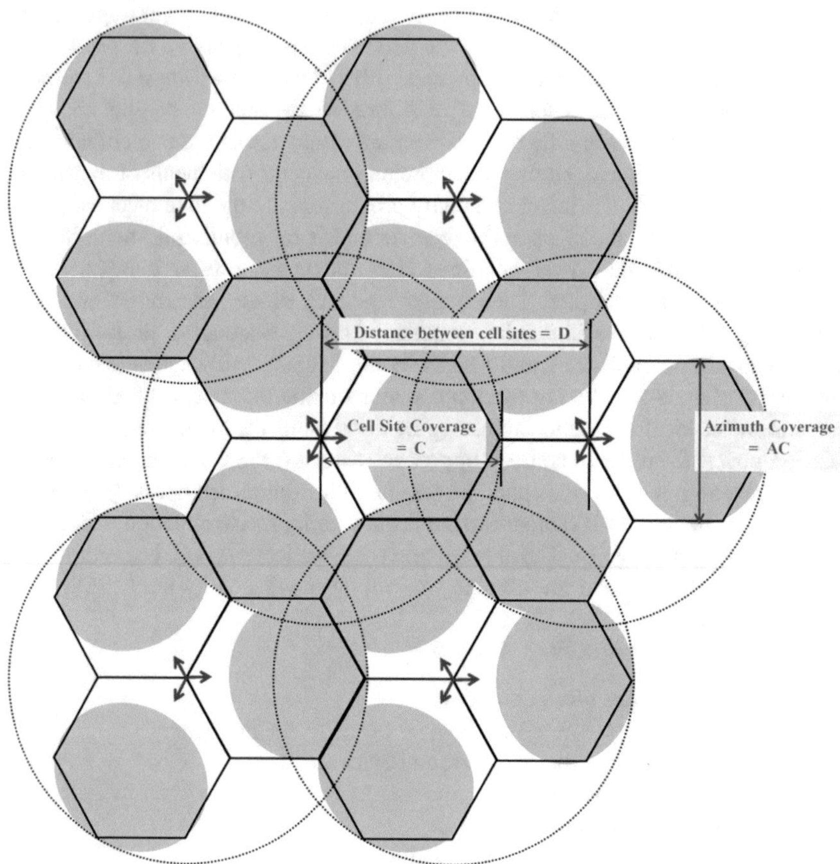

Fig. 5 Three-sector cell site configuration

will have the highest received power and the received power of the sector will decrease with the traffic load reduction.

The monitored data will help to detect the errors automatically. The parameters such as the inapt antenna azimuths, unseemly antenna tilts, and radio frequency power can be corrected automatically after evaluating the feedbacks. These flaws are caused due to human errors such as the inept cell site survey and inept physical parameter settings. After the analysis of the monitored feedbacks, the errors such as overshooting cells, call drop due to congestion in some specific hot spot areas, and coverage holes [13] due to untoward antenna parameter settings will be mitigated automatically by altering the antenna tilts, azimuth, and the radio frequency power. The traffic loads of sectors will be monitored, and the relative traffic load-based antenna array tilt optimization will be implemented to ameliorate the key performance indicators. The expression for the calculation of the cell site coverage (C) which is further derived by using the equations given in [1] is given below:

$$C=\left\{\left(\frac{(AH)}{\tan\left((AT)+\left(\frac{VB}{2}\right)\right)}\right)+\left(\frac{2\times(\sin(VB))\times(AH)}{((\cos(VB))-(\cos(2\times AT)))}\right)\right\} \quad (1)$$

The expression for the azimuth cell site coverage (AC) which is further derived from [1] is given below:

$$AC=\left\{\frac{2\times\left(\tan\left(\frac{HB}{2}\right)\right)\times(AH)}{\sin(AT)}\right\} \quad (2)$$

3 Numerical Results

In this section, the advantages and challenges of using barometric pressure sensor-based altimeters will be analyzed. Furthermore, the effects of unseemly antenna tilt (AT) and antenna azimuth will be analyzed, and the solutions will be proposed.

3.1 Barometric Altimeters

According to the US standard atmosphere [14], the pressure at a specific altitude can be calculated by using the equation given below:

$$p=p_0\left[1-\frac{\text{Altitude}}{44330}\right]^{5.255} \quad (3)$$

P = the air pressure at the altitude in the unit of mbar
P_0 = the standard atmosphere, 1013.25 mbar
Altitude = the height above the sea level in meters.

The antenna height above ground level can be detected by monitoring the relative pressure difference of two altitudes and the temperature. Although this device is not accurate as the laser meter [1] is, but the accuracy of the device can be meliorated by increasing the sensitivity of pressure sensors. Furthermore, the micro-electromechanical systems have enabled a paradigm shift in the field of the navigation and the control by increasing the accuracy of the modern atmospheric pressure sensor-based altimeters [3, 4] and tilt sensors [11, 12].

From Eq. (3), we can say that pressure difference between altitude A and altitude B in Fig. 6 can be calculated by using:

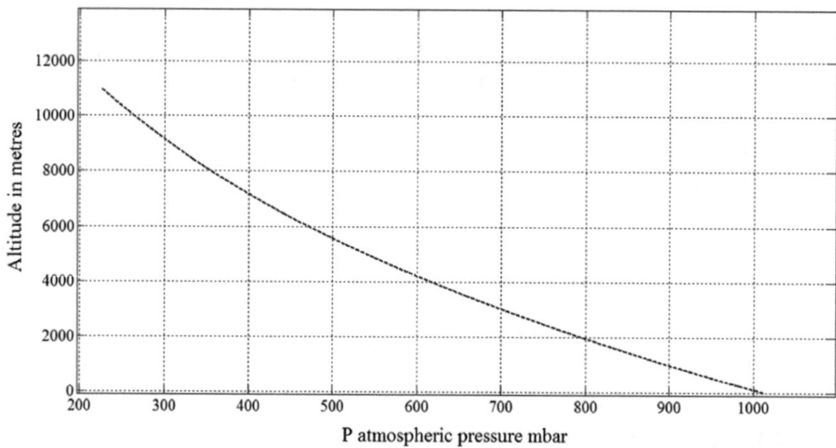

Fig. 6 Atmospheric pressure versus altitude using Eq. (3)

$$P_A - P_B = p_0\left(\left(\left[1-\frac{\text{AltitudeA}}{44330}\right]^{5.255}\right)-\left(\left[1-\frac{\text{AltitudeB}}{44330}\right]^{5.255}\right)\right) \tag{4}$$

Although laser meters [1] are accurate, the laser rays may be obstructed by the objects which may cause errors in measurements of the antenna heights above the ground level. The pressure of the atmosphere decreases with the increasing altitude, and therefore the accuracy of barometric altimeter may reduce at the higher altitudes. The pressure of the atmosphere also changes with the temperature; therefore, temperature sensors are also important to trace the atmospheric pressure changes. The laser meter is therefore a more accurate device for the measurement of the antenna height above ground level, whereas MEMS-based barometric altimeters [3, 4] are more reliable.

3.2 *Case Analysis of Inapt Radio and Antenna Parameter Settings*

We are considering the monitored feedbacks as AH = 30 m, D = 600 m, HB = 65° and based on that, the effect of antenna tilts on the footprints is evaluated (Figs. 7, 8 and 9).

3.2.1 Overshooting Cells

Overshooting cells are caused when the cell site coverage (C) or azimuth coverage (AC) or both exceed the cell site distance D. These circumstances are responsible

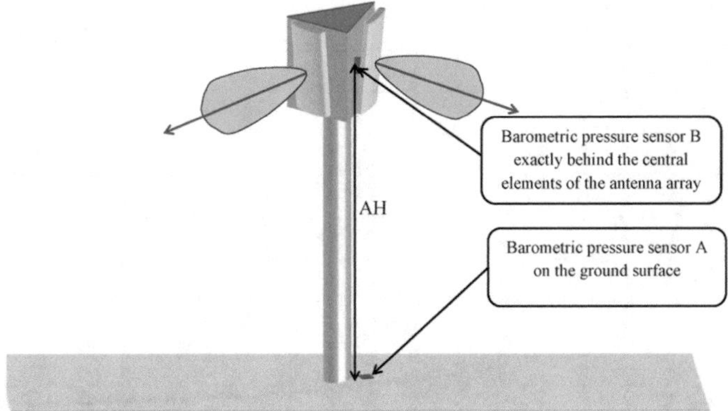

Fig. 7 Barometric pressure sensor-based altimeters for antenna height above the ground level measurements

for the physical cell identity (PCI) overshoot in LTE and pilot pollution in 3G. In our simulation, we are considering D = 600 m and AH = 30 m; therefore, from the results in Fig. 8, we observe that the overshooting cells are caused when the antenna down tilt angles are less than or equal to 6° and vertical beam widths are greater than or equal to 6.25°. Overshooting cells are also caused when the antenna down tilt angles are less than or equal to 6.5° and vertical beam widths are greater than or equal to 7.26°. To eliminate overshooting cells, the antenna down tilt angles must be increased accordingly.

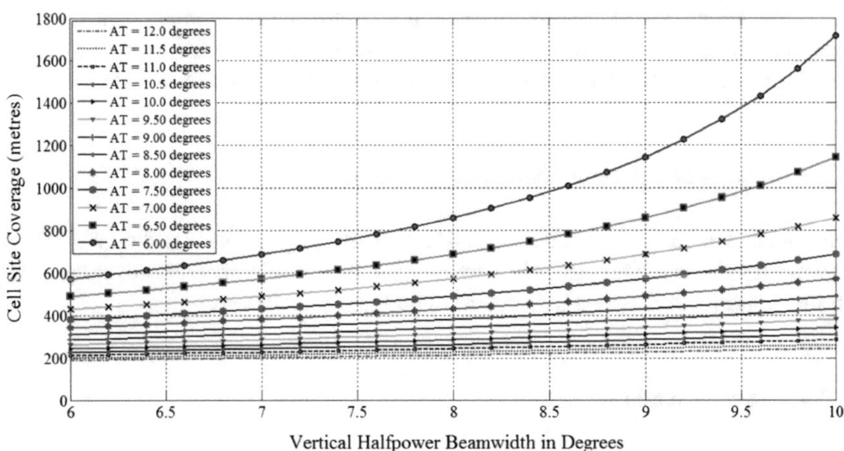

Fig. 8 Cell site coverage (C) variation of the antenna array due to (AT) tilt control

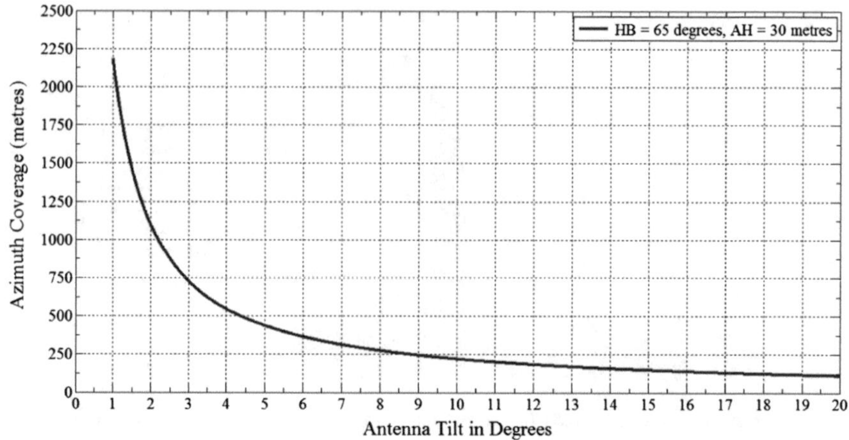

Fig. 9 Azimuth coverage (AC) variation of the antenna array due to tilt (AT) control

3.2.2 Coverage Holes, Overlap Planning, and Capacity

By considering the above conditions and from the results in Figs. 8 and 9, we observe that when the antenna down tilt angles are greater than or equal to 9.5°, the coverage holes are generated in the mobile radio network, thereby degrading the handoff performance due to the absence of coverage and footprint overlaps, whereas larger footprint overlaps [15, 16, 1] will enhance the handoff performance, thereby reducing the overall capacity of the network. Relative traffic load-based automatic antenna tilt optimization with footprints overlaps control can control coverage and capacity of the mobile radio network. In smart automatic traffic load balancing technique, the lightly loaded cell will reduce the antenna down tilt angle and heavily loaded cell will increase the antenna down tilt angle by constantly monitoring the traffic loads of the sector antennas in the mobile radio network.

3.2.3 Effects of Unseemly Antenna Azimuths

The antenna azimuth should be controlled properly as given in Fig. 5 because if the antenna azimuths are not according to the plan, then it will generate coverage holes in one region and increase unnecessary handoffs in another region.

3.2.4 Control Actions

Once the cell site parameters are monitored, then the algorithm at the operation support subsystem (OSS) will generate control signals to mitigate the inapt radio and antenna parameter settings according to the solutions of the cases we have stated earlier.

4 Conclusions

The numerical results indicate that the proposed close-loop control system can enable a paradigm shift in the field of fully automated self-organizing and energy-efficient network, thereby reducing the OPEX and CAPEX of the service provider. It was observed that for the accurate measurement of the antenna height above the ground level, the pressure sensor with a very high sensitivity is required.

Acknowledgements This work was partially supported by the Government of Odisha, India under the scheme Orissa Youth Innovation Fund-2016 on March 05, 2016.

References

1. Lahiry, A., Datta, A., Tripathy, S.: Active and entire candidate sector channel utilization based close loop antenna array amplitude control technique for UMTS and CDMA networks to counter non uniform cell breathing. In: Berretti, S., Thampi, S., Dasgupta, S. (eds.) Intelligent Systems Technologies and Applications. Advances in Intelligent Systems and Computing, vol. 385, pp. 33–44. Springer, Cham (2016)
2. Khanafer, R.M., Solana, B., Triola, J., Barco, R., Moltsen, L., Altman, Z., Lázaro, P.: Automated diagnosis for UMTS networks using Bayesian network approach. IEEE Trans. Veh. Technol. **57**(4), 2451–2461 (2008)
3. Bevermeier, M., Walter, O., Peschke, S., et al.: Barometric height estimation combined with map-matching in a loosely coupled Kalman-filter. In: 7th Workshop on Positioning, Navigation and Communication, pp. 129–132 (2010)
4. Lin, C.E., Huang, W.C., Hsu, C.W., et al.: Electronic barometric altimeter in real time correction. In: 27th IEEE Digital Avionics Systems Conference, 26–30 October, pp. 30–34 (2008)
5. Andrades, A.G., Barco, R., Serrano, I.: A method of assessment of LTE coverage holes. EURASIP J. Wirel. Commun. Netw., 01–12 (2016)
6. Hapsari, W.A., Umesh, A., Iwamura, M., Tomala, M., Gyula, B., Sébire, B.: Minimization of drive tests solution in 3GPP. IEEE Commun. Mag. **50**(6), 28–36 (2012)
7. Johansson, J., Hapsari, W.A., Kelley, S., Bodog, G.: Minimization of drive tests in 3GPP release 11. IEEE Commun. Mag. **50**(11), 36–43 (2012)
8. Barco, R., Lazaro, P., Munoz, P.: A unified framework for self-healing in wireless networks. IEEE Commun. Mag. **50**(12), 134–142 (2012). https://doi.org/10.1109/MCOM.2012.6384463
9. Choi, S., Kim, S., Yoon, Y., Allen, M.G.: A magnetically excited and sensed MEMS-based resonant compass. In: IEEE International Magnetics Conference (INTERMAG), San Diego, California, pp. 595–595 (2006)
10. Li, M., Rouf, V.T., Thompson, M.J., Horsley, D.: Three-axis Lorentz-force magnetic sensor for electronic compass applications. J. Microelectromech. Syst. **21**(4), 1002–1010 (2012)
11. Li, B., Zhang, H., Zhong, J., Honglong, C.: A Mode localization based resonant MEMS tilt sensor with a linear measurement range of 360. In: IEEE MEMS 2016, Shanghai, pp. 938–941 (2016)
12. Zou, X., Thiruvenkatanathan, P., Seshia, A.A.: A highresolution micro-electro-mechanical resonant tilt sensor. Sens. Actuators A **220**, 168–177 (2014)
13. Galindo-Serrano, A., Sayraç, B., Jemaa, S.B., Riihijärvi, J., Mähönen, P.: Automated coverage hole detection for cellular networks using radio environment maps. In: IEEE, 9th

International Workshop on Wireless Network Measurements (WiOpt), Tsukuba Science City, Japan, pp. 35–40 (2013)
14. U.S. Standard Atmosphere. U.S. Government Printing Office, Washington, D.C. (1976). https://ntrs.nasa.gov/archive/nasa/casi.ntrs.nasa.gov/19770009539.pdf
15. Lahiry, A., Tripathy, S., Datta, A.: W-CDMA busy hour handoff optimization using OMC-R controlled remote electronic variable tapered planar array. In: 3rd IEEE International Conference on Communication and Signal Processing, Melmaruvathur, pp. 031–035 (2014)
16. Lahiry, A., Datta, A., Maiti, S.: Improved self optimized variable antenna array amplitude tapering scheme to combat cell size breathing in UMTS and CDMA networks. In: 2nd IEEE International Conference on Signal Processing and Integrated Networks, Noida, pp. 077–082 (2015)

Tumour Classification in Graph-Cut Segmented Mammograms Using GLCM Features-Fed SVM

C. A. Ancy and Lekha S. Nair

Abstract Mammograms are customarily employed as one of the reliable computer-aided detection (CAD) techniques. We propose an efficient modified graph-cut (GC) segmented, grey-level co-occurrence matrix (GLCM)-based support vector machine (SVM) technique, for classification of tumour. In this work, SVM classification was carried out in single-view mammograms, subsequent to preprocessing, GC segmentation and GLCM feature extraction. Segmentation of pectoral muscles was done first, followed by segmentation of tumour, using kernel space mapped normalized GCs. We believe this process is the first of its kind used in mammograms. A suitably large number of features were extracted from GLCM, using Haralick method, which in turn increased the training efficiency. The proposed method was tested on 322 different mammograms from Mammographic Image Analysis Society (MIAS) and hence successfully verified to provide efficient results. High accuracy rates were achieved by combining best methods at each stage of diagnosis.

Keywords Gamma expansion · Normalized graph cuts · Mammograms
Haralick method · GLCM · SVM

1 Introduction

Breast tumour is one of the dominant types of cancer observed in women. Among various scanning methodologies, digital mammography is most commonly used because mammograms help detect tumours, even when the lumps are small. Tumours are of two types: benign and malignant tumours. Benign tumours occur normal in appearance and do not spread to other parts, whereas malignant tumours are dispersive and can divide uncontrollably, and thus, hazardous to health. The process to

C. A. Ancy (✉) · L. S. Nair
Department of Computer Science and Engineering, Amrita School of Engineering,
Amrita Vishwa Vidyapeetham, Amrita University, Amritapuri, India
e-mail: aancieantony93@gmail.com

L. S. Nair
e-mail: lekhaas@gmail.com

V. Bhateja et al. (eds.), *Intelligent Engineering Informatics*, Advances in Intelligent
Systems and Computing 695, https://doi.org/10.1007/978-981-10-7566-7_21

detect mammary tumour involves four steps: mammogram preprocessing, segmentation of masses i.e. region of interest (ROI), feature extraction from the segmented part and classification using machine learning algorithms. In order to reduce the number of misclassification errors, error-free acquisition of results at each stage of the process is of prime concern. This study was performed by efficient adoption of GC algorithms, to segment tumours, by considering the parameters pertinent to weight, type of cut and energy minimization criteria. The research also encompassed scrutiny of maximum number of Haralick texture features from GLCM, to facilitate the training process. Thus, when tumour region was detected, SVM could proficiently classify the captured images into benign or malign types.

2 State of the Art

Several researches have been carried out, towards development of different types of CAD systems, where mammogram preprocessing is a crucial step. In 2011, Somasundaram et al. [1] proposed a new automatic method for enhancement of medical images based on gamma correction. Segmentation follows preprocessing, and among different methods for ROI segmentation, GC segmentation is an age-old method, but less employed in mammograms, due to its complexity in computing the segmentation. In 2009, Saidin et al. [2] came up with density-based segmentation in mammograms, using GC method. In this work, segmentation was done semi-automatically, where the user needed to mark the seed labels for segmentation. Presence of pectoral muscles can give false positive results. So, pectoral muscles need to be removed prior to segmenting tumour region. In 2012, Abdellatif et al. [3] proposed eigenvector segmentation, to automatically detect pectoral muscle boundary in mammograms. Their work successfully segmented pectoral muscles but does not speak about segmentation of masses. In 2015, Angayarkanni et al. [4] used dynamic GC segmentation in mammograms, which can efficiently segment the tumour by iteratively using GC algorithms. Feature extraction was performed to compute the features that characterize the suspicious region. Bellotti et al. [5] used texture features from GLCM to characterize the ROI. Among different machine learning classifiers, SVM family is widely employed for classification of tumours because of its consistent, reliable performance. Moayedi et al. [6] in 2010 used SVM family for classification of masses. They obtained an accuracy of 96.6% for successive enhancement learning (SEL)-weighted SVM, 91.5% for support vector-based fuzzy neural network (SVFNN) and 82.1% for kernel SVM.

3 Proposed Method

The main objective of this paper was to integrate a set of promising, proven methods on each module of image processing, which leads to reliable classification results.

Image preprocessing is done using gamma expansion. Preprocessed mammograms are then segmented in two stages: segmentation of breast from background using morphological operations and segmentation of tumour within the breast, using normalized graph cuts. Next, texture features are extracted from these segmented portions, by GLCM method. Using Haralick method from GLCM, a total of 56 features are obtained. These features are fed into the SVM for classification. SVM could accurately classify the tumour into benign and malignant types. A block diagram for the proposed method is shown in Fig. 1.

3.1 Segmentation of Mammograms

Initially, preprocessing of mammogram is done using gamma expansion [1], which could increase the image-contrast, preserving the brightness. Preprocessed images are fed into segmentation phase. Segmentation of tumour is the most intricate task in this work, intended to detect the suspicious regions within mammograms. Segmentation is performed in two stages. First stage involves segmentation of breast from

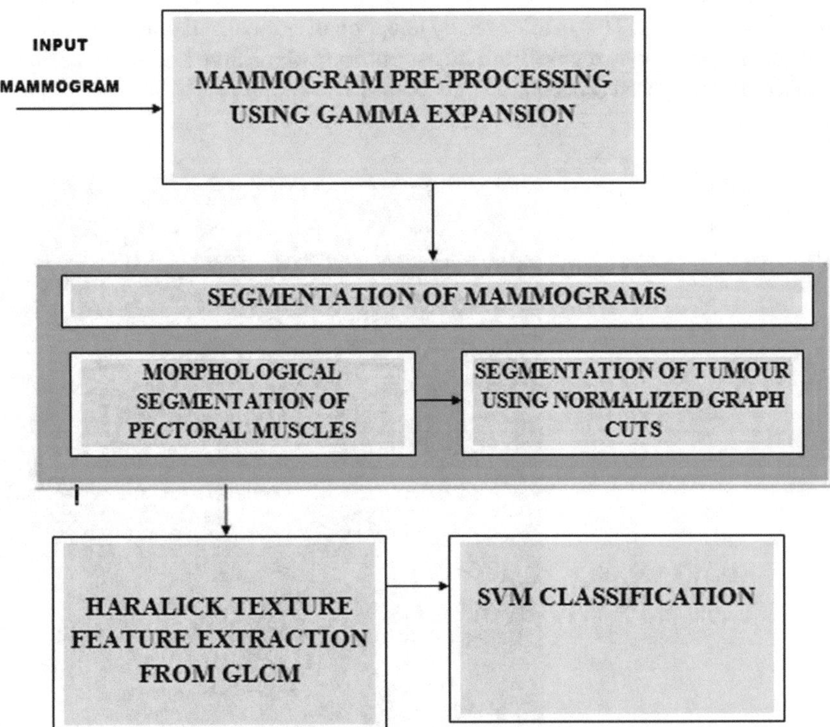

Fig. 1 Block diagram for the proposed work

background, as the pectoral muscles, which can give false positive results, need to be segmented. Morphological closing and image gradient is utilized to carry this out. Figures 2 and 3 show original mammograms, whereas Fig. 4 shows the segmented grey scale of tumour. Figure 5 graphically shows the segmentation of pectoral muscles from mammograms. Second stage involves segmentation of masses, which are in fact, the tumours. These were segmented using normalized GC algorithms [7]. As image data is highly convoluted, Kernel tricks to map the image data were applied implicitly, using a mapping function, to efficiently adopt graph cuts in mammograms, into data of higher-dimensional space [8]. Kernel radial basis function (RBF) is employed as mapping function. RBF kernel function is defined as:

$$K(y, z) = exp(-\left\|y - z\right\|^2 / \sigma^2) \quad (1)$$

Based on density, using unsupervised graph-cut formulations, multi-region segmentation of mammograms is carried out in higher-dimensional space. Perusing Mercers theorem [8], datacost and smoothcost do not have to be computed explicitly, as the dot product in the feature space is sufficient for data term computation, which is kernel induced, as a function of the region parameters, a kernel function and the image. The α/β swap algorithm for energy minimization of graph cuts has been developed in C++ by Boukov et al. [9], using the multi-label MRF optimization software library. This is followed by adoption of normalized graph cut, in order to segment the tumour region from breast portion. Algorithm 1 explains pertinent details of the proposed procedure.

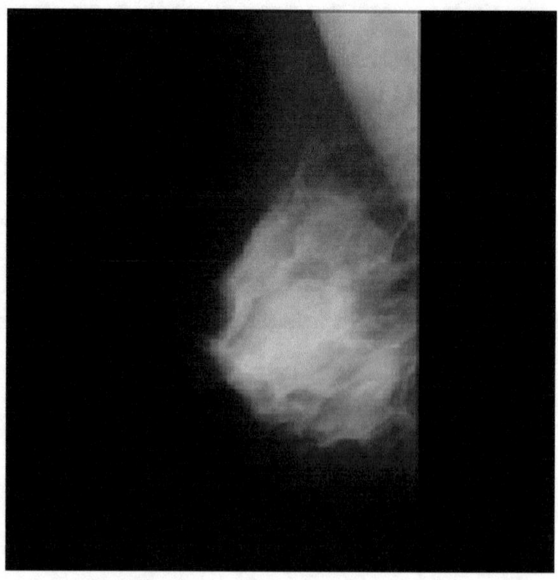

Fig. 2 Original mammograms

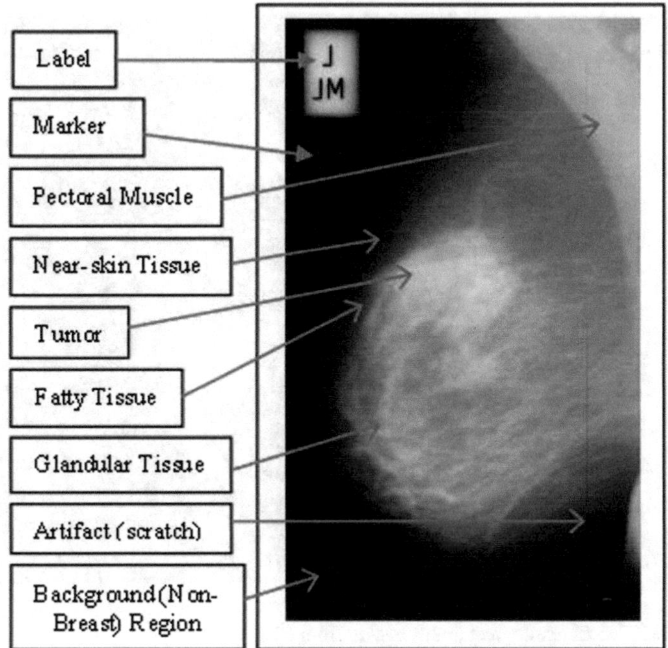

Fig. 3 Different parts of breast tissue

Algorithm 1

```
Proposed Segmentation Algorithm
  begin
     Initialize the datacost and smoothcost to apply the graph cuts.
     Begin iteration to achieve minimum energy criteria for graphcuts.
     repeat for every iteration
          Image is mapped to kernel space using rbf kernel function.
          Initialize datacost and smoothcost of the image to define
          segmentation functional.
          Calculate datacost and smoothcost.
          Optimize the segmentation functional.
          Find the normalized cut.
     until energy optimizes.
end.
```

3.2 Texture Feature Extraction

Feature extraction is an imperative step, as the number and values of features play a key role in the training of classifier. Higher number of features are more helpful in the training phase. Among different feature types, texture features are widely employed to capture visual content of images. GLCM is frequently applied to calculate texture

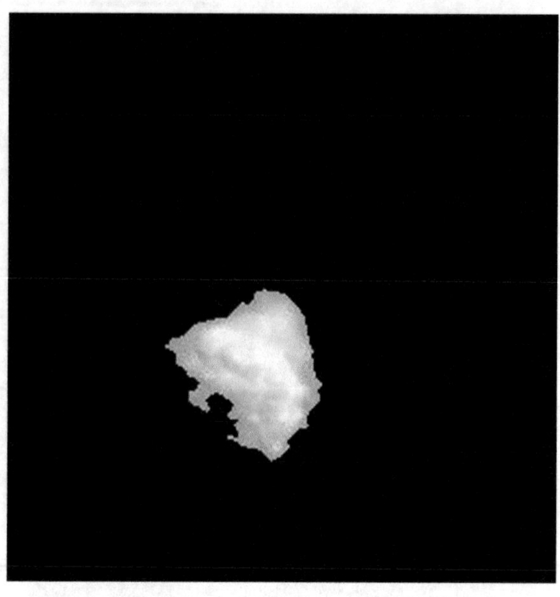

Fig. 4 Segmented grey scale of tumour

Table 1 Formulas of texture features extracted from GLCM

GLCM texture features	Formula
Inverse difference moment	$\sum \frac{P(i,j)}{1+(i-j)^2}$
Entropy	$\sum P(i,j) * [-lnP(i,j))]$
Energy	$\sum P^2(i,j)$
Correlated coefficient	$\sum \frac{P(i,j)*(i-\mu_x)*(j-\mu_y)}{\delta_x \delta_y}$
Contrast	$(i,j)^2 * P(i,j)$
Angular second moment	$\sum_i \sum_j p(i,j)^2$
Sum of squares	$\sum_i \sum_j (i-\mu)^2 p(i,j)$
Sum average	$\sum_{i=2}^{2N_g} ip_(x+y)(i)$
Sum variance	$\sum_{i=2}^{2N_g} (i-f_\delta)^2 P_(x+y)(i)$
Sum entropy	$-\sum_{i=2}^{2N_g} P_(x+y)(i) P_(x+y)(i)$
Difference variance	$\sum_{i=0}^{N(g-1)} i^2 P_(x-y)(i)$
Difference entropy	$-\sum_{i=2}^{2N_g} P_(x+y)(i) P_(x+y)(i)$
Info.measure of correlation 1	$\frac{HXY-HXY_1}{maxHX,HY}$
Info.measure of correlation 2	$(1-exp[-2(HXY_2 HXY)])^(1/2)$

features as it considers the spatial relationship of pixels in relative locations. Haralick function, from cytometry toolbox, employing GLCM features as basis, can compute up to 14 sets of features from a single-axis direction. From four different axes, we obtained a total of 56 feature values. The texture features taken from GLCM are tabulated in Table 1:

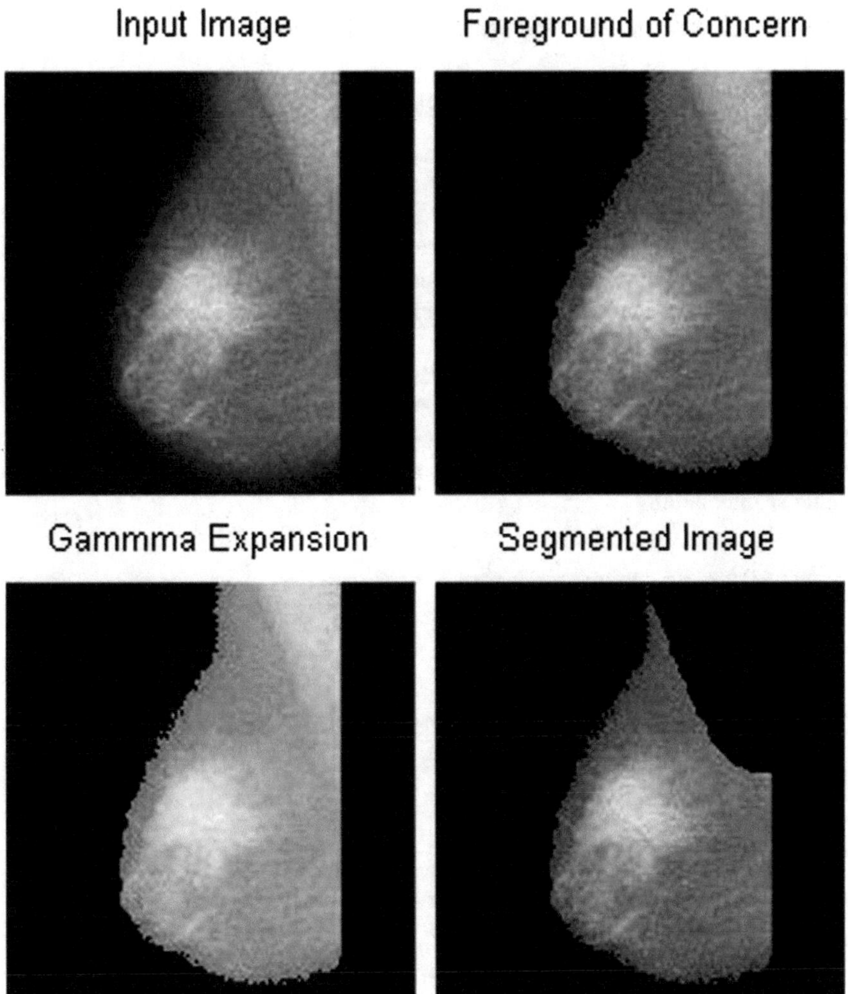

Fig. 5 Segmentation of pectoral muscles

3.3 Classification of Tumour

Appropriate number of texture features extracted from the segmented tumour is fed into SVM for training purposes. We employ linear SVMs for our classification. When a new test data comes, based on the supervised learning it receives, it can efficiently classify the tumour into benign and malignant types. SVM converts the two-class classification problem into a higher-dimensional space and hence finds the best hyperplane, thereby protecting the data points from misclassification and providing good classification performance.

4 Test Results and Performance Evaluation

4.1 Experimental Data and its Performance Evaluation

The verification followed by performance evaluation for the proposed work was carried out on benchmarked datasets, from MIAS database. For experimental analysis, 322 samples of single-view mammograms were taken for studies. Among the 322, 118 were tumorous and 104 non-tumorous. Tumour samples contains 63 as benign type and 55 malignant type. The proposed method worked well with all these samples.

Figure 6 shows the edge detection of mammograms, based on different densities, using Canny edge and different greyscale shades. Figure 7 shows the cropped portion of segmented greyscale tumour. The segmentation accuracy of the proposed method was checked using dice-similarity coefficient and was found to be 0.97 which ensures better segmentation.

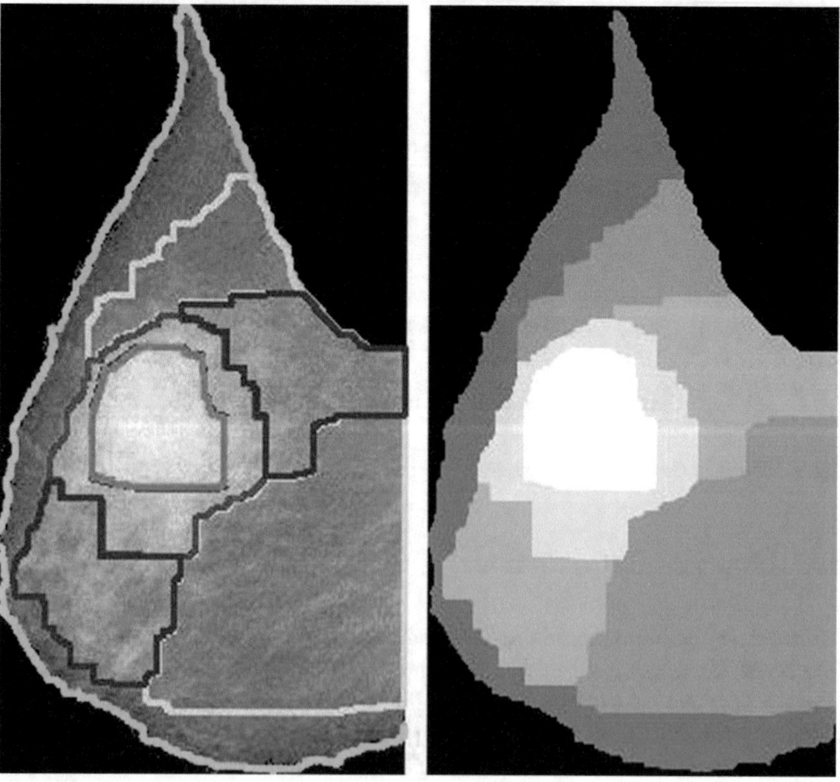

Fig. 6 Edge detection based on different densities

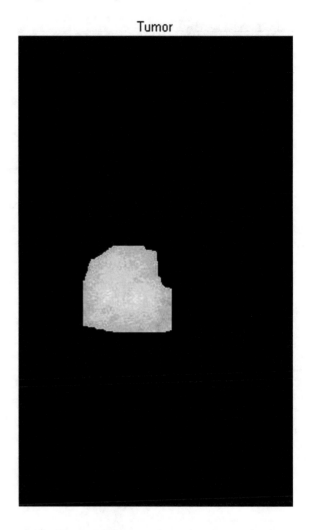

Fig. 7 Segmented tumour from normalized GCs

From the classifier results, Tp, Tn, Fp and Fn ratios were found first. Based on their values, five evaluation metrics: precision, recall (sensitivity), specificity, accuracy and F1 scores were computed. The computed metrics were plotted to generate the receiver operating characteristic (ROC) curve. The area under the curve was calculated. Area under the ROC curve is obtained to be 0.97, which reveals that classifier returns good values, as shown in Fig. 8. Table 2 shows the name, equation, meaning and obtained values of the performance metrics employed.

Table 2 Efficiency check

Metrics	Equations	Meaning	Values obtained
Accuracy	$\frac{T_n+T_p}{T_n+T_p+F_n+F_p}$	Gives closeness to the true value	0.9811
Specificity	$\frac{T_n}{T_n+F_p}$	Gives true negative rate	0.9643
Sensitivity	$\frac{T_p}{T_p+F_n}$	Gives false negative rate	0.9871
Precision	$\frac{T_p}{T_p+F_p}$	Gives false positive rate	0.9615
$F_1 score$	$\frac{2*P*R}{P+R}$	Gives test accuracy	0.9804

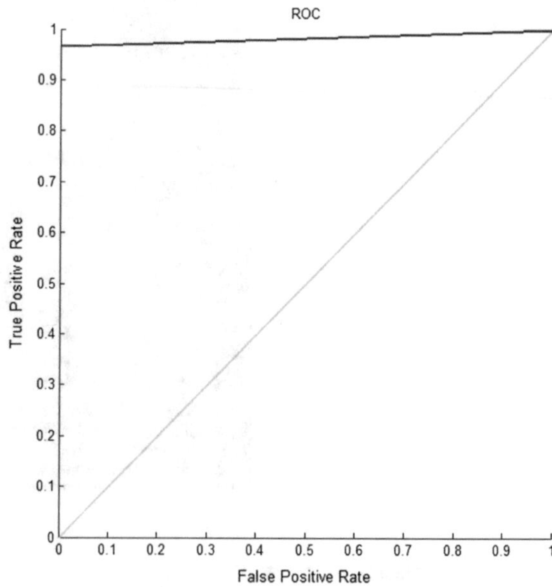

Fig. 8 ROC curve

4.2 Comparison with Existing Works

In Table 3, a comparison of different works was made with the proposed method. All the authors had used different combinations of SVM classifier. Still, high values for the performance metrics reveal that the proposed method outperforms all other works. Combining best methods had also ensured less time-consuming classification results.

Table 3 Comparison of proposed method with existing works

Detection method employed	Authors	Efficiency obtained	Efficiency of proposed system
Modified SVM	Ramin Nateghi et al. (2005)	F1 score = 0.77	F1 score = 0.98
SVM	F. Eddaudi et al. (2011)	Area under ROC = 0.95	Area under ROC = 0.97
SVM-KNN classifier	Li.Rong, Sunyuan (2011)	Accuracy = 0.97	Accuracy = 0.9811
Morphological segmented GLCM-based SVM	Ancy C A, Lekha S Nair (2017)	Accuracy = 0.81	Accuracy = 0.9811

5 Conclusion

Experimental results and comparisons reveal that the combined technique of image processing and machine learning, employed in our research, could efficiently classify the tumour type with 98.11% accuracy. As future enhancement, additional modifications can be tried out in parameters chosen for weight updating, type of cuts and energy minimization criteria chosen for graph-cut algorithms.

References

1. Somasundaram, K., Kalavathi, P.: Medical image contrast enhancement based on gamma correction. Int. J. Knowl. Manag. e-Learn. **3**(1), 15–18 (2011)
2. Saidin, N., et al.: Density based breast segmentation for mammograms using graph cut techniques. In: TENCON Region 10 Conference, pp. 663–667. IEEE (2009)
3. Abdellatif, H., et al.: K2. Automatic pectoral muscle boundary detection in mammograms using eigenvectors segmentation. In: Radio Science Conference (NRSC), 10 Apr 2012, pp. 633–640. IEEE (2012)
4. Angayarkanni, S.P., et al.: Dynamic graph cut based segmentation of mammogram. SpringerPlus **4**(1), 591 (2015)
5. Bellotti, R., et al.: A completely automated CAD system for mass detection in a large mammographic database. Med. Phys. **33**(8), 3066–3075 (2006)
6. Georgsson, F.: Differential analysis of bilateral mammograms. Int. J. Pattern Recognit. Artif. Intell. **17** (2003)
7. Shi, J., et al.: Normalized cuts and image segmentation. IEEE Trans. Pattern Anal. Mach. Intell. **22** (2000)
8. Salah, M.B., et al.: Multiregion image segmentation by parametric kernel graph cuts. IEEE Trans. Image Process. **20**(2) (2011)
9. Boykov, Y., et al.: An experimental comparison of min-cut/max-flow algorithms for energy minimization in vision. IEEE Trans. PAMI **26**(9) (2004)
10. Nair, J.J., et al.: A robust non local means maximum likelihood estimation method for Rician noise reduction in MR images. In: Proceedings of the IEEE International Conference on Communication and Signal processing, ICCSP (2014)

11. Rajan, J., Kannan, K., Kaimal, M.R.: Smoothening and sharpening effects of theta in complex diffusion for image processing. In: Seventh International Conference on Advances in Pattern Recognition (2009)
12. Nair, J.J., Bhadran, B.: Denoising of SAR images using maximum likelihood estimation. In: Proceedings of the IEEE International Conference on Communication and Signal processing, ICCSP (2014)
13. Priya, S.U., Nair, J.J.: Denoising of DT-MR images with an iterative PCA. Procedia Comput. Sci. **58**, 603–613 (2015)

Improving Feature Selection Using Elite Breeding QPSO on Gene Data set for Cancer Classification

Poonam Chaudhari and Himanshu Agarwal

Abstract This paper focuses on feature set selection on microarray gene expression for cancer classification. The gene expression data are incommensurate when it comes to number of genes and number of samples. This imbalance makes it important to study feature selection algorithms from the complex gene expression data. We conducted a research with quantum particle swarm optimization with elitist breeding (EBQPSO) on gene data sets. To the best of our knowledge, the exploration of the elitist is not taken into account for deep searching and classification applications with genetic data sets. Our contribution in this paper is to use EBQPSO algorithm on gene data sets for classification of cancer. The algorithm is tested with supervised and unsupervised learning approach, viz. support vector machine, J48 and neural network. The results show that EBQPSO outperforms particle swarm optimization (PSO) and quantum particle swarm optimization (QPSO) algorithms in terms of precision and recall value.

Keywords EBQPSO · Elitist breeding · Precision · Recall

1 Introduction

According to the report published by World Health Organization, cancer is the reason for death in every 1 out of 6 individuals which has claimed 8.8 million lives in 2015. The reports further state that 30–40% of cancers could be prevented if diagnosed in initial stage. This accentuates the need for early prediction of the disease. For diagnosis, various evaluation tests, blood sampling and biopsy are required to be done. Pathologists perform enumerable test, often including immunochemical tissue stains, to complete a diagnosis.

In order to overcome the limitations of the conventional techniques, advanced techniques such as molecular and genomic techniques have evolved. The analysis

P. Chaudhari (✉) · H. Agarwal
Symbiosis Institute of Technology, Pune, India
e-mail: poonam.chaudhari@ges-coengg.org

V. Bhateja et al. (eds.), *Intelligent Engineering Informatics*, Advances in Intelligent Systems and Computing 695, https://doi.org/10.1007/978-981-10-7566-7_22

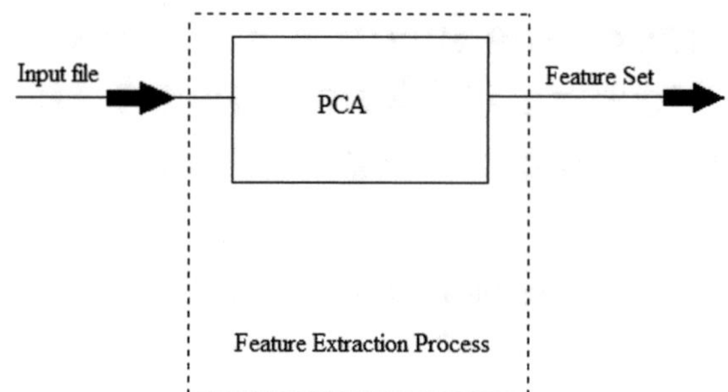

Fig. 1 Block diagram for principal component analysis

protocols are fine-tuned to offer the best balance between prognostic power and feasibility in daily clinical routine. Authors Gu et al. [1] worked on molecular morphology and stated that personalized medicine demands personalized pathology which can be easily met only with the amalgamation of genetic data.

Genomic techniques aim at the collective characterization and quantification of genes. A concise definition is given in nature journal [2]: a genetic database is a collection of one or more sets of genetic data (genes, gene products, variants and phenotypes). High-dimensional data, when used for the classification task, essentially need data preprocessing. Along with data cleansing, retaining important features, understanding the intricate relationships among the features and removing excess features can all improve the results. Dimension reduction is segregated as (a) feature extraction and (b) feature selection (Figs. 1 and 2).

We propose to make use of quantum particle swarm optimization with elite breeding on gene data set to generate a strong feature set which can be then used for classification purpose. As per the understanding of the subject, EBQPSO has not been used with gene data for classification task. The algorithm is tested on four different types of cancer data sets (lung, colon, leukaemia and cervical cancer) with three feature selection algorithms (PSO, QPSO and EBQPSO) and three classification algorithms (SVM, J48 and NN). The convergence time of EBQPSO is the lowest, thus boosting time efficiency. The algorithm also improves the classification task by enhancing the accuracy.

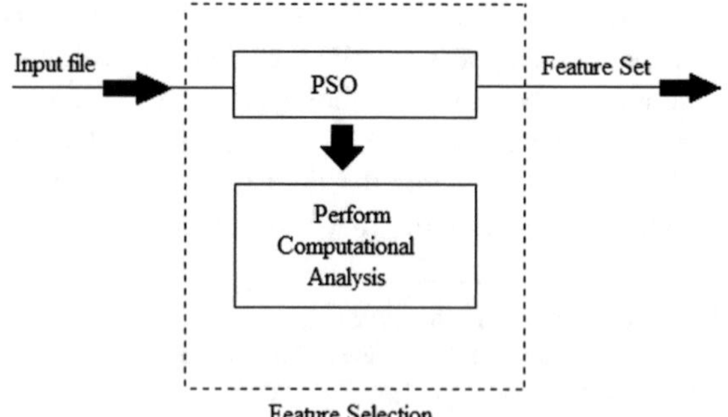

Fig. 2 Block diagram for particle swarm optimization

2 Related Works

2.1 Principal Component Analysis (PCA)

The primary job of PCA is to represent the data in lower dimension by removing the redundant features. This target is achieved by finding the orthogonal principal component. Prof. Long et al. [3] in his research work stated that components are selected based on the interpretation of the eigenvalues and eigenvectors of the autocorrelation matrix with discrete spectral representations. In the research work conducted by Hira and Gillies [4], the results elaborated linear algorithm, i.e. PCA and nonlinear algorithm, i.e. locally linear embedding and Laplacian Eigenmaps for feature selection. The work concluded that the over fitting issues persist and need to be resolved. According to a study published by author Jonathan Shlens in a tutorial to principal component analysis, it is difficult to extract independent statistical components by PCA until we make an implicit assumption that data should follow Gaussian distribution. Kernel PCA [5] rectifies this limitation to some extent. The subspace is computed with the help of dot products between vectors in the subspace. The function also generates a weight matrix [6]. The reconciliation of linear data to subspace vectors can lead to gross errors. These aggregate corruptions affect the performance of canonical PCA [7]. To overcome the drawback, inductive robust PCA was introduced. The algorithm is taught to solve the minimization problem in polynomial time [7]. However, some components with a low variance are thrown out. These thrown out components are crucial when we consider a real-time application. However, author Malaya K. Nath, in his research study with pattern recognition, concluded that the algorithm leads to the time delay [8].

2.2 Particle Swarm Optimization

PSO is a metaheuristic approach which operates in huge space of candidate solutions. It works towards optimizing the problem by iteratively improving feasible solutions with respect to a particular parameter. Convergence and biases are the prime factors of PSO [9]. The PSO algorithm and its parameters must be chosen so as to properly balance between exploration and exploitation to avoid premature convergence to a local optimum yet still ensures a good rate of convergence to the optimum [9]. The parameters are to be selected in such a way that the swarm converges to a point and also stops from further divergence of the swarm's particles. The research outputs by author Cleghorn [10] stated that parameter boundaries did not affect the area of convergence of the swarm. Shift in the boundaries creates

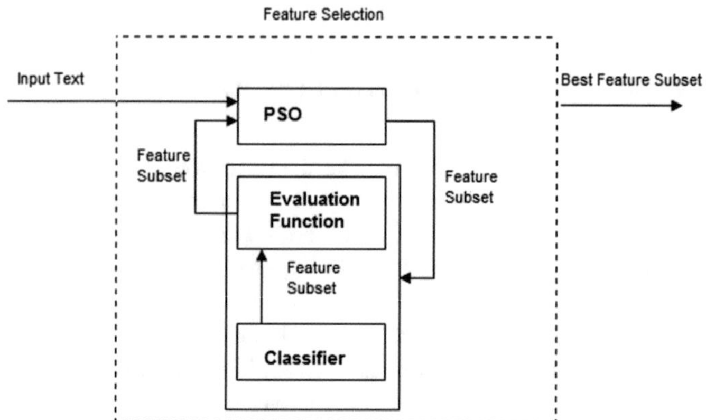

Fig. 3 Feature selection through particle swarm optimization

biases in the results, and hence, variants in PSO were needed. In the research work undertaken by Yang et al. [11], variants of PSO were explored along with the use of each variant for respective application domains (Fig. 3).

Authors Shinde and Gunjal [12] in their paper have made a comparison of PSO with other feature selection algorithms. They proposed an algorithm which searched for optimal feature subsets based on a well-formulated differential measure. Researcher Ahmad [13] gave a clear classification among linear and nonlinear algorithms with supervised and unsupervised techniques. Author Fernandes in [14] has stated the importance of feature extraction for better accuracy in classification. Researcher Dr. Vikrant Bhateja in [15] has emphasized the importance of neural networks in medical applications. The features were extracted using image analysis techniques which reduced the size of the data set and improved the accuracy.

Since then, multiple variations of PSO have been studied in genetic data analysis task. Author Enrique Alba has implemented geometric PSO with SVM as fitness function [16]. The results state that GPSO yields improved accuracy. Author D. Ramyachitra has proposed a variant of PSO named as interval-based PSO [17]. Author Maolong Xi has extended the research work with binary PSO [18]. The algorithm states that the efficiency of BPSO is better than PSO as the next position in the swarm is not influenced by the current position of the particle. Author Jing Zhao has experimented with quantum PSO [19]. He suggested improvement in the algorithm as cooperative algorithm, CBQPSO.

Based on quantum mechanics and trajectory analysis of PSO [20], Sun et al. stated that we need to consider, position and distance. Jun Sun concluded that QPSO is a global convergent technique which considers less number of parameters, thus improving the convergence rate [21].

3 Feature Selection Algorithm: EBQPSO

In the iterative process of finding personal best and global best, many a times the good solution sets would get overruled. To conserve the best, an elitist pool was introduced. Yang [11] was the one to predict that probing the elitist will surely produce some additional important information which would be crucial for searching the optimal solution.

Our algorithm makes use of elite pool set. EBQPSO is used as a strong feature selection step before applying basic classification algorithms like SVM, J48 and neural network, which in return would add to classification accuracy (Fig. 4).

The algorithm uses the elitists generated in the evolutionary processes to create new sub-swarms through the proposed breeding scheme. In the elitist breeding scheme, an elitist pool consists of personal best particles and global best particle and then uses the transposon operators to enhance the diversity of solutions. Moreover, the mechanism of updating the elitists with the new bred individuals with better fitness also provides a more efficient and precise search guidance for the swarm.

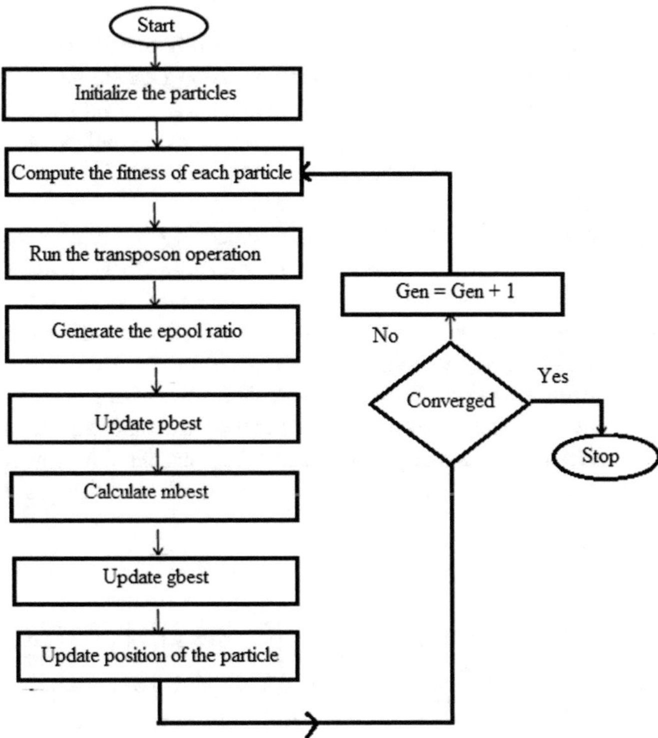

Fig. 4 Flow chart for EBQPSO [11]

A transposon is made of consecutive genes in each chromosome. The major types of transposon operators are (a) cut and paste (b) copy and paste. Size of transposon operator is decided by jumping percentage, whereas the jumping rate decides the probability of activation of the transposon operator.

However, transposon operator faces boundary constraints and might generate invalid individual. To overcome the boundary violation problem, the position vector of each particle is normalized. The conversion equation is defined as follows [11]:

$$\mathrm{conv}(x) = x - \mathrm{lb}(x)/\mathrm{ub}(x) - \mathrm{lb}(x) \quad (1)$$

After the transposon operations, the value of each gene should be restored back to its corresponding positional value in the search space according to the equation as follows [11]:

$$\mathrm{restore}(x) = \mathrm{conv}(x) \times (\mathrm{ub}(x) - \mathrm{lb}(x)) + \mathrm{lb}(x) \quad (2)$$

4 Experimental Results and Discussions

The data are taken from NCBI repository [22]. Four different types of cancer data sets have been considered, namely lung cancer, colon cancer, blood cancer and cervical cancer. The proposed model tries to classify the data into classes, namely cancerous and non-cancerous. Cleansing of data, removing null values and handling of redundant data are taken care in preprocessing techniques to convert the data into normalized form (Tables 1, 2, 3, 4 and 5).

The proposed algorithm EBQPSO is implemented in Java along with PSO and QPSO. The prime focus was to evaluate the efficiency of the feature selection algorithms. Hence, standard classifiers were used. Secondly, the motive was to evaluate the feature selection algorithms for supervised as well as unsupervised learning algorithms. Hence, the output was computed on the following algorithms: SVM, J48 and neural network classifier. Each data set was inputted to three different feature selection techniques, and each intermediate result was then inputted to three different classifiers. Accuracy alone cannot be considered as a good measure for evaluation as most of the data sets have two output classes. Binary classification

Table 1 Data set with sample size

Type of cancer	Genomic spatial event (GSE)	Geo data set (GDS)	No of samples
Lung	GSE series4115	GDS2771	192
Colon	GSE series11223	GDS3268	202
Leukaemia	GSE series4619	GDS2118	66
Cervical	GSE series3578	GDS3017	156

Table 2 Precision value of lung cancer data sets using PSO, QPSO and EBQPSO with the help of SVM, J48 classifier and NN classifier

Lung cancer	SVM	J48	NN
PSO	0.68	0.67	0.72
QPSO	0.77	0.75	0.79
EBQPSO	0.81	0.79	0.83

Table 3 Precision value of colon cancer data sets using PSO, QPSO and EBQPSO with the help of SVM, J48 classifier and NN classifier

Colon cancer	SVM	J48	NN
PSO	0.72	0.74	0.74
QPSO	0.79	0.78	0.79
EBQPSO	0.84	0.81	0.85

Table 4 Precision value of leukaemia data sets using PSO, QPSO and EBQPSO with the help of SVM, J48 classifier and NN classifier Table

Leukaemia	SVM	J48	NN
PSO	0.61	0.62	0.65
QPSO	0.69	0.71	0.73
EBQPSO	0.73	0.78	0.81

Table 5 Precision value of colon cancer data sets using PSO, QPSO and EBQPSO with the help of SVM, J48 classifier and NN classifier

Cervical cancer	SVM	J48	NN
PSO	0.69	0.70	0.71
QPSO	0.78	0.76	0.78
EBQPSO	0.84	0.81	0.86

could lead to accuracy paradox. Hence, for unbiased evaluation, we have considered precision, recall and f-measure (Figs. 5, 6, 7, 8, 9 and 10).

The lung cancer data set is initially inputted to PSO for feature selection. The output is then used for classification using SVM. The precision value is 0.68.

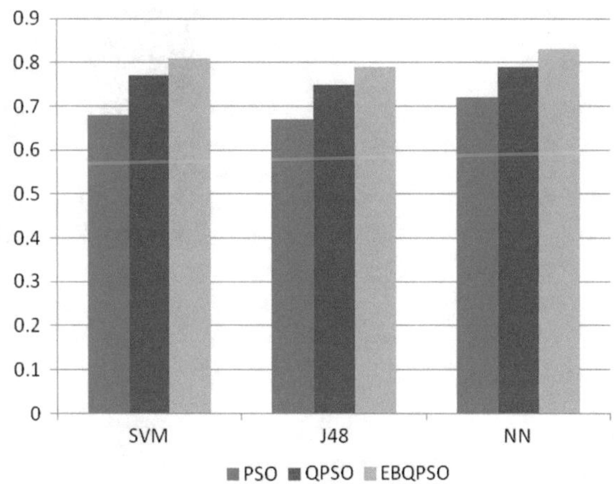

Fig. 5 Graph for precision value of lung cancer data set using SVM, J48 and NN

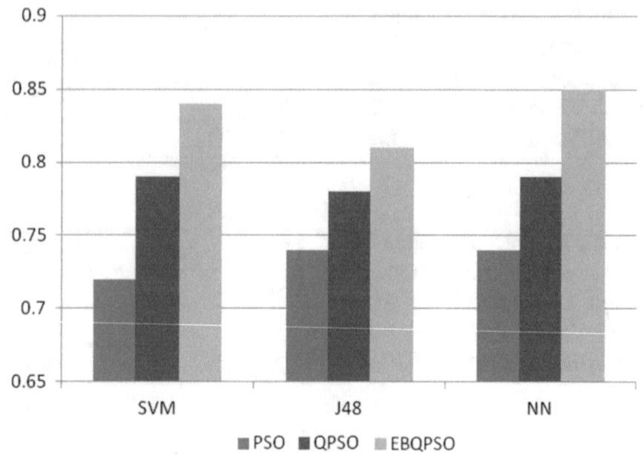

Fig. 6 Graph for precision value for colon cancer data set using SVM, J48 and NN

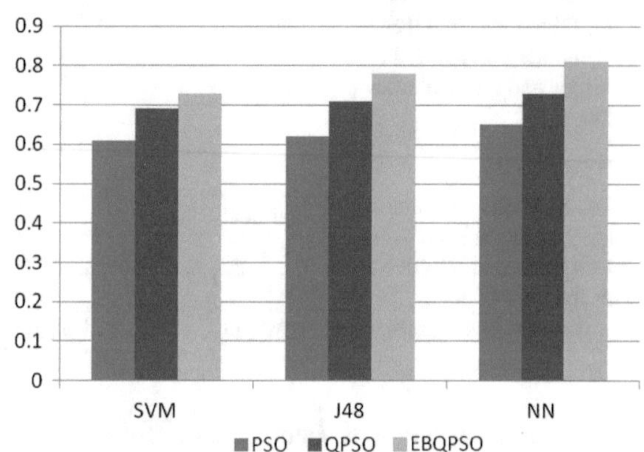

Fig. 7 Graph for precision value for leukaemia data set using SVM, J48 and NN

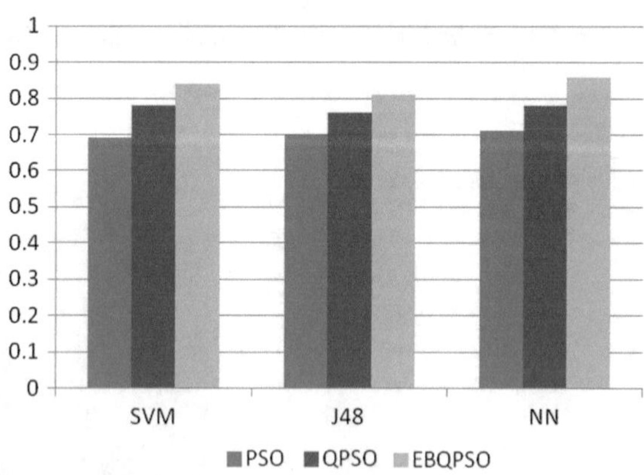

Fig. 8 Graph for precision value for cervical cancer data set using SVM, J48 and NN

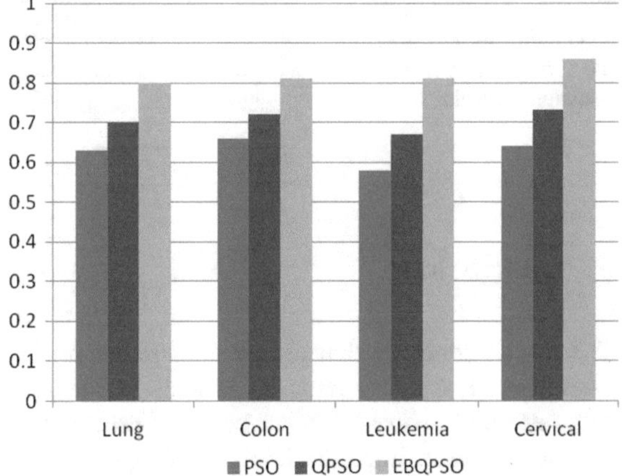

Fig. 9 Graph for f-measure value for cancer data sets with features selected through PSO, QPSO and EBQPSO

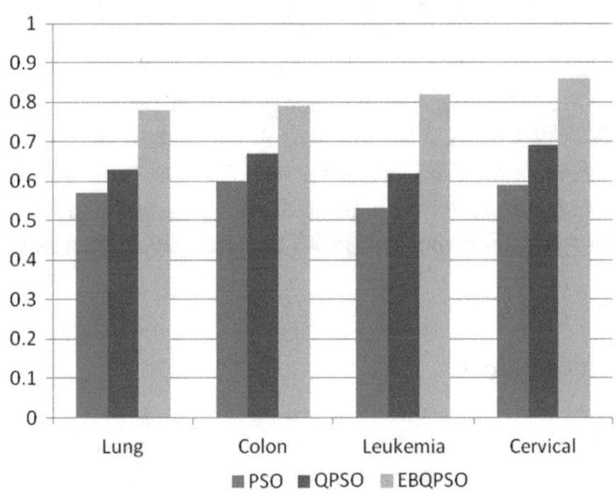

Fig. 10 Graph for recall value of cancer data sets with features selected through PSO, QPSO and EBQPSO

This value increases as we input the same lung cancer data set to EBQPSO and then provide it to SVM again. The precision value of later combination is 0.81. Similarly, the model is run again for the remaining two classification algorithms. The final result shows substantial increase when the classification algorithm is inputted with the feature selection set generated by EBQPSO (Tables 6 and 7).

Table 6 F-measure value of neural network classifier for the four data sets with feature selection algorithms

F-measure values	PSO	QPSO	EBQPSO
Lung	0.63	0.70	0.80
Colon	0.66	0.72	0.81
Leukaemia	0.58	0.67	0.81
Cervical	0.64	0.73	0.86

Table 7 Recall value of neural network classifier for the four data sets with feature selection algorithms

Recall values	PSO	QPSO	EBQPSO
Lung	0.57	0.63	0.78
Colon	0.60	0.67	0.79
Leukaemia	0.53	0.62	0.82
Cervical	0.59	0.69	0.86

5 Conclusions

The results conclude that precise feature set selection plays a crucial role in the classification task. Our contribution includes the use of EBQPSO algorithm on gene data sets for classification of cancer. This breeding strategy acts on the elitists of the swarm to escape from the likely local optima and helps to perform search efficiently. It improves the search efficiency of QPSO, and the mechanism of updating the elitists with the new bred individuals with better fitness also provides a more efficient and precise search guidance for the swarm. Our algorithm has outperformed with better global search capability and convergence rate. Our results state that the feature selection set generated by elitist breeding quantum PSO improves the classification accuracy with supervised as well as unsupervised learning algorithms. EBQPSO performs well for data sets with different sizes. As all the tests were performed on same software and hardware along with same parameters, we can conclude that EBQPSO consistently improves the result.

References

1. Gu, J., Taylor, C.R., Phil, D.: Practicing pathology in the era of big data and personalized medicine. Appl. Immunohistochem. Mol. Morphol., 1–9 (2014)
2. Genetic databases. www.nature.com/articles/
3. Long, D.W., Brown, M., Harris, C.: Principal components in time-series modeling. In: Control Conference (ECC) (2015)
4. Hira, Z.M., Gillies, D.F.: A review of feature selection and feature extraction methods applied on microarray data. Department of Computing, Imperial College London (2015)
5. Scholkopf, B., Smola, A., Muller, K.R.: Kernel principal component analysis. J. Neural Comput., 1299–1319 (1998)
6. Dongxiao, N., Xihua, G.: Application of Gauss Radial Kernel function principal component analysis model in the industrial enterprise's wastewater treatment. In: Industrial Electronics and Applications, pp. 1–12. IEEE (2007)
7. Bao, B.-K., Liu, G., Xu, C., Yan, S.: Inductive Robust principal component analysis. IEEE Trans. Image Process., 3794–3800 (2012)
8. Nath, M.K.: Independent component analysis of real data. In: Advances in Pattern Recognition, pp. 266–279. IEEE (2009)
9. Trelea, I.C.: The particle swarm optimization algorithm: convergence analysis and parameter selection. Inf. Process. Lett., 317–325 (2003)
10. Cleghorn, C.W.: Particle swarm convergence: standardized analysis and topological influence. In: Swarm Intelligence Conference, pp. 134–145 (2014)

11. Yang, Z.-L.,Wu, A., Min, H.-Q.: An improved quantum-behaved particle swarm optimization algorithm with elitist breeding for unconstrained optimization. Comput. Intell. Neurosci., 674–686 (2015). Hindawi Publishing Corporation
12. Shinde, P.L., Gunjal, B.L.: Particle swarm optimization best feature selection method for face images. Int. J. Sci. Eng. Res., 2229—5518 (2012)
13. Ahmad, I.: Feature selection using particle swarm optimization in intrusion detection. Int. J. Distrib. Sensor Netw. Special Issue Enabling Technol. Next-Gener. Sensor Netw. Prospects (2015)
14. Fernandes, S.L., Chakraborty, B., Gurupur, V.P., Prabhu, G.A.: Early skin cancer detection using computer aided diagnosis techniques. J. Integr. Design Process Sci., 33–43 (2014)
15. Gautam, A., et al.: An improved mammogram classification approach using back propagation neural network. Data Eng. Intell. Comput., 369–376 (2018). Springer
16. Alba, E., Garcia-Nieto, J., Jourdan, L., Talbi, E.-G.: Gene selection in cancer classification using PSO/SVM and GA/SVM hybrid algorithms. IEEE (2007)
17. Ramyachitra, D., Sofia, M., Manikandan, P.: Interval-value based particle swarm optimization algorithm for cancer-type specific gene selection and sample classification. Genomics (2015). Elsevier
18. Xi, M., Sun, J., Liu, L., Fan, F., Wu, X.: Cancer feature selection and classification using a binary quantum-behaved particle swarm optimization and support vector machine. Comput. Math. Methods Med. (2016). Hindawi Publishing Corporation
19. Zhao, J., Sun, J., Xu, W.: A binary quantum-behaved particle swarm optimization algorithm with cooperative approach. IJCSI Int. J. Comput. Sci. Issues (2013)
20. Sun, J., Feng, B., Xu, W.: Particle swarm optimization with particles having quantum behavior. In: Proceedings of the IEEE Congress on Evolutionary Computation, pp. 325–331 (2004)
21. Sun, J., Lai, C.H., Wu, X.J.: Particle Swarm Optimization: Classical and Quantum Perspectives. CRC Press (2012)
22. https://www.ncbi.nlm.nih.gov

Epileptic Seizure Mining via Novel Empirical Wavelet Feature with J48 and KNN Classifier

M. Thilagaraj and M. Pallikonda Rajasekaran

Abstract In this paper, we are providing the application of options using empirical wavelet transform (EWT) and J48 decision tree model for electroencephalogram (EEG) signal classification. The features were extracted from EEG signal based on the empirical wavelet transform. Empirical wavelet transform (EWT) decomposes the EEG signal in the form of the intrinsic mode functions (IMFs) which is an AM–FM signal. The statistical values were extracted from the decomposed signals resulting in the EWT features. The extracted features were classified using classifiers J48 and KNN classifier. The proposed J48 model achieved higher accuracy rates than that of the KNN algorithm.

Keywords Epilepsy • Wavelet features • KNN classifier • J48 classifier

1 Introduction

A brain–computer interface (BCI) is a substitute method for correspondence and control framework in which the messages or orders do not rely upon the cerebrum's ordinary movement of fringe nerves and muscles [1]. This is made possible with the use of sensors which can monitor the physical processes that occur within the brain to measure the thought from the brain. In these systems, the manipulation of the brain activity can be produced by using motor movements and the signals can be used to control computers or communication devices [2, 3]. EEG signals are noninvasive, inexpensive, and compared with other signals like MEG, PET, and FMRI [4].

M. Thilagaraj (✉)
Department of Instrumentation and Control Engineering,
Kalasalingam University, Anand Nagar, Krishnankoil, India
e-mail: thilagaraj@klu.ac.in

M. Pallikonda Rajasekaran
Department of Electronics and Communication Engineering,
Kalasalingam University, Anand Nagar, Krishnankoil, India
e-mail: m.p.raja@klu.ac.in

V. Bhateja et al. (eds.), *Intelligent Engineering Informatics*, Advances in Intelligent Systems and Computing 695, https://doi.org/10.1007/978-981-10-7566-7_23

The voltage of the EEG signal can be determined by its amplitude. In the frequency spectrum range of the EEG, the frequency range is between the ultraslow to ultrafast frequency components.

The frequency range lies between 0.1 and 100 Hz for the classification purpose. These calculations are for discovery of epileptic seizures which rely upon the distinguishing proof of different examples, for example an increase in amplitude or EEG flattening. The algorithms have been developed which is based on wavelet features, amplitude relative to the activity, and spatial context [5]. We have analyzed the Normal and Epileptic EEG Waveform signals of human using Empirical wavelet transform (EWT), and their abilities are detected by the frequency in the EEG. The extracted features were taken, and these elements are utilized to check the sections in the nearness or nonappearance of epileptic seizures. Many soft computing techniques have been evolved for the classification of features in EEG signals [6]. Nonlinear deterministic dynamics were found in the seizure activity in the EEG time series [7]. The features from EEG signals were extracted using principal component analysis and classified with neural network [8]. Linear feature extraction methods were extracted and classified by using KNN classifier which includes decision rules [9]. Feed-forward neural network was implemented using wavelet transform for classification of EEG signals [10].

In this work, we propose a new technique for EEG classification based on J48 classifier and K-nearest neighbor, which are used as statistical tools for classification. The remaining paper is structured in the form as follows. Sect. 2.1 deals with the EEG signal data description. Section 2.2 deals with the feature extraction using Empirical Wavelet Transform methods which is applied to the EEG signals. Section 2.3 describes the classification using J48 decision tree, Sect. 2.4 deals with the classification using KNN algorithm, and Sect. 3 shows the simulation and results. In Sect. 4, conclusion is discussed.

2 Methodologies

2.1 Data Selection

The electroencephalogram signal dataset which is available in public on the market in University (http://www.meb.unibonn.de/epileptologie/science/physik/eegdata.html) was used. The entire dataset consists of five sets (A, B, C, D, and E), and every set contains one hundred electroencephalogram signals with 4097 samples and lasts for 23.6 s with a frequency of 173.61 Hz. Sets A and B are taken from surface electroencephalogram recordings of five healthy volunteers with eyes open and closed. Samples in both groups C and D have been taken from the hippocampus formation of the opposite hemisphere, the epileptogenic zone of the brain. The dataset E contains solely seizure activity that is chosen from all recording sites exhibiting attack activity. The flowchart of the proposed methodology for detection of epileptic data is shown in Fig. 1.

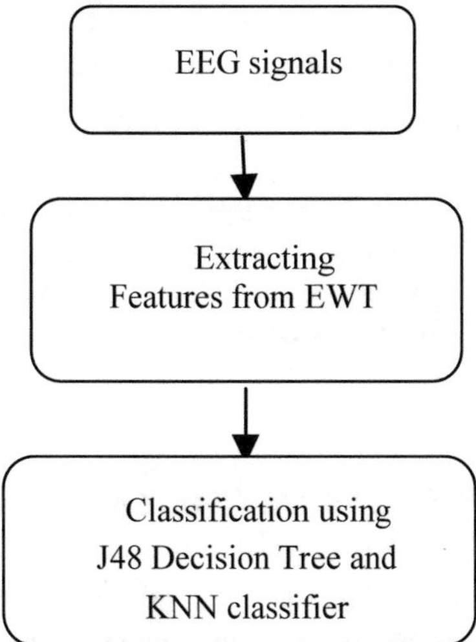

Fig. 1 Flowchart of the proposed technique

2.2 *Feature Extraction Methods*

This method is adopted to retract some special information from a given input EEG signal. So in general, we transform the signal from a domain to related information for the purpose of analysis. A moment request on the change is that the first capacity can be incorporated, i.e., reproduced from its changed state. The empirical wavelet transform is exhibited, and its most imperative elements are examined.

Empirical wavelet transform is one type of wavelets which is used to produce the processed signal for feature extraction. Let Fourier point of view is taken into consideration, this construction is used for equating with band-pass filters. Other way to develop the potential adaptability and the performance of the filters and their support depends on which the information gathered in the spectrum of the verified signal that is located. To determine the clarity, only the true signals were taken into account. But there are some following reasons that can be easily explained by imaginary signal through constructing various types of filters with both the positive and negative frequencies. A general Fourier axis with 2π value is considered. To fulfill the Shannon criteria, we limit our explanation to $\omega \in [0, \pi]$. By considering the Fourier support in the range from 0 to π by sectioning into N bordering fragments. ω_n is the range between each segment (where $\omega_0 = 0$ and $\omega_N = \pi$). Each section is indicated as $\Lambda_n = [\omega_{n-1}, \omega_n]$; then, it is easy to see that SNn = 1 $\Lambda n = [0, \pi]$. Centered on each n, we define a transition phase Tn of width $2\tau n$. The empirical wavelet acts as a band-pass filter for all Λn. We are forced to divide the Fourier spectrum for analyzing the signal. Our goal is to divide the spectrum. This shows that we need a number of $N + 1$ boundaries, and to find $N - 1$ extra boundaries in the region, we need 0 and π. Again, we are in need to find the local maxima through the spectrum, and they listed in the order of decreasing manner in which 0 and π are in exclusion. Next, we consider revising algorithm in which M acts as maxima. Two cases are produced as follows:

- M ≥ N: The process of algorithm shows the maxima number of division, only the N − 1 maxima in the boundary acts as first.
- M < N: The signal has fewer modes, then we can keep all the detected maxima, and the N value is reset. Now, with this set of maxima plus 0 and π is equipped, we characterize the limits ω n of each section as the inside between the two back to back maxima.

2.3 J48 Decision Tree Classifier

A decision tree is one of the predictive machine learning models which are used to fix the target values based on the dependent variable of new samples through various attributes of the available data. The nodes from the internal of a decision tree denote with different attributes; the branches between the nodes are used to show the possible values from the observed samples through the attributes. Next, with the help of the terminal nodes, we can determine the final value (classification) of the variable. The additional options of J48 area unit accounting for missing worth within the continuous attribute value vary and are used for derivation of rules, etc., within the maori hen data processing tool, J48 is an open supply form of Java implementation of the C4.5 formula. The woodhen tool is employed to offer variety of choices connected with the tree pruning. In different case of the potential through the fitting pruning that is employed as a tool for precising the tree. The opposite algorithms say that he classification is needed to perform recursively with each single leaf is pure to urge these the classification of the info ought to be as excellent as attainable. This formula will be wont to generate the foundations from identity of that information. The target is more and more generalization of a choice tree till it we have a tendency to obtain the equilibrium of flexibility and accuracy.

2.4 KNN Classifier

The *k*-nearest neighbor algorithm (*k*-NN) is one type of pattern recognition and a nonparametric method which is simple among all the machine learning algorithms that are used for the classification of the dates based on the training. KNN classification is based on the voting of the data which are majority in its neighbors and K-stands as positive integer which is typically small in nature. When object is considered as the class of that single nearest neighbor, then the value of $k = 1$. The other way to assign the value of k is based on the object property value with the average values of *k*-nearest neighbors. The process can be calculated by knowing the weighting scheme that is of $1/d$ where d is the distance of the neighbor.

From the set of object, the neighbors are taken with the help of doing the training step of algorithm. The *k*-nearest neighbor algorithm is suitable for certain boundary and acts as sensitive to the data which are local in structure. By measuring the location of the nearest neighbor in the instance space, the unknown neighbor can be classified. The most accurate result is obtained by choosing the value of k which must be greater than 1. If the value of k is higher, then the output is smoother and acts as less sensitive.

3 Results and Discussion

The dataset contains of 100 series data in each set and in each series is consist of 4096 dataset samples. The original EEG signal for set A and set E is shown in Figs. 2 and 3, respectively. In this process, each series is composed of 256 discrete data windowed by a rectangular window. After the process of downsampling method for the 100 series set, a total number of 1600 vectors are taken from each set. The down-sampled EEG signal for set A and set E is shown in Figs. 4 and 5, respectively. The training and testing datasets were obtained by 3200 vectors (1600 data from each one set). For training, 1600 data (800 data from each set) are used and other 1600 data (800 data from each set) are used for testing.

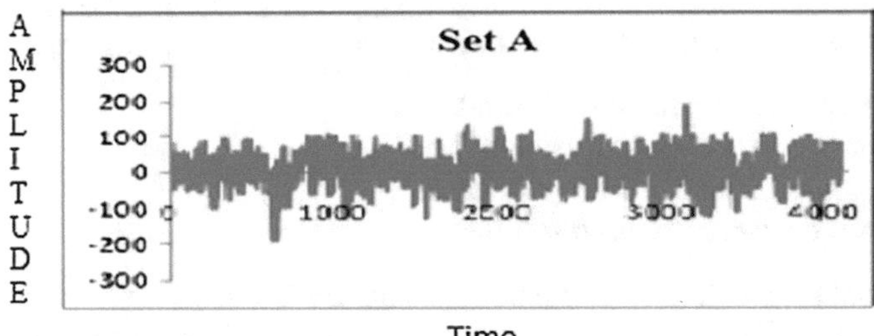

Fig. 2 Set A before downsampling

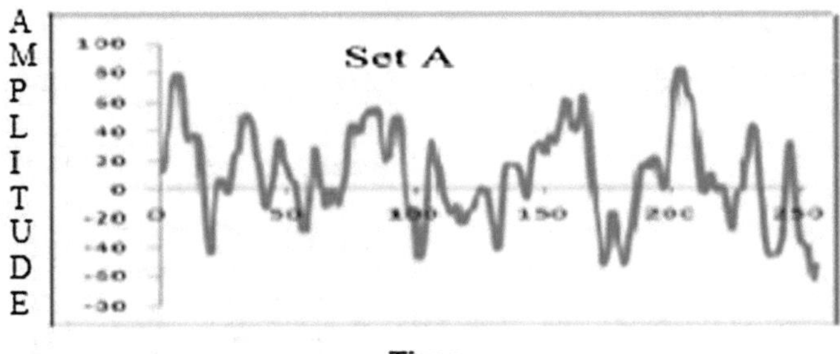

Fig. 3 Set A after downsampling

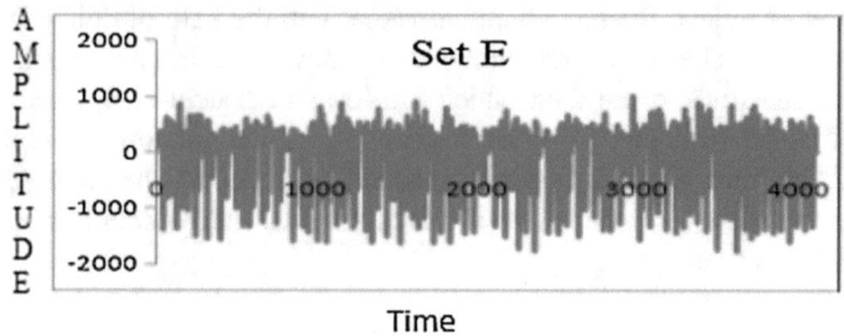

Fig. 4 Set E before downsampling

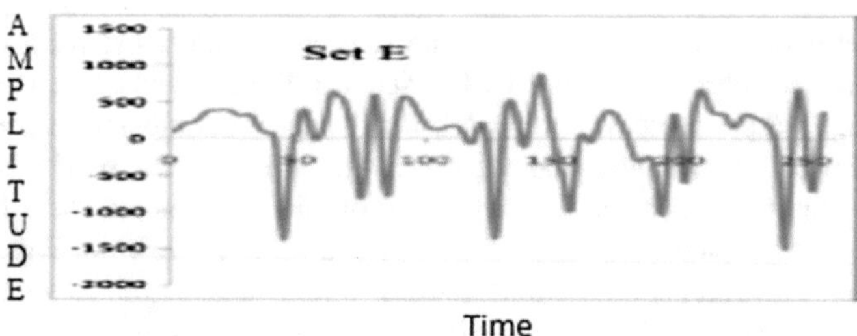

Fig. 5 Set E after downsampling

The extracted statistical features derived from the EEG signals are as follows:

(1) Maximum coefficients in each sub-band.
(2) Minimum coefficients in each sub-band.
(3) Mean of coefficients in each sub-band.
(4) Standard deviation for each sub-band.

The results of extracted features by using EWT for two sets are shown in Tables 1 and 2. The outcome of the classifier which is determined from the taken data is extracted in the statistical features is a J48 decision tree with 0.25 as confidence factor. Henceforth, the processed EWT wavelet coefficients of point of

Table 1 Features extracted using "EWT" for set A and set E

Dataset	Features extracted	EWT
Set A	Maximum coefficient	57.2027
	Minimum coefficient	−75.92
	Mean	−0.5937
	Standard deviation	8.0126
Set B	Maximum coefficient	−71.25
	Minimum coefficient	58.6354
	Mean	−0.4996
	Standard deviation	8.7565

Table 2 Performance comparison of KNN and J48 classifier

Dataset	Performance metrics	KNN	J48
Set A	Mean absolute error	0.4902	0.4788
	Root mean square error	0.6924	0.5183
	Classification accuracy	94	96.2
Set B	Mean absolute error	0.5293	0.5087
	Root mean square error	0.7201	0.5122
	Classification accuracy	94.2%	96.2%

interest and estimation of the EEG signs are utilized as the element information relates to the signs. From the decomposed EWT wavelets the parameters like maximum coefficient, minimum coefficient, mean and standard deviation are extracted. The extracted values were displayed in Table 1. The classification of KNN classifier depends on the category of the mean absolute error (MAE), root mean square error (RMSE), and classification accuracy (CA) of the taken data, and the results were displayed in Table 2.

4 Conclusion

The signals from several human subjects are collected. The collected EEG signals were identified to be normal or abnormal, i.e., the true label of the signals collected. The preprocessing is applied to the input signal. The input signal is decomposed using wavelets, and the filtering is applied. Features are extracted from the signals based on the EWT algorithm. The features were extracted for all the signals in the database. The extracted test signal features and the train signals are sending through the True labels passed into the classifier. The classifier finds whether the input signal is normal or abnormal. The proposed EWT feature with J48 algorithm shows better performance compared to the KNN classifier.

References

1. Wolpaw, J.R., Birbaumer, N., McFarland, D.J., Pfurtscheller, G., Vaughan, T.M.: Brain-computer interfaces for communication and control. Clin. Neurophysiol. **113**, 767–791 (2002)
2. Fisher, R.S., van Emde Boas, W., Blume, W., Elger, C., Genton, P., Lee, P., et al.: Epileptic seizures and epilepsy: definitions proposed by the International League against Epilepsy (ILAE) and the International Bureau for Epilepsy (IBE). Epilepsia **46**(4), 470–472 (2005)
3. Sornmo, L., Laguna, P.: Bioelectrical Signal Processing in Cardiac and Neurological Applications, 1st edn. Elsevier Store, Amsterdam (2005)
4. Cunningham, P., Delany, S.J.: K-Nearest neighbor classifiers. Technical Report. D UCD-CSI-2007-4, Artificial Intelligence Group, Dublin

5. Polat, K., Günes, S.: Classification of epileptic from EEG using a hybrid system based on decision tree classifier and fast Fourier transform. Appl. Math. Comput. **187**(2), 1017–1026 (2007)
6. Fisher, R.S., van Emde Boas, W., Blume, W., Elger, C., Genton, P., Lee, P., et al.: Epileptic seizures and epilepsy: definitions proposed by the International League against Epilepsy (ILAE) and the International Bureau for Epilepsy (IBE). Epilepsia **46**(4), 470–472 (2005)
7. Andrzejak, R.G., Lehnertz, K., Mormann, F., et al.: Indications of nonlinear deterministic and finite dimensional structures in time series of brain electrical activity: dependence on recording region and brain state. Phys. Rev. E **64**(6), 061907 (2001)
8. Lekshmi, S.S., Selvam, V., Pallikonda Rajasekaran, M.: EEG signal classification using principal component analysis and wavelet transform with neural network. In: Proceedings of International Conference on Communication and Signal Processing, India, 3–5 April 2014
9. Bhuvaneswaria, P., Satheesh Kumar, J.: Influence of linear features in nonlinear electroencephalography (EEG) signals. Elsevier Procedia Comput. Sci. **47**, 229–236 (2015)
10. Hazarika, N., Chen, J.Z., Tsoi, A.C., Sergejew, A.: Classification of EEG signals using the wavelet transform. In: Proceedings of 13th International Conference on Digital Signal Processing, India, vol. 59(1), pp. 89–92 (1997)

Gallbladder Shape Estimation Using Tree-Seed Optimization Tuned Radial Basis Function Network for Assessment of Acute Cholecystitis

V. Muneeswaran and M. Pallikonda Rajasekaran

Abstract In this paper, computerized scheme for automatic volume estimation of inflamed gallbladder in ultrasound images has been investigated. Diagnosis of acute cholecystitis at an early stage is an arduous task as the difference between normal shape and inflamed gallbladder shape cannot be visualized in ultrasound images. This paper comes out with an unsupervised machine learning algorithm—tree-seed optimization algorithm tuned radial basis function network for segmentation of gallbladder in ultrasound images. Tree-seed optimization algorithm, which optimizes function and parameters in real values, is a population-based stochastic search algorithm. Prior to the classification, speckle reduction and feature extraction process were successfully used. These features are then used in classification process to define the gallbladder and non-gallbladder regions. The proposed optimized classifier system is evaluated in real-time clinical datasets with cholecystitis and cholelithiasis. The inherent differentiation of the proposed intelligent classifier is analyzed using standard evaluation parameters. Comparison with expert decisions provides further evidence that the optimally tuned radial basis function network has important implications for diagnosis of acute cholecystitis.

Keywords Tree-seed optimization · Gallbladder segmentation
Radial basis function network

V. Muneeswaran (✉)
Department of Instrumentation and Control Engineering, Kalasalingam University,
Krishnankoil 626126, Tamil Nadu, India
e-mail: munees.klu@gmail.com

M. Pallikonda Rajasekaran
Department of Electronics and Communication Engineering, Kalasalingam University,
Krishnankoil 626126, Tamil Nadu, India
e-mail: m.p.raja@klu.ac.in

V. Bhateja et al. (eds.), *Intelligent Engineering Informatics*, Advances in Intelligent Systems and Computing 695, https://doi.org/10.1007/978-981-10-7566-7_24

1 Introduction

Ultrasound imaging modality is a conventional medical diagnostic imaging technique that is used to visualize the structure of soft tissue structures like gallbladder. Noteworthy advancements in the identification of acute cholecystitis have occurred with advancements in imaging technology. The greater part of cases of acute calculous cholecystitis can consistently be established in an uncomplicated fashion with the use of ultrasound or computed tomography. Less recurrently, MRI and/or Tc-HIDA scans may be used to aid in the analysis. It should be well thought-out, the first imaging alternative for all assumed cases of acute cholecystitis [10]. However, in clinical perspective, radiologist analyzes the abdominal ultrasound image for disorders including the presence of gallstones and cholecystitis which may likely to contain inaccuracy to some extent. Certain patient factors and clinical settings can significantly confound and delay the diagnosis of acute cholecystitis, increasing the risk of complication. Thus, the extraction of gallbladder shape using supporting software gives light to the way of computer-aided diagnosis that results in earlier detection of lesions and sludges with reduced inaccuracy. The four modes of examination in ultrasound are amplitude mode, motion mode, brightness mode, and Doppler mode. Ultrasound brightness mode scan becomes the useful diagnostic imaging modality as it converts the amplitude of ultrasound signals reflected from the patient body to a grayscale image. As the ultrasound scanners give the output of distinguishable 256 shades of gray, accurate echogenicity is possible to visualize the pathological lesions. It is also capable of displaying the 2D cross section of internal organ using its acoustic impedance.

Radial basis function networks are considered to be a powerful tool for classification as they tend to have an inherent ability of learning complex and nonlinear classes. Though RBFN performs to a satisfactory level during the course of classification, conventional classifiers based on the static radial basis function network have not performed well. The steepest descent method used for training RBFN has some deficiencies. In particular, the steepest descent method gets trapped into local minima, thus making it to greatly depend on the initial settings. In order to overcome these disadvantages, optimized radial basis function network is proposed which designs RBF network in an optimal way using tree-seed optimization algorithm.

The proposed computer-aided diagnosis (CAD) system uses tree-seed optimization algorithm for classifying gallbladder region and non-gallbladder region. It consists of three major stages: preprocessing of ultrasound images, feature extraction, and classification of true gallbladder region. A tuned RBFN is used as tool for classification of discriminative input features. A large number of performance evaluation metrics have been used to evaluate the performance of the proposed system.

2 Literature Review

2.1 Related Works

Few notable works in the context of gallbladder segmentation are discussed in this section. Ciecholewski [5] suggested an algorithm for shape extraction of gallbladder in ultrasound images using active contour model inhibiting motion equation model. Area error rate produced by ACM was about 22 and 19% for lithiasis and folds, respectively, which could further be elevated. Boundary detection using probabilistic data association filter (PDAF) with interacting multiple model (IMM) [1] recommended by P. Abolmaesumi and M. R. Sirouspour is pertinent to handle more severe cases of echo dropouts during the process of shape extraction in ultrasound images. But the selection of proper values for the process and measurement noise covariances necessitates manual intervention in the shape extraction process.

There are many works in the literature for solving the problem of RBFN training and its structure optimization. Most of them have been used in optimizing the center, width, and weight of the RBFN. A framework for using trained RBFN in practical application has been discussed by Li et al. [12]. This approach gives greater flexibility in designing RBFN using modified ant colony optimization, but still a great deal of disagreement between prediction errors remains unsolved. Fathi and Montazer [7] have proposed improvement in RBF learning technique using particle swarm optimization algorithm. These algorithms have the limitation of utilizing disproportionate run length time for exploring better network parameter. A combination of orthogonal least square algorithm and genetic algorithm (OLS–GA) for RBFN training was suggested by Chang [3]. The usage of this algorithm is confined to wind power forecasting and requires large computational time as it involves hybrid algorithms. Shen et al. [18] introduced artificial fish swarm algorithm for fine tuning of RBFN network parameters. This methodology encompasses artificial fish swarm algorithm to detect optimal center, width, and weight values of RBFN. This parameter selection technique requires larger time for processing as it involves three biological behaviors of fish swarm. Ayala and Coelho et al. [2] suggested a system identification model using modified RBFN, which is associated with the training of RBFN using cuckoo search optimization. Architectural parameters of RBFN produced by this optimization technique require enhancement. Gholam Ali Montazer and Davar Giveki rendered a new approach of training RBFN using particle swarm optimization improved with orthogonal least square algorithm for image [13].

2.2 Tree-Seed Optimization

Tree-seed optimization (TSA) algorithm deals with three main processes such as random tree location, searching seeds, and selection of best solution. Random tree location refers to location of trees within the search space, searching with seeds elaborates the location search operation by a seed based on the information from control parameter search tendency (ST). Selection of best solution is inhibited based on the objective function of the problem. Finally, substitution process takes place by the replacement of trees. The sequential steps adopted in TSA algorithm are given as follows [8, 9]:

1. Instantiating the number of population (N), control parameter(ST), termination condition X, and sequence step x.
2. **Random trees location**
   Inference to the value of minimum and maximum values assumed for a given search space, generate initial location of trees randomly using Eq. (1)

$$T = L + rand(H - L) \tag{1}$$

   The random trees were also checked for fitness using the problem-specific objective function. A sample best solution for minimization process could be represented as

$$T_{best} = minf(T_i); i = 1, 2, 3, N \tag{2}$$

3. **Searching seeds**
   Searching seeds for a tree is formulated using a control parameter ST in the range [0 1]. Based on the rand value chosen in calculation of initial trees, the seed for the trees can be calculated using one of the two following updating rules

$$S_{((\forall rand<ST))} = T + (T_{best} - T_r) \tag{3}$$

$$S_{((\forall rand>ST))} = T + (T - T_r) \tag{4}$$

4. **Comparison of seeds with tree**
   The seed values are compared to the trees for better optimization performance.
5. Test for termination: If $x < X$, go to Step 3. Continue the seed searching procedure till the problem gets optimal solution.

3 Optimized RBF Based on Tree-Seed Optimization

Radial basis function networks are tri-layered structure that offers better performance in real-time classification problems. The activation function in RBFN differs from the other neural networks, and in the hidden layer, they use input stimulus-responsive functions [14]. As the name suggests, it uses $\phi(x)$ as the radial basis function. In most of the cases, Gaussian function is used and it is expressed in Eq. (5) below

$$\phi(x) = \exp[-\frac{(\|x - C_i\|)^2}{(2\sigma_i)^2}] \tag{5}$$

where i denotes the hidden node and it ranges from $i = 1, 2, \ldots M$.

By choosing suitable values for center c_i, width σ_i, and weight w_{ij} between hidden node i and output node j, the output of the RBFN $\hat{y}$ can be calculated as

$$\hat{y}_{ij} = \sum_{j=1}^{N} \sum_{i=1}^{M} w_{ij} * \exp[-\frac{(\|x - C_i\|)^2}{(2\sigma_i)^2}] \tag{6}$$

As discussed in Sect. 2.1, optimization techniques can be used in tuning the parameters of RBFN. The procedural steps for tuning RBFN using tree-seed algorithm are given below

1. Initialize the basic RBFN by building the input vectors and application of target vectors.
2. Determine the error goal *err* with all the available random network parameters. With proper determination of error goal, the network parameters of RBFN can be optimized.
3. Use TSA optimization for optimal selection of network parameters of RBFN. This process gets initialized by considering the network parameters as trees. Initialize the number of iteration sequence x (total number of iterations are x). With known search space, the initial random trees were generated using Eq. (1).
4. Iterate over the seed searching process for all random tree value produced with exploitation of control parameter (*ST*) and update the seed values using Eq. (3) or Eq. (4)
5. Set $x = x + 1$ and assign the parameters updated from Eq. (3) or Eq. (4) as the parameters of RBFN if it tends to minimize the *err* value as in Eq. (11). Thus the seed value substitutes tree value, the best solution from TSA will be noted as T_{best} as in Eq. (2).
6. If $x < X$, start from step 3, else take the tree T_{best} as the final solution of the optimization problem.

The above-said procedures indicate the reassessment of RBFN network parameters using TSA algorithm. Thus, TSA-tuned RBF network has paved the way to produce efficient network parameters for gallbladder shape extraction.

4 Proposed Model for Gallbladder Shape Estimation

4.1 Preprocessing

Ultrasound image displays a coarse-grained appearance, which disguises the minute changes of gray levels and makes the process of intolerance between original gray level and noise a tiresome process. Thus in most cases, the preliminary task for ultrasound image processing especially in medical diagnosis and therapeutic decision making is speckle reduction, which makes the sequential steps simplified [15, 16].

A modification in the strategy followed by Latifoglu [11] for speckle reduction in ultrasound images is adopted by replacing the optimization algorithm as shown in Fig. 1. The role of optimization in this speckle reduction process is to find optimal filter coefficients for a 2D finite impulse response filter. Replacement of artificial bee colony (ABC) algorithm with TSA provides a useful estimate of filter coefficients for speckle reduction. The speckle reduction process involves the use of the simulated test image of size 128 ∗ 128 pixels. This simulated test image was corrupted with speckle noise at a desirable level. Using the mean square error between the filter output and simulated test image as in Eq. (7) as fitness function, at each iteration of TSA, the filter coefficients are updated. Fig. 2 shows the filter coefficients produced by ABC algorithm and tree-seed optimization algorithm.

$$MSE = \frac{1}{KL} \sum_{k=0}^{K-1} \sum_{l=0}^{L-1} [I_{org}(k,l) - I_{Noisy}(k,l)]^2 \quad (7)$$

These filter coefficients produced by tree-seed optimization will be used for speckle reduction in ultrasound images

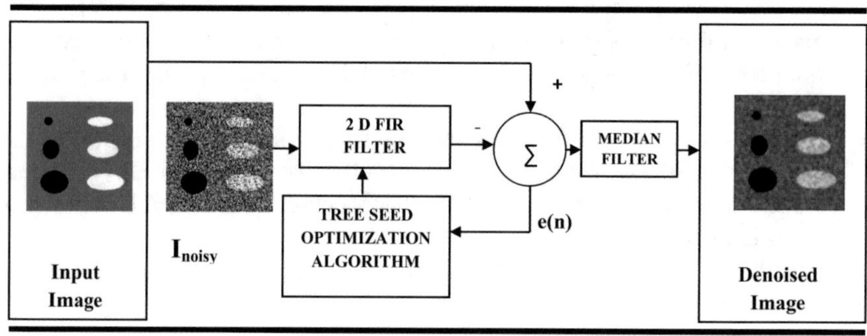

Fig. 1 Block diagram for preprocessing of ultrasound images

(a)

a_{00} 0.1032	a_{01} 0.1047	a_{02} 0.1012
a_{10} 0.1063	a_{11} 0.1110	a_{12} 0.1059
a_{20} 0.1006	a_{21} 0.1043	a_{22} 0.1024

(b)

a_{00} 0.1085	a_{01} 0.1143	a_{02} 0.1075
a_{10} 0.1042	a_{11} 0.1152	a_{12} 0.1137
a_{20} 0.1097	a_{21} 0.1101	a_{22} 0.1069

Fig. 2 **a** Filter coefficients produced by ABC algorithm; **b** filter coefficients produced by tree-seed optimization

4.2 Feature Extraction

Point features can be used for the extraction of region of gallbladder using pixels, which in turn will reflect the gallbladder length and width. Feature extraction based on the gallbladder region is the extraction of set of image features from the region of interest (ROI). Features to be taken from the ROI include major axis length, minor axis length, and area. These features will greatly help in differentiating pixels into classes such as gallbladder region and non-gallbladder region. For reducing the computational complexity, only about a hundred dominant pixel values were considered for feature extraction from both gallbladder region and non-gallbladder region. Prior to the feature extraction process, a single-level thresholding process is applied to the speckle-reduced image for selecting the ROI. A process of image enhancement becomes necessary in this case as the edges of the gallbladder region are non-uniform and unclear. Histogram equalization is almost preferred in such cases [6]. Histogram equalization transforms the gray levels in an image using the transformation given in Eq. (8)

$$b = T(a) \tag{8}$$

Given an image $f_{(i,j)}$, the suitable histogram-equalized image $g_{(i,j)}$ can be calculated as follows:

$$g_{(i,j)} = floor((L-1)\sum_{n=0}^{f_{(i,j)}} P_n) \tag{9}$$

where P_n represents the transformation applied to all values of $f_{(i,j)}$, and L represents the number of gray levels in the image.

An order filtering process is adopted following the histogram equalization process. Adopting the strategy followed in Hough transform [4], local maxima points in all directions for pixel in ROI are obtained. The (x, y) coordinates of all local maxima point are accumulated for all eight directions. An approximate measure of the center of all the eight local maxima points can be found by sum and average process. Then the Euclidean distance d between all the successive (x, y) coordinates the center can be obtained using the Eq. (10)

$$d = \sqrt{((x_1 - x_2)^2 + (y_1 - y_2)^2)} \tag{10}$$

where (x_1, x_2) and (y_1, y_2) represent coordinates of center of all local maxima points and individual local maxima points of pixels in successive directions. Thus, connecting all the points will give a shape measure such as the perimeter of the gallbladder shape. Thus, the feature extraction process will come out with eight local maxima points and a shape measure.

4.3 Gall Region Detection

In the present study, the concurrent ability of the radial basis function network in nonlinear mapping of gallbladder region and the global optimization capability of TSA are fully used to forecast the gallbladder structure. During the training process, the feature vector extracted from 30% of the total preprocessed images according to the rule of thumb was taken for training set. Based on the optimal parameters determined and feature vectors as input, an estimate of the gallbladder shape is found iteratively by the evaluation of fitness function. The fitness function for the above process is estimated using Eq. (11)

$$E(b,a) = \frac{1}{2}\{\frac{1}{m}\sum_{i=1}^{m} d(a_i,b) + \frac{1}{n}\sum_{j=1}^{n} d(a,b_j)\} \tag{11}$$

where a and b are two sets of contour points, and contour points in set a are obtained from the contour of a gold standard image. In the context of image segmentation, manually segmented images are usually referred as gold standard images. The set of contour points of the extracted shape resulting from the proposed method is stored in set b. Every contour point a_i, b_j has specific coordinates (x, y). Due to minimum loss during the training process, a post-processing step becomes inevitable. Post-processing includes holes filling process. Thus, the extracted gallbladder shape is found which will be greatly helpful in accessing inflammation in gallbladder.

5 Interpretation and Validation of Results

The dataset for the implementation of the proposed methodology is composed of 180 images which were collected from private diagnostic center *VijayScans* in Rajapalayam city of Tamil Nadu, India. Images were obtained from patients particularly affected with cholelithiasis and both acalculous and calculous cholecystites. Extraction of the accurate shape in the gallbladder becomes difficult due to the presence of calcification in it. 1.a and 2.a of Fig. 3 shows the 2 input images from the dataset. Gold standard images obtained from segmenting the gallbladder region by an expert radiologist and are named as ground truth images as shown in 1.b and 2.b of Fig. 3 are used for comparison. It is visually evident that the preprocessing process removes speckle noise and acoustic shadows in the input images and can be seen in 1.c and 2.c of Fig. 3. Features taken from the preprocessed image will be effective mentor for the segmentation process. A final segmented result of the gallbladder region without superfluous information present in the input image is obtained after the post-processing process and is shown in 1.d and Fig. 5.d of Fig. 3. Values of performance indicators such as sensitivity, specificity, and accuracy for a set of five input images taken from the database are shown in Table 1. Methodology proposed by Shao [17] was also taken up into consideration and was modified for

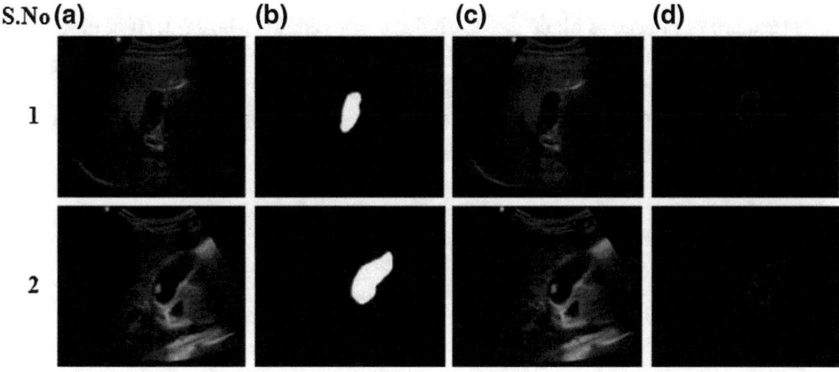

Fig. 3 **a** Input images; **b** corresponding ground truth images; **c** preprocessed images; **d** segmented images using TSA-based RBFN

Table 1 Comparison of evaluation metrics for gallbladder segmentation

	PSO-based RBFN			TSA-based RBFN		
Images	Metrics			Metrics		
	Sensitivity	Specificity	Accuracy	Sensitivity	Specificity	Accuracy
Image 1	0.721381	0.752910	0.810210	0.856482	0.835712	0.899210
Image 2	0.790826	0.783556	0.781421	0.878541	0.834576	0.832475
Image 3	0.744201	0.721067	0.761257	0.814544	0.875315	0.895246
Image 4	0.772813	0.751461	0.801702	0.857632	0.819251	0.823547
Image 5	0.781257	0.765728	0.772130	0.847620	0.803028	0.825860

gallbladder segmentation for comparative analysis and to demonstrate the efficacy of the proposed segmentation. Higher values of sensitivity, specificity, and accuracy stand as a proof for the effectiveness of the proposed methodology, and it is reflected in Table 1. The mean value of the area occupied by gallbladder found using the proposed methodology was about 19700 pixels which were very closer to the values of mean area occupied by gallbladder calculated from the ground truth image segmented using experts.

6 Conclusion and Future Work

In this paper, a classification based on the computer-aided decision support scheme is followed for detecting the gallbladder inflammation. The recommended TSA-based RBFN is capable of processing abdominal ultrasound images. When compared with PSO-based RBFN technique, TSA-based RBFN reduces the difficulty in discrimination process between gallbladder and non-gallbladder regions to a greater extent.

The presence of sludge is identified with the proposed TSA-based RBFN algorithm; it also rules out PSO-based RBFN training algorithm; and it is justified using both performance evaluation procedures in Table 1 and also in 1.d and 2.d of Fig. 3. Thus, the proposed algorithm proves to be an efficient technique for gallbladder region segmentation, and it can be a technique for clinical significance.

Acknowledgements The authors thank Vijay Scans, Rajapalayam, Tamil Nadu, for supporting the research by providing ultrasound images and necessary patient information. Also we thank the Department of ECE, Kalasalingam University, Tamil Nadu, India, for permitting to use the computational facilities available in Centre for Research in Signal Processing and VLSI Design which was set up with the support of the Department of Science and Technology (DST), New Delhi, under FIST Program in 2013 (Reference No: SR/FST/ETI-336/2013 dated November 2013).

References

1. Abolmaesumi, P., Sirouspour, M.R.: An interacting multiple model probabilistic data association filter for cavity boundary extraction from ultrasound images. IEEE Trans. Med. Imaging **23**(6), 772–784 (2004)
2. Ayala, H.V.H., Coelho, L.: Multiobjective Cuckoo Search Applied to Radial Basis Function Neural Networks Training for System Identification. IFAC Proc. **47**(3), 2539–2544 (2014)
3. Chang, W.Y.: An RBF neural network combined with OLS algorithm and genetic algorithm for short-term wind power forecasting. J. Appl. Math., 9 (2013). 10.1155/2013/971389. Article ID 971389
4. Chu, A., Sehgal, C.M., Greenleaf, J.F.: Use of gray value distribution of run lengths for texture analysis. Pattern Recogn. Lett. **11**(6), 415–419 (1990)
5. Ciecholewski, M.: Gallbladder boundary segmentation from ultrasound images using active contour model. In: Fyfe, C., Tino, P., Charles, D., Garcia-Osorio, C., Yin, H. (eds.) Intelligent Data Engineering and Automated Learning IDEAL 2010. IDEAL 2010. LNCS, vol. 6283 , pp. 63–69. Springer, Berlin, Heidelberg (2010)
6. Das, N., Sarkar, R., Basu, S., Kundu, M., Nasipuri, M., Basu, D.K.: A genetic algorithm based region sampling for selection of local features in handwritten digit recognition application. Appl. Soft Computing. **12**(5), 1592–1606 (2012)
7. Fathi, V., Montazer, G.A.: An improvement in RBF learning algorithm based on PSO for real time applications. Neurocomputing **111**, 169–176 (2013)
8. Kiran, M.S.: TSA: Tree-seed algorithm for continuous optimization. Expert Syst. Appl. **42**(19), 6686–6698 (2015)
9. Kiran, M.S.: An Implementation of Tree Seed Algorithm (TSA) for Constrained Optimization. In: Lavangnananda, K., Phon-Amnuaisuk, S., Engchuan, W., Chan, J.H. (eds.) IES 2015. PALO, vol. 5, pp. 189–197. Springer, Cham (2016)
10. LaRocca, C.J., Hoskuldsson, T., Beilman, G.J.: The use of imaging in gallbladder disease. In: Eachempati, S., Reed II, R. (eds.) Acute Cholecystitis, pp. 41–53. Springer, Cham (2015)
11. Latifoglu, F.: A novel approach to speckle noise filtering based on Artificial Bee Colony algorithm: An ultrasound image application. Comput. Methods Program. Biomed. **111**(3), 561–569 (2013)
12. Li, J., Liu, X., Jiang, H., Xiao, Y.: Melt index prediction by adaptively aggregated RBF neural networks trained with novel ACO algorithm. J. Appl. Polym. Sci. **125**, 943–951 (2012). 10.1002/app.35688
13. Montazer, G.A., Giveki, D.: An improved radial basis function neural network for object image retrieval. Neurocomputing **168**, 221–233 (2015)

14. Muneeswaran, V., Pallikonda Rajasekaran, M.: Performance evaluation of radial basis function networks based on tree seed algorithm. In: Proceeding of the 2016 International Conference of Circuit Power and Computing Technologies, pp. 1-4. IEEE Explore (2016)
15. Muneeswaran, V., Pallikonda Rajasekaran, M.: Analysis of particle swarm optimization based 2D FIR filter for reduction of additive and multiplicative noise in images. In: Arumugam, S., Bagga, J., Beineke, L., Panda, B. (eds) Theoretical Computer Science and Discrete Mathematics. ICTCSDM 2016. Lecture Notes in Computer Science, vol. 10398, Springer, Cham (2017)
16. Muneeswaran, V., Pallikonda Rajasekaran, M.: Beltrami-regularized denoising filter based on tree seed optimization algorithm: an ultrasound image application. In: Satapathy, S., Joshi, A. (eds.) Information and Communication Technology for Intelligent Systems (ICTIS 2017) - Vol. 1. ICTIS 2017. Smart Innovation, Systems and Technologies, p. 83, Springer, Cham (2018)
17. Shao, Y., Chen, Q., Jiang, H.: RBF neural network based on particle swarm optimization. In: Zhang, L., Lu, B.L., Kwok, J. (eds.) Advances in Neural Networks—ISNN 2010. LNCS, vol. 6063, pp. 169–176. Springer, Berlin, Heidelberg (2010)
18. Shen, W., Guo, X., Wu, C., Wu, D.: Forecasting stock indices using radial basis function neural networks optimized by artificial fish swarm algorithm. Knowl. Based Syst. **24**(3), 378–385 (2011)

Materialized View Selection Using Backtracking Search Optimization Algorithm

Anjana Gosain and Kavita Sachdeva

Abstract Selecting the materialized views optimally is very important in designing a data warehouse and is NP-hard problem. Various evolutionary algorithms exist in literature for the appropriate selection of materialized views. In this paper, we have examined the application of backtracking search optimization algorithm (BSA), for selecting the materialized views in data warehouse. According to our experiments, the results obtained by our proposed backtracking search optimization-based materialized view selection algorithm (BSMVSA) are superior to those found using particle swarm optimization and genetic algorithm. The solution obtained by BSMVSA greatly reduces the total cost within the storage constraint.

Keywords Aggregation · Dimension · Measure · Heuristic Evolutionary

1 Introduction

A data warehouse (DW) is a data repository that provides an integrated environment for reporting, analyzing, and supporting queries which requires complex aggregations of huge amounts of historical data [1, 2]. It is a big challenge to operate and manage such an integrated data store in a cost-effective way. Materialized views are the intermediate results stored in a data warehouse [3] that avert

A. Gosain · K. Sachdeva (✉)
University School of Information and Communications Technology, GGS Indraprastha University, New Delhi 110078, India
e-mail: kavitasachdeva4@gmail.com

A. Gosain
e-mail: anjana_gosain@hotmail.com

V. Bhateja et al. (eds.), *Intelligent Engineering Informatics*, Advances in Intelligent Systems and Computing 695, https://doi.org/10.1007/978-981-10-7566-7_25

accessing the original data sources and thus increase the efficiency of the queries posed to a data warehouse. Optimal selection of materialized views is NP-hard problem [4] and is needed to design a data warehouse effectively. Recently, lot of attention is given to solve this problem. Researchers have given various frameworks and algorithms to deal with this problem. Data cube and lattice [5–13], MVPP [14–17] and AND-OR view graphs [18–21] are various frameworks that exist in literature. Lattice framework captures dependency among aggregate views and is used in building the data cubes with multiple dimensions. MVPP is a global query processing plan for the complete query set and exploits the existence of common sub-expressions. AND-OR view graph is used to express all the possible execution plans for evaluating a query in the query set. Using the above frameworks, various algorithms like heuristic-based algorithms [11, 20], greedy algorithms [5, 8, 18, 22], evolutionary algorithms such as genetic algorithm [9, 10, 14, 15, 21, 23, 24] and simulated annealing [13, 16, 17]have been presented for view selection. However, heuristic-based algorithms cannot compute a perfect solution within the acceptable time due to the complex nature of problem. Greedy algorithms are highly problem-dependent and are susceptible to poor local minima, while evolutionary algorithms (EA) work on randomly selected multiple solutions simultaneously to find out the optimum most solution and can be applied on various problems. Various evolutionary algorithms like particle swarm optimization (PSO) [25, 26], bee colony optimization (BCO) [27], ant colony optimization (ACO) [28], and differential evolution (DE) [29]have been used in context of materialized view selection. Backtracking search optimization algorithm (BSA) comes under the category of evolutionary algorithm and aids in solving various optimization problems [30]. With a simple structure, it is effective, fast, and has strategies for generating trial populations by taking favor of the experiences gained from previous generations and gives BSA very powerful capabilities for exploring and exploiting the population [30]. In this paper, we have implemented BSA on lattice framework to find an optimal set of views within the space constraint, thereby minimizing the total cost of query processing. We have compared our results with genetic algorithm and PSO to prove the effectiveness of BSMVSA.

This paper is organized as follows: Definition and mathematical model of materialized view selection problem is given in Sect. 2. Section 3 gives the brief review of backtracking search optimization algorithm (BSA), while Sect. 4 presents BSA-based view selection in a stepwise manner. Experimental results under different space constraints, along with comparative performance analysis with other algorithms, are discussed in Sect. 5. Section 6 ends up with conclusions.

2 Materialized View Selection

2.1 Problem Definition

Materialized view selection [31] problem can be illustrated as follows: Given some storage space X and query set Z, the problem is to determine a set of materialized views V that reduce the total query processing cost of the selected views, under the constraint that the total space occupied by selected views, V, is less than X. View selection is an important decision in the effective design of data warehouse. We have used the lattice framework [5] in determination of materialized view selection as lattice framework captures and models dependencies among aggregate views. Group-by clause [32] characterizes a data cube (DC). A path exists between the two cubes ci and cj if there is a dependency relation $c_i \leq c_j$, exists between the two cubes c_i and c_j, implying if a query can be resolved from c_i, then it can also be resolved by using cube c_j. 2^N data cubes are possible in the lattice framework, for a fact relation having N dimensions. For example, sales transactions in a data warehouse system (taken from TPC-H star schema benchmark [33]) have three dimensions, part (P), supplier (S), and customer (C), and in fact depicts the sale of parts(P) from suppliers(S) to customers(C). This fact relation will generate $2^3 = 8$ data cubes in the lattice as shown in Fig. 1.

Each level of the lattice has data cubes which are aggregated at same level of aggregation along different dimensions. The number beside each cube indicates its size (i.e., number of rows). The top cube, i.e., psc, is the base cuboid at the lowest level of aggregation. Bottom cuboid has highest level of aggregation. If a query can be directly answered from cube (*, s,*) then it can also be answered from any of its parent cubes (p, s,*), (*, s, c), or (p, s, c) by summarizing the data along some dimensions [26].

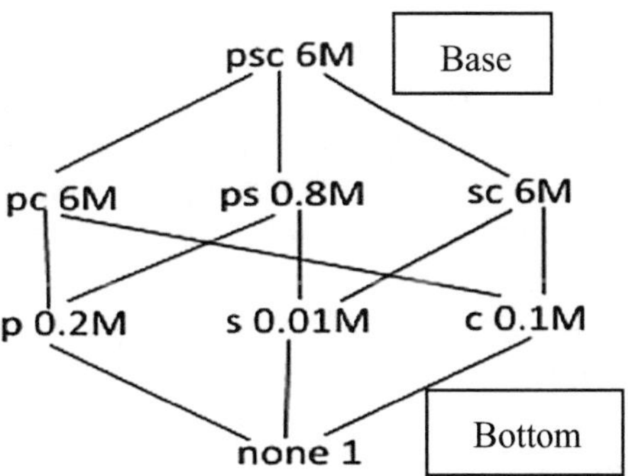

Fig. 1 Lattice framework with 8 possible data cubes [5]

2.2 *Mathematical Model of Materialized View Selection Problem*

Selection of materialized views is crucial for designing an efficient data warehouse. It aims at reducing the query response time by selecting an optimal set of materialized views within the storage space and cost considerations, to help accelerating the entire data warehouse. We have used the lattice framework, where cubes correspond to the views to be selected, whenever a query is invoked by user. Query invoking frequency corresponds to the cube invoking frequency. We have followed the linear cost model as proposed in [5], for evaluating the cost of answering query. This query answering cost is equivalent to the no. of rows to be accessed in the corresponding cube for the query. So, the materialized view selection (MVS) using lattice framework [9] can be defined as follows: Given a cube-lattice X, having a set of s cubes $B = (b_1, b_2, \ldots, b_s)$, set of y user queries $U = (u_1, u_2, \ldots, u_y)$, a set of query frequency values $W = (w_{u1}, w_{u2}, \ldots, w_{uy})$, set of update frequency values $CF = (g_{b1}, g_{b2}, \ldots, g_{bs})$ of the cubes in B, constrained by storage space T. Our objective function is to select a set of cubes (views) P to minimize the cost function defined in Eq. 1, under the space constraint $\sum_{b \in P} |B| \leq T$ [9].

$$\sum_{i=1}^{k} wui * R(ui\ , P) + \sum_{b \in P} gb * M(b\ , P) \tag{1}$$

where $R(u_i, P)$ and $M(b, P)$ depict the cost to evaluate query u_i and the maintenance cost of cube b, with reference to the set of materialized cubes (views) P.

3 Brief Review of Backtracking Search Optimization Algorithm

Backtracking search optimization algorithm (BSA) is one of the most popular evolutionary algorithms, for finding solution to global optimization problems. It has a simple structure, i.e., fast and efficient in solving various problems. BSA retains a memory in which it caches a population from previous generation and uses the past generated solutions while searching for solutions with superior fitness values. It has strategies for generating trial populations by taking favor of the experiences gained from previous generations and gives BSA very powerful capabilities for exploring and exploiting the population [30]. Random mutation and non-uniform crossover strategy of BSA produces very useful trial populations in each generation and enhances its ability to solve problems. It can be described by partitioning its functions into five major processes: initialization, selection-I, mutation, crossover, and selection-II [30]. Algorithm 1 presents the pseudocode for BSA.

Algorithm1: Pseudocode for BSA [30]

```
Input: ObjFun, N, D, maxcycle, mixrate, low1:D, up1:D
Output: globalminimum, globalminimizer
// rnd ~ U(0,1), rndn ~ N(0,1), w=rndint(.), rndint(.) ~ U(1,.) | w ε (1,2,3,……,.)
//INITIALIZATION
function bsa(ObjFun,N,D,maxcycle,low,up)
globalminimum = inf
for i from 1 to N do
for j from 1 to D do
P i,j =rnd.(upj – lowj) + lowj //Initialization of population,P
oldP i,j = rnd.(upj – lowj) + lowj //Initialization of oldP
end
fitnessPi = ObjFun(Pi) //Initial fitness values of P
end
for iteration from 1 to maxcycle do
//Selection – I
if (a < b | a,b ~ U(0,1)) then oldP := P end
oldP := permuting (oldP) // permuting arbitrary changes in positions of two
 Individuals in oldP
Generation of Trial-Population
// Mutation
mutant= P+3.rndn.(oldP - P)
//CROSSOVER
map1:N,1:D = 1 //Initial-map is an N-by-D matrix of ones.
if (c < d | c,d ~ U(0,1)) then
for i from 1 to N do
map i,u(1:⌈mixrate · rnd · D⌉) = 0 | u = permuting((1,2,3,…,D))
end
else
for i from 1 to N do, map i,randi(D) = 0 , end
end
T := mutant
for i from 1 to N do
for j from 1 to D do
if map i,j = 1 then T i,j := P i,j
end
end
// Boundary Control Mechanism
for i from 1 to N do
for j from 1 to D do
if (T i,j < lowj) or (T i,j > upj) then
T i,j = rnd.(upj – lowj) + lowj
end
end
end
end
// SELECTION-II
fitnessT = ObjFnc (T)
for i from 1 to N do
if fitnessTi < fitnessPi then
fitnessPi := fitnessTi
Pi := Ti
end
end
fitnessPbest = min(fitnessP) | best ε {1,2,3,…,N}
if fitnessPbest < globalminimum then
globalminimum := fitnessPbest
globalminimizer := Pbest
// Export globalminimum and globalminimizer
end
end
```

4 BSA-Based Materialized View Selection (BSMVSA)

The steps of the BSA-based algorithm for materialized view selection are elaborated below.

Step1: Population initialization: In this step, a set of views V is created constrained by the rule that the total space occupied by V is less than S as shown in (2) and the objective function, i.e., cost function of (1) is defined as

$$V_{ab} \sim U(x, y, z) \tag{2}$$

where b = 1, 2, … z, a = 1, 2, …, y, x represents the number of cubes (views), z corresponds to the problem dimension, and y represents the number of user queries. Each V_a is a target individual in the view set V. U is the uniform distribution. Fitness function, of the initial population (set of views V), fitness V_a is initialized as the objective function.

Step2: Selection-I: Selection-I of BSMVSA gives the historical population oldV according to (3). It is used for directing the search toward the set of

$$oldV_{ab} \sim U(x, y, z) \tag{3}$$

The historical population oldV can be redefined at the start of each iteration by the 'if-then' rule shown in (4):

$$\text{if } m < n \text{ then } oldV := V | m, n \sim U(0, 1), \tag{4}$$

After determining oldV, (5) is used to alter the order of views in oldV randomly.

$$oldV := permuting(oldV) \tag{5}$$

Step3: Mutation: BSMVSA's mutation process gives the basic form of trial set of views,M using (6)

$$M = V + R.\,(oldV - V) \tag{6}$$

In this, R manages the amplitude of (oldV − V). The value of R = 3.rndn.

Step4: Crossover: During the crossover, BSMVSA generates the final form of the trial set of views S with the help of two steps as shown in Algorithm1 (Lines 15–21). The boundary control mechanism in (Algorithm1, lines 29–34) is used to reconstruct the individuals.

Step5: Selection-II: In this stage, the trial set of views S_i's having better fitness values than the analogous set of views V_i's are used to update the V_i's.

5 Experimental Results

We have implemented BSMVSA using MATLAB and conducted experiments by running it over standard test data sets of TPC-H star schema benchmark [33]. To test and validate the effectiveness of BSMVSA, we have compared it with PSO and genetic algorithm. The main parameters of BSMVSA are as follows: population size k = 50, problem dimension d = 3, and mix rate = .5. We have calculated the cost function on varying the dimensions of lattice, query invoking frequency, and cube invoking frequency to show the effectiveness of BSA for materialized view selection.

5.1 Considering Different Space Constraints

If unlimited space is provided to store materialized views, then all the cubes can be materialized to attain minimum query processing cost. But providing unlimited space is infeasible, so we considered different cases of space constraints as 5, 10, 15, 20, 25, 30, 35, and 40% to examine the performance of the proposed algorithm in selecting views under different space constraints. It is from Fig. 2 that aimlessly increasing the storage space cannot reduce the cost. According to the test data set, the optimal and effective storage space in terms of cost is about 20% of the total views.

5.2 On Varying the Dimensions of Lattice

We have considered three cases, i.e., by using (a) three dimensions (b) four dimensions, and (c) five dimensions, respectively. We have used the part of TPC-H benchmark [33]. For three dimensions, we chose Product p, Customer c, and

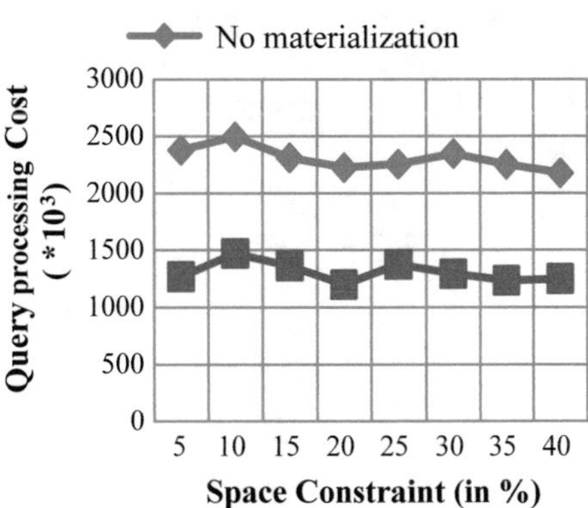

Fig. 2 Results of BSMVSA under different space constraints

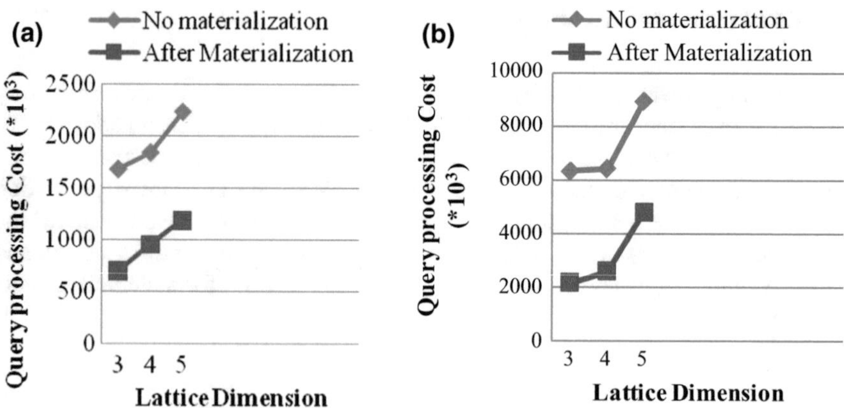

Fig. 3 **a** Results of BSMVSA on varying lattice dimensions keeping the query frequency and update frequency uniform **b** Results of BSMVSA on varying lattice dimensions with random query frequency and update frequency

Supplier s from the test set. For the fourth dimension, we included Time t dimension from the benchmark. For considering five dimensions, we added an additional dimension, Location l. The results of the experiments on changing the number of dimensions of lattice are depicted in Fig. 3a, b with uniform frequency sets and random frequency sets, respectively. It has been observed that on an exponential rise in the number of views, the processing cost of query increases linearly.

5.3 *On Increasing the Number of User Queries*

To examine the performance distribution of query processing cost achieved, we conducted an experiment on increasing the user queries. Keeping the query invoking frequency and cube invoking frequency uniform or random, the difference between query processing cost without materialization and query processing cost with materialization decreases on increasing the number of user queries as shown in Fig. 4a, b. This apparently depicts that our proposed algorithm, BSMVSA is scalable.

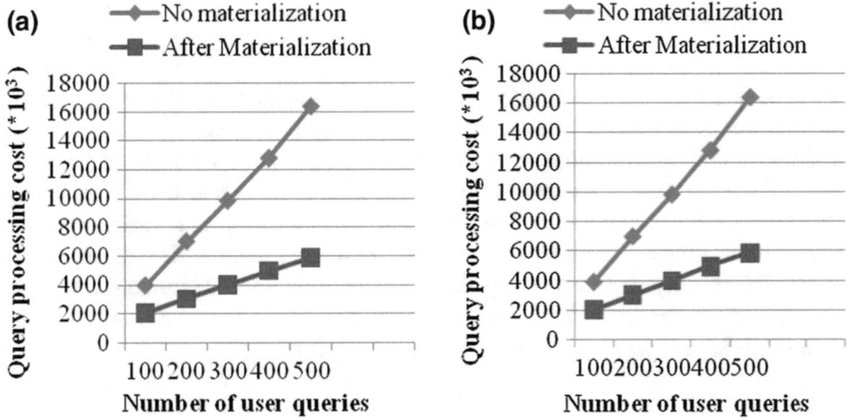

Fig. 4 **a** Number of user queries versus query processing cost (keeping query frequency and update frequency uniform) **b** Number of user queries versus query processing cost (with random query frequency and update frequency)

5.4 Comparison with PSO and Genetic Algorithm

BSMVSA yields much better results than PSO and genetic algorithm, inspite of storage and invoking frequency of data set used. Thus, BSMVSA is a better option than PSO and GA in selecting materialized views with lower query processing cost (Fig. 5).

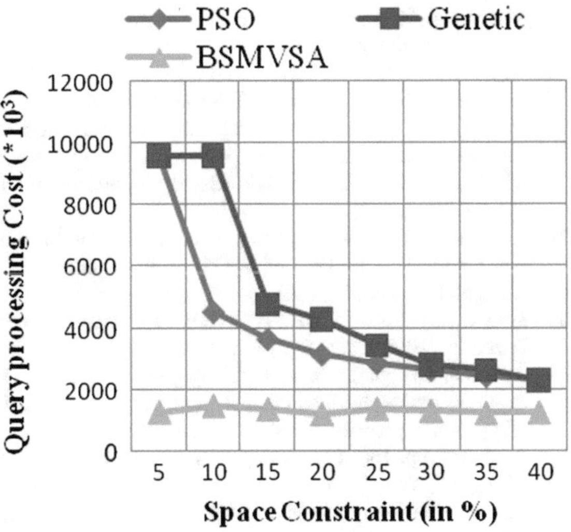

Fig. 5 Comparative results using genetic, PSO, and BSMVSA

6 Conclusion

In this study, we have implemented BSA for selection of materialized views using lattice framework. Experiments were conducted using TPC-H benchmark data set on different lattice dimensions, on different frequency sets, and considering different cases of storage space. According to the experimental results, BSMVSA always generates a superior solution. The total cost of processing and maintaining the queries approaches a minimum when space is approximately 20% of size of all views. This clearly shows that any further increase beyond this amount does not notably reduce the cost. To prove its effectiveness, over other view selection methods, it is compared with PSO and genetic algorithm.

References

1. Han, J., Kamber, M.: Data Mining: Concepts and Techniques. Morgan Kaufman, San Francisco, CA, USA (2001)
2. Morse, S., Isaac, D.: Parallel Systems in the Data Warehouse. Prentice Hall, Upper saddle River, NJ, USA (1998)
3. Jain, H., Gosain, A.: A comprehensive study of view maintenance approaches in data warehousing evolution. ACM SIGSOFT Soft. Eng. Notes **37**(5) (2012)
4. Gupta, H., Mumick, I.S.: Selection of views to materialize under a maintenance cost constraint. In: Proceedings of the 7th International Conference on Database Theory, pp. 453–470. Springer (1999)
5. Harinarayan, V., Rajaraman, A., Ullman, J.D.: Implementing data cubes efficiently. In: Proceedings of the 1996 ACM SIGMOD International Conference on Management of Data, Montreal, Que, Canada, pp. 205–216 (1996)
6. Yang, D.L., Huang, M.L., Hung, M.C.: Efficient utilization of materialized views in a data warehouse. In: Advances in Knowledge Discovery and Data Mining, pp. 393–404. Springer, Berlin, Heidelberg (2002)
7. Yu, X., J., Yao, X., Choi, C.-H., Gou, G.: Materialized view selection as constrained evolutionary optimization. IEEE Trans. Syst. Man Cybern. Part C: Appl. Rev. **33**(4) (2003)
8. Vijay, K.T.V., Haider, M.: Materialized views selection for answering queries. In: Data Engineering and Management, pp. 44–51. Springer, Berlin Heidelberg (2012)
9. Lin, W.Y., Kuo, I.C.: A genetic selection algorithm for OLAP data cubes. Knowl. Inf. Syst. **6** (1), 83–102 (2004)
10. Lawrence, M.: Multiobjective genetic algorithms for materialized view selection in OLAP data warehouses. In: Proceedings of the 8th Annual Conference on Genetic and Evolutionary Computation. ACM (2006)
11. Gou, G., Xu Yu, J., Lu, H.: A*search: an efficient and flexible approach to materialized view selection. IEEE Trans. Syst. Man Cybern. Part C: Appl. Rev. **36**(3), 411–425 (2006)
12. Talebian, S.H., Sameem, A.K.: Using genetic algorithm to select materialized views subject to dual constraints. In: International Conference on Signal Processing Systems. IEEE (2009)
13. Vijay, K.T.V., Kumar, S.: Materialized view selection using simulated annealing. In: Big Data Analytics, pp. 168–179. Springer, Berlin, Heidelberg (2012)
14. Zhang, C., Yao, X., Yang, J.: An evolutionary approach to materialized views selection in a data warehouse environment. IEEE Trans. Syst. Man Cyberne. Part C: Appl. Rev. **31**(3), 282–294 (2001)

15. Horng, J.-T., Chang, Y.-J., Liu, B.-J.: Applying evolutionary algorithms to materialized view selection in a data warehouse. Soft. Comput. **7**(8), 574–581 (2003)
16. Derakhshan, R., Dehne, F., Korn, O., Stantic, B.: Simulated annealing for materialized view selection in data warehousing environment. In: Databases and Applications (2006)
17. Derakhshan, R., Dehne, F., Korn, O., Stantic, B.: Parallel simulated annealing for materialized view selection in data warehousing environments. In: Algorithms and Architectures for Parallel Processing, pp. 121–132. Springer, Berlin, Heidelberg (2008)
18. Gupta, H., Mumick, S.: Selection of views to materialize in a data warehouse. IEEE Trans. Knowl. Data Eng., 24–43 (2005)
19. Mami, I., Coletta, R., Bellahsene, Z.: Modeling view selection as a constraint satisfaction problem. In: Database and Expert Systems Applications. Springer, Berlin, Heidelberg (2011)
20. Tamiozzo, A.S., Ale, J.M.: A solution to the materialized view selection problem in data warehousing. In: XX Congreso Argentino de Ciencias de la Computación (Buenos Aires, 2014)
21. Horng, J.-T., Chang, Y.-J., Lin, B.-J., Kao, C.-Y.: Materialized view selection using genetic algorithms in a data warehouse system, evolutionary computation, 1999. In: CEC 99 Proceedings of the 1999 Congress on. vol. 3. IEEE (1999)
22. Vijay, K.T.V., Ghoshal, A.: A reduced lattice greedy algorithm for selecting materialized views. In: Information Systems, Technology and Management. pp. 6–18. Springer, Berlin, Heidelberg (2009)
23. Wang, Z., Zhang, D.: Optimal genetic view selection algorithm under space constraint. Int. J. Inf. Technol. **11**(5), 44–51 (2005)
24. Talebian, S.H., Sameem, A.K.: Using genetic algorithm to select materialized views subject to dual constraints. In: International Conference on Signal Processing Systems. IEEE (2009)
25. Sun, X., Ziqiang, W.: An efficient materialized views selection algorithm based on PSO. In: Proceeding of the International Workshop on Intelligent Systems and Applications, ISA 2009, Wuhan, China (2009)
26. Gosain, A., Heena: Materialized cube selection using particle swarm optimization algorithm. In: 7th International Conference on Communication, Computing and Virtualization, Elsevier (2016)
27. Vijay Kumar, T.V., Arun, B.: Materialized view selection using improvement based bee colony optimization. Int. J. Softw. Sci. Comput. Intell. 7(4) (2015)
28. Song, X., Gao, L.: An ant colony based algorithm for optimal selection of materialized view. In: International Conference on Intelligent Computing and Integrated Systems (ICISS) (2010)
29. Vijay Kumar, T.V., Kumar, S.: Materialized view selection using differential evolution. Int. J. Innov. Comput. Appl. **6**(2) (2014)
30. Civicioglu, P.: Backtracking search optimization algorithm for numerical optimization problems. Appl. Math. Comput. **219**, 8121–8144 (2013)
31. Gupta, H.: Selection of views to materialize in a data warehouse. In: Proceedings of the 6th International Conference on Database Theory, pp. 98–112. Springer (1997)
32. Gray, J., Layman, A., Bosworth, A., Pirahesh, H.: Data cube: a relational aggregation operator generalizing group-by, cross-tabs and subtotals. Data Min. Knowl. Discovery **1**(1), 29–53 (1997)
33. O'Neil, P.E., O'Neil, E.J., Chen, X.: The star schema benchmark (SSB). Pat (2007)

Fractal Coding for Texture, Satellite, and Gray Scale Images to Reduce Searching Time and Complexity

Sandhya Kadam and Vijay Rathod

Abstract Fractal coding techniques are time-consuming and complex. The proposed Grover's quantum search algorithm (QSA) reduces the computational complexity in searching mechanism and achieves square root speedup over classical algorithms in an unsorted database. The quantum fidelity can be calculated to reduce minimum matching error between a given range block and its corresponding domain block. The proposed system is implemented for texture, satellite, and grayscale images for different sizes of range and domain blocks. The results are compared and displayed to reduce the complexity in the searching mechanism. The comparative analysis of existing methods and proposed algorithm has been carried out using performance parameters as compression ratio (CR), computational complexity and PSNR.

Keywords Compression ratio · Computational complexity · Fractal image compression · Grover's quantum search · PSNR · Quantum representation

1 Introduction

M. Barnsley proposed fractal coding technique in 1987 [1]. FIC uses the feature of self-similar structure. It is a lossy digital image compression technique used to encode the image which reduces the storage space. The main principle of the fractal transform coding is to use redundancies in an image that can be efficiently exploited by means of block self-affine transformations. The fractal decoding process is faster as compared to other image compression methods. In order to reduce the computational cost of FIC, a lot more FIC algorithms were proposed such as wavelet

S. Kadam (✉)
Faculty of Engineering, Pacific Academy of Higher Education and Research Center, Udaipur, Rajasthan, India
e-mail: kadam.sandhya@gmail.com

V. Rathod
St. Xavier's Technological Institute, Mumbai, India

V. Bhateja et al. (eds.), *Intelligent Engineering Informatics*, Advances in Intelligent Systems and Computing 695, https://doi.org/10.1007/978-981-10-7566-7_26

transform based FIC (wavelet-FIC) using Haar wavelet [2], hybrid method [3] which combines discrete cosine transform with FIC (DCT-FIC), fuzzy systems [4], neural networks can be combined [5] with fractal. The performance of FIC can be improved based on fractal dimension [6] and different size of range and domain blocks [7] which depends on variance. The image quality of FIC can be assessed by using modified PSNR which is based on HVS characteristics [8]. Here, weighted sum of different distortions as error, structural, and edge of the image is considered to assess the quality of an image. It is efficient and simple to calculate and has high correlation with HVS. Fractal theory can also be applied to video compression [9].

Different approaches for implementation of fractal image compression on medical images [10] are also studied, and results of implementation using MATLAB as well as FPGA are compared to improve the encoding speed. The results of fractal methods are comparable to JPEG compression for medical images [11]. Finally, NP-hard problem was detected to find optimal fractal code [12]. Therefore, quantum approach is proposed [13, 14] to reduce the complexity in searching mechanism.

2 Fractal Coding Using Quantum Approach

Image processing can be carried out using quantum computations [15]. However, one of the difficulties with quantum-based FIC method is that, when the domain-range block size is reduced to improve compression ratio, it tends to be computationally expensive in searching similarities. Therefore, it is interesting to consider the methods under examination from the point of view of reducing computational complexity in searching self-similarities. A possible approach could be the use of Grover's search with quantum representation. Therefore, system is proposed using Grover's search. Grover's algorithm performs quantum computing. This search principle is useful in reducing the search complexity in FIC to $O(\sqrt{N})$ steps. For representation of classical images as normalized quantum states, a novel method is proposed.

Geometric transformations can also be applied to quantum images [16]. The 'qubit lattice' representation was incorporated as SQR by Yuan [17] for infrared images. The position and color information in images in quantum approach are suggested by Li [18]. Sun [19] expanded FRQI for RGB color. A literature survey is elaborated in [20, 21].

3 Proposed System

The quantum-based FIC is shown in Fig. 1. A search is made for the desired element in Ω. Consider a quantum circumstances for a given quantum database with

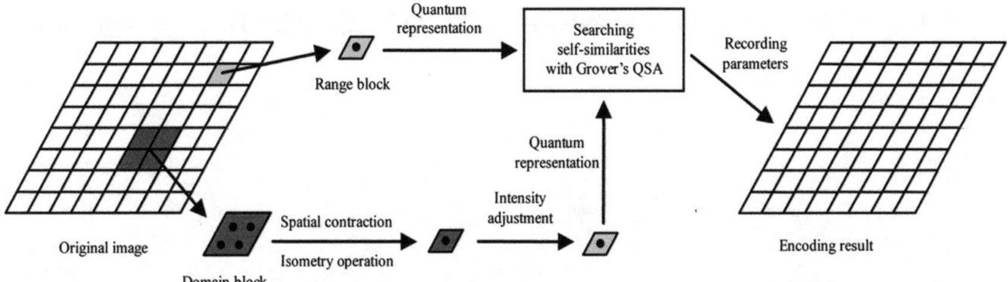

Fig. 1 Quantum-based FIC

N states X1, X2 … XN. G-QSA can find the expected state through a series of three steps.

Step 1: Initialization.

$$|\psi\rangle = \left(N^{-\frac{1}{2}}, N^{-\frac{1}{2}}, \ldots, N^{-\frac{1}{2}}\right)^{\mathrm{T}}. \quad (1)$$

Step 2: Repeat the following two-step unitary operation $\sqrt{N}$ times.

$$D_{ij} = \begin{cases} -1 + (2/N), & \text{if } i = j \\ 2/N, & \text{otherwise} \end{cases}. \quad (2)$$

Step 3: Measurement of the resulting state.

Since G-QSA has three steps of quantum mechanism and calculation to find the desired item, quantum operations are essentially just linear transformations in a two-dimensional space and can be interpreted by 2 × 2 matrices. The first step is achieved by using the Walsh–Hadamard transformation. Further, the first state of the unitary operation can be obtained by a conditional phase gate. In second phase of the unitary operation, states are inverted about their mean. This can be carried out by the product of a Walsh–Hadamard transform W, a conditional phase shift. The third step is a measurement of the state.

4 Experimental Results

A set of standard test images from the USC-SIPI image database [22] is used for simulation. The GUI designed for fractal compression method is shown in Fig. 2. Two image sets, consisting of textures and satellite images, are used. The proposed algorithms are simulated using MATLAB R2012a on Intel(R) Core i5 2.5 GHz PC. The original and reconstructed satellite images are shown in Fig. 3. The results are given for texture images in Table 1. It is seen, from Table 2, the compression ratio CR is high in satellite image for quantum algorithm as compared to DCT, since this type of image is more based on fractal geometry. The decoded image quality measured as

PSNR is good in quantum algorithm as there is no loss of detailed coefficients as in DCT.

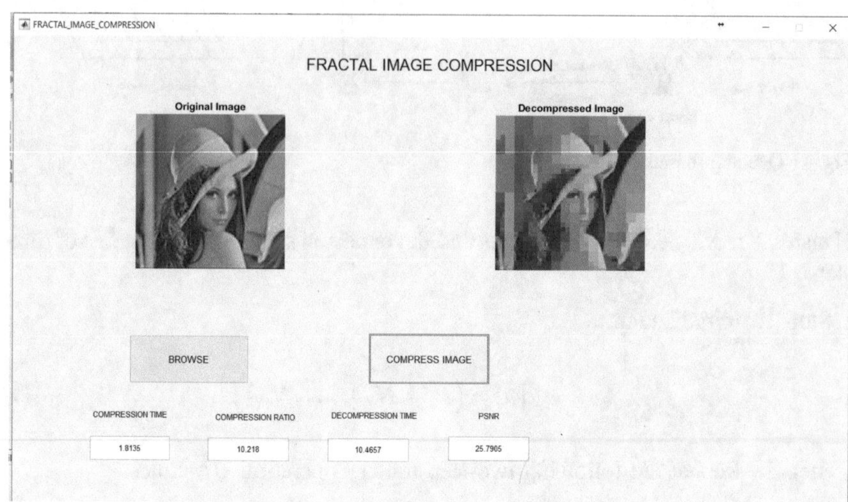

Fig. 2 GUI designed for fractal compression method

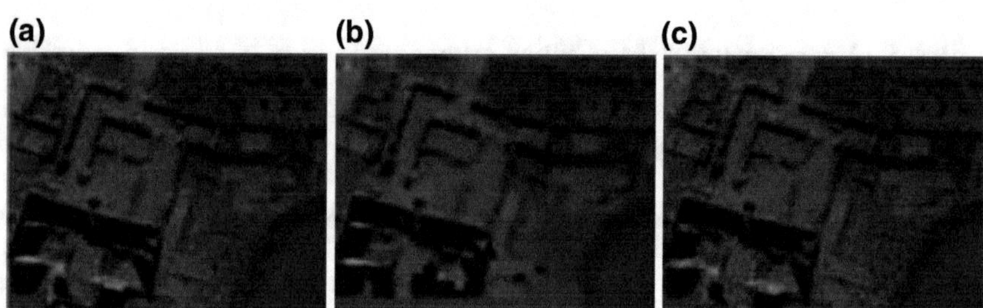

Fig. 3 **a** Original image **b** reconstructed image range size 16 × 16 **c** reconstructed image range size 8 × 8

Table 1 Performance comparison of quantum algorithm for texture image set

Input image	Compression factor		PSNR (dB)		Complexity (in computations)	
Range size	16 × 16	8 × 8	16 × 16	8 × 8	16 × 16	8 × 8
Image 1	38.59	39.70	35.25	37.53	13010	59534049
Image 2	37.90	38.62	34.37	35.93	13011	59534305
Image 3	37.68	37.82	34.43	36.16	13010	59533793
Image 4	38.10	38.83	34.24	36.04	13010	59533921
Image 5	37.54	38.82	34.25	35.76	13011	59534305
Image 6	37.67	39.11	34.30	36.88	13011	59534177

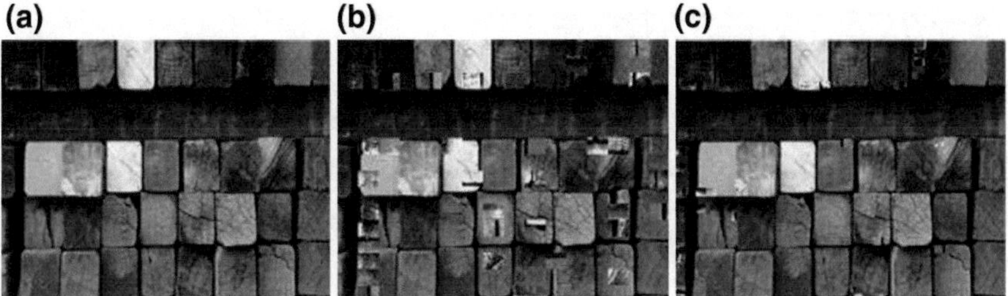

Fig. 4 **a** Original image **b** reconstructed image range size 16 × 16 **c** reconstructed image range size 8 × 8

Table 2 Performance comparison of quantum algorithm for satellite image set

Input image	Compression factor		PSNR (dB)		Complexity (in computations)	
Range size	16 × 16	8 × 8	16 × 16	8 × 8	16 × 16	8 × 8
Image 1	39.46	40.29	40.26	39.32	13010881	59534049
Image 2	41.59	42.75	37.08	37.16	13011137	59534305
Image 3	28.53	30.72	36.72	36.85	13010625	59533793
Image 4	29.56	30.87	38.40	37.29	13010753	59533921
Image 5	39.72	42.41	34.29	32.57	13011137	59534305
Image 6	39.45	39.80	39.20	39.24	13011009	59534177

The original and reconstructed texture images are shown in Fig. 4. It is seen that for texture images, compression ratio is less that satellite images. So proposed method is more suitable for satellite images to achieve more compression.

The Fig. 5 shows reduction in complexity of Grover's search as compared to quantum algorithm before Grover's search Tables 3 and 4 gives complexity before Grover's search and complexity after Grover's search.

Tables 5, 6, and 7 give compression ratio, compression time(s), and PSNR for comparison of proposed method with DCT, quantum, DCT-based FQT, and FQTH. Performance analysis of proposed method with different methods DCT, quantum, DCT-based FQT, and FQTH is shown in Figs. 6, 7, and 8. We have seen that even though DCT gives less compression ratio, it is computationally efficient over other traditional techniques. So, the comparison between quantum algorithm and existing computationally efficient DCT algorithm helps to understand the improvement of computational efficiency over the existing best algorithms. The results show that the required speedup is achieved by the proposed algorithm using quantum superposition of states. Especially for the fractal oriented satellite image, the compression

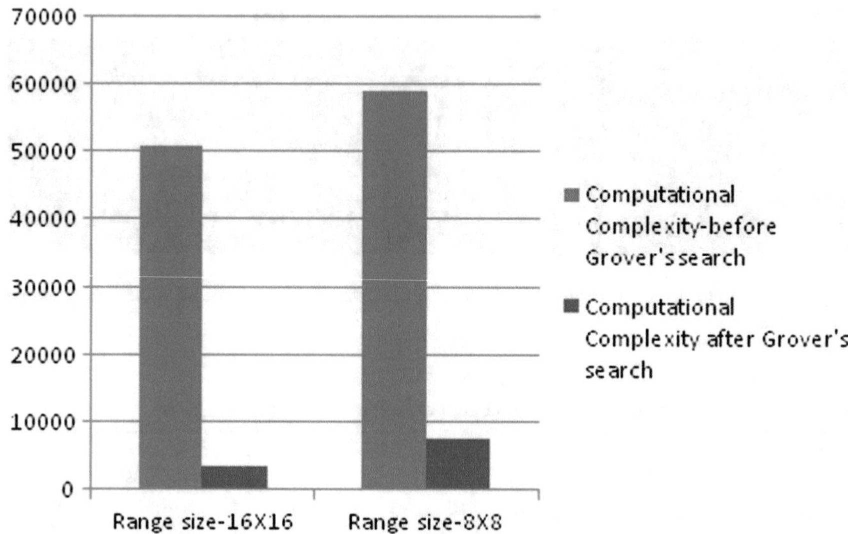

Fig. 5 Computational complexity before and after Grover's search

Table 3 Complexity before Grover's search

Complexity	Quantum range (16 × 16)	Quantum range (8 × 8)
Quantum representation	50881	59105
Searching similarities	12960000	59474944

Table 4 Complexity after Grover's search

Complexity	Quantum range (16 × 16)	Quantum range (8 × 8)
Full search	12960000	59474944
Grover's search	3584	7680
Grover's search (theoretical)	3600	7712

Table 5 Comparison of performance parameters as compression factor for DCT, proposed quantum method, DCT-based fractal quad tree and fractal quad tree with Huffman coding

Input image	Compression factor			
	DCT	Quantum	DCT-based FQT	FQTH
Satellite	24.93	28.53	27.89	25.65
Texture	19.57	19.12	22.58	17.83
Lena	11.79	8.86	37.44	10.12

Table 6 Comparison of performance parameters as PSNR for DCT, proposed quantum method, DCT-based fractal quad tree, and fractal quad tree with Huffman coding

Input image	PSNR(dB)			
	DCT	Quantum	DCT-based FQT	FQTH
Satellite	35.43	36.71	34.32	27.74
Texture	32.13	36.87	38.36	28.05
Lena	34.32	37.13	34.01	25.69

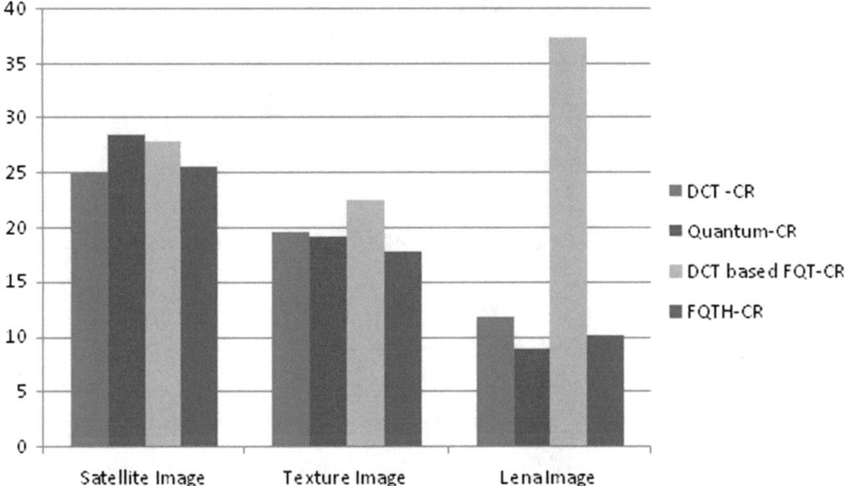

Fig. 6 Performance analysis of compression ratio for satellite, texture, and lena image for different compression methods

time (CT) is reduced from 9.4 to 3.35 s. To increase the compression ratio further, the algorithm is run with different sizes of range and domain blocks. Figures 3 and 4 give the compressed images of single satellite and texture image, respectively, for two different domain and range block sizes.

To reduce the search complexity specified in Table 3, Grover's search is adopted along with the quantum FIC algorithm which reduces complexity. Fig. 7 depicts how Grover's search is deployed to search and code the single fractal part from the total blocks available. Table 4 shows the amount of reduction in computations with Grover's algorithm (Table 7).

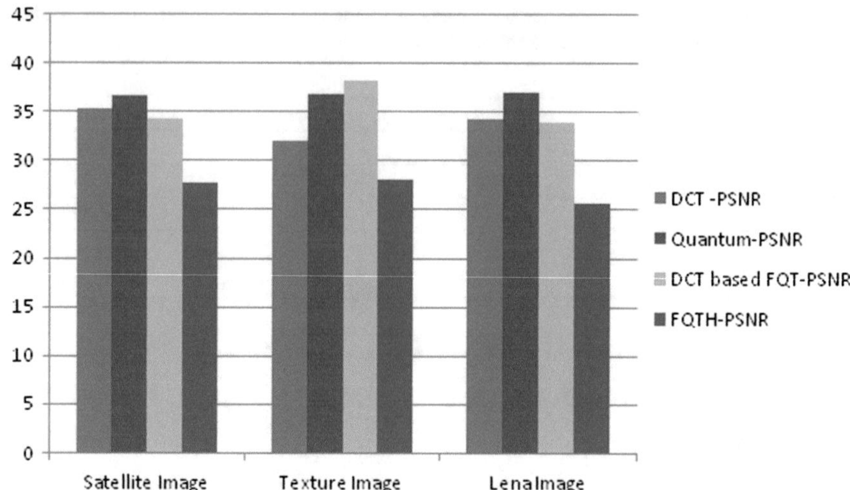

Fig. 7 Performance analysis of PSNR for satellite, texture, and lena image for different compression methods

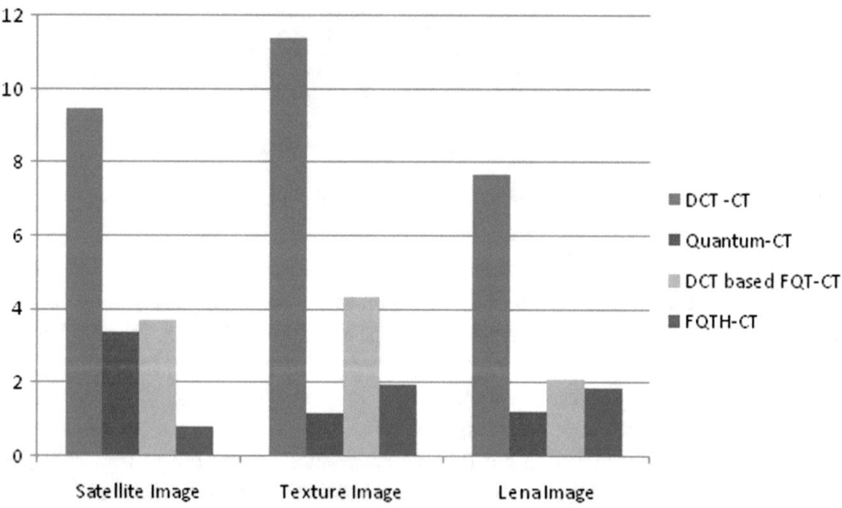

Fig. 8 Performance analysis of compression time(s) or satellite, texture, and lena image for different compression methods

Table 7 Comparison of performance parameters as compression time for DCT, proposed quantum method, DCT-based fractal quad tree, and fractal quad tree with Huffman coding

Input image	Compression time (s)			
	DCT	Quantum	DCT-based FQT	FQTH
Satellite	9.45	3.35	3.69	0.78
Texture	11.42	1.17	4.30	1.95
Lena	7.69	1.20	2.08	1.85

5 Conclusion and Future Scope

Quantum approach to fractal image compression has been examined and sought to improve it by formulating the search approach using Grover's algorithm. The experiment shows that the proposed representation on the algorithm without Grover's search is able to provide good compression factor and better PSNR for compressing natural, texture, and satellite images. In order to increase compression factor further, the simple modification to the algorithm like decreasing range and domain block size significantly had an impact on the compression results. But the search complexity in the algorithm remained as a drawback because it results in high computational and time requirements of encoding part. Grover's search algorithm is used along with quantum FIC, which has reduced searching complexity. It can be seen that the quantum FIC algorithm along with Grover's search is more powerful and outperforms the existing algorithms for satellite images. As a future work, different partitioning schemes and entropy encoding for the fractal codes can be implemented to further improve the compression. Also, the proposed method can be applied to medical images to achieve more compression.

References

1. Fisher, Y.: Fractal Image Compression (1995)
2. Chaudhari, R.E., Dhok, S.B.: Wavelet transformed based fast fractal image Comp. In: International Conference on Circuits, Systems, Communications and Information Technology Applications, pp. 64–69 (2014)
3. Kadam, S., Rathod, V.: DCT with quad tree and Huffman coding for color images. Int. J. Comput. Appl. **173**(9), 33–37 (2017)
4. Nodehi, A., Mznah, G.S.: Intelligent fuzzy approach for fast fractal image compression. EURASIP J. Adv. Signal Process, 1–9 (2014)
5. Mahalaxmi, G.V.: Implementation of image compression using fractal image compresssion and neural network for MRI images. In: IEEE International Conference on Information Science (ICIS), pp. 60–64 (2016)
6. Al-saidi, N.M.G., Ali, A.H.: Towards enhancing of fractal image compression via block complexity. In: IEEE Annual Conference on New Trends in Information and Communication Technology Applications (NTICT), pp. 246–251 (2017)

7. Abdul, N., Salih, J.: Fractal coding technique based on different block size. In: Al-Sadeq International Conference on Multidisciplinary in IT and Communication Science and Applications (AIC-MITCSA), pp. 1–6 (2016)
8. Gupta, P., Srivastva, P., Bhardwaj, S., Bhateja, V.: A modified PSNR metric based on HVS for quality assessment of color images. In: International Conference on Communication and Industrial Application, pp. 1–4 (2011)
9. Zhu, S., Zhang, S., Ran, C.: An improved inter-frame prediction algorithm for video coding based on fractal and H.264. In: IEEE Early Access, p. 1 (2017)
10. Padmashree Rohini, S., Padma, N.: Different approaches for implementation of fractal image compression on medical images. In: IEEE International Conference on Electrical, Electronics, communication and optimization techniques, pp. 66–72 (2016)
11. Padmashree, S., Nagpadma, R.: Comparative analysis of JPEG compression and fractal image compression for medical images. Int. J. Eng. Sci. Technol., 1847–1853 (2013)
12. Rahul, M., Hartenstein, H.: Optimal fractal coding is NP-HARD. In: Proceedings IEEE, Data Compression Conference, pp.. 261–270 (1997)
13. Amin, Q., Ali, N., Ali, A., Nodehi, S.: Square function for population size in quantum evolutionary algorithm and its application in fractal image compression. In Sixth International Conference on Bio-Inspired Computing: Theories and Applications, pp. 3–8 (2011)
14. Yang, Y., Bai, G., Chiribella, G.: Masahito Hayashi: compression for quantum population coding. In IEEE International Symposium on Information Theory(ISIT), pp. 1973–1977 (2017)
15. Songlin, D., Yaping, Y., Yide, M.: Quantum-accelerated fractal image compression-an interdisciplinary approach. IEEE Signal Process. Lett. **22**(4) (2015)
16. Hirota, K., Le, P.Q., Lliyasu, A.M., Dong, F.: Strategies for designing geometric transformations on quantum images. Theor. Comput. Sci. **412**, 1406–1418 (2011)
17. Yuan, S., Mao, X., Xue, Y., Xiong, Q.: A compare: SQR: a simple quantum representation of infrared images. Quantum Inf. Process. **13**(6), 1353–1379 (2014)
18. Li, H.-S., Qingxin, Z., Lan, S., Shen, C.-Y., Zhou, R., Mo, J.: Image storage, retrival, compression and segmentation in a quantum system. Quantum Inf. Process. **12**, 2269–2290 (2013)
19. Sun, B., Lliyasu, A., Yan, F., Dong, F., Hirota, K.: An RGB multi channel representation for images on quantum computers. J. Adv. Comput. Intell Inform. **17**(3), 404–417 (2013)
20. Caraiman, S., Manta, V.: Image representation and processing using ternary quantum computing. Adapt. Nat. Comput. Algorithms **7824**, 366–375 (2013). Springer, Berlin
21. Lliyasu, A.M.: Towards secure and efficient image and video processing applications on quantum computers. Entropy **15**(8), 2874–2974 (2013)
22. USC-SIPI Image Database. http://sipi.usc.edu/database

Hybrid Genetic Algorithm–Differential Evolution Approach for Localization in WSN

P. Srideviponmalar, V. Jawahar Senthil Kumar and R. Harikrishnan

Abstract Nature-inspired algorithms have the characteristics to learn and decide and to be adaptable, intelligent, and robust, and so they can be used for solving complex problems. This paper deals with one such algorithm named hybrid genetic algorithm–differential evolution for localization in wireless sensor network. This algorithm is used to estimate the position of sensor node. A novel hybrid algorithm is analyzed, designed, and implemented. This algorithm provides better accuracy and is simple to implement.

Keywords Genetic algorithm · Wireless sensor network · Sensor node
Anchor node · Differential evolution

1 Introduction

Internet of things and other intelligent monitoring techniques need sensors to be embedded with the devices under control. The sensor node localization would reduce the signal traffic which otherwise can lead to huge traffic of signals for location estimation and location update [1]. In most of the applications, the sensed information becomes meaningful only with the addition of location parameter [2].

P. Srideviponmalar (✉)
Faculty of Information and Communication Engineering, Anna University, Chennai, India
e-mail: sridevi_ponmalar@yahoo.co.in

V. Jawahar Senthil Kumar
Department of Electronics and Communication Engineering,
Anna University, Chennai, India
e-mail: veerajawahar@gmail.com

R. Harikrishnan
School of Electrical and Electronics Engineering,
Sathyabama University, Chennai, India
e-mail: rhareish@gmail.com

V. Bhateja et al. (eds.), *Intelligent Engineering Informatics*, Advances in Intelligent Systems and Computing 695, https://doi.org/10.1007/978-981-10-7566-7_27

Depending upon the application, the sensor nodes are connected in a topology to form wireless sensor networks (WSNs). The WSNs are used for applications such as forest fire detection, road traffic management, healthcare monitoring, animal habitat monitoring, precision agriculture, disaster management, military surveillance, and environmental monitoring [3].

The locations of sensor nodes can be predicted by analytical method. These methods are complex, and process becomes tedious with scalability of the network. So for this type of complex problems, nature-inspired algorithms are discussed in the literature. Because of the intelligence, robustness, flexibility, and simple implementation, nature-inspired algorithm is preferred [4].

Global positioning system (GPS) can be used for sensor node position detection. But GPS technique would consume more power from energy-constraint WSN. Moreover, every node has to connect with GPS which is costly [5]. This paper deals with a nature-inspired hybrid algorithm known as hybrid genetic algorithm–differential evolution localization algorithm (GADELA). This is a range-based distributed algorithm. A centralized algorithm requires sensor information to be transmitted to a centralized process. Therefore, the energy required compared to distributed algorithm is more. In distributed algorithm, the sensor location information need not be transmitted to a centralized process. Thus, the algorithm is energy efficient as the process is done locally [6]. In range-based algorithm, the distance is calculated by using radio signal strength indicator. The need for any extra hardware is not required for calculating the distance which is required for location estimation technique [7].

2 Localization Problem Modeling

A total of 50 nodes are taken within the solution space, in which 10 are the landmark nodes for which the location information in known priori and the remaining 40 are sensor nodes for which the location has to be found. Using trilateration method, the location of 40 sensor nodes is estimated by using the 10 landmark nodes as shown in Fig. 1.

This type of estimation follows two steps: First step is the distance calculation, and the next one is the position estimation. The distance calculation is carried out by using the Eq. (1):

$$\mathrm{d_{ist}} = \sqrt{\left((\mathrm{y_{est}} - \mathrm{y_{landmark}})^2 + (\mathrm{z_{est}} - \mathrm{z_{landmark}})^2\right)} \tag{1}$$

The position estimation is carried out by using Eq. (2):

Fig. 1 Trilateration method

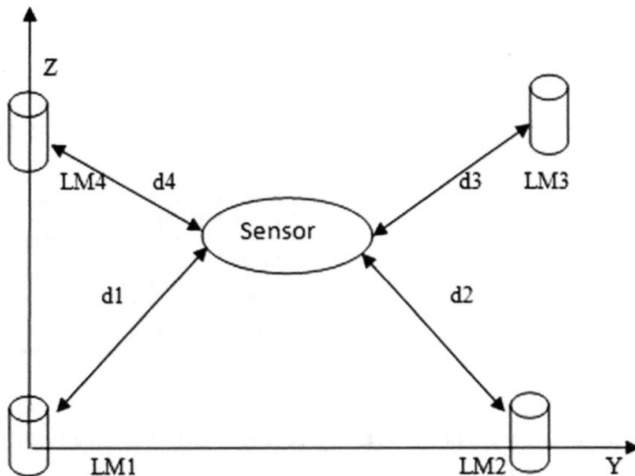

$$f(y_u, z_u) = \left[\sqrt{(y_{est} - y_{landmark})^2 + (z_{est} - z_{landmark})^2} - d_{ist}\right]^2 \quad (2)$$

The average position estimation is calculated by using Eq. (3):

$$f(y_u, z_u) = \left[\sqrt{(y_{est} - y_{landmark})^2 + (z_{est} - z_{landmark})^2} - d_{ist}\right]^2 / N \quad (3)$$

y_{act}, z_{act} is the actual y and z positions of sensor node, respectively.
y_{est}, z_{est} is the estimated y and z positions, respectively, of the sensor node which does not know the location.
$y_{landmark}, z_{landmark}$ is the anchor or landmark of y and z positions, respectively.
N is the total sensor nodes.

3 Hybrid Genetic Algorithm–Differential Evolution (GADE) Localization Algorithm

Genetic algorithm (GA) solves nonlinear, nonconvex operation problems. It uses the nature's concept of survival of fittest. For the localization of sensor node problem, four stages of optimization are carried out. The localization error is optimized, so that optimal localization information is estimated. The four stages are initialization, selection, crossover, and mutation. Initialization of sensor node population is carried out first. Genetic algorithm performs better selection

Table 1 Hybrid GADELA design parameters

Parameters (localization)	Tuned values
Maximum no. of iterations	100
Space size	100
No. of landmarks	10
No. of unknown node	40
No. of total node	50
No. of vector	20
Crossover constant (GA)	0.3
Scaling factor, S (DE)	0.9

operation and crossover operation. So these operations are carried out by genetic algorithm [8].

Differential evolution performs better mutation. So the mutation operation is carried out by differential evolution (DE). DE has the ability to solve non-continuous and non-differential real-world problems. DE performs better on mutation operation [9]. It mutates vector with other vectors within the given population which are randomly selected. Mutation operation leads to global optimization. So hybrid genetic algorithm and differential evolution would give better optimal location of sensor nodes. So this idea is implemented in this GA-based DE algorithm.

Following is the design procedure for hybrid genetic algorithm–differential evolution localization algorithm:

Step 1 Control variables of localization of sensor nodes are selected as genes of a chromosome.
Step 2 Create initial population of sensor nodes.
Step 3 Find fitness of sensor node location information (chromosomes) by using the localization function.
Step 4 Select location information using Roulette wheel selection for mating.
Step 5 Perform crossover operation as in genetic algorithm.
Step 6 Perform mutation operation as in differential evolution.
Step 7 Select new population of location of sensor nodes for the next generation.
Step 8 Repeat steps 4, 5, 6, and 7 for stopping the condition.
Step 9 Print location estimate.

Design consideration for hybrid GADELA is recorded in Table 1. The flowchart of hybrid GADELA is shown in Fig. 2.

4 Result Analysis

Hybrid GADELA is used to estimate the location. The outputs of hybrid GADE localization algorithm are shown in Figs. 3, 4, 5, 6, and 7.

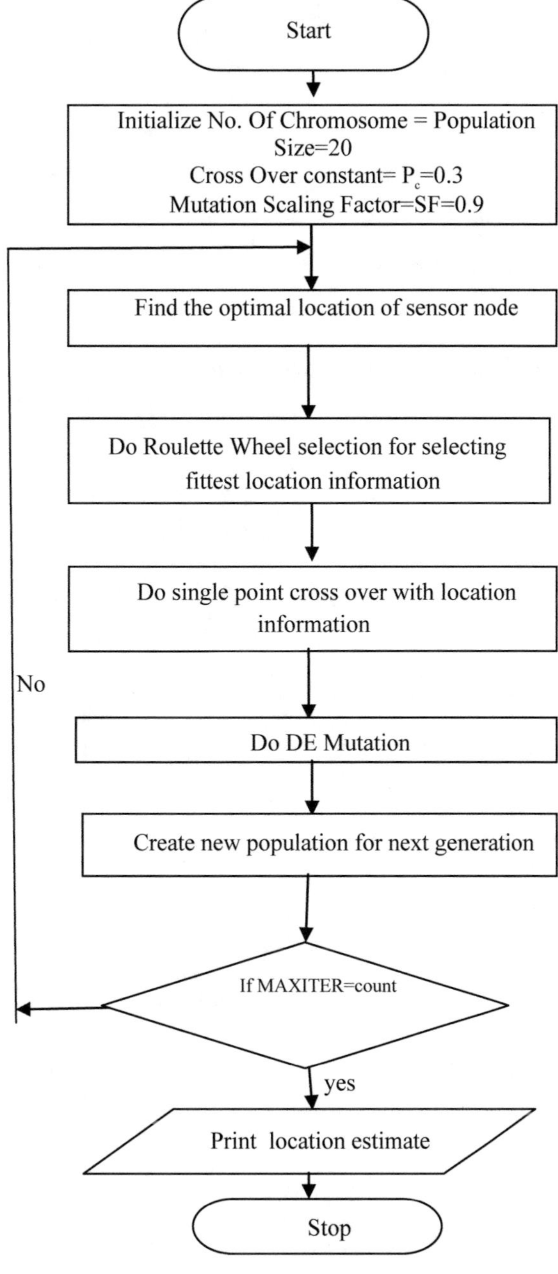

Fig. 2 Flowchart of hybrid GADE localization algorithm

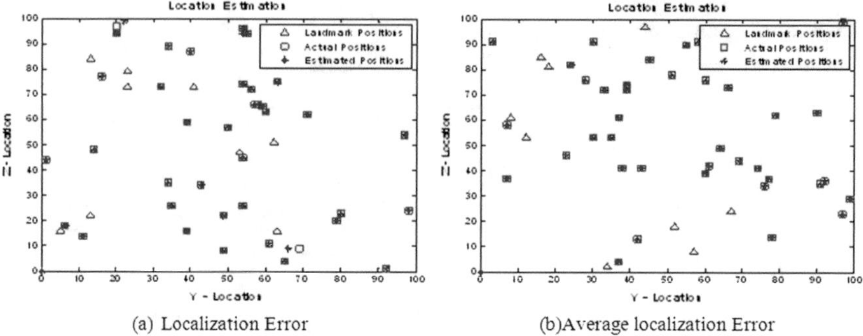

(a) Localization Error (b)Average localization Error

Fig. 3 Output of GADELA for 100 population vector size

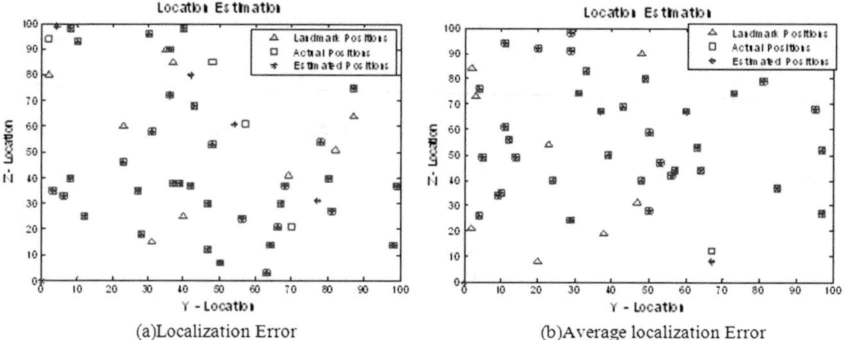

(a)Localization Error (b)Average localization Error

Fig. 4 Output of GADELA for 80 population vector size

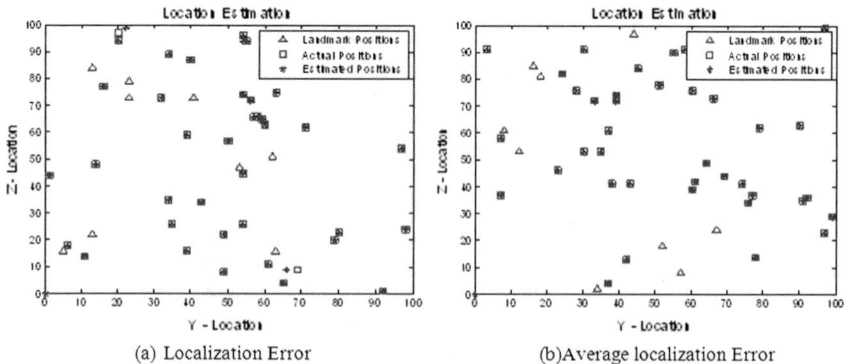

(a) Localization Error (b)Average localization Error

Fig. 5 Output of GADELA for 60 population vector size

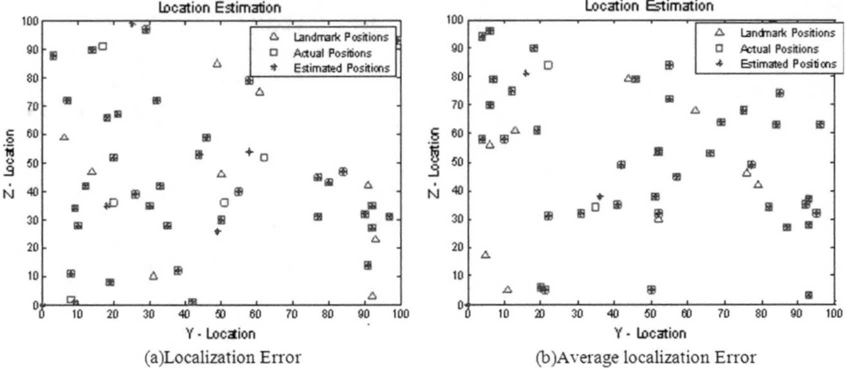

(a)Localization Error (b)Average localization Error

Fig. 6 Output of GADELA for 40 population vector size

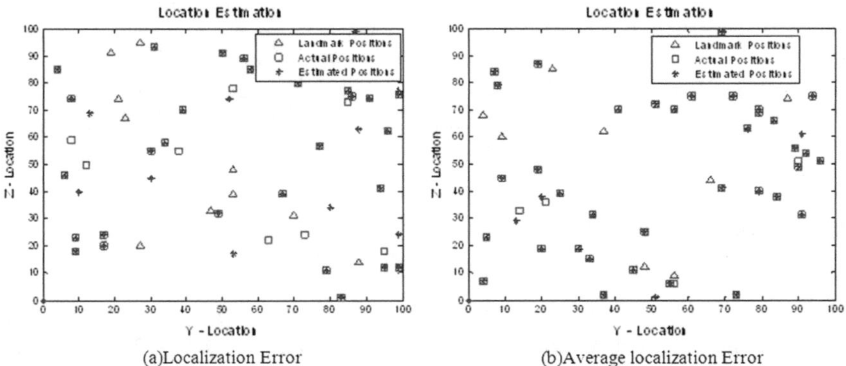

(a)Localization Error (b)Average localization Error

Fig. 7 Output of GADELA for 20 population vector size

The output of GADELA in terms of location error and average localization error is shown in Figs. 3, 4, 5, 6 and 7 for a population vector size of 100, 80, 60, 40 and 20 respectively.

The performance analysis of hybrid GADE localization algorithm with localization function is presented in Table 2. It presents the accuracy and time complexity of the hybrid algorithm for various population vector sizes under consideration with localization function. Similarly, Table 3 shows the accuracy and time complexity of the hybrid algorithm for various population vector sizes under consideration with average localization function. By comparison, it is found that the accuracy increases and the time complexity performance is better with the increase in population vector size. And also the hybrid GADE localization algorithm performs better with accuracy and time complexity if it uses average localization function instead of simple localization function.

Table 2 Performance analysis of hybrid GADE localization algorithm with localization function

No. of population vector	Hybrid GADE localization algorithm with localization function	
	Accuracy	Time complexity
20	78	28.456
40	82	26.456
60	84	17.678
80	90	12.438
100	94	10.002

Table 3 Performance analysis of hybrid GADE localization algorithm with average localization function

No. of population vector	Hybrid GADE localization algorithm with average localization function	
	Accuracy	Time complexity
20	88	15.564
40	94	14.987
60	96	12.845
80	98	10.765
100	100	8.789

5 Conclusion

The hybrid GADELA is analyzed, designed, and implemented. In general, the algorithm shows better accuracy and it has better time complexity. The algorithm performs well in terms of accuracy and time complexity as the population vector size is increased. Further by using average localization function, the performance is further enhanced to get better accuracy and time complexity. The performance comparisons are provided. A careful selection of design parameters and better hybrid techniques might lead to improved performance and estimation accuracy.

References

1. D'Ore, S., Galluccio, L., Morabito, G., Palazzo, S.: Exploiting object group localization in the internet of things: performance analysis. IEEE Trans. Veh. Technol. **64**(8), 3645–3656 (2015)
2. Singh, S.P., Sharma, S.C.: Range free localization techniques in wireless sensor network: a review. Procedia Comput. Sci. **57**, 7–16 (2015). Science Direct
3. Mao, G., Fidan, B., Anderson, B.D.: Wireless sensor network localization techniques. Comput. Netw. **51**, 2529–2553 (2007). Science Direct
4. Raghavendra Kulkarni, V., Förster, A., Venayagamoorthy, G.K.: Computational intelligence in wireless sensor networks: a survey. IEEE Commun. Surv. Tutor. **13**(1), 68–96 (2011)
5. Halders, S., Ghosal, A.: A survey on mobility-assisted localization techniques in wireless sensor networks. J. Netw. Comput. Appl. 82–94 (2016)

6. Shi, Q., He, C., Chen, H., Jiang, L.: Distributed wireless sensor network localization via sequential greedy optimization algorithm. IEEE Trans. Signal Process. **58**(6), 3328–3340 (2010)
7. Salman, N., Ghogho, M., Kemp, A.H.: Optimized low complexity sensor node positioning in wireless sensor networks. IEEE Sens. J. **14**(1), 39–46 (2014)
8. Grefenstette, J.J.: Optimization of control parameters for genetic algorithms. IEEE Trans. Syst. Man Cybern. **16**(1), 122–128 (1986)
9. Vaisakh, K., Srinivas, L.R.: Differential evolution approach for optimal power flow solution. J. Theor. Appl. Inform. Technol. 261–268 (2008)

Local Tetra Pattern-Based Fruit Grading Using Different Classifiers

Ramanpreet Kaur, Mukesh Kumar and Mamta Juneja

Abstract Agriculture is an integral part of economic development, and thus, it becomes essential to lift the impact factor of agriculture development. In past years, researchers had introduced many nondestructive image processing technique to grade the food products. These techniques ensure the quality of food products, are consistent, and save the labor time as well. Many dedicated systems or techniques are available for grading particular type of fruit; therefore, there is need to devise common technique to grade various type of fruits. This paper introduces the common feature extraction method which uses local tetra pattern to grade fruits. In this research, we graded guava fruit into four categories (unripe, ripe, overripe, and defected). The performance of the proposed method is evaluated and compared using ensemble classifiers and compared using accuracy and error rate. The experimental results showed the highest accuracy of 93.8% by Subspace Discriminant Ensemble classifier. The proposed method can be easily adapted for any other spherical fruit or vegetable.

1 Introduction

Agriculture is the ruling occupation in India. Directly or indirectly, two-third of population is dependent on agriculture. It is not just a source of livelihood but a way of life. In India, it is an integral part of economic development and hence provides highest contribution to national income. Thus, it becomes essential to lift the impact factor of agriculture development. India is a front-runner in many fruits and vegetables. It ranks second in fruit and vegetable production after China. As per the

R. Kaur (✉) · M. Kumar · M. Juneja
UIET, CSE Department, Panjab University, Chandigarh, India
e-mail: cheema8raman@gmail.com

M. Kumar
e-mail: mukesh_rai9@pu.ac.in

M. Juneja
e-mail: mamtajuneja@pu.ac.in

V. Bhateja et al. (eds.), *Intelligent Engineering Informatics*, Advances in Intelligent Systems and Computing 695, https://doi.org/10.1007/978-981-10-7566-7_28

national horticulture database published by National Horticulture Board during 2014–15, India produced 86.602 million metric tons of vegetables. The area under cultivation of fruits stood at 6.110 million hectares while vegetables were cultivated at 9.542 million hectares. Therefore, it becomes essential to boost the production and productivity of vegetables and fruits in the country. Along with increasing their productivity, it is equally essential to label the quality of fruits and vegetables before dealing them out. Labeling goods are done manually by humans by observing the external quality feature like color, shape, size. Manual inspection is time-consuming, tiresome, and inconsistent as well since it is human nature to become inattentive after a period of time. Thus, automatic grading system of goods gives accurate, consistent, and uniform outputs, resulting in saving time and assisting the economic development. For determining the quality of fruits, external and internal quality features are taken into consideration. Some of the internal quality factors are sweetness, aroma, taste, sourness, nutritive value like minerals, vitamins and external quality factors are color, size, shape, texture, surface defects.

In this research, we proposed a method of classifying guava fruit into four categories, namely unripe, ripe, overripe, and defected. The extracted color, texture, shape, and size values of guava fruit are used in different classifiers. The paper of organization is as follows Sect. 2 briefly explains the related work, Sect. 3 deals with the materials and methods, and Sect. 4 describes the proposed methodology. In Sect. 5 results are discussed and Sect. 6 is conclusion and future scope.

2 Related Work

The design requirements for grading different fruits vary from fruit to fruit. Building a dedicated system focused on particular product or fruit has been the aim of many researchers. Kondo [1], Gay et al. [2] built common fruit grading and classification system but dedicated system available sort particular type of fruit only. Performance of different grading system depends upon the quality factors taken into consideration. Quality factors used by farmers are external and internal quality factors as mentioned earlier in introduction section. For grading of fruits, nondestructive and standardize technique is required; hence, computer vision-based fruit grading comes into play. The captured images of fruit are converted into digital images. Various image processing techniques are applied on digital images to extract features which are then used in classification.

To automate the apple grading, Unay et al. [3] extracted color, shape, size, and texture features from segmented multispectral images followed by specific segmentation of defective part and then categorizing fruit. From all the features extracted, total of 67 features are selected various classifiers used for classifying fruit into respective categories are nearest neighbor classifier (K-NN), support vector machine (SVM), C4.5, linear discriminant classifier. For grading fruits into different categories, two approaches were used (i) direct approach where fuzzy K-NN achieved highest of 83.5% recognition rate and (ii) cascading approach consists of SVM (85.6%) followed by fuzzy K-NN. From direct approach and cascading approach, direct approach is preferred with limited computational resources and cascading approach is better choice if significant accuracy is required. Suresha et al. [4] segmented apples using threshold value from HSV images, and then average green and red components are used in classifying apples using SVM classifier, achieving 100% accuracy. M. Khojastehnazhand developed lemon grading system consists of two CCD cameras, lighting system, two capture cards, computer, and mechanical parts. Color value was extracted from HSI image by averaging the hue component, and volume was calculated by dividing the image into number of different sectors. Blasco et al. [5] detected peel defects in citrus fruits by using region selecting, growing, and merging algorithm. It involves selecting the seeds from appropriate region of interest (ROI), and the selected ROI grows iteratively by addition of neighboring pixels that satisfy the criterion. After region growing, images were segmented into many regions. Among many segmented area, largest area was determined which was assumed to be sound skin of fruit. But this approach fails if area of defected skin comes out to be the largest area. An expert measured the performance of proposed system and accuracy of detecting defects was 94%. Arakeri and Lakshmana [6] developed computer vision-based tomato grading system. Median filter was applied on captured images to eliminate the noise and reflections. Resulting images were segmented using Otsu's method, and features extracted from individual channel of RGB image were mean, standard deviation, skewness. Extracted texture features from GLCM matrix were contrast, homogeneity, energy, and correlation. For optimal feature selection, sequential forward selection (SFS) was applied. Classification using multilayer neural network obtained accuracy of 100% for defective/nondefective and 96.47% for ripe/unripe tomatoes. Mhaski et al. [7] introduced tomato grading system using Raspberry Pi. Captured images of tomato fruit are converted into HSV images. Red, green, and yellow masks from HSV images are calculated, and maximum of these three masks defines the ripeness stage of tomato. Shape and size are estimated by dividing image into contours, and the biggest of all the contours estimated is the fruit size. K-mean clustering algorithm detects the defects in tomato images. Prabha and Kumar [8] determined the banana fruit maturity by the use of color intensity algorithm obtained from the image histogram. Banana region was segmented from the background using a threshold value. Maturity level of banana was calculated using mean values, and variance determines the smoothness texture. The number of pixels in banana region is the area, and the number of pixels in boundary region defines the perimeter. Color value algorithm achieved 99.1% accuracy and 85% accuracy was obtained for size value algorithm.

Ji et al. [9] measured the ripeness level of banana with the use of grade color chart of banana given by ASDA. The RGB image of banana was converted into CIE XYZ images, followed by separating colors into four clusters. The colorfulness (C), lightness (L), and hue (H) for every pixel were calculated and then values are averaged for further use. In this research, comparison between spectrophotometry and digital imaging was made and showed that digital imaging gives improved results and is more flexible and consistent. Hu et al. [10] used computer vision-based size determination technique for banana. The five-point method algorithm along with Euclidean distance was used to estimate the indicators of banana size Sanaeifer et al. [11] predicted banana quality indices using color features. For extracting color information, background was removed from image and the resulted image was converted into HSV and l*a*b color spaces. From banana surfaces, red, blue, green, saturation, hue, intensity, lightness, and a*, b* components were determined and saved for further analysis. Correlation between color information and quality indices is analyzed. Capability of support vector regression (SVR) and artificial neural network (ANN) was evaluated, and SVR showed better results. To automate mango ripeness evaluation using image processing, Ramya and Anitha [12] cropped the mango area and segmented using Otsu's method. After that, the segmented image is divided into three regions, namely equator, apex, and stalk regions. The average value of three regions for every color channel of RGB image is determined. K-NN algorithm calculates the Euclidean distance and classifies mangoes using the nearest distance with accuracy of 93%.

Al Ohali [13] designed and implemented grading system for data fruit. In this research, various external quality factors are defined and the shape of date fruit is estimated from edge tracking operator. Date with high intensity is considered as better quality, and intensity is measured by area covered by edges divided by total fruit area. Defects in date fruit (bird flicks) are estimated by use of brightness value of pixel, and bruises are determined from the shape. After various features were extracted, fruit is classified using back propagation neural network and obtained 80% accuracy. Dutta et al. [14] classified grapes into pesticide treated and untreated grapes. Region of interest from the captured images is segmented. Eighteen features are extracted from segmented image ín frequency domain using haar filter. Features extracted are used in support vector machine classifier, and results showed the 100% accuracy. Sardar [15] classified guava into different categories, updated Hassu algorithm was used, which considers the number of pixels belonging to specific color intensity. The different range of values obtained from algorithm are observed and are assigned to each category. The value generated by an input was matched to existing specified range, and category was assigned appropriately.

The dedicated fruit grading systems and techniques are available because of variation in characteristics. Thus, there is need to devise a common grading system or techniques that can grade variety of fruits. In this research, we proposed a common technique of grading fruits.

3 Materials and Methods

Samples of guava are collected from local market; the category of each sample is decided from survey of 10 people. Every person after looking at the guava image label it as ripe, unripe, overripe, and defected, then maximum votes a sample gets for particular label is named as the same. Sample images were captured through camera, and MATLAB R2015a was used for processing of images. Images are captured by a NIKON S2600 camera, and its specification are detailed below

- 14.0 megapixels
- Lens-5x zoom
- Image sensor-1/2.3-in
- Interface-high-speed USB
- Type-CCD

Guava fruit was placed at white background to simplify the task of segmentation, and camera was located at distance of 12–15 cm from guava and vertically at angle of (manually oriented the side of fruit). Images are captured during daylight, and image size was set to 640 * 480 pixel, ISO sensitive-400 and format-JPG. A total of 113 images are collected.

4 Proposed Methodology

The proposed method uses image processing technique and various classifiers to categorize the guava fruit into four categories, namely ripe, unripe, overripe, and defected. This involves segmenting the fruit image from background and extracting color, size, shape, and texture features. Table 1 gives the brief information about the parameter used for various features (detailed explanation of LTrPs is given in Sect. 4.5). Figure 1 represents the flowchart of the proposed method.

Table 1 Parameters used for various features' extraction

Features extracted	Parameter used
Color value extraction	Mean
	Median
	Variance
	Standard deviation
Shape and size value extraction	Area
	Perimeter
	Roundness
	Equidistant
	Circularity
	Major axis
	Minor axis
Texture features' extraction	Local tetra patterns (LTrPs) [16]

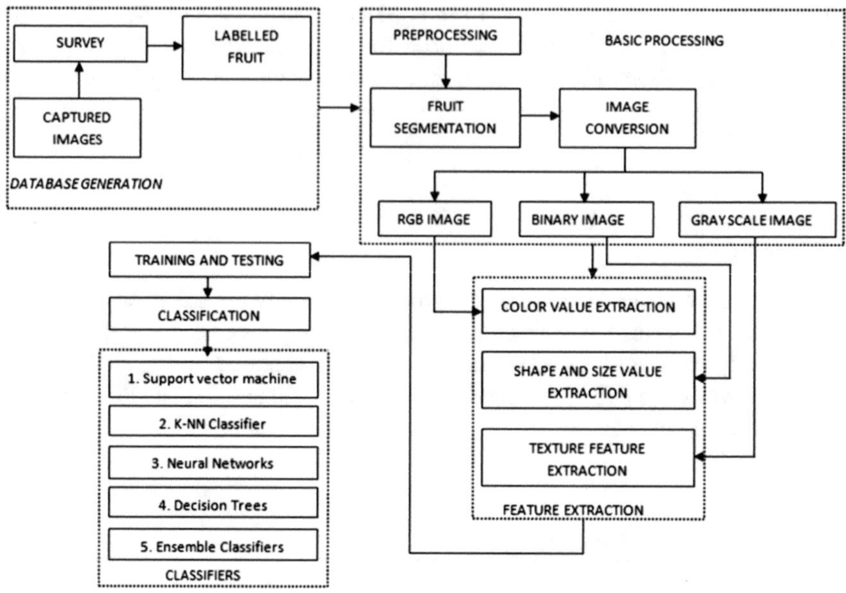

Fig. 1 Flowchart of the proposed method

4.1 Database Creation

Samples of guava are collected from local market; the category of each sample is decided from survey of ten people. Every person after looking at the guava image label it as ripe, unripe, overripe, and defected, and then maximum votes a sample gets for particular label is named as same. Images are captured by a NIKON S2600 camera, *and* guava fruit is placed on white background to simplify the task of segmentation. Captured images are used for further processing.

4.2 Basic Processing

(i) *Preprocessing*—To eliminate the noise and reflection, median filter was applied to the input image. (ii) *Fruit segmentation*—Step 1 Convert the processed RGB image to grayscale image. Step 2 Conversion of grayscale image to binary image using threshold value, where binary image contains 1's in the fruit region and 0's in the background. Step 3 Binary image obtained is multiplied with each color channel of original processed RGB image and then concatenated together to get the guava region named as segmented image. (iii) *Image conversion*—Binary image is obtained in step 2, segmented image is the RGB image of guava fruit in step 3 and grayscale image is obtained from the conversion of segmented image into grayscale image. These three images are used further in feature extraction.

4.3 Feature Extraction

4.3.1 Color value extraction

Color of guava determine the ripeness level, thus is the primary feature. From segmented image, mean and median (for each color channel of RGB), variance and standard deviation are calculated.

$$\text{Mean}(\text{m}) = \frac{\sum_u \sum_v I(u, v)}{R{*}S} \quad (1)$$

$$\text{Standard deviation} = \sqrt{\frac{\sum_u \sum_v (I(u, v) - m)^2}{R{*}S}} \quad (2)$$

$$\text{Variance} = \frac{\sum_u \sum_v (I(u, v) - m)^2}{R{*}S} \quad (3)$$

I(u, v) is the image, and (R * S) is the image size.

4.3.2 Shape and size value extraction

From the binary image, *area* is counting the number of pixels having value 1, and perimeter counts the number of pixels having value 1 at the edges of guava region.

$$\text{circularity} = \frac{perimeter \wedge 2}{area} \quad (4)$$

$$Diameter = \frac{major\ axis + minor\ axis}{2} \quad (5)$$

$$Equidistant = \frac{4{*}area}{3.14} \quad (6)$$

$$Roundness = \frac{4{*}3.14{*}area}{perimeter \wedge 2} \quad (7)$$

Total ten geometric features are computed.

4.3.3 Texture features

For computing texture features, local tetra pattern proposed by Murala et al. [16] encode connection between a given pixel and its neighbors.

LTrPs uses gray scale image and traced the spatial structure with the use of the centre pixel's direction R_c. For calculating direction of centre pixel R_c in image M, we need to determine M_{0° and M_{90°

$$\text{where } M_{0^\circ}(R_c) = M(R_h) - \mathrm{M}(R_c) \tag{8}$$

$$M_{90^\circ}(R_c) = M(R_v) - \mathrm{M}(R_c) \tag{9}$$

R_h and R_v are vertical and horizontal neighbor of centre pixel R_c. Direction calculation of $R_c$is as follows

$$M'_{dir}(R_c) = \begin{cases} 1, & M_{0^\circ}(R_c) \geq 0 \& M_{90^\circ}(R_c) \geq 0 \\ 2, & M_{0^\circ}(R_c) < 0 \& M_{90^\circ}(R_c) \geq 0 \\ 3, & M_{0^\circ}(R_c) < 0 \& M_{90^\circ}(R_c) < 0 \\ 4, & M_{0^\circ}(R_c) \geq 0 \& M_{90^\circ}(R_c) < 0 \end{cases} \tag{10}$$

$$\begin{aligned} = \{ & g_3(M'_{dir}(R_c), M'_{dir}(R_1)), g_3(M'_{dir}(R_c), M'_{dir}(R_2)), \\ & g_3(M'_{dir}(R_c)), M'_{dir}(R_p))\}|_{p=8} \end{aligned} \tag{11}$$

$$\text{where } g_3(M'_{dir}(R_c), M'_{dir}(R_p)) = \begin{cases} 0, & M'_{dir}(R_c), M'_{dir}(R_p) \\ M'_{dir}(R_p) & \text{else} \end{cases}$$

For every pixel in image, M is converted into 1, 2, 3, 4 values.

Now, LTrPs^2 (R_c) is calculated as

From Eqs. 10 and 11, for every pixel, 8-bit tetra pattern is obtained. For every direction of center pixel, we divided patterns into four parts, followed by conversion into three binary patterns for each part as shown in

$$\begin{aligned} & LTrP^2|_{direction=2,34} = \sum_{q=1}^{q} 2^{(q-1)} \times g_4(LTrP^2(R_c)|_{direction=2,34} \\ & \text{where } g_4(LTrP^2(R_c)|_{direction=\phi} = \begin{cases} 1, & \text{if } LTrP^2(R_c) = \phi \\ 0. & \text{else} \end{cases} \end{aligned} \tag{12}$$

Now, LTrPs^2 (R_c) is calculated as $\begin{cases} 0, & M'_{dir}(R_c), M'_{dir}(R_p) \\ M'_{dir}(R_p) & \text{else} \end{cases}$

Thus, for total of four directions, 12 (4 * 3) binary patterns are obtained. For magnitude determination of every pixel, 13th binary pattern (MP) is included as follows.

For magnitude determination of every pixel, 13th binary pattern (MP) is included as follows.

$$T_{M'_{(R_p)}} = \sqrt{(M'_{0^\circ}(R_p))^2 + (M'_{90^\circ}(R_p))^2} \tag{13}$$

$$\mathrm{MP} = \sum_{q=1}^{q} 2^{(q-1)} \times f_1 T_{M'(R_p)} - T_{M'(R_c)}|_{q=8} \tag{14}$$

Total 13 binary patterns for every pixel are generated; these 13 patterns are used to construct histogram. From combined histogram, feature vector of length 2^q is constructed. In this research, feature vector (FV) of known sample of unripe, ripe, overripe, and defected guava fruit is constructed. From four saved FVs, every input image' FV is subtracted and the result is summed up (negative values are taken as positive). Thus for each input, we get four summed up values (i.e., for unripe, ripe, overripe, and defected category) and these values are used for further analysis.

4.4 Training and Testing

Data comprised of 22 features extracted from each image. In this study, we have evaluated the performance of the proposed method on dataset having three different percentages of training and testing data.

4.5 Classification

All the features extracted are used in classification of guava fruit into four categories, namely unripe, ripe, overripe, and defected. Ensemble classifiers are used for classification. A classification ensemble is a predictive model composed of a weighted combination of multiple classification models. In general, combining multiple classification models increases the predictive performance. It combines a set of trained weak learner models and data on which these learners were trained. The goal of ensemble classifiers to be more accurate than individual classifiers that make up the whole new ensemble classifier. In this research, the classification of guava by ensemble classifiers outperformed from all other classifiers is used, thus proving that ensemble classifiers can often be more accurate than using individual classifiers. It can predict ensemble response for new data by aggregating predictions from its weak learners. It stores data used for training, can compute re-substitution predictions, and can resume training if desired. Various ensemble classifiers are Boosted trees, Bagged trees, Subspace discriminant, Subspace K-NN, and RVS boosted trees.

5 Results and Discussion

This section presents the results produced by different classifiers on dataset, comprising of different percentage of training and testing data. Accuracy and error rate are evaluated and plotted as shown in figures. Boosted trees, Bagged trees, Subspace

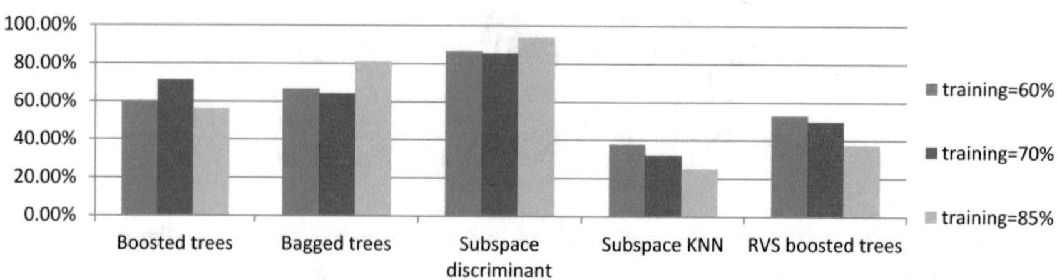

Fig. 2 Accuracy of different ensemble classifiers

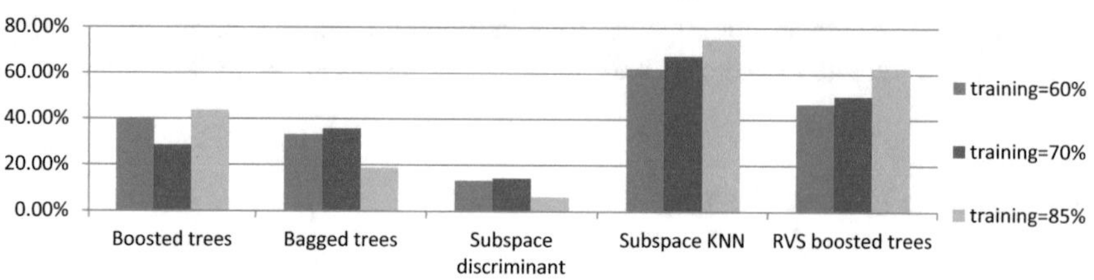

Fig. 3 Error rate of different ensemble classifiers

Table 2 Accuracy of ensemble classifiers

Name of classifier	Types	Training = 60%	Training = 70%	Training = 85%
		Testing = 40%	Testing = 30%	Testing = 15%
1. Ensemble classifiers	Boosted trees (%)	60.0	71.4	56.3
	Bagged trees (%)	66.7	64.3	81.3
	Subspace discriminant (%)	86.7	85.7	93.8
	Subspace KNN (%)	37.8	32.1	25.0
	RVS boosted trees (%)	53.3	50.0	37.5

Discriminant, Subspace K-NN, and RVS boosted trees are used and Subspace Discriminant gave highest accuracy out of all ensemble classifiers. It gave 93.8% accuracy and 6.2% error rate. Figures 2 and 3 present the results produced by various ensemble classifiers, Tables 2 and 3 summarizes the accuracy and error rate.

6 Conclusion and Future Scope

This paper proposes method that grades the guava fruit into four categories (unripe, ripe, overripe, and defected) based on color, shape, size, and texture features. For texture features, we have successfully implemented local tetra patterns. Features

Table 3 Error rate of ensemble classifiers

Name of classifier	Types	Training = 60%	Training = 70%	Training = 85%
		Testing = 40%	Testing = 30%	Testing = 15%
2. Ensemble classifiers	Boosted trees (%)	40.0	28.6	43.7
	Bagged trees (%)	33.7	33.7	19.7
	Subspace discriminant (%)	13.7	14.3	6.2
	Subspace KNN (%)	62.8	67.9	75.0
	RVS boosted trees (%)	46.7	50.0	62.5

extracted are classified using various Ensemble classifiers. Experimental results showed 93.8% accuracy which is achieved by Subspace Discriminant Ensemble classifier followed by Bagged trees. Overall Subspace Discriminant Ensemble classifier showed better results for different percentage of training and testing data. Future scope may be to decrease the computational time for calculating local tetra patterns. The local tetra patterns involve four directions, and it may be extended to eight directions considering diagonals as well. The proposed method can be adapted for other spherical fruits or vegetables.

Declaration

The work under consideration only and contains survey (grading of fruit images) for the purpose of empirical validation of the proposed system. We hereby declare that due permissions have been taken from the expert committee formulated by UIET, Panjab University, Chandigarh, India.

References

1. Kondo, N.: Fruit grading robot. In: International Conference on Advanced Intelligent Mechatronics, pp. 1366–1371 (2003)
2. Gay, P., Berruto, R.: Innovative techniques for fruit color grading (2002). http://www.deiafa.unito.it
3. Unay, D., Gosselin, B., Kleynen, O., Leemans, V., Destain, M.-F., Debeir, O.: Automatic grading of Bi-colored apples by multispectral machine vision. Comput. Electron. Agric. **75**, 204–212 (2011)
4. Suresha, M., Shilpa, N.A., Soumya, B.: Apples grading based on SVM classifier. Int. J. Comput. Appl. 0975–8878 (2012)
5. Blasco, J., Alexios, N., Molto, E.: Computer vision detection of peel defects in citrus by means of a region oriented segmentation algorithm. J. Food Eng. **81**, 535–543 (2007)
6. Arakeri, M.P., Lakshmana: Computer vision based fruit grading system for quality evaluation tomato in agriculture industry. In: 7th International Conference on Communication, Computing and Virtualization, vol. 79, pp. 426–433 (2016)
7. Mhaski, R.R., Chopade, P.B., Dale, M.P.: Determination of ripeness and grading of tomato using image analysis on Rasberry Pi. In: International Conference on Communication, Control and Intelligent Systems (2015). 978-1-4673-7541-2

8. Prabha, D.S., Kumar, J.S.: Assessment of banana fruit maturity by image processing technique. J. Food Sci. Technol. **52**(3), 1316–1327 (2015)
9. Ji, W., Koutsidis, G., Luo, R., Hutchings, J., Akhtar, M., Megias, F., Butterworth, M.: A Digital Imaging Method for Measuring Banana Ripeness, vol. 38(3), pp. 364–374. Wiley Periodicals (2012)
10. Hu, M.-H., Dong, Q.-L., Malakar, P.K., Liu, B.-L., Jaganathan, G.K.: Determining banana size based on computer vision. Int. J. Food Prop. **18**(3), 508–520 (2015)
11. Sanaeifar, A., Bakhshipour, A., Guardia, M.: Prediction of banana quality indices from color features using support vector regression **148**, 54–61 (2016)
12. Ramya, M., Anitha Raghavendra: Ripeness evaluation of mangoes using Image processing. Int. J. Eng. Sci. Comput. **6**(7) (2016)
13. Al Ohali, Y.: Computer vision based date fruit grading system: design and Implementation. J. King Saud Univ.–Comput. Inf. Sci. **23**, 29–36 (2013)
14. Dutta, M.K., Sengar, N., Minhas, N., Sarkar, B., Goon, A., Banerjee, K.: Image processing based classification of grapes after pesticide exposure. LWT-Food Sci. Technol. **72**, 368–376 (2016)
15. Sardar, H.: Fruit quality estimation by color for grading. Int. J. Model. Optim. **4**(1) (2014)
16. Murala, S., Maheshwari, R.P., Balasubramanian, R.: Local tetra patterns: a new feature descriptor for content-based image retrieval. IEEE Trans. Image Process. **21**(5), 2874–2886 (2012)

An Integrated Approach Incorporating Nonlinear Dynamics and Machine Learning for Predictive Analytics and Delving Causal Interaction

Indranil Ghosh, Manas K. Sanyal and R. K. Jana

Abstract Development of predictive modeling framework for observational data, exhibiting nonlinear and random characteristics, is a challenging task. In this study, a neoteric framework comprising tools of nonlinear dynamics and machine learning has been presented to carry out predictive modeling and assessing causal interrelationships of financial markets. Fractal analysis and recurrent quantification analysis are two components of nonlinear dynamics that have been applied to comprehend the evolutional dynamics of the markets in order to distinguish between a perfect random series and a biased one. Subsequently, three machine learning algorithms, namely random forest, boosting and group method of data handling, have been adopted for forecasting the future figures. Apart from proper identification of nature of the pattern and performing predictive modeling, effort has been made to discover long-rung interactions or co-movements among the said markets through Bayesian belief network as well. We have considered daily data of price of crude oil and natural gas, NIFTY energy index, and US dollar-Rupee rate for empirical analyses. Results justify the usage of presented research framework in effective forecasting and modeling causal influence.

I. Ghosh (✉)
Department of Operations Management, Calcutta Business School, Kolkata, India
e-mail: fri.indra@gmail.com

M. K. Sanyal
Department of Business Administration, University of Kalyani, Kalyani, India
e-mail: manas_sanyal@klyuniv.ac.in

R. K. Jana
Indian Institute of Management Raipur, 492015 Raipur, CG, India
e-mail: rkjana1@gmail.com

V. Bhateja et al. (eds.), *Intelligent Engineering Informatics*, Advances in Intelligent Systems and Computing 695, https://doi.org/10.1007/978-981-10-7566-7_29

1 Introduction

Predictive modeling has been an active area of interdisciplinary research that has garnered lot of attention and tremendous surge among the academicians and practitioners. It is a very challenging task to predict future movements in the presence of random, nonlinear, and sporadic components in historical data. The task becomes even tougher in real life if the series under consideration is sensible to shocks from endogenous events. It is essential to extract and understand the nature of hidden pattern in historical observations in order to make future predictions. If the discovered pattern is found to have dominant linear trend, then forecasting exercise becomes relatively simpler. However, in real life, cases such as financial time markets are largely characterized by superimposition of linear and nonlinear trend, periodic, random parts. Hence, a simple attempt of fitting a standard linear parametric model may not yield good forecasts always. Literature is replete with the development of plethora of sophisticated predictive modeling frameworks in the area of stock market forecasting, weather forecasting, computer network traffic forecasting, and performance forecasting in mechanical systems. Tao [1] proposed a hybrid framework incorporating wavelet neural network (WNN) and genetic algorithm (GA) for prediction of daily, weekly, and other interval-based exchange rates. Rather et al. [2] made a comparative study of predictive modeling on the performance of autoregressive moving average (ARMA), exponential smoothing (ES), and recurrent neural network (RNN) in forecasting returns of stock prices of TCS, BHEL, Wipro, Axis Bank, Maruti, and Tata Steel. Ramasamy et al. [3] attempted to predict wind speeds of different locations (Bilaspur, Chamba, Kangra, Kinnaur, Kullu, Keylong, Mandi, Shimla, Sirmaur, Solan, and Una location) in the Western Himalayan Indian state of Himachal Pradesh adopting artificial neural network (ANN)-based framework. Yin et al. [4] studied fractal nature of Chinese gold market applied both single fractal model and multifractal detrended fluctuation analysis (MFDFA) model and identified the key responsible factors for multifractal characteristics of the said market. It is also important to delve into interrelationships among multiple variables to assess the impact of one on other as most of the times univariate forecasting models fail to address this issue. Majority of the existing research work follow the conventional approach of designing a standalone research model based on a typical discipline, viz. econometric model, computational intelligence, machine learning, and nonlinear dynamics. However, it is very hard to find a study presenting a hybrid research framework comprising Machine Learning in conjugation with other techniques. We have attempted to void this gap by presenting a novel nonlinear dynamics and machine learning-based models to carry out predictive analytics and discovering causal interconnections in time series observations. For experimentation process to test the effectiveness of the presented framework, daily closing prices of crude oil, NIFTY energy index, US dollar-Rupee rate, and natural gas have been selected for predictive modeling and assessing causal interrelationships.

Rest of the article is structured as follows. Section 2 narrates the nature and key statistical characteristics of the dataset. Empirical investigations through the lenses of nonlinear dynamics frameworks, i.e., fractal modeling, recurrence plot (RP), and recurrence quantification analysis (RQA), have been elucidated in Sect. 3. Subsequently, machine learning models are summarized, and results of predictive performance are discussed in Sect. 4. Analysis of causal interactions using Bayesian belief network (BBN) has been summarized in Sect. 5. Finally, we conclude the article in Sect. 6 by highlighting the overall contribution of the work and discussing the future scopes.

2 Data Description

We have compiled the daily closing price data of crude oil, NIFTY energy index, US dollar-Rupee rate, and natural gas during January 2013 to May 2017, from 'Metastock' data repository. Key statistical properties of these four series are mentioned in Table 1.

It is amply evident that none of the time series abides by normal distribution according to Jarque–Bera and Shapiro–Wilk tests. Both augmented Dickey–Fuller (ADF) and Phillips–Perron (PP) tests indicate that stationary assumption does not hold for any of these series. Further autoregressive conditional heteroscedasticity—Lagrange multiplier test—strongly suggests the presence of conditional heteroscedasticity embedded in behavioral patterns of crude oil price, NIFTY energy index, US dollar-Rupee rate, and oil natural gas price which justifies the utilization of nonlinear dynamics and machine learning in modeling. Figure 1 depicts the temporal movements of these four time series.

Table 1 Statistics of crude oil, NIFTY energy index, US dollar-Rupee rate, and natural gas

Measures	Crude oil	NIFTY energy	US dollar-Rupee rate	Natural gas
Mean	4297.9669	8796.373	62.991262	204.632108
Median	3639.5000	8466.850	63.263500	201.750000
Maximum	7347.0000	12406.85	68.773500	386.200000
Minimum	1844.0000	7960.413	53.020000	111.700000
Std. Dev.	1432.4941	1107.722	3.882737	43.529406
Jarque–Bera	89.6476***	63.4264***	77.7584***	58.3287***
Shapiro–Wilk Test	0.89905***	0.90139***	0.91388***	0.98622***
ADF test	−1.725#	−1.3615#	−2.4604#	−2.1483#
PP test	−4.743#	−6.0227#	−15.166#	−12.167#
ARCH LM test	413.256***	373.442***	109.114**	554.756***

***Significant at 1% level of significance, **Significant at 5% level of significance, *Significant at 10% level of significance, #Not Significant

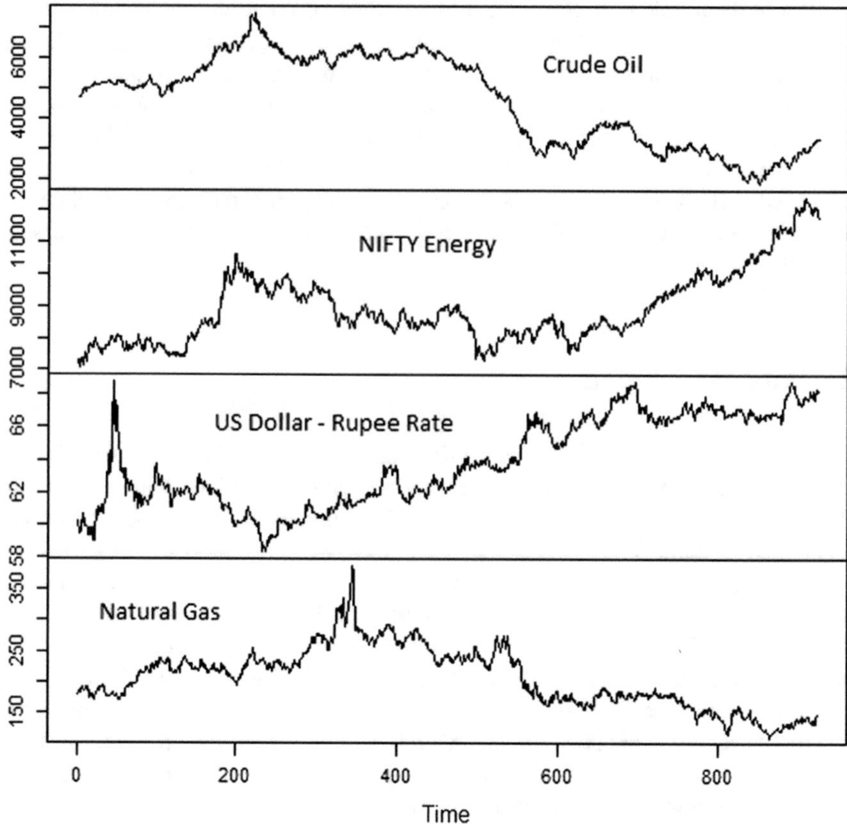

Fig. 1 Temporal movements of four financial time series

3 Empirical Analysis of Pattern

Initially to understand the underlying evolution dynamics, fractal modeling and recurrence analysis, tools of nonlinear dynamics, have been deployed. First, we describe the principle and findings of fractal modeling framework based on Mandelbrot's rescaled range analysis.

3.1 Rescaled Range (R/S) Analysis

On the basis of pioneer work of Hurst [5] Mandelbrot and Wallis [6] developed the principle of R/S analysis to distinguish between pure random time series and a biased random one. Two indices, namely Hurst exponent (H) and fractal dimensional index (FDI), are estimated applying R/S analysis to identify the presence of long-range (memory) dependence and short-range (memory) dependence. The key steps of R/S analysis are narrated below.

Step 1: The time series (R_N) of length N is initially segregated into f groups of continuous subseries of length n.
Step 2: The mean (E_f) of each subseries $R_{j,f}$ (j = 1, 2, …., n) is then determined.
Step 3: The cumulative deviation of individual elements from the mean of respective subseries is calculated as

$$X_{j,f} = \sum_{i=1}^{j} (R_{i,f} - E_f) \tag{1}$$

Step 4: The range is computed as

$$R_f = max\,(X_{j,f}) - min\,(X_{j,f}) \tag{2}$$

Step 5: Subsequently, the standard deviation (S_f) of individual subseries is determined using the following equation.

$$S_f = \sqrt{(1/n) \sum_{j=1}^{n} (R_{j,f} - E_f)^2} \tag{3}$$

Step 6: Finally, the mean value of the rescaled range for all subseries is calculated using Eq. 4.

$$(R/S)_n = (1/A) \sum_{f=1}^{F} (R_d/S_d) \tag{4}$$

3.2 *The Hurst Exponent*

The R/S statistic and Hurst coefficient (H) are asymptotically associated as

$$(R/S)_n = C{*}n^H \tag{5}$$

where C is a constant.

The value of Hurst exponent (H) is estimated via ordinary least squares (OLS) regression over a sample of increasing time horizons on the following equation.

$$\log\ (R/S)_n = \log C + H\ \log n \tag{6}$$

Magnitude of H varies in the range [0, 1]. Theoretically, it can be inferred that an estimated H value of exactly 0.5 implies that the series purely follows an independent and identically distributed (IID) Gaussian random walk model. On the counterpart, if the magnitude of H differs from 0.5, then the series is said to possess fractional Brownian motion (FBM) or biased random walk characteristics.

3.3 Fractal Dimensional Index (FDI)

It is a quantifiable measurement index that assists in comprehending the dynamics of any chaotic system. Fractals can be thought of as objects with self-similar components. Any financial time series can be thought of as a series of fractals that appear similar across a range of time periods. Through estimation of FDI, dynamics of time series observations can be understood. The FDI and H are related by Eq. 7.

$$FDI = 2 - H \tag{7}$$

Since the range of H is [0, 1], the values of FDI vary in the range of [1, 2]. For time series data of financial markets, FDI value closer to 1 infers trending market in one direction, i.e., presence of long-range dependence. This long-range dependence is also termed as 'JOSEPH'S EFFECT' and is drawn from the Old Testament Prophecy of seven years of plenty followed by seven years of famine. Alternatively, when FDI value is closer to 2 corresponds to short-range dependence which is also termed as 'NOAH'S EFFECT'. Table 2 displays the results of fractal investigation.

In brevity, it can clearly be inferred that all four series follow fractional Brownian motion. Hence, random walk hypothesis is rejected. Strong presence of trend component, as apparent from values of H and FDI, encourages the subsequent attempt of forecasting future figures through advanced machine learning. Although H value closer to 1 implies dominance of trend component, it does not

Table 2 Results of fractal analysis

Series	Hurst exponent	FDI	Nature of pattern	Effect	Random walk hypothesis
Crude oil	0.893	1.107	JOSEPH'S EFFECT	PERSISTENT	REJECTED
NIFTY energy	0.885	1.115	JOSEPH'S EFFECT	PERSISTENT	REJECTED
US dollar-Rupee rate	0.877	1.123	JOSEPH'S EFFECT	PERSISTENT	REJECTED
Natural gas	0.880	1.119	JOSEPH'S EFFECT	PERSISTENT	REJECTED

necessarily point to linear trend. To gain further deeper insights about the temporal evolution, empirical analyses have been made using recurrence plots (RP) and recurrence quantification analysis (RQA).

3.4 Recurrence Plot (RP)

The approach of fitting linear dynamical system to model time series exhibiting irregular oscillations and randomness hardly results in any success. Alternatively, the said behavior can be captured through low-dimensional deterministic chaos. Deterministic dynamic systems representing nonlinear or chaotic systems can be surmised by the recurrence of state vectors, i.e., states with large temporal distances are often tend to be in close proximity to state space. The magnitude of recurrence largely influences the overall predictability and complexity of the underlying dynamics. Recurrence plot (RP), a highly decorated graphical tool, can effectively evaluate the degree of recurrence in dynamical systems in a high-dimensional phase space. It was developed numerically by Eckman [7] in 1987. However, mere visualization through RP may not always provide deeper insights. To overcome the shortcoming, Webber and Zbilut [8] developed recurrence quantification analysis (RQA) to add quantitative depiction to standalone RP.

3.5 Phase Space Reconstruction

It is a process of recreation of topological equivalent snapshot of behavior of any multidimensional system in a single manifest variable using Taken's Delay-Embedding theory [7]. For a given one-dimensional time sequence {x1, x2, …., xn}, phase space profile of it can be reconstructed by identifying proper dimensionality (D) and time delay unit (τ). It is done by taking D times values, each shifted by a delay of τ from x to construct a particular coordinate in D dimensional phase space. Mathematically, according to delay method the first coordinate of the reconstructed state space trajectory, V1 is expressed as

$$V_1 = \left(x_1, x_{1+\tau}, x_{1+2\tau}, \ldots., x_{1+(D-1)\tau}\right) \tag{8}$$

It is possible to maximally construct n-(D-1)τ numbers of coordinates which are arranged in a matrix form as shown below:

$$V = \begin{pmatrix} V_1 \\ V_2 \\ . \\ . \\ V_{n-(D-1)\tau} \end{pmatrix} \tag{9}$$

Once the matrix is generated, the recurrence plot can be obtained by executing the following steps.

Step 1: Calculate the Euclidean norm between elements of matrix V in pairwise manner applying Eq. 10.

$$D_{ij} = \left\| V_i - V_j \right\| \tag{10}$$

Step 2: A recurrence matrix (R_{ij}) of size N is calculated as

$$R_{ij} = \theta(\varepsilon - D_{ij}) \tag{11}$$

where $\theta()$ is a Heaviside function and ε is a predefined threshold parameter.
Step 3: Two-dimensional graphical representation of the recurrence matrix by substituting a black dot to the value 1 and a white to value 0 results in the recurrence plot.

3.6 *Recurrence Quantification Analysis (RQA)*

It is a nonlinear data analysis technique that provides deeper insights beyond the capacity of RPs by classifying the nature of dynamical systems through computations of some manifest measures. It basically attempts to quantify the magnitude and duration of recurrence observed in state space trajectory.

3.6.1 Recurrence Rate

It measures the percentage of recurrent points in RP. Higher value of RR indicates the presence of periodicity while smaller value implies chaotic behavior. It is calculated as

$$RR = \frac{1}{N^2} \sum_{i,j=1}^{N} R_{ij} \tag{12}$$

3.6.2 Determinism Rate

It is the percentage of recurrence points that construct additional diagonal lines parallel to main diagonal. It accounts for overall deterministic nature of a system. RP of chaotic systems tends to possess none or diagonals of marginal length in contrast to periodic or quasi-periodic systems dynamics that are characterized by a mixture of short and long diagonals parallel to the main diagonal. It is computed as:

$$DET = \frac{\sum_{l=l_{min}}^{N} lp(l)}{\sum_{i,j}^{N} R_{ij}} \tag{13}$$

where p(l) is the probability of the length l of the diagonal lines.

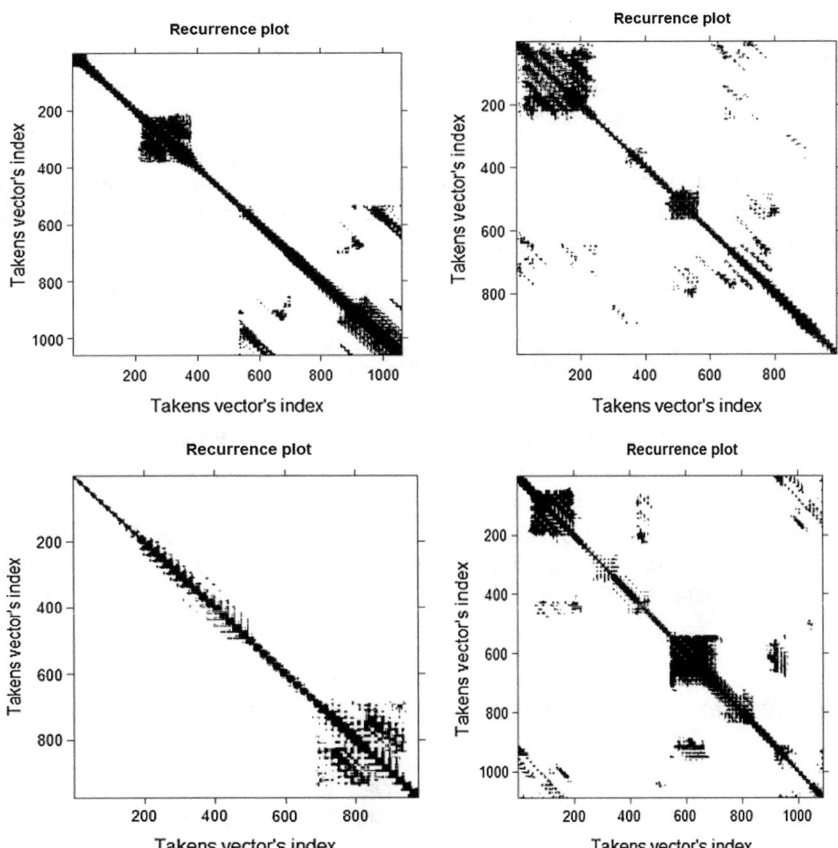

Fig. 2 Recurrence plots of respective time series

Table 3 Results of RQA

Time Series	%REC	%DET	%LAM
Crude oil (%)	5.45	93.95	96.02
NIFTY energy (%)	3.80	87.60	91.93
US dollar-Rupee rate (%)	3.04	87.31	93.58
Natural gas (%)	5.09	92.90	95.61

Laminarity (LAM): It measures the amount of recurrence points that form vertical hypersurface and is estimated as:

$$LAM = \frac{\sum_{v=v_{min}}^{N} vp(v)}{\sum_{v=1}^{N} vp(v)} \tag{14}$$

where p(v) is the probability of the length v of the vertical lines, and v_{min} is the minimum length of vertical line.

It can be noticed that none of the RPs do not exhibit complete deterministic behavior as shown in Fig. 2. Presence of parallel line to main diagonal, disjoint areas, and small surfaces rather suggest the existence of higher order deterministic chaos. For better interpretation, RQA has been applied and the findings are summarized in the following table (Table 3).

It can be observed that the recurrence rates of four financial series are higher than that of chaotic series. High values of rate determinism and laminarity justify deterministic characteristics and predictable nature of time series which in turn validate the findings of fractal modeling

4 Machine Learning-Based Modeling

Since the findings of empirical investigation through nonlinear dynamics framework rule out the random walk hypothesis hint at deterministic dynamics, machine learning models are employed to discover the governing pattern. Three machine learning techniques, namely random forest (RF), boosting, and group method of data handling (GMDH), have been used for predictive analytics. Brief overview of these three methods is elucidated below.

4.1 Random Forest (RF)

RF was developed by Breiman [9] back in 2001. It is an ensemble-based machine learning tool that is predominantly used in predictive modeling problems. It can perform both classification and regression tasks based on the nature of its base models. In general, decision trees for classification and regression are selected as base learners of RF. Lately, it has garnered lot of attention worldwide due to its high precision and flexibility in solving complex pattern recognition tasks [10, 11]. The base learners, i.e., the decision trees are constructed based on randomly selected subset of available input features in training dataset. The assignment of class label of an unknown instance (for classification) or prediction of continuous outcome (regression) is performed using simple majority voting or averaging strategy.

4.2 Boosting

Similar to RF, boosting is also an ensemble-based predictive modeling technique where individual training samples are assigned weight values [12]. A series of k number of classifiers/regression modelers are utilized and iteratively trained to recognize the pattern. Once the learning operation of a particular classifier/ regression modeler is completed, the weights are updated to allow the subsequent learner to focus more on the training samples where misclassification/error rates were comparatively higher. The final outcome of boosting is the weighted outcome of individual base learners. These weights are decided based on the predictive accuracy of respective base modelers. In this work, AdaBoost (Adaptive Boosting) [13] has been adopted. Cortes et al. [14] applied AdaBoost M1 algorithm for prediction of European corporate failure in multiclass classification framework. Leshem and Ritov [15] developed machine learning-based intelligent system for traffic flow management system

4.3 Group Method of Data Handling (GMDH)

It is a variant of neural network that applies a series polynomial function of combinations of attributes to learn complex manner in a heuristic self-organizing mode. Originated by Ivakhenko [16], GMDH can effectively carry out modeling of univariate time series by identifying relational form among the time lags. The basic Ivakhenko polynomial for n attributes is expressed as:

$$W = f(x_1, x_2, \ldots\ldots, x_n) = \alpha_0 + \sum_{i=1}^{n} \alpha_i x_i + \sum_{i=1}^{n}\sum_{j=1}^{n} \beta_{ij} x_i y_j + \sum_{i=1}^{n}\sum_{j=1}^{n}\sum_{k=1}^{n} \gamma_{ijk} x_i y_j z_k + \ldots \tag{15}$$

where W is the response variable, coefficients of variables are represented by α_0, α_i, β_{ij},γ_{ijk}. The predictors are represented by x_i, y_j, z_k, etc. For forecasting in univariate framework, the above equation can be expressed as:

$$W = f(x_1, x_2, \ldots.., x_n) = \alpha_0 + \sum_{i=1}^{n} \alpha_i x_i + \sum_{i=1}^{n}\sum_{j=1}^{n} \beta_{ij} x_i x_j + \sum_{i=1}^{n}\sum_{j=1}^{n}\sum_{k=1}^{n} \gamma_{ijk} x_i x_j x_k \quad (16)$$

Parameters are learned through regularized least square estimation process. Literature reports successful use of GMDH in various real-life problems [17, 18]. Implementation of the algorithm has been carried out using *'gmdh'* package of R.

To evaluate the performance of predictive modeling, four quantitative measures, namely mean squared error (MSE), Nash–Sutcliffe coefficients (NSC), index of agreement (IA), and Theil inequality (TI), are computed. These measures are calculated as:

$$MAPE = \frac{100}{N} \sum_{i=1}^{N} \left| \frac{Y_{act}(i) - Y_{pred}(i)}{Y_{act}(i)} \right| \quad (17)$$

$$NSC = 1 - \frac{\sum_{i=1}^{N} \{\mathrm{Y_{act}(i)} - \mathrm{Y_{pred}(i)}\}^2}{\sum_{i=1}^{N} \{\mathrm{Y_{act}(i)} - \overline{Y_{act}}\}^2} \quad (18)$$

$$IA = 1 - \frac{\sum_{i=1}^{N} \mathrm{Y_{act}(i)} - \mathrm{Y_{pred}(i)}^2}{\sum_{i=1}^{N} \{|\mathrm{Y_{pred}(i)} - \overline{Y_{act}}| + |\mathrm{Y_{act}(i)} - \overline{Y_{act}}|\}^2} \quad (19)$$

$$TI = \frac{\left[\frac{1}{N}\sum_{i=1}^{N} (\mathrm{Y_{act}(i)} - \mathrm{Y_{pred}(i)})^2\right]^{1/2}}{\left[\frac{1}{N}\sum_{i=1}^{N} \mathrm{Y_{act}(i)}^2\right]^{1/2} + \left[\frac{1}{N}\sum_{i=1}^{N} \mathrm{Y_{pred}(i)}^2\right]^{1/2}} \quad (20)$$

where $\mathrm{Y_{act}(i)}$ and $\mathrm{Y_{pred}(i)}$ are actual and predicted outcomes. Values of MAPE must be as low as possible to indicate efficient prediction; ideally, a value of zero signifies no error. NSC value of 1 corresponds to a perfect match between model and observations whereas value less than zero occurs when the observed mean is a better predictor than the modeler. The range of IA lies between 0 (no fit) and 1 (perfect fit). TI values should be close to 0 for good prediction while a value close to 1 implies no prediction at all.

The datasets of individual time series have been partitioned into training (80%) and test (20%) datasets. As the algorithms are sensible to different parameters, viz. number of base learners and degree of polynomial, ten experimental trials have been designed by varying the respective parameters. Arithmetic average values of obtained performance indicators on these trials have been computed to judge the forecasting performance. Results of predictive performance for individual models are shown in the following tables (Tables 4, 5, and 6).

It can clearly be observed that for all three models, NSE and IA values are close to one and values of MAPE and TI are quite low on both training and test datasets.

Table 4 Predictive performance of RF

Series	MAPE	NSE	IA	TI
Training dataset				
Crude oil	0.6879	0.9989	0.9992	0.0044
NIFTY energy	0.7045	0.9914	0.9987	0.0047
US dollar-Rupee rate	0.6227	0.9992	0.9998	0.0036
Natural gas	0.7237	0.9876	0.9973	0.0041
Test dataset				
Crude oil	0.6811	0.9991	0.9993	0.0044
NIFTY energy	0.7946	0.9919	0.9990	0.0045
US dollar-Rupee rate	0.6144	0.9993	0.9998	0.0035
Natural gas	0.7178	0.9884	0.9977	0.0041

Table 5 Predictive performance of boosting

Series	MAPE	NSE	IA	TI
Training dataset				
Crude oil	0.6861	0.9990	0.9993	0.0041
NIFTY energy	0.6985	0.9927	0.9988	0.0046
US dollar-Rupee rate	0.6163	0.9994	0.9999	0.0034
Natural gas	0.7188	0.9879	0.9975	0.0040
Test dataset				
Crude oil	0.6856	0.9990	0.9993	0.0041
NIFTY energy	0.6972	0.9927	0.9988	0.0046
US dollar-Rupee rate	0.6149	0.9994	0.9999	0.0034
Natural gas	0.7174	0.9879	0.9975	0.0040

Table 6 Predictive performance of GMDH

Series	MAPE	NSE	IA	TI
Training dataset				
Crude oil	0.7126	0.9893	0.9885	0.0061
NIFTY energy	0.7517	0.9813	0.9869	0.0072
US dollar-Rupee rate	0.6834	0.9905	0.9891	0.0049
Natural gas	0.7789	0.9786	0.9842	0.0065
Test dataset				
Crude oil	0.6878	0.9927	0.9926	0.0057
NIFTY energy	0.7405	0.9845	0.9899	0.0069
US dollar-Rupee rate	0.6671	0.9919	0.9910	0.0046
Natural gas	0.7537	0.9834	0.9883	0.0061

Hence, inference can be drawn that RF, boosting, and GMDH can effectively be used for forecasting future movements crude oil, NIFTY energy, exchange rate, and natural gas.

5 Causal Analysis

To accomplish the secondary objective of this research, to explore the causal interactions among the four financial time series, Bayesian belief network (BBN) has been adopted to accomplish the task.

Bayesian Belief Network (BBN): It represents the joint probability distribution of a set of interlinked variables in a graphical manner [19]. It specifies a set of conditional independence assumptions and conditional probabilities that govern the variables. Unlike classical naive Bayes classifier, BBN allows the flexibility to impose the conditional independence condition to a subset of variables. BBN comprises of a set of node representing the stochastic variables and arcs, connecting the nodes which account for direct causal interrelationships among the variables. A properly learned BBN can capture the true interrelationship among set of variables and represent it in a directed acyclic graph (DAG). In most of the occasions, traditional econometric models, viz. Granger causality analysis, co-integration test, vector autoregression (VAR), and vector error correction method (VECM) have

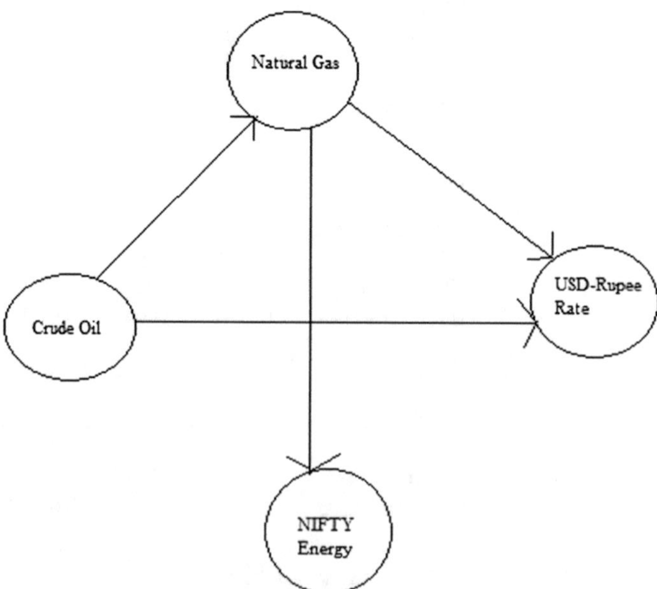

Fig. 3 Causal interrelationship

been deployed to analyze causal influence as reported in literature. The limitations of these methods are well known. On the other hand, BBN is capable of discovering nonlinear association network in a complete nonparametric framework. We have used *'bnlearn'* package of R for simulation. The following figure displays the causal influence network. It is found that crude oil has significant impact on price of natural gas and USD-Rupee rate while price of natural gas influences US dollar-Rupee rate and NIFTY energy (Fig. 3).

6 Concluding Remarks

This paper presents a novel research framework for predictive and empirical analysis. Findings implicate its efficacy in overall modeling and discovering hidden causal interactions. In the context of financial markets, results are extremely useful for different market players who are actively involved in trading, portfolio balancing, resource allocation, etc. Although the research model has been applied to financial time series, it can easily be extended for pattern recognition in other domains too. In the future, statistical comparative analysis of predictive performance of machine learning models can be conducted and a new ensemble can be designed incorporating many other state-of-the-art predictive modeling algorithms. Also, nonlinear causality models can be applied in parallel to construct the influence network for validating the findings of BBN.

References

1. Tao, H.: A wavelet neural network model for forecasting exchange rate integrated with genetic algorithm. IJCSNS Int. J. Comput. Sci. Netw. Sec. **6**, 60–63 (2006)
2. Rather, A.M., Agarwal, A., Sastry, V.N.: Recurrent neural network and a hybrid model for prediction of stock returns. Expert Syst. Appl. **42**, 3234–3241 (2015)
3. Ramasamy, P., Chandel, S.S., Yadav, A.K.: Wind speed prediction in the mountainous region of India using an artificial neural network model. Renew. Energy **80**, 338–347 (2015)
4. Yin, K., Zhang, H., Zhang, W., Wei, Q.: Fractal analysis of gold market in China. Rom. J. Econ. Forecast. **16**, 144–163 (2013)
5. Hurst, H.E.: Long-term storage capacity of reservoirs. Trans. Am. Soc. Civ. Eng. **116**, 770–808 (1951)
6. Mandelbrot, B., Wallis, J.: Noah, Joseph and operational hydrology. Water Resour. Res. **4**, 909–918 (1968)
7. Eckmann, J.P., Kamphorst, S.O., Ruelle, D.: Recurrence plot of dynamical system. Europhys. Lett. **4**, 4973–4977 (1987)
8. Webber, C.L., Zbilut, J.P.: Dynamical assessment of physiological systems and states using recurrence plot strategies. J. Appl. Physiol. **76**, 965–973 (1994)
9. Breiman, L.: Random forests. Mach. Learn. **45**, 5–32 (2001)
10. Xie, Y., Li, X., Ngai, E.W.T., Ying, W.: Customer churn prediction using improved balanced random forests. Expert Syst. Appl. **36**, 5445–5449 (2009)

11. Lariviere, B., Van den Poel, D.: Predicting customer retention and profitability by random forests and regression forests techniques. Expert Syst. Appl. **29**, 472–484 (2005)
12. Schapire, R.E., Singer, Y.: Improved boosting algorithms using confidence-rated predictions. Mach. Learn. **37**, 297–336 (1999)
13. Collins, M., Schapire, R.E., Singer, Y.: Logistic Regression, AdaBoost and Bregman distances. Mach. Learn. **48**, 253–285 (2002)
14. Cortes, E.A., Martinez, M.G., Rubio, N.G.: Multiclass corporate failure prediction by Adaboost. M1. Int. Adv. Econ. Res. **13**, 301–312 (2007)
15. Leshem, G., Ritov, Y.: Traffic flow prediction using Adaboost Algorithm with random forests as a weak learner. Int. J. Math. Comput. Phys. Electr. Comput. Eng. **1**, 1–6 (2007)
16. Ivakhnenko, A.G.: The group method of data handling—a rival of the method of stochastic approximation. Soviet Autom. Control **13**, 43–55 (1968)
17. Zhang, M., He, C., Liatsis, P.: A D-GMDH model for time series forecasting. Expert Syst. Appl. **39**, 5711–5716 (2012)
18. Najafzadeh, M., Barani, G.H., Azamathulla, H.M.: GMDH to predict scour depth around a pier in cohesive soils. Appl. Ocean Res. **40**, 35–41 (2013)
19. Scutari, M.: Learning Bayesian Networks with the bnlearn R Package. J. Stat. Softw. **35**, 1–22 (2010)

Adaptive Channel Equalization Using Decision Directed and Dispersion Minimizing Equalizers Trained by Variable Step Size Firefly Algorithm

Archana Sarangi, Shubhendu Kumar Sarangi and Siba Prasada Panigrahi

Abstract This paper signifies to present a design methodology for equalization of nonlinear channels for weights adaptation. Adaptive algorithms such as PSO, FFA, and VSFFA-based channel equalizer aimed to minimize inter-symbol interference associated with broadcast channel. In this paper, we implemented various channel equalizers such as decision directed equalizer, dispersion minimizing equalizer using PSO, FFA, and VSFFA which are principally derivative-free optimization tools. These algorithms are appropriately used to update weights of equalizers. Accomplishment of proposed diverse channel equalizers are evaluated in terms of mean square error (MSE) and BER plots and assessments are made using evolutionary algorithms applied to equalizers. It is observed that proposed equalizer-based adaptive algorithms, mostly VSFFA trained equalizers, offer improved performance so far as accurateness of reception is taken into account.

Keywords Particle swarm optimization · Adaptive channel equalizer Firefly algorithm · Variable step size firefly algorithm · Decision directed equalizer · Dispersion minimizing equalizer

1 Introduction

In nonlinear adaptive signal processing areas such as system identification, nonlinear and non-stationary signal prediction, equalization, and noise cancelation, nonlinear adaptive filtering practices have acknowledged extensive curiosity in contemporary years. To realize requirement, voluminous scholars have done research to its advances. Innumerable varieties of nonlinear scheme methodologies are projected to build and estimate nonlinear systems.

A. Sarangi (✉) · S. K. Sarangi
ITER, Siksha 'O' Anusandhan University, Bhubaneswar, India
e-mail: archanasarangi24@gmail.com

S. P. Panigrahi
Department of Electrical Engineering, VSSUT, Burla, India

V. Bhateja et al. (eds.), *Intelligent Engineering Informatics*, Advances in Intelligent Systems and Computing 695, https://doi.org/10.1007/978-981-10-7566-7_30

In contemporary digital communication schemes, broadcast of advanced data over medium is limited by inter-symbol interference (ISI) instigated by distortion in broadcast medium. Higher data broadcast over mediums with acute distortion can be realized by using inverse filter in receiver that stabilizes distortion. Since medium is time fluctuating and unidentified in strategized point due to discrepancies in broadcast standard, we require an adaptive equalizer that offers benefit over time-fluctuating medium and endeavors to recover broadcast symbols. Mostly task of equalizer is to reorganize corrupted data series from set of received symbols with inverse modeled transmission medium. In case of acute channel distortion, non-linear equalizers provide superior advantages as compared to linear ones. Also, linear equalizer is not applicable for non-minimum phase mediums, which are demanding for equalization. The most recurrently used structure of equalizer is transversal adaptive filter with gradient-based algorithms.

An adaptive filter has feature of self-adjustment, its parameters to remove the effects of distortion on input signal. It can be coined as filter which fine-tunes its transfer function with respect to optimization algorithm motivated by error signal. The center of work is to advance and examine diverse equalizers, which can be self-adaptive with evolutionary algorithms, so as to reduce error. Above mentioned algorithms diverge in means that they adapt their coefficients, irrespective of how they accomplish convolution procedures. To resolve ISI complications, voluminous processes and structures are projected. Remedy to diminish consequence of ISI is channel equalization. Adaptive algorithms are utilized to modernize factors of equalizer when medium is unidentified and time fluctuating. Initiation of coefficients is carried by conveying a skilling system from broadcaster to receiver and hence tracked by decision directed approach, dispersion minimizing approach for reception of facts. With numerous processes and structures recommended, total effort is profitable to produce decision directed equalization, dispersion minimizing equalization.

The main aim of equalizer is to eradicate inter-symbol interference and additive noise as much as conceivable. Inter-symbol interference occurs with scattering of communicated symbols due to depressive feature of medium, which outcomes in overlay of neighboring symbols. Adaptive equalizers undertake medium is time fluctuating and attempt to project equalizer whose coefficients are changing in time conferring to variation in medium, and eradicate ISI and noise at each time. Equalization is utilized to eradicate inter-symbol interference produced due to constrained bandwidth of broadcast medium. When medium is band limited, symbols conveyed through will be disseminated. Also, multipath transmission in wireless infrastructures grounds interference at receiver. Thus, equalizers are utilized to spot frequency response of equalizer system (Fig. 1).

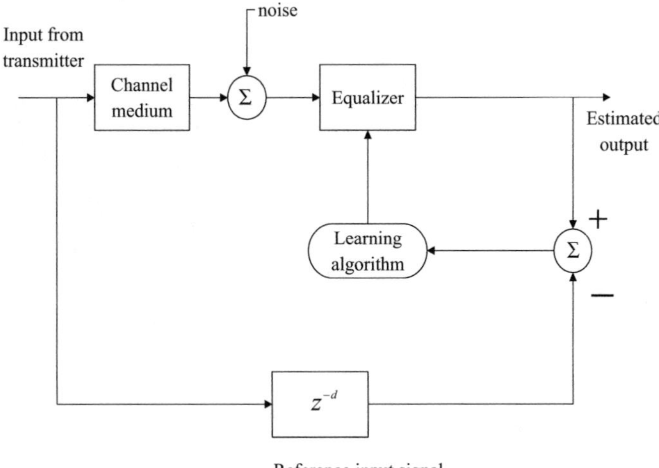

Fig. 1 Channel equalization system

2 Design Formulation of Equalizers

2.1 Decision Directed Equalizer

Blind channel equalization is recognized as self-recovery process. Principles to improve unidentified input categorization to unidentified medium rely solely on probabilistic and geometrical possessions of input class. The word blind is utilized in such disparity as it achieves equalization without referring any signal. As a substitute, blind equalizer trusts on acquaintance of signal assembly and its indicators to accomplish equalization. Equalizers whose actions rely on this principle are decision directed and dispersion minimization equalization. This DDE means that receiver conclusions are utilized to produce error signal. Decision directed equalizer tuning is effectual in pursuing inactive disparities in medium response.

Equalizer constraints can be improved without use of training data. This methodology supports to recover medium ability as well as lessen cost. Seeing state in which some technique has fashioned equalizer situation that releases eye of medium. All judgments are faultless but equalizer constraints may not be at optimum point. So output of decision device is precise copy of deferred source. For source of binary ± 1 and decision device that is sign operator, deferred source retrieval error can be figured as sign {yy[k]} – yy[k] where yy[k] is equalizer output and sign {yy[k]} equivalents deferred source $s[k-\delta]$ (i.e., conveyed signal delayed by δ). This translates LMS equation to decision directed least mean square (LMS) as,

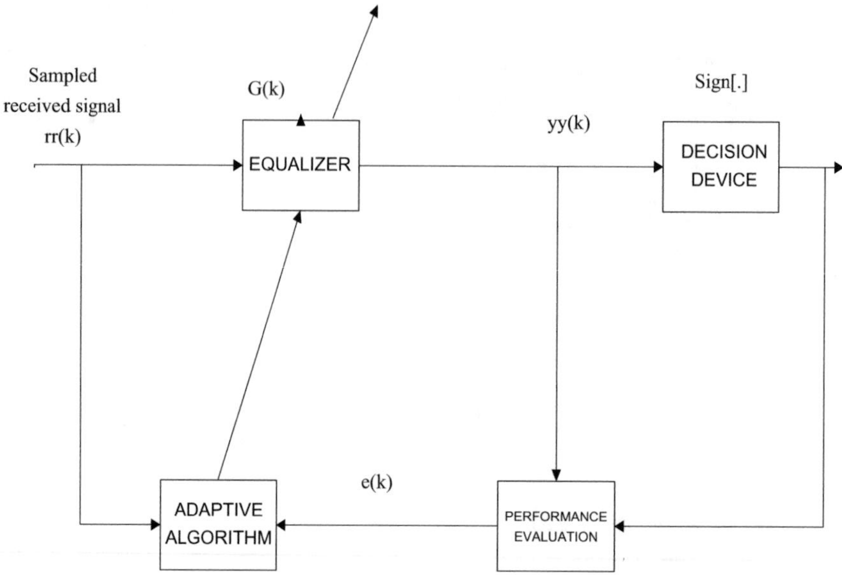

Fig. 2 Decision directed adaptive linear equalizer

$$G(k+1) = G_i(k) + \mu(sign(yy(k)) - yy(k))rr(k-i) \tag{1}$$

Practically that source [k] does not seem in above equation. So no training signal is obligatory for its employment and decision directed LMS equalizer adaptation law is titled as "blind" equalizer. The rudimentary regulation of thumb is that 5% (or so) decision errors can be abided before decision directed LMS flops to assemble appropriately. Initialization utilized switches taps at zero except for one in middle that initiates at unity, title das "center-spike" initialization [1] (Fig. 2).

2.2 Dispersion Minimizing Equalizer

This segment of equalizer contemplates alternative enactment function with additional kind of blind equalization. Perceived that for source of binary ±1, square of source is recognized while specific values of source are not. Thus, $SS^2(k) = \gamma = 1$ for all k. In specific, consider,

$$J_{DMA} = \frac{1}{4} avg(\gamma - yy^2(k)) \tag{2}$$

It measures dispersion of equalizer outcome about its anticipated square value. Accompanying adaptive element for modernizing equalizer coefficients is,

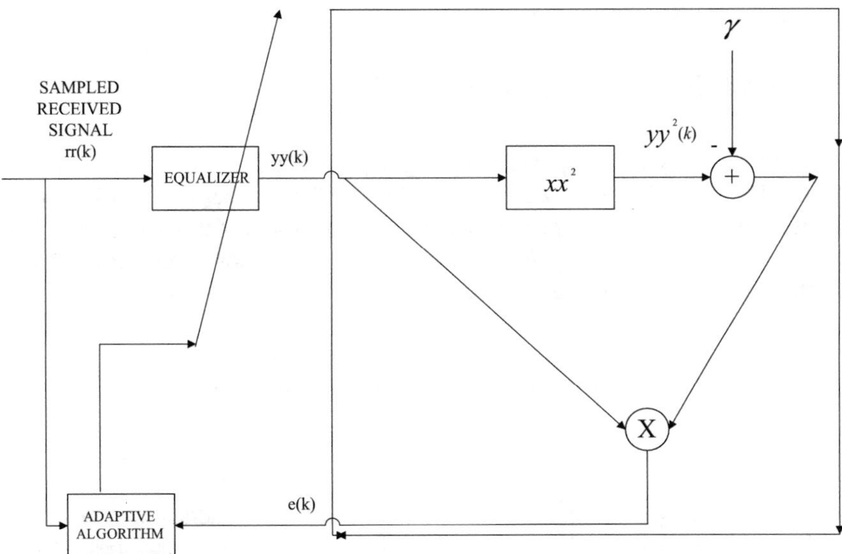

Fig. 3 Dispersion minimizing adaptive linear equalizer

$$G_i(k+1) = G_i(k) + \mu \frac{d\,J_{DMA}}{d\,G_i}\bigg|\; G_i = G_i(k) \tag{3}$$

Dispersion minimization equalization (DM) for blindly adapting coefficients of linear equalizer is

$$G_i(k+1) = G_i(k) + \mu avg(1 - yy^2(k))yy(k)rr(k-1) \tag{4}$$

Without averaging operation, it will become (Fig. 3)

$$G(k+1) = G_i(k) + \mu avg(1 - yy^2(k))yy(k)rr(k-1) \tag{5}$$

While DM equalizer characteristically may converge to anticipate response from inferior starting than decision directed, it is not as enthusiastic as proficient LMS. For a specific delay, squared retrieval error surface being sloped along gradient by LMS is unimodal, i.e., having one minima [1].

So, no issue where exploration is ready, it discovers anticipated sole minima, allied with δ utilized in figuring recover error. Dispersion performance function is multimodal with distinct minima agreeing to dissimilar attained deferrals and polarities. It is perceived that +1's are swapped with −1's has identical value at optima. Thus, convergent deferral and polarity attained relied on initialization utilized. Typical initialization for DM equalizer is single nonzero spike positioned nearby center of equalizer.

3 Optimization Techniques

3.1 Particle Swarm Optimization (PSO)

PSO algorithm significantly depends on group behavior of birds, etc., with their natural habits flocking. In PSO, a plentiful particle propagates everywhere in solution space to find out supreme solution. They also stare at best elements in their path [2]. So particles think about their individual finest solutions with excellent solution lay down so far. PSO was exactly modeled as follow:

$$\begin{aligned} velc(t+1) = w.velc(t) + p_1.rand.(pbest(t) - posc(t)) \\ + p_2.rand.(gbest(t) - posc(t)) \end{aligned} \tag{6}$$

$$posc(t+1) = posc(t) + velc(t+1) \tag{7}$$

where t, velc(t), posc(t), pbest(t) are existing number of iterations, velocity of particles, position of particles, and personal finest position of particle, respectively. Also, gbest(t) is global best position and p_1, p_2 are acceleration coefficients.

3.2 Firefly Algorithm (FFA)

The fireflies flashing is an exceptional sight in summer sky in tropical and temperate areas. The crucial characteristic of flashes is to engaged mating partners and entreat potential prey.

Basic concepts for FFA are:

(1) A firefly may be engaged to another in spite of their group.
(2) For a couple of blinking fireflies, lighter firefly will follow brighter one. But if brighter firefly is not accessible, then it travels irregularly [3].

For global minima, brilliance can only inversely proportionate to estimation of working function. Firefly algorithm is centered on two crucial concepts, one is discrepancy in light intensity and other is initiation of attraction of other fireflies. Attractiveness of firefly is determined by its brightness which is by some means connected to working function. Brightness of firefly at fixed spot x is preferred as the G(x). For any certain medium, light absorption coefficient γ, light intensity G fluctuates with distance r as,

$$G = G_0 e^{-\gamma r^2} \tag{8}$$

Here G_0 is actual radiant intensity at r = 0.

Distance among any two fireflies i and j can be specified as,

$$r_{ij} = \left\|x_i - x_j\right\| = \sqrt{\sum_{k=1}^{D} (x_{i,k} - x_{j,k})^2} \tag{9}$$

where $x_{i,k}$ is kth element of ith firefly.

The motion of ith firefly is occupied to one more striking jth firefly and is evaluated by,

$$x_i = x_i + \beta_0 e^{-\gamma r_{ij}^2}(x_j - x_i) + \alpha(rand - \frac{1}{2});\ \beta_0\ ;\text{is}\ \ +\text{ve} \tag{10}$$

3.3 Variable Step Size Firefly Algorithm (VSFFA)

In normal firefly, size of step α is constant. It cannot actually track search mechanism. But searching for more area in solution space, larger step size is required, but it is not supportive for convergence to global optimum. If size of step has a small value, effect is divergent. Accordingly, size of step α has a huge influence on exploration and convergence of algorithm. It would be beneficial to stability capability of global exploration and local exploitation about presenting situation [4].

Due to this factor the size of step should modify dynamically with a new pattern as follows:

$$\begin{aligned}&\alpha(existing\_iteration) = 0.4/(1 + \exp(0.015* \\ &(existing\_iteration - \max generation)/3))\end{aligned} \tag{11}$$

Sequence of execution as follows:

Step 1 Create primary population of fireflies.
Step 2 Estimate intensity of light for particular firefly.
Step 3 Determine size of step α;
Step 4 Using the modified step size update position of fireflies.
Step 5 Jump to step 2 if stopping criterion is achieved.

4 Result Analysis

This paper utilizes standard nonlinear channel for testing successfulness of evolutionary algorithms. From simulation results, parameters utilized for algorithms are specified as population size = 30, α = 0.2, β = 1, ϒ = 1. For superior analysis of three algorithms, different MSE curves and BER plots, i.e., MSE of three

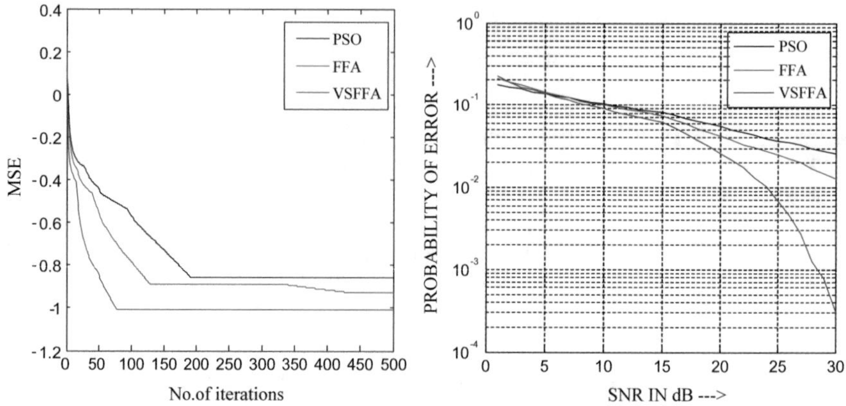

Fig. 4 MSE and BER for decision directed equalizer

algorithms with respect to 500 no. of iterations, BER with respect to SNR plots, were considered.

Transfer function of channel utilized for equalization is,

$$H(z) = 0.2600 + 0.9300\,z^{-1} + 0.2600\,z^{-2} \tag{12}$$

Decision Directed Equalizer Above Fig. 4 specifies combination of three algorithms for Decision directed equalizer for given channel. The value of MSE for PSO trained DD equalizer was found out to be −0.8606 with 197 no. of iterations. The value of MSE for FFA trained DD equalizer was found out to be −0.9284 with 433 no. of iterations. The value of MSE for VSFFA trained DD equalizer was found out to be −1.009 with 84 no. of iterations.

Dispersion Minimizing Equalizer (Fig. 5)

BER Performance All equalizer structures were competent for certain iterations, and their weights are intended and restored. Thereafter, to compute bit error rate, restored weights are given into equalizer network which can be adapted through optimization algorithms. Rely on new received ones, equalizer approximates conveyed symbol. This progress was frequent for diverse values of additive noise having SNR = 30 dB. In order to illustrate efficiency of equalizers under different channel nonlinearities, we have plotted BER with ranging SNR = 0 to 30 dB, We have stated channel for plotting SNR vs BER plots and BER's are compared as per algorithms trained.

- BER values for PSO, FFA, and VSFFA trained decision directed equalizer at 10 dB are $10^{-0.985}$, $10^{-1.0028}$, and $10^{-1.046}$ and at 30 dB are $10^{-1.57}$, $10^{-1.88}$, and $10^{-3.52}$, respectively. Hence, we noticed that at SNR(dB) = 10, VSFFA trained decision directed equalizer illustrates better BER performance than that of PSO trained decision directed equalizer. Similarly, when we compare it at SNR

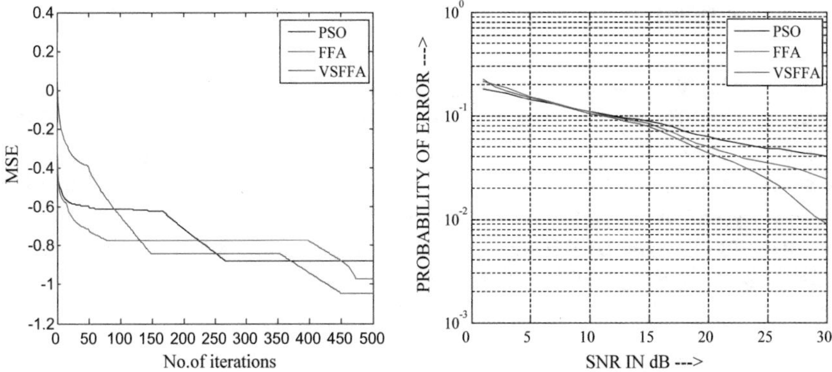

Fig. 5 MSE and BER for dispersion minimizing equalizer

(dB) = 30, VSFFA trained decision directed equalizer illustrates better BER performance than that of PSO trained decision directed equalizer. When BER values for both dB's are compared, VSFFA trained equalizer (at SNR = 30 dB) outperforms VSFFA trained equalizer (at SNR = 10).

- BER values for PSO, FFA, and VSFFA trained dispersion minimizing equalizer at 10 dB are $10^{-0.962}$, $10^{-0.968}$, and $10^{-0.982}$ and at 30 dB are $10^{-1.4}$, $10^{-1.62}$, and $10^{-2.051}$, respectively. Hence, we noticed that at SNR(dB) = 10,VSFFA trained dispersion minimizing equalizer illustrates better BER performance than that of PSO trained dispersion minimizing equalizer. Similarly, when we compare it at SNR(dB) = 30, VSFFA trained dispersion minimizing equalizer illustrates better BER performance than that of PSO trained dispersion minimizing equalizer. When BER values for both dB's are compared, VSFFA trained equalizer (at SNR = 30) outperforms VSFFA trained equalizer (at SNR = 10).

Simulation results up to SNR = 30 dB is carried out with channel conditions. The above plots signify comparison of MSE performance of three algorithms, particle swarm optimization, firefly algorithm, and variable step size firefly algorithm applied to different equalizers. It can be revealed that PSO algorithm reliably acts poorest, in that it sometimes exhibits higher rate of convergence, with less MSE as compared to other algorithms trained equalizers. Also, concluded that VSFFA algorithm reliably accomplishes convergence with high MSE, with least sensitive to variations noticed that, for low channel distortion, enactment of VSFFA is very close by to optimum solution. Therefore, performance of equalizers is likened by scheming BER plots. It can be noticed that, with less noisy channel circumstances, PSO- and FFA-based equalizers accomplish almost same. However, with high noisy channel circumstances, VSFFA-based equalizer outperforms than other two algorithms trained equalizers.

5 Conclusion

This work implies applicability of gradient-free optimization practices such as PSO, FFA, and VSFFA for error minimization. Assessment of PSO, FFA, and VSFFA trained channel equalizers are carried out and hence compared. It is established that VSFFA-based channel equalizers yield better MSE values as compared to MSE yielded by rest other algorithms but convergence rate varies from one equalizer to another equalizer and BER analysis for VSFFA algorithm trained equalizer yielded better results as compared to other algorithms trained equalizer.

References

1. Akaneme, E.S.A., Onoh, G.N.: Algorithms for evaluating adaptive equalization in wireless communication systems. Eur. J. Eng. Technol. **3**(2) (2015). ISSN 2056-5860
2. Kennedy, J., Eberhart, R.C.: Particle swarm optimization. In: Proceedings of the IEEE International Conference on Neural Networks, vol. 4, pp. 1942–1948 (1995)
3. Yang, X.S.: Firefly algorithms for multimodal optimization. In: Proceedings of the 5th International Conference on Stochastic Algorithms Foundations and Applications, vol. 5792, pp. 169–178. LNCS Springer (2009)
4. Yu, S., Zhu, S., Ma, Y., Mao, D.: A variable step size firefly algorithm for numerical optimization. Appl. Math. Comput. **263**, 214–220 (2015)

A Study on Wrapper-Based Feature Selection Algorithm for Leukemia Dataset

M. J. Abinash and V. Vasudevan

Abstract In many fields, big data play a predominant role such as in research, business, biological science, and many other fields in our day-to-day activities. It is mainly the voluminous amount of structured, semi-structured, and unstructured data. It is the base from the data mining. So the knowledge discovery using this data is very difficult. Bioinformatics is an interdisciplinary of biology and information technology; the gene expression or the microarray data are analyzed using some softwares. These gene data are grown higher and higher, so the analyze and the classification are more difficult among these growing big data. So we focus on analyzing these data for cancer classification. The proposed work discusses the SVM-based wrapper feature selection for cancer classification. The cancer dataset are applied in two feature selection algorithms, and among them, the wrapper-based SVM method is made best for feature selection for cancer classification.

Keywords Gene · Feature selection · Support vector machines (SVM)

1 Introduction

Big data is the tremendous amount of data; it can in the form of audio, video, spreadsheets, biological data, etc. The data are in the structured such as tables, spreadsheets, semi-structured such as audio, video, radio frequency identification etc., unstructured such as XML data, mail data. These data gave challenges to the data researches, data scientists for storing, capturing, analysis, retrieval, classification, etc.

M. J. Abinash (✉) · V. Vasudevan
Department of Information Technology, Kalasalingam University, Virudhunagar, Tamil Nadu, India
e-mail: mj.abinash@gmail.com

V. Vasudevan
e-mail: vasudevan_klu@yahoo.co.in

V. Bhateja et al. (eds.), *Intelligent Engineering Informatics*, Advances in Intelligent Systems and Computing 695, https://doi.org/10.1007/978-981-10-7566-7_31

Bioinformatics is the interdisciplinary of science that combines the statistical, computer science, and the biological data for analyzing and for interpreting using some software tools and methods. While analyzing the biological data i.e., gene expression data is more difficult for classification because the curse of dimensionality may be occur. Using gene expression data, is essential in cancer classification and these gene data are analyzed by the data analyzers, statisticians, researchers in their laboratory. The feature selection is needed to select the relevant features for classifying the data. The previous methods have some more drawbacks in feature selection such as less accuracy, time complexity, computational cost. In this paper, we overcome this to get the accurate features for cancer classification.

A wrapper-based support vector machine (SVM) is proposed for selecting the relevant features. The gene expression data or the microarray data are in the row x column matrix of expression levels. The rows represent genes or samples, and the columns represent the gene attributes or variables. It may be in the ordinal or numerical. The microarray data are arranged in the form of matrix [1]. Machine Learning is a technique that gives a prior knowledge base to the computers without being explicitly programmed frequently. Nowadays, machine learning algorithms are used in cancer diagnosis. Last few years, the researchers working on cancer diagnosis were focused on improvement of classification accuracy [2]. The main aim of this machine learning technique is the classification, prediction, etc. Before executing the process of classification, feature selection is essential in selecting the relevant features.

The prediction of cancer based on microarray is expressed here [3]. There are three methods for selecting the features they are as follows: filter, wrapper, and embedded, and these are mentioned in [4]. R. Kohavi et al. explained the wrapper methods being used for feature selection [5]. The induction algorithm is associated with wrapper methods for selecting feature subsets. Here it acts as a black box. The induction algorithm acts on the internal training datasets, which removed different sets in the data. The evaluation of highest data is chosen. Some examples for wrapper feature selection methods are forward selection, backward elimination, recursive feature elimination (RFE). One of the machine learning algorithms SVM is constructing a new model based on the hyperplanes and dividing the data into two points; then the distance between the decision hyperplanes and the classes are maximized to the boundary [6]; from these, the possible output will be found out. SVM can be used in vast range of classification problems including medical diagnosis, prediction of protein functions, and some diagnosis such as text, speech [7].

Breast cancer is the most popular disease; it is of two types, basal and luminal where it originate in the breast. When compare to luminal, basal type cancer cells are more lethal and their prognosis is worse [8]. A novel predictive model was proposed [9] for breast cancer prediction using support vector machines (SVM). It includes the modification of wrapper based forward selection using SVM and the entire process was implemented with the help of R tool. In the proposed method, the leukemia datasets are applied in two feature selection methods to get the relevant variable subset for classification.

2 Related Works

Hanaa Salem et al. explained the implementation of the feature selection process proceeds by the filter method called information gaining (IG); the next wrapper method genetic algorithm (GA) is carried out for the feature reduction purpose. Then finally, the genetic programming (GP) is used to classify the cancer cells to predict whether the given set is caused by cancer disease or not [10]. Kononenko. I. et al. explained the overview of machine learning algorithms and their applications in medical diagnosis [11]. Feldman B et al. mentioned that big data provide characteristics such as variety, velocity, veracity, and it gives importance to the healthcare for biomedical research [12].

As patient medical record increases at a higher rate, automated genomic analysis [13] are very important during the medical decision process. Mouhamadou et al. proposed a modified method called RFE-SVM-based feature selection for classification, with local search operators to improve the quality [14]. I. Guyon, J. Weston et al. mentioned the problem of selecting genes from the DNA microarray data and they attempted to use SVM based Recursive Feature Elimination (RFE) in classification to attain better performance [15]. J. C. H. Hernandez, B. Duval et al. mentioned that the support vector machines are based on generalized linear classifiers, and it belong to the supervised learning technique, and acts as the maximum margin classifiers [16]. M. Raza, I. Gondal et al. used these two methods, multivariate permutation test (MPT) and significant analysis of microarray (SAM) to select the relevant genes among the breast cancer data for prediction [17]. Gautam et al. [18] proposed the new classification approach using back propagation artificial neural network (BP-ANN) for separating the mammogram where it checks whether the class is normal or abnormal; it shows better accuracy compared to the existing techniques.

From the above-related works, we find out many the issues in the efficient feature selection. It has so many algorithms to predict the features for the cancer dataset. In that, we find out the best feature selection method to predict the features compared with the correlation and SVM-based wrapper methods.

3 Proposed Work

The important part of the feature selection is the variable selection or attribute selection. After collecting the raw data, the essential features are necessary to increase in the performance. So the features can be discrete, nominal, or it may be continuous. In general, features are represented as

Relevant—The output of the features has the greatest impact, and it cannot be suspected by their halt.

Irrelevant—These irrelevant features cannot have the greatest impact on the output, and in these, the random values are generated.

Redundant—In this feature, same information is presented in various places. When neglecting the halt, there may be the problem of selecting subset. The attention will be focused on the input variables in the learning algorithm.

Selecting the important features is essential among all the features in the feature selection mechanism. Consider all features of this method are same, some features may be redundant or irrelevant. In machine learning, the attributes selection is based on the variable subset selection or the feature selection. From the raw data, the selected features are applicable in the learning algorithm for building a model. The reduced number of the dimensions among the features led to the best subset that contributes the highest accuracy. So we reject the unimportant dimensions among the features. In the preprocessing stage, we need to avoid the curse of dimensionality; it can be done through feature selection and feature extraction. Minimal feature subset is based on the problem domain to discover the feature selection from the original features.

3.1 Proposed Methodology

Fig. 1 discusses the work flow diagram for proposed methodology; it shows that the raw data get preprocessed through feature selection algorithms such as correlation and SVM, and after, that it selects the relevant features for future classification. The abundance of noisy, irrelevant and inconsistent features in the input data necessitates the feature selection process for collecting the relevant information from it. The data subset selection gives the benefits of removal of the irrelevant features. All the list of attribute subsets are tested for feature selection, for example a set of 'n' attributes have a 2^n subsets in most cases, which will make it infeasible for model recognition, statistical information's and some fields in the communes of data mining, so the feature selection process plays a vital role.

Basically, the feature selection process acquires the following main methods for attribute selection algorithm, i.e., filter, wrapper, and the embedded.

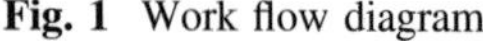
Fig. 1 Work flow diagram

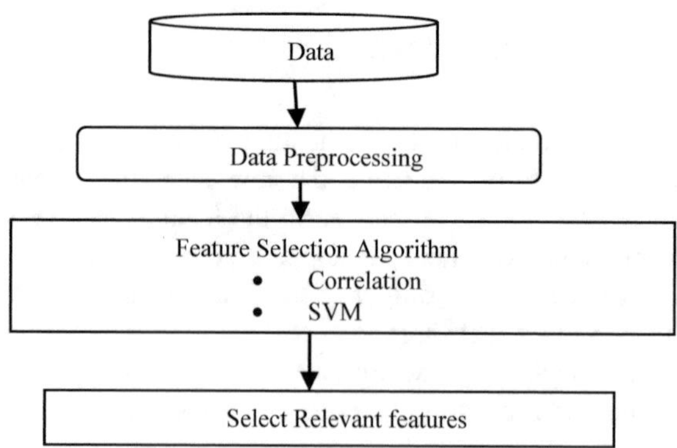

Fig. 2 It shows the working process of the filter method using ranking method to select the relevant features from the huge amount of the dataset

3.2 Filter Methods

By categorizing the feature selection methods, these filter methods are relatively independent in classification. Fisher's discriminant criterion method helps in calculating the distance between mean values and find the combination of features of the two or more classes it slightly related to correlation coefficient method. The implementation process is done by statistical tests (T test, F test) or mutual information. Correlation between features cannot be considered in isolation in earlier evaluation in filter-based methods. We make minimum repetition in the dataset to select the relevant features.

MRMR (Minimum Redundancy Maximum Relevance) method enhances the maximum relevance along with minimum redundancy criteria for choosing the additional features and already identified features are maximally dissimilar. The representative power of the feature set is expanded through the MRMR, so it boosts their generalization properties (Fig. 2).

3.3 Wrapper Methods

Wrapper methods are considered as the black box because it uses the classifier based on the predictive power to score the feature subsets. In machine learning community, the wrapper method based on SVM has been widely studied. Here we use backward feature elimination, support vector machine recursive feature elimination (SVM-RFE) method to avoid the insignificant features from the subset recursively. It is one of the wrapper method applied on the cancer dataset. The main objective of the function is to select the features based on the ranks and eliminate the features based on the bottom ranked. This elimination function is used to find the linear kernel variance (Fig. 3).

Fig. 3 Shows the working process of the wrapper method, it wrapped around the classifier model

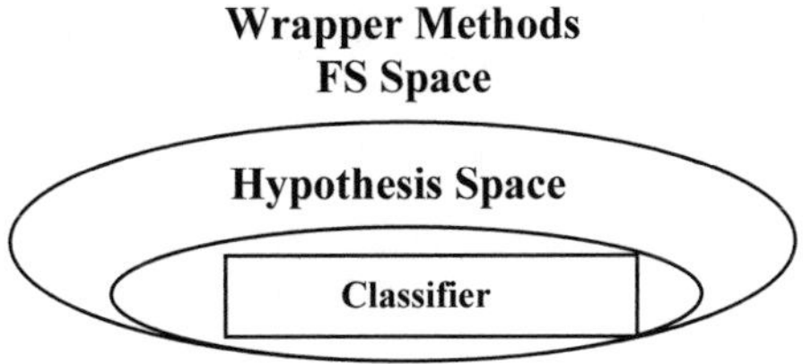

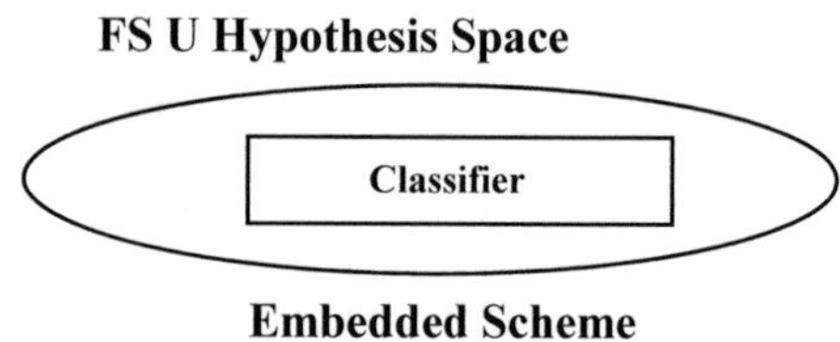

Fig. 4 Integration of the logical and machine learning algorithms

3.4 Embedded Scheme

It uses both the combination of the filter and wrapper method qualities. And the feature selection algorithm (either explicit or implicit) has its own inducer. The logical conjunctions determine an example of this embedding to induce the methods (Fig. 4).

4 Correlation-Based Feature Selection (CFS)

According to the degree of redundancy, CFS searches feature subsets among the features. The individual subsets of features are highly correlated based on the evaluator but have low inter-correlation with the class. Conditional entropy is a numeric measure used by the subset evaluators; the classes have the highest correlation, to guide the search iteratively and add the features. Information gain is the example of univariate filters, it does not make interactions between features and it is overcome by multivariate filters for e.g. CFS. It evaluates the degree of redundancy for every feature subset. Correlation between subset of attributes can be used to estimate the correlation coefficient and class; the features are as inter-correlations between them. It grows with the correlation between the classes and features. The best feature subset can be determined by CFS, thus the growing inter-correlation will be decreased. The CFS has the following strategies to find the best feature subset by searching as forward selection, backward elimination, bidirectional search, best-first search, and genetic search used by Eq. (1) [19]

$$r_{zc} = \frac{\overline{kr_{zi}}}{\sqrt{k + k(k-1)r_{ii}}} \tag{1}$$

where r_{zc} is the correlation between the summed feature subsets and the class variable, k is the number of subset features, r_{zi} is the average of the correlation between the subset features and the class variable, and r_{ii} is the average inter-correlation between subset features.

5 Wrapper-Based SVM

Due to limited computing resources, the performance of the training classifiers is not fit for the full set of data because there is an abundance of training data. All the time, the training data are stored to main memory like support vector machines. With the small number of subsets the training data work well. Majority of research community raises a question of "For training the classifier how much training data set has to be taken into an account?". We are going to propose a simple procedure effectively and also get the good results and discussion. The linear support vector machines (SVMs) are first trained in our strategy to create initial classifiers based on the subset of training data.

In SVM there are many classifiers. Each classifier has a hyperplane, separating 'positive' and 'negative,' and also represented by the normal and a constant. Elimination of features was done in the second step in order to achieve a specified level of data sparsity such that the weight becomes normal i.e., close to zero. Sparsity is defined by the average numbers of features that have been discarded by the feature selection process. After the feature selection step, the features are retained finally and the documents of all the data training set are represented. We classify the test data based on the SVM classifier in reduced future space. The proposed method provides improved data sparsity and better classification results. We have explored other conventional feature selection techniques to provide a valid comparison with the proposed SVM-based feature selection. Final Classification results are evaluated based upon the density of the training datasets. Hence SVM can be used for wrapper approach, as it provides efficient results in classification and regression tasks. For that, our future classification model will use the modified SVM algorithm for classification purpose. The pseudocode for the wrapper-based SVM approach is implemented as follows:

Algorithm: Wrapper based Feature Selection using Support Vector Machine
INPUT: J * K matrix, where J is no. of attributes and K are the samples
OUTPUT: Accuracy in feature selection of the provided dataset

```
Begin
normalize job
mice():
    x=apply the function(x)
    x1=mice function applied on x
    x=binding the x to complete the normalization
train and tune():
    a=train function carried over x & x$classes using the method SVMLinear
prediction();
    b=generating the confusion matrix to predict from the a
END
```

Table 1 Gene expression data matrix—leukemia

Samples	Gene 1	Gene 2	…	Gene m-1	Gene m
Sample 1	25.8	615.9	…	2068.5	308.6
Sample 2	203.7	858.6	…	2337.5	611.7
;	;	;	;	;	;
Sample n -1	226.2	942.4	…	2392.5	2567
Sample n	203.9	926.8	…	3989.5	548.8

6 Results and Discussion

6.1 Dataset Description

For our experiment, we attained the dataset from the UCI machine learning repository http://archive.ics.uci.edu/ml/datasets [20]. Gene expression matrix shown in Table 1 is obtained by transforming the microarray data images. The rows in the matrix represent samples, and the columns denote the genes. Every cell signifies the expression values of gene under a specific condition. The matrix generally contains large volume of data; therefore, data mining methods are used to find interesting pattern from such matrices. Information attained via microarray data analysis is the most important for classification of phenotype.

Table 2 shows the particulars of gene expression data applied in the experiment. The datasets contain gene expression matrix vectors taken from microarray data from a number of patients (Table 3).

6.2 Correlation

Correlation algorithm is implemented to the dataset to select the relevant features. Before the feature selection, the dataset that contains some missing values are normalized using mice [R package]. From the dataset index, the training and the testing values are created using the normalization process which inturn has attained the accuracy of 0.9426 and is shown in Fig. 5. The accuracy attained is 0.9426.

Table 2 Details of microarray dataset

Dataset	Total samples	No. of genes	Class labels	Class wise samples
Leukemia	172	17129	Acute Lymphoblastic Leukemia (ALL)	147
			Acute Myeloid Leukemia (AML)	125

Table 3 Gene's description for leukemia

Gene number	Accession number	Symbol	Gene description
4189	5114	X16667_at	HOXB3 homeobox B3
5104	388	Z28339_at	AKR1D1 aldo-ketoreductase family 1, member D1
6493	17001	X91911_s_at	GLIPR1 GLI pathogenesis-related 1
6281	29823	M31211_s_at	MYL6B myosin, light chain 6B, alkali, smooth muscle, and non-muscle
6373	6152	M81695_s_at	ITGAX integrin, alpha X (complement component 3 receptor 4 subunit)
6855	11633	M31523_at	TCF3 transcription factor 3
4965	9795	Y08200_at	RABGGTA Rab geranylgeranyltransferase, alpha subunit
2043	6563	M57710_at	LGALS3 lectin, galactoside-binding, soluble, 3
3451	7763	U58681_at	NEUROD2 neuronal differentiation 2
4894	7600	X98411_at	MYO1F myosin IF

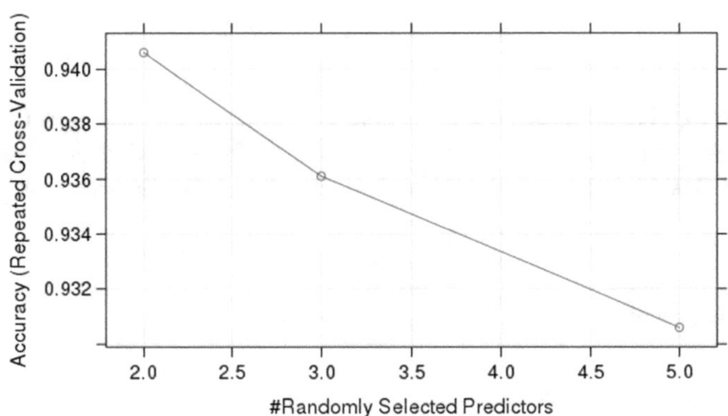

Fig. 5 Accuracy measured in correlation

6.3 Wrapper-Based Support Vector Machine

The number and the input variables are affected to performance of supervised learning algorithms. Feature selection techniques are used to overcome the affected problems, find the subset of the variable, and also get the better result than the previous feature selections. From the dataset index, the train and the test values are created using the normalization process which will provide accuracy value of 0.9713 and is shown in Fig. 6. The accuracy is 0.9713.

Table 4 shows the accuracy for two algorithms (correlation and SVM). Here, the kappa statistics is a statistical measure; it measure the interconnected categorical

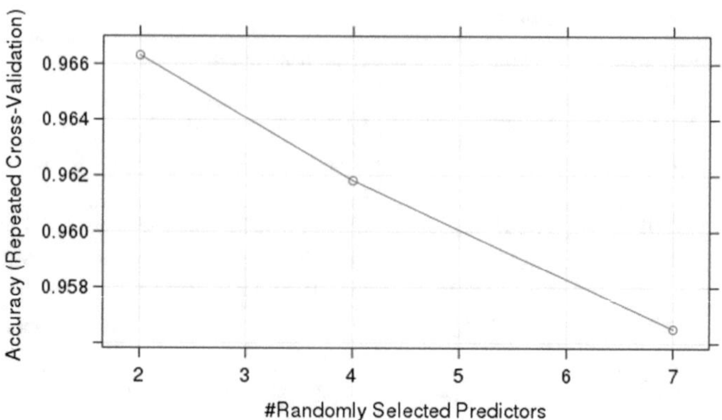

Fig. 6 Accuracy measured in wrapper-based SVM

Table 4 Results from RStudio for leukemia dataset

Algorithm	Accuracy	Kappa statistics	Balanced accuracy	Class
Correlation	0.9426	0.8729	0.9364	Positive
SVM	0.9713	0.9364	0.9682	Positive

values through balanced accuracy. In that, SVM shows the best result is used to select the relevant features of the Leukemia dataset for classification. Based on the SVM, the 50 relevant gene features are selected for future classification purpose.

7 Conclusion

In this paper, feature selection using wrapper-based SVM is highly attainable for cancer classification in high-dimensional datasets. It shows better performance when compared to the filter methods on many real datasets. The two feature selection algorithms, correlation and wrapper-based SVM, are used to analyze the leukemia gene dataset. And the evaluation of the feature selection is measured by kappa statistics with balanced accuracy. From our method, we find that many features are irrelevant and the relevant features give the high inaccurate result. So the gene description is mentioned above. So our proposed feature selection method focuses on cancer classification using SVM machine learning algorithm.

References

1. Tuimala, J., Laine, M.: DNA Microarray Data Analysis, 2nd edn. PicasetOy, Helsinki (2005)
2. Cruz, J.A., Wishart, D.S.: Applications of machine learning in cancer prediction and Prognosis. Cancer Informat (2006)
3. Michiels, S., Koscielny, S., Hill, C.: Prediction of cancer outcome with microarrays a multiple random validation strategy. Lancet **365**, 488–492 (2005)
4. Pang-Ning, T., Steinbach, M., Kumar, V.: Introduction to data mining (2006)
5. Kohavi, R., John, G.H.: Wrappers for feature subset selection. Artif. Intell. **97**, 273–324 (1997)
6. Platt, J.C., Cristianini, N., Shawe-Taylor, J.: Large margin DAGs for multiclass classification, pp. 547–53 (1999)
7. Duda, R.O., Hart, P.E., Stork, D.G.: Pattern classification, 2nd edn. Wiley, New York (2001)
8. Karplus, A.: Machine learning algorithms for cancer diagnosis. Santa Cruz County Science Fair (2012)
9. Kim, W., Kim, K.S., Lee, J.E., Noh, D.Y., Kim, S.W., Jung, Y.S.: Development of novel breast cancer recurrence prediction model using support vector machine. J Breast Cancer **15**, 230–238 (2012)
10. Salem, H., Attiya, G., El Fishway, N.: Classification of human cancer diseases by gene expression profiles. Appl. Soft Comput. **50**, 124–134 (2016)
11. Kononenko, I.: Machine learning for medical diagnosis: history, state of the art and perspective. Artif. Intell. Med. **23**, 89–109 (2001)
12. Feldman, B., Martin, E.M., Skotnes, T.: Big data in healthcare hype and hope. Dr. Bonnie 360 (2012). http://www.west-info.eu/files/big-data-in-healthcare.pdf
13. IBM: Large Gene interaction Analytics at University at Buffalo, SUNY (2012). http://public.dhe.ibm.com/common/ssi/ecm/en/imc14675usen/IMC14675USEN.PDF
14. Samb, M.L., Camara, F., Ndiaye, S., Slimani, Y., Esseghir, M.A.: A novel RFE-SVM-based feature selection approach for classification. Int. J. Adv. Sci. Technol. **43** (2012)
15. Guyon, I., Weston, J., Barnhill, S., Vapnik, V.: Gene selection for cancer classification using support vector machines. Mach. Learn. **46**, 389–422 (2002)
16. Hernandez, J.C.H., Duval, B., Hao, J.K.: SVM based local search for gene selection and classification of microarray data, in BIRD, pp. 99–508 (2008)
17. Raza, M., Gondal, I., Green, D., Coppel, R.L.: Feature selection and classification of gene expression profile in hereditary breast cancer. In: Hybrid Intelligent Systems, Fourth International Conference on Kitakyushu, Japan (2004)
18. Gautam, A.: An improved mammogram classification approach using back propagation neural network. In: Data Engineering and Intelligent Computing, pp. 369–376. Springer, Singapore (2018)
19. Cui, Y., Jin, J.S., Zhang, S., Luo, S., Tian, Q.: Correlation-based feature selection and regression. Part I, LNCS 6297, pp. 25–35. Springer-Verlag Berlin Heidelberg (2010)
20. Uci machine learning repository. http://archive.ics.uci.edu/ml/datasets

Analysis and Prototype Sequences of Face Recognition Techniques in Real-Time Picture Processing

G. D. K. Kishore and M. Babureddy

Abstract The present-day implementation and demand of some real-time image processing applications are to be increased in various recent technologies. Some of these techniques are related to biometric applications such as fingerprint identification, face recognition, iris scan implementation, and speech identification; instead of all these techniques, face identification is an emerging concept in biometric applications with video summarization and surveillance, computer interaction with human being, face identification, and image databases in various applications to categorize face classification. In recent days, face identification is a rapid technology, which is a criminal forensic analysis, access control policy and prison dimensionality privacy with recent approaches. So, we have gone through formalized different approaches and techniques with feasible development of various biometric applications. We also discuss different classification and clustering techniques to extract all features of faces for identification in real-time applications. We also give all comprehensiveness with critical analysis in calculation of face detection in general implementation. This survey gives traditional methods to automatic face identification formulation to increase the performance in face identification. We also investigate review of implementation in data mining classification, clustering, and feature extraction methods with parameters which consist of face identification such as facial expression, variation, and facial illumination in biometric application developments.

Keywords Face identification · Principal component analysis
Independent component analysis
Authentication facial databases and industrial neural networks

---

G. D. K. Kishore (✉) · M. Babureddy
Department of Computer Science, Krishna University, Machilipatnam, Andhra Pradesh, India
e-mail: kishore.galla1@gmail.com

V. Bhateja et al. (eds.), *Intelligent Engineering Informatics*, Advances in Intelligent Systems and Computing 695, https://doi.org/10.1007/978-981-10-7566-7_32

1 Introduction

In biometric technology, face identification and recognition is an essential module defined by different researchers in embedded systems to detect pattern recognition and in system-oriented applications. Some of the data mining-oriented machine learning applications and system graphics were also involved in face classification and identification in real-time applications. More number of advanced security and forensic real-time applications require the usage of face recognition and detection technologies. In biometric applications, face recognition is an effective and important people pervasive system to define implementation of human tasks with human computational in face recognition application systems. Computational evolutionary model gives practical data with respect to practical surveillance with human access control applications in implementation of human–computer interface and image database management systems and identification of unwanted human activities and so on. Faces also play an independent activity in social network face sharing applications with convening to identify emotion and other feature with preferable operations. Human identifies and recognizes thousands and lacks of faces in their lifetime with familiar faces appeared in large number of years and decades in separate formation of face recognition. Because of dynamic changes in face recognition and feature extraction manual effort to detect faces from different dimensional based images. So in face recognition, face detection is an essential part to identify automatic face identification in some image-oriented applications. Face detection is not only a straightforward procedure for various image appearances such as image orientation, image occlusion, and image orientation with different facial expression in image acquisition. So the main aim of face recognition is a specific object classification for digital images with different facial expressions. However, some of the face detection calculations, approaches, and techniques were introduced to handle feature extraction and pattern classification. So in this paper, we give brief survey regarding face recognition techniques with different facial features and also discuss classification clustering, feature extraction, and facial object class methodologies in relevant image processing.

2 Related Work

This section describes different author's opinion regarding face recognition procedures and methodologies. In this area, we temporarily review some of the technological innovation that has been used for experience identification. Normally, experience identification techniques continue by finding the experience in picture, related to consequence of calculating and decreasing interpretation, spinning with different dimensions. Given a stabilized picture, the functions, either global or regional, are produced and compacted in a lightweight experience with identification which can then be held in a data source or a smart card and in contrast to face representations produced at later times.

a. **Appropriate Techniques**: Any object face identification is related to some other Web processing applications and stocks rich common literary works with many of them. Mainly, object face identification is based upon face recognition. For identification of encounters in an image and video, object face monitoring is compulsory, possibly in 3D with evaluation of the head cause [1]. This normally brings to evaluation of the person's attention [2] and evaluation of look [3] which is essential in computer with human interaction for knowing the objective in particular conversational interfaces to monitor visuals effectively. Relatively there was a heavy work of person is monitoring and activity identification [4] which are essential books for face monitoring and for which face identification is an excellent resource of information. Latest reports have also started to concentrate on face appearance research either to infer effective state [5] or for generating personality-animated graphics, particularly in MPEG-4 compression [6]. Excellent visible conversation is also growing in different image processing external events.
b. **Data Classification**: Basically, classification refers to categorization of different attributes to improve the execution of decision-making process in BI and collect data from different outsourcing Web sites or publicly available ULRs in real-time data processing. Classification recognizes and processes data based on entities for easy identification in the real-time data stream evaluation. There are many applications of classifications such as Component Naïve Bays (CNB) and Support Vector Machine (SVM) that give better performance and easy handling of classifiers with efficient generalization abilities and parameter processing. Finally, classification performs great speedups in terms of accuracy of categorization as well as generalization ability with different sequences.
c. **Data Clustering**: Clustering is an exploratory data analytical tool that gives better solutions for classification problems. Primarily, clustering is the task of grouping a set of same objects in a cluster or similar cluster groups. The main task of clustering is to extract relevant data for statistical data analysis that is used in many data analytic fields, which includes machine learning, pattern recognition, information retrieval, image data analysis, and big data informatics. Clustering is not an automatic problem orientation to achieve iterative progress in the knowledge discovery of attributes and also achieve interactive multi-objective optimization that involves training and testing sequences. Clustering is the most important data mining methodology to be used in BI in order to manage marketing and customer relation management, telecommunication, and large associations of data transaction organizations

Normally, before acknowledging an experience, it must be located in the picture. In some supportive techniques, experience identification is a constraint to every user with different semantics. Some techniques use a mixture of skin touch and experience structure to identify the place with experience and use a picture chart to allow encounters of various sizes to be recognized. Progressively, some techniques are to detect and identify different object faces, which have not fully organized [5]. Hints such as activity and person identification can be used [1] to localize encounters for identification. Typically, interpretation, range, and in-plane spinning for the head are approximated at the same time, along with rotation in depth when this is considered.

Some of the great experience is considered for identification by an automated system. Presently, more number of methods are under growth, which is most applicable. A significant improvement in techniques is whether to signify the overall look of the experience or the geometry. Brunelli and Poggio [7] have compared these two techniques, but eventually most techniques nowadays use a mixture of both overall look and geometry. Geometry is used to evaluate difficulty in permutation of detection of human objects with any precision, particularly from a single still picture, but provides more sturdiness against cover-up and aging. Required data is easily extracted from an experience picture, but is more susceptible to trivial distinction, compulsory from cause, and appearance changes. In exercise, for most reasons, even human-related techniques must calculate some geometrical factors in order to obtain a 'shape-free' reflection that is outside of appearance and causes artifacts [1, 8]. This is acquired by finding experience attractions and bending the experience to a canonical fairly neutral cause and appearance. Facial features are also important for geometrical techniques and for anchoring local representations. These are the main related concepts discussed in face recognition systems.

3 General Procedure for Face Identification

In this section, we discuss general procedure to define face recognition in image processing. Module description is as follows:

a. **Picture databases**: Because of the lack of here we are at finishing this venture, we desire to use whatever exterior sources are available. In particular, we would like to use current experience details as source for our tests. There is a huge variety of databases available, such as the AR Face Databases [2] and the Yale Face Databases B [9]. The AR Face Databases contain 4000 images of 126 topics (70 men and 56 women). The pictures are front opinions and function modifications in lighting, facial expression, and cover-up. Accessibility to the database must be asked via e-mail. The Yale Face Databases B contain images of 10 topics under 576 mixtures of 9 go orientations and 64 lighting circumstances and are publicly accessible on the Web sites.
b. **Foundation set construction**: The primary prevent plan for our tests has been caved determine 1. We will begin with an exercise set of faces and execute PCA on this set. This will lead to the groundwork set of eigenfaces. Its encounters will be extracted from a pre-existing face database that yet to be recognized

(see resources). A coaching set of K pictures each containing M × N pixels will normally generate a set of MN eigenfaces. This basis set will be calculated once and used as an ordinary for all subsequent tests.

c. **Truncating the foundation set**: In exercise, it is ineffective to use the groundwork set that is MN in variety. To pack the details and then create the recognition process more effectively, we will truncate the foundation set. By purchasing the foundation set by the scale of the eigenvalues, we can determine which of the foundation pictures play a role the least details to the breaking down. Truncating these basis pictures will lead to image pressure and a more effective way to shop and procedure considerable levels of details.
d. **Variations in experience images**: After developing the groundwork using PCA on the coaching set of encounters, we will develop a collection of the PCA coefficients for all the encounters we have acquired. We will then analyze the sturdiness of the coefficients to recognize changed editions of the encounters in the collection. These modifications will be a kind of interference or crime of the very first image. The alterations we offer and examine are preservative white-colored Gaussian interference (AWGN) and changes in lighting circumstances. To test the encounters with AWGN, we will imitate AWGN with zero mean and different differences in MATLAB and then add this to the very first experience. To analyze the consequence of lighting, we will use experience pictures taken under different illumination conditions than the coaching pictures.
e. **Nonlinear preprocessing techniques**: We will also examine the efficiency of our face identification program with three different nonlinear preprocessing techniques. We recommend to use advantage recognition, average filtration, and comparison improvement. We will implement the preprocessing to the coaching set and to all pictures saved in the collection. We will then assess the efficiency of our facial identification program by analyzing pictures that have also gone through the same preprocessing. We hypothesize that advantage recognition will enhance the efficiency of the program when the pictures are changed by illumination, since lighting modifications can be showed as low regularity multiplicative interference. We hypothesize that average filtration will enhance the efficiency of the program when the pictures are changed by AWGN. We hypothesize that regional comparison improvement could enhance the efficiency of the program under changes in lighting.
f. **Face identification algorithm**: After implementing the appropriate preprocessing, if any, we will then use an experience identification criteria to find out whether the face connected to anyone in our collection. The collection will include the PCA coefficients of each experience along with the eigenfaces produced by the coaching set. The feedback experience image will be decomposed using the eigenfaces, and a set of PCA coefficients will be recognized. The lowest squared range between the feedback coefficients and all of the other places of coefficients in the collection will be measured. The collection image with the least lowest squared distance between its PCA coefficients and the feedback image coefficients will be the calculation of the identification. If the least minimum squared range between the PCA coefficients between the

calculated and the feedback image is below a certain threshold, we determine that quality image has been recognized. Otherwise, we determine that quality image does not coordinate any identity stored in the library.

4 Techniques Used in Face Recognition

Traditionally, more number of approaches and techniques were used for detection of faces in image processing applications to some secure applications in biometric applications. First, basically face recognition appears in manual to detect faces in different images. In other words, because of independent and random picture changes artificial face recognition and automatic face recognition. The following are the techniques used in the face recognition:

a. **Artificial Face Recognition**: Generally, there are three stages of synthetic experience recognition system, mainly experience reflection, experience recognition, and experience recognition. Face reflection is the first phase in synthetic experience recognition and handles the reflection and modeling of encounters. The reflection of an experience decides the subsequent methods of recognition and recognition. In the entry-level recognition, an experience class should be recognized by general qualities of all the encounters, whereas in the subordinate-level recognition, specific features of eyes, nasal area, and mouth have to be sent to each individual experience. There are different techniques for experience recognition, which can be categorized into three categories: template-based, feature-based, and appearance-based.
   Face recognition is second phase in synthetic experience recognition and it handles segregation of a experience in a analyze picture and to separate it from the staying field and background. Several techniques have been suggested for experience recognition. In one of the strategy, elliptical exerciser structure of human go has been utilized [4]. This technique discovers the summarize of the go by using one of the advantage recognition plan, viz. Canny's advantage sensor and an ellipse is fit to indicate the face area in the given picture. But the problem with this technique is that it can be applied only to front experience opinions. Another strategy of experience recognition manipulates pictures is face space [7]. Facial pictures are not changed when they are estimated into the experience area, and the forecasts on the non-face pictures appear quite different. This is the technique used in discovering the face areas. For each face area, the range between the local sub-images and experience area is measured. This measure is used as an signal of the faceness. The result obtained by determining the range from the experience area at every point in the picture is called as face map. Low principles represent the existence of a face area in the given picture.
   Face recognition is the third phase in synthetic experience recognition, and it is performed at the subordinate level. Each experience in the analyze data source is

compared with each experience in it experience data source and then regarded as a successful recognition if there is a match between the analyze picture and one of the experience pictures in it data source. The category process may be affected by several factors such as scale, cause, lighting, face appearance, and disguise.

b. **Principal Component Analysis (PCA)**: This component is also known as Karhunen–Loeve strategy, one of the favored techniques for function choice and dimension decrease. Identification of individual human relations was first done by Turk and Pentland [1], and renovation of individual encounters was done by Kirby and Sirovich [2]. Excellent strategy, known as eigenface strategy, describes a function space which decreases the attribute formation of the original data. This decreased information section is used to define face identification. But inadequate discerning energy within the category and huge calculations are the well-known common issues in PCA strategy. This limitation is get over by line discriminant analysis (LDA). LDA is appropriate prominent methods for feature choice to look at centered methods [2]. But many LDA-centered experience recognition program has first used PCA to decrease measurements, and then, LDA is used to improve the discerning energy of feature choice. This is because LDA has the small example dimension condition in selection of datasets which are larger with different examples per category for good discerning functions removal. LDA straightly led to poor extraction of discerning functions. In the proposed strategy [9], Gabor narrow is used to filter frontal experience pictures and PCA is used to decrease the dimension of strained function with different vector presentations in reliable streams. The activities of appearance-centered mathematical methods such as PCA, LDA, and ICA are examined and evaluated for the recognition of shaded encounters pictures in [10]. Based on the above considerations, PCA is better than the LDA and ICA with different illuminations. LDA is better than ICA and LDA is most applicable that PCA and ICA on different occlusions, but PCA is partial to occlusions in evolutions of LDA and ICA and is used to decrease strategy in [8] and for modeling appearance relations in [11]. An determinative and discriminate of PCA & LDA processed in [12]. These methods focus on challenging issue for high-dimensional data with and without processing and understanding the data with time. The proposed PCA and LDA are very efficient in storage usage and effective with fundamental factors. Those algorithms give excellent experience face identification and achievements with extremely identification calculations. Appearance methods and techniques and modified PCA and local protecting projects (LPP) are mixed in [13] and provide face identification for decision to detect feature extraction formations.

c. **Template-Based Methods**: To identify a experience in a new picture, first the head summarize, which is pretty continually approximately elliptical exerciser, is recognized using filtration, edge sensors, or silhouettes. Then, the shapes of local facial features are produced in the same way, taking advantage of knowledge of face and feature geometry. Lastly, the connection between features extracted from the feedback picture and predetermined saved layouts of face and face functions is calculated to determine whether there is experience present in the picture. Design-related techniques based on predetermined layouts are understanding of range, form, and pose variations. To deal with such modifications, deformable template methods have been suggested, which design experience geometry using elastic designs that are permitted to convert, range, and move. Model parameters may include not only form, but also strength information of face functions as well.

d. **Gabor Wavelet**: For enhancing knowledge recognizable proof serious capacity vectors obtained Gabor wavelet change of front experience pictures joined together with ICA in [1]. Gabor capacities have been perceived as extraordinary compared to other portrayals for encounter recognizable proof. As of late, Gabor wavelets have been generally utilized for encounter reflection by encounter recognizable proof researchers [2, 3, 5, 14, 15], in light of the fact that the popcorn bits of the Gabor wavelets are the same 2D-responsive field data of the warm-blooded animal cortical straightforward cells, which shows reasonable highlights of spatial region and arrangement selectivity. Past takes a shot at Gabor capacities have likewise shown astounding results for encounter distinguishing proof. Commonplace strategies incorporate the Powerful Link Structure (PLS) [2], Flexible Bunch Chart Related (FBCR) [15], Gabor Fisher Classifier (GFC) [14], and AdaBoosted GFC (AGFC) [3]. Gabor capacities are likewise utilized for step recognizable proof and sex distinguishing proof as of late [15, 7]. In this paper, it was watched that however Gabor stages are comprehension of territorial adjustments and they can separate between styles with precisely the same; i.e., they give more data about the territorial picture capacities. Along these lines, the Gabor stages can work similarly well with the sizes, as long as its affectability to irregularity and territorial alterations can be paid deliberately. Over the errand, essayists recommended to mean experience pictures utilizing the Local Gabor Binary Points (LGBP), which unites Gabor extents with Local Binary Point (LBP) proprietor [4]. Enhanced results were obtained when contrasted and the LBP and the GFC. Since encounter reflection with LGBP focused on local histograms, which were obtuse to local changes [1], similarly provincial histograms of LGBP can be utilized to decrease the comprehension of Gabor stages to local adjustments. By advancement Gabor organizes through LBP and territorial histograms, an extremely astonishing recognizable proof rates practically identical with those of Gabor sizes based systems were acquired, which demonstrates productivity of Gabor arranges in the class of various experiences. A novel technique for expulsion of facial capacities was recommended in [2] fixated on Gabor wavelet impression of experience pictures and bit minimum pieces style criteria. The trial results

Fig. 1 Haar-based classification to face detection

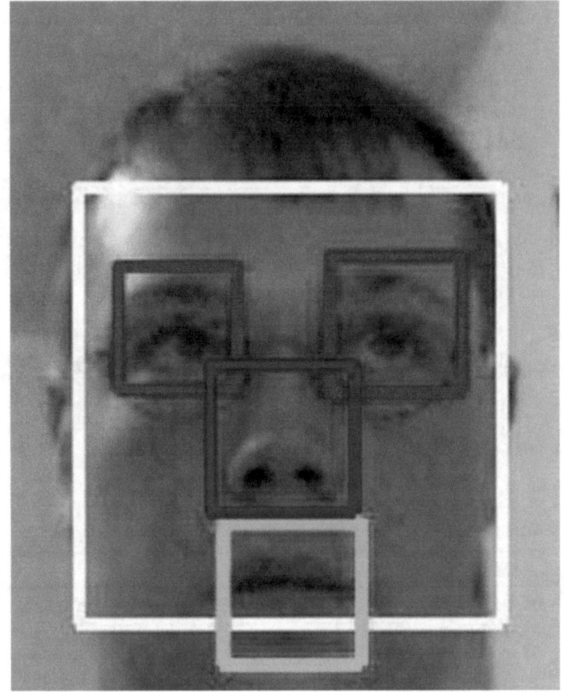

focused on XM2VTS [7] and ORL [7] information source demonstrated that Gabor focused part minimum pieces style approach beats work evacuation systems, for example, PCA, LDA, Kernel PCA or General Discriminate Analysis (GDA) and additionally mix of these methods with Gabor portrayals of experience pictures. A strategy is introduced in [7] by which high-force work vectors bought the Gabor wavelet alteration of front experience pictures combined with ICA for upgraded encounter distinguishing proof.

e. **HAAR Classifier**: The primary foundation for Haar classifier item recognition is the Haar-like functions. These functions, rather than using the intensity principles of a pixel, use the modified contrast values between nearby rectangular-shaped categories of p. The contrast differences between the pixel categories are used to determine comparative mild and eye shadows. Two or three adjacent categories with a comparative comparison difference type a Haar-like function. Haar-like functions are used to identify an image. Haar functions can simply be scaly by improving or decreasing the dimensions of the pixel team being analyzed. This allows functions to be used to identify things of various dimensions. General procedure of the HAAR is shown in Fig. 1.

f. **Support Vector Classifier**: Support Vector Machine (SVM) is the most useful method in classification resolutions. Basic example is experiencing identification. SVM consecutively provides when the function features interpreting examples has lost records. A category criterion that has efficiently been used in this structure is the all-known SVM [5], which can apply to the unique overall look for space with available space of it acquired after implementing an element removal technique [6, 7, 15]. The benefits of SVM classifier over conventional sensory system are that SVMs can accomplish better generalization efficiency.

5 Comparison of Approaches in Face Recognition

See (Table 1).

Table 1 Comparison of different face recognition approaches

Technique	Description
Artificial face recognition	This technique discovers the summarize of the go by using one of the advantage recognition plan, viz. Canny's advantage sensor and an ellipse is fit to indicate the face area in the given picture. But the problem with this technique is that it can be applied only to front experience opinions
Principle component analysis	This strategy is one of the favored methods for function choice and dimension decrease. Identification of individual faces using PCA was first done by Turk and Pentland. But inadequate discerning energy within the category and huge calculations are the well-known common issues in PCA strategy
Template-based methods	To identify a experience in a new picture, first the head summarize, which is pretty continually approximately elliptical exerciser, is recognized using filtration, edge sensors, or silhouettes. In parameter facing and other features in data evaluation, template-based methods do not maintain proper strength in face recognition
Gabor wavelet	For improving experience identification intense function vectors purchased Gabor wavelet modification of front experience pictures are relevant to correspond with ICA in picture retrieval. Gabor functions have identified better representations for experience identification. Gabor wavelet modification of front experience pictures is mixed together with ICA for enhanced experience identification
HAAR classifier	The primary foundation for proposed HAAR classifier item recognition is the HAAR-like functions. HAAR functions can simply be scaly by improving or decreasing the dimensions of the pixel team being analyzed
Support vector classifier	SVM is one of the most effective and useful approach to appear classification issues. However, SVM cannot provide when the function vectors interpreting examples have losing records. The benefits of SVM classifier over conventional sensory system are that SVMs can accomplish better generalization efficiency

6 Major Challenges in Face Recognition

Nowadays, based on data presentation and dynamic changes in picture calculations with respect to facial features and expressions, the following challenges are required to do effective face recognition in real-time image processing applications.

Lighting Pushed—For better evaluation, experience identification frameworks inside systems have accessed different levels, and experience identification in outside environments continues to be as a complicated subject to the consequence of differences in circumstances, and it causes to perform dynamic changes in the experience overall look and is one of the most complicated circumstances in a sensible face identification program that needs to be obtained.

Face pose—In a monitoring program, you are mostly installed to an area where individuals cannot reach to you. Increasing a digital high place, the encounters are thought by some position level. This is the simplest situation in town monitoring programs. The complex situation is that individuals normally pass through you in perspective. They cannot access away from the digital camera. Regulators cannot limit individual's behaviors in public venues. However, earliest techniques have 20–35° position restrictions to recognize an experience. Acknowledging encounters from more perspectives is another task.

Face expression—Face appearance is cognitive problem to match up against position and illumination, but it affects the experience. Face identification results have impact on the identification rate by 1–10%, and a experience with large has a good laugh which has an impact as more as 30% since having a fun experience changes the face overall look and distorts the connection of sight, nasal area, and oral cavity.

Face aging—Face identification techniques use either geometric methods or attribute-based techniques or natural techniques. Based on them do not fix the aging problem. Almost all of them give an age patience as long as twenty decades after preprocessing. Human objects between decades cannot be identified since experience appearance changes fast. Face overall look becomes constant after teenage decades. Identification criteria that can recognize faces for everyone are not available.

Dynamic background—It is easier to learn a face when the qualifications are constant or individual, but issues arise when the qualifications are shifting or powerful. Multiple-face single-experience identification is easy in comparison with several experiences so it is also a big task of this type.

7 Conclusion

In years of decade, face recognition is the essential formation from different research authors in real-time biometric applications. Pattern recognition and facial expression detection in computer-oriented applications are effective data management schema for processing face recognition. There are more number of approaches introduced to face recognition effectively with different features. In this paper, we discuss related fields in face recognition with different authors' definitions. And we also discuss basic method to detect face recognition in different real-time applications. Finally, we discuss major challenges to detect face recognition in real-time application. Further improvement of our work is to introduce some of advanced feature extraction and face recognition approaches for privacy-oriented applications.

References

1. Rautaray, S.S., Agrawal, A.: Real time multiple hand gesture recognition system for human computer interaction. Int. J. Intell. Syst. Appl. **5**, 56–64 (2012). https://doi.org/10.5815/ijisa.2012.05.08
2. Song, F., et al.: A multiple maximum scatter difference discriminant criterion for facial feature extraction. IEEE Trans. Syst. Man Cybern. Part B Cybern. **37**(6), 1599–1606 (2007)
3. Ravi, S., Nayeem, S.: A study on face recognition technique based on Eigenface. Int. J. Appl. Inf. Syst. (IJAIS), **5**(4) (2013). ISSN: 2249-0868 (Foundation of Computer Science FCS, New York, USA)
4. Al-Ghamdi, B.A.S., Allaam, S.R., Soomro, S.: Recognition of human face by face recognition system using 3D. J. Inf. Commun. Technol. **4**, 27–34
5. Rath, S.K., Rautaray, S.S.: A survey on face detection and recognition techniques in different application domain. Int. J. Mod. Educ. Comput. Sci. **8**, 34–44 (2014) (Published Online August 2014 in MECS)
6. Jafri, Rabia, Arabnia, Hamid R.: A survey of face recognition techniques. JIPS **5**(2), 41–68 (2009)
7. Viola, Paul, Jones, Michael J.: Robust real-time face detection. Int. J. Comput. Vis. **57**(2), 137–154 (2004)
8. Zhang, C., Zhang, Z.: A survey of recent advances in face detection. Technical report, Microsoft Research (2010)
9. Bowyer, Kevin W., Chang, Kyong, Flynn, Patrick: A survey of approaches and challenges in 3D and multimodal 3D + 2D face recognition. Comput. Vis. Image Underst. **101**(1), 1–15 (2006)
10. Xiaoguang, L.: Image analysis for face recognition. Personal notes, 5 May 2003
11. Zhao, W., et al.: Face recognition: a literature survey. ACM Comput. Surv. (CSUR) **35**(4) 399–458 (2003)
12. Abate, A.F., et al.: 2D and 3D face recognition: a survey. Pattern Recognit. Lett. **28**(14) 1885–1906 (2007)
13. Colombo, Alessandro, Cusano, Claudio, Schettini, Raimondo: 3D face detection using curvature analysis. Pattern Recognit. **39**(3), 444–455 (2006)

14. Bhele1, S.G., Mankar, V.H.: A review paper on face recognition techniques. Int. J. Adv. Res. Comput. Eng. Technol. (IJARCET) **1**(8) (2012)
15. Scheenstra, A., Ruifrok, A., Veltkamp, R.C.: A survey of 3D face recognition methods. In: Audio-and Video-Based Biometric Person Authentication. Springer Berlin Heidelberg (2005)

Closed-Set Text-Independent Automatic Speaker Recognition System Using VQ/GMM

Bidhan Barai, Debayan Das, Nibaran Das, Subhadip Basu and Mita Nasipuri

Abstract Automatic speaker recognition (ASR) is one type of biometric recognition of human, known as voice biometric recognition. Among plenty of acoustic features, Mel-Frequency Cepstral Coefficients (MFCCs) and Gammatone Frequency Cepstral Coefficients (GFCCs) are used popularly in ASR. The state-of-the-art techniques for modeling/classification(s) are Vector Quantization (VQ), Gaussian Mixture Models (GMMs), Hidden Markov Model (HMM), Artificial Neural Network (ANN), Deep Neural Network (DNN). In this paper, we cite our experimental results upon three databases, namely Hyke-2011, ELSDSR, and IITG-MV SR Phase-I, based on MFCCs and VQ/GMM where maximum log-likelihood (MLL) scoring technique is used for the recognition of speakers and analyzed the effect of Gaussian components as well as Mel-scale filter bank's minimum frequency. By adjusting proper Gaussian components and minimum frequency, the accuracies have been increased by 10–20% in noisy environment.

Keywords ASR · Acoustic Feature · MFCC · GFCC · VQ · GMM · MLL Score

1 Introduction

Automatic speaker recognition (ASR) system was first introduced by Pruzansky et al. [7]. There are two primary tasks within the *speaker recognition* (SR), namely *speaker identification* (SI) and *speaker verification* (SV). This paper concerns with

B. Barai (✉) · D. Das · N. Das · S. Basu · M. Nasipuri
Jadavpur University, Kolkata 700032, India
e-mail: bidhanb@research.jdvu.ac.in

D. Das
e-mail: debayan.157@gmail.com

N. Das
e-mail: nibaran@cse.jdvu.ac.in

S. Basu
e-mail: subhadip@cse.jdvu.ac.in

M. Nasipuri
e-mail: mnasipuri@cse.jdvu.ac.in

V. Bhateja et al. (eds.), *Intelligent Engineering Informatics*, Advances in Intelligent Systems and Computing 695, https://doi.org/10.1007/978-981-10-7566-7_33

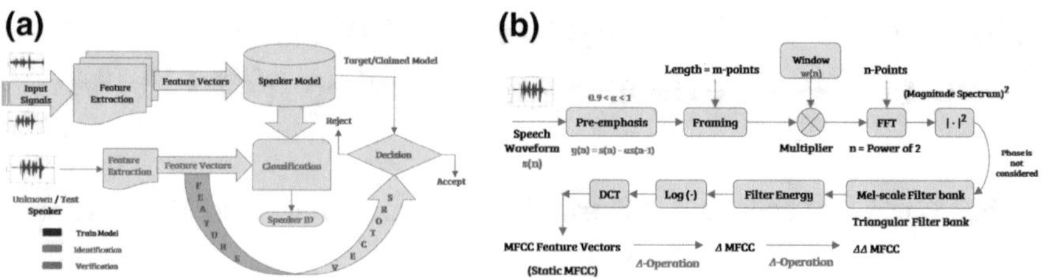

Fig. 1 Block diagram **a** for SR and **b** for MFCC feature extraction

SI and we have used the terms SI and SR synonymously. A general block diagram for SI and SV is shown in Fig. 1a. The SR can also be classified as *text-dependent* and *text-independent* recognition and further divided into *open-set* and *closed-set* identification.

An immense number of features are invented, but at present day the features that are popularly used for robust SR are *Linear Predictive Cepstral Coefficient* (LPCC) and *Perceptual Linear Predictive Cepstral Coefficient* (PLPCC) [9], *Gammatone Frequency Cepstral Coefficient* (GFCC) [11], *Mel-Frequency Cepstral Coefficient* (MFCC), combination of *MFCC and phase information* [6], *Modified Group Delay Feature* (MODGDF) [5], *Mel Filter Bank Energy-Based Slope Feature* [4], *i-Vector* [3], *Bottleneck Feature of DNN* (BF-DNN). In some cases to increase robustness, combined features are developed by fusion of some of these robust features. Some of combined features are LPCC+MFCC, MFCC+GFCC, PLPCC+MFCC+GFCC. The state-of-the-art methods for modeling/classification are *Vector Quantization* (VQ) [10], *Hidden Markov Model* (HMM), *Gaussian Mixture Model* (GMM) [8], *GMM-Universal Background Model* (GMM-UBM) [1], *Support Vector Machine* (SVM), *Deep Neural Network* (DNN) and hybrid models like *VQ/GMM, SVM/GMM, HMM/GMM*. Among these, the hybrid modelbreak HMM/GMM is very useful for SR in noisy environment because HMM isolates the speech feature vectors from the noisy feature vectors and then estimates the multivariate probability density function using GMM in the feature space.

2 Feature Extraction

The first step of SR is feature extraction, also known as *front-end processing*. It *transforms(maps)* the raw speech data into the *featurespace*. The features like MFCC and GFCC are computed using frequency domain analysis and *Spectrogram*. In our experiment, the MFCC feature is used for SR. The block diagram for extracting MFCC feature is shown in Fig. 1b. The computation of MFCC is discussed briefly as follows:

Pre-emphasis: The speech signal is passed through a HPF to increase the amplitude of high frequency. If $s(n)$ is the speech signal, then it is implemented as $\tilde{s}(n) = s(n) - \alpha s(n)$, where $0.9 < \alpha < 1$. Generally, the typical value of α is 0.97.

Framing: To compute MFCC, short time processing of speech signal is required. The whole speech signal is broken into overlapping frames. Typically, 25–60 ms frame is chosen with the overlap of 15–45 ms.

Window: For *Short Time Fourier Transform* (STFT) for $x(n)$, where $x(n)$ be a short time frame, we must choose a window function $h(n)$. A typical window function $h(n)$ is given by

$$h(n) = \beta - (1 - \beta)\, cos\left(\frac{2\pi n}{N - 1}\right) \tag{1}$$

where N is the window length. Here, $\beta = 0.54$ for Hamming window and $\beta = 0.5$ for Hanning window.

DFT and FFT: The *Discrete Fourier Transform* (DFT) for the windowed signal is computed as $X(\omega, n) = \sum_{m=-\infty}^{\infty} x(m)h(n - m)e^{-j\omega n},\ where\ 0 \le n \le N$. For discrete STFT, continuous $X(\omega, n)$ is sampled with N (length of windowed signal) equal points in frequency (ω) as $X(k, n) = X(k) = X(\omega, n)|_{\omega=\frac{2\pi}{N}k},\ where\ 0 \le k \le N$. The graphical display of $|X(k, n)|$ as color intensity is known as *Spectrogram.* Fortunately, two previous equations can be simplified with the help of *Fast Fourier Transform* (FFT) as $X(k) = \mathcal{FFT}\{x(n)h(n)\}, where\ 0 \le k \le N$. To facilitate *FFT*, we must make N as power of 2. To do so, it is required to pad zeros with the frame to make frame length a nearest power of 2 if N is not a power of 2, otherwise zero padding is not required.

Magnitude Spectrum: The squared magnitude spectrum is computed as $S(k) = |X(k)|^2$, where $0 \le k \le N$

Mel-Scale Filter Bank: In Mel scale, n_B number of overlapping triangular filters are set between $M(f_{min})$ and $M(f_{max})$ to form a filter bank. The relation between Mel scale (mel) and Linear scale (Hz) is given by

$$M(f) = 1127 log_e\left(1 + \frac{f}{700}\right) \tag{2}$$

where f in *Hz* and $M(f)$ in *mel*. A filter in filter bank is characterized by start, center, and end frequencies, i.e., $M(f_s)$, $M(f_c)$, and end $M(f_e)$, respectively. Using inverse operation of (2), we can compute f_s, f_c, and f_e using the following equation:

$$f = 700(e^{\frac{M(f)}{1127}} - 1) \tag{3}$$

where f in *Hz* and $M(f)$ in *mel*. Next, we map the frequencies f_s, f_c, and f_e to the corresponding nearest FFT index numbers given by f_{bin}^s, f_{bin}^c, and f_{bin}^e, respectively, which are called FFT bins by using the following equation:

$$f_{bin} = \lfloor \frac{(N+1)f}{F_s} \rfloor, \qquad f = f_s, f_c, f_e \tag{4}$$

Here, F_s is the sampling frequency of the speech signal. The filter weight is maximum at center bin f^c_{bin} which is 1, and zero weight is assumed at start and end bins, f^s_{bin} and f^e_{bin}. The weights are calculated as follows:

$$H_m(k) = \begin{cases} 0 & \text{if } k < f^s_{bin} \\ \frac{k - f^s_{bin}}{f^c_{bin} - f^s_{bin}} & \text{if } f^s_{bin} \le k \le f^c_{bin} \\ \frac{f^e_{bin} - k}{f^s_{bin} - f^c_{bin}} & \text{if } f^c_{bin} \le k \le f^e_{bin} \\ 0 & \text{if } k > f^e_{bin} \end{cases} \tag{5}$$

Filter Energy: The filter bank is set over the squared magnitude spectrum $S(k)$. For each filter in the filter bank, the filter weight is multiplied with the corresponding $S(k)$ and summed up all the products to get the *filter energy*, denoted by $\{\tilde{S}(k)\}_{k=1}^{k=n_B}$. Taking *logarithm*, we get *log energies*, $\{log(\tilde{S}(k)\}_{k=1}^{k=n_B}$.

DCT: To perform *Discrete Cosine Transform* (DCT), the following operation is carried out.

$$C_n = \sum_{k=1}^{D} (\log \tilde{S}(k)) cos(n(k - \frac{1}{2})\frac{\pi}{D}), \qquad n = 1, 2, \ldots\ldots, D \tag{6}$$

Here, $D = n_B$ is the dimension of the vector C_n which is called MFCC vector.

3 Speaker Model

The models that are used frequently in SR are Linear Discriminative Analysis (LDA), Probabilistic LDA (PLDA), Gaussian Mixture Model (GMM), GMM-Universal Background Model (GMM-UBM), Hidden Markov Model (HMM), Artificial Neural Network (ANN), Deep Neural Network (DNN), Vector Quantization, Dynamic Time Warping (DTW), Support Vector Machine (SVM). GMM is the most popular model used in SR. These models are used to build speaker templates. Score domain compensation aims to remove handset-dependent biases from the likelihood ratio scores. The most prevalent methods include H-norm, Z-norm, and T-norm.

3.1 Vector Quantization (VQ)

It is used as a preliminary method for clustering data, so that the process of Vector Quantization(VQ) can be applied more suitably. The grouping is done by minimizing *Euclidean* distance between vectors. If we get V number of vectors after the feature extraction phase, then after VQ we will get K vectors where $K < V$. This set of K vectors is called *codebook* which represents the set of centroids of the individual clusters. In the modeling section, the GMM model is built upon these K vectors.

3.2 Gaussian Mixture Model (GMM)

Let for jth speaker there are K number of quantized feature vectors of dimension D, viz. $\mathcal{X} = \{\boldsymbol{x}_t \in \mathbb{R}^D : 1 \leq t \leq K\}$. The GMM for jth speaker, λ_j, is the weighted sum of M component D-variate Gaussian densities where mixture *weights* w_i $\{i = 1 \ to \ M\}$ must satisfy $\sum_{i=1}^{M} w_i = 1$. Hence, the GMM model λ_j is given by $p(\boldsymbol{x}_t|\lambda_j) = \sum_{i=1}^{M} w_i \mathcal{N}(\boldsymbol{x}_t; \boldsymbol{\mu}_i, \boldsymbol{\Sigma}_i)$ where $\mathcal{N}$ $(\boldsymbol{x}_t;\ \boldsymbol{\mu}_i, \boldsymbol{\Sigma}_i)$ $\{i = 1 \ to \ M\}$ are D-variate Gaussian density functions given by

$$\mathcal{N}(\boldsymbol{x}_t; \boldsymbol{\mu}_i, \boldsymbol{\Sigma}_i) = \frac{1}{(2\pi)^{D/2}|\boldsymbol{\Sigma}_i|^{1/2}} e^{-\frac{1}{2}(\boldsymbol{x}_t-\boldsymbol{\mu}_i)' \boldsymbol{\Sigma}_i^{-1}(\boldsymbol{x}_t-\boldsymbol{\mu}_i)} \tag{7}$$

with mean vector $\boldsymbol{\mu}_i \in \mathbb{R}^D$ and covariance matrix $\boldsymbol{\Sigma}_i \in \mathbb{R}^{D\times D}$. $(\boldsymbol{x}_t - \boldsymbol{\mu}_i)'$ represents the transpose of vector $(\boldsymbol{x}_t - \boldsymbol{\mu}_i)$. The GMM model for jth speaker λ_j is parameterized by weight w_i, mean vector $\boldsymbol{\mu}_i$, and covariance matrix $\boldsymbol{\Sigma}_i$. Hence, $\lambda_j = \{w_i, \boldsymbol{\mu}_i, \boldsymbol{\Sigma}_i\}$.

These three parameters are computed with the help of EM algorithm. In the beginning of the EM iteration, the three parameters are required to initialize per Gaussian component. Initialization could be absolutely random, but in order to converge faster one can use *k-means* clustering algorithm also. A block diagram for GMM is shown in Fig. 2.

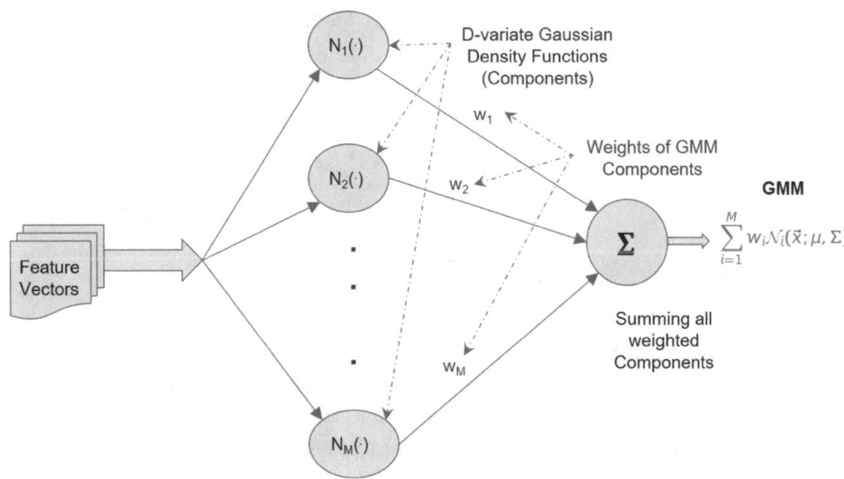

Fig. 2 A block diagram of GMM

3.2.1 Maximum Likelihood (ML) Parameter Estimation (MLE):

The aim of the EM algorithm is to re-estimate the parameters after initialization, which give the maximum likelihood (ML) value given by

$$p(\mathcal{X}|\lambda_j) = \prod_{t=1}^{K} p(\boldsymbol{x}_t|\lambda_j) \tag{8}$$

The EM algorithm begins with an initial model λ_0 and re-estimate a new model λ in such a way that it always provides a new λ for which $p(\mathcal{X}|\lambda) \geq p(\mathcal{X}|\lambda_0)$.

To estimate the model parameters, mean vector $\boldsymbol{\mu}_i$ is initialized using k-means clustering algorithm and this mean vector is used to initialize covariance matrix $\boldsymbol{\Sigma}_i$. w_i is assumed to $1/M$ as its initial value. In each EM iteration, three parameters are re-estimated according to the following three equations to get the new model λ_{new}.

$$w_i = \frac{1}{K}\sum_{t=1}^{K} \mathcal{P}(i|\boldsymbol{x}_t, \lambda_j) \tag{9}$$

$$\boldsymbol{\mu}_i = \frac{\sum_{t=1}^{K} \mathcal{P}(i|\boldsymbol{x}_t, \lambda_j)\boldsymbol{x}_t}{\sum_{t=1}^{K} \mathcal{P}(i|\boldsymbol{x}_t, \lambda_j)} \tag{10}$$

$$\boldsymbol{\Sigma}_i = \frac{\sum_{t=1}^{K} \mathcal{P}(i|\boldsymbol{x}_t, \lambda_j)(\boldsymbol{x}_i - \boldsymbol{\mu}_i)(\boldsymbol{x}_i - \boldsymbol{\mu}_i)'}{\sum_{t=1}^{K} \mathcal{P}(i|\boldsymbol{x}_t, \lambda_j)} \tag{11}$$

The iteration continues until a suitable convergence criteria holds. For the covariance matrix $\boldsymbol{\Sigma}_i$, only diagonal elements are taken and all off-diagonal elements are set to zero. The probability $\mathcal{P}(i|\boldsymbol{x}_t, \lambda_j)$ is given by

$$\mathcal{P}(i|\boldsymbol{x}_t, \lambda_i) = \frac{w_i \mathcal{N}(\boldsymbol{x}_t; \boldsymbol{\mu}_i, \boldsymbol{\Sigma}_i)}{\sum_{j=1}^{M} w_j \mathcal{N}(\boldsymbol{x}_t; \boldsymbol{\mu}_j, \boldsymbol{\Sigma}_j)} \tag{12}$$

4 Speaker Identification with MLL Score

Let there are S speakers $\mathcal{S} = \{1, 2, 3, ..., \mathrm{S}\}$ and they are represented by the GMM's $\lambda_1, \lambda_2, \lambda_3, ..., \lambda_S$. Now, the task is to find the speaker model with the maximum *posteriori* probability for the set of feature vectors $\mathcal{X}$ of test speaker. Using minimum error Bayes' decision rule, the identified speaker is given by

$$\hat{S} = \arg\max_{k \in S}(\sum_{t=1}^{K} log(p(\mathbf{x}_t|\lambda_k))) \tag{13}$$

Here, $\hat{S}$ is the identified speaker and kth speaker's log-likelihood (LL) score is given by $\sum_{t=1}^{K} log(p(\mathbf{x}_t|\lambda_k))$ [8]. The identified speaker $\hat{S}$ has the maximum log-likelihood (MLL) score.

5 Experimental Results and Discussion

We conducted the SR experiment extensively over the three databases, namely IITG Multi-Variability Speaker Recognition Database (IITG-MV SR), ELSDSR, and Hyke-2011. The IITG-MV SR database contains recorded speech from five recording devices, namely digital recorder (D01), Headset (H01), Tablet PC (T01), Nokia 5130c mobile (M01), and Sony Ericsson W350i mobile (M02), in noisy environment. However, ELSDSR and Hyke-2011 contain clean speech; i.e., noise level is very low and the speeches are recorded with a microphone. The sampling frequency for D01, H01, T01 is 16 kHz, for M01, M02 is 8 kHz, and for ELSDSR, Hyke-2011 is 8 kHz. We chose frame size about 25 ms and overlap about 17 ms, i.e., frameshift is $(25 - 17) = 8$ ms for 16 kHz speech signal and 50 ms frame size and about 34 ms overlap; i.e., frameshift is $(50 - 34) = 16$ ms for 8 kHz speech signal. The pre-emphasis factor α is set to 0.97. To compute *FFT* 512-point, FFT algorithm is used. For *mel*-scale frequency conversion, maximum and minimum linear frequencies are $f_{min} = 0, 300$ Hz and $f_{max} = 5000$ Hz. The frequency, f_{min}, has significant effect on the accuracy of ASR. Number of triangular filters in filter bank is $n_B = 26$ which produces 26 MFC coefficients, and among them first 13 MFCC are chosen to create MFCC feature vector of dimension $D = 13$. The accuracy rate for the mentioned databases is shown in Table 1. In VQ, we consider 512 clusters, to reduce large number of vectors, upon which GMM is built using 5 EM iteration.

It is clearly shown that the accuracy rate is low for noisy speech as compared to the clean speech. This is because the noise level distorts the frequency spectrum of the signal considerably and vectors in the feature space are shifted and distorted from the original vectors. All the databases show the highest accuracy with vector dimension equal to 13, no. of Gaussian components equal to 32, and accuracy degrades beyond this limits. Another observation is that the bandwidth of the filters in the filter bank in linear scale (Hz) also influences the accuracy rate. The SR under mismatch and reverberant conditions are more challenging tasks, because in these cases, performance of SR system degrades drastically. Other important issues for SR are language dependency and device mismatch. It has been seen that the accuracy rate degrades if there is a mismatch of language between training and testing data. Specially, for the device mismatch between training and testing data, the accuracy rate degrades drastically. Though GMM shows satisfactory accuracy rate, HMM is more robust than GMM and provides better result in environmental mismatch condition. Hybrid

Table 1 Recognition accuracy for databases IITG-MV SR, Hyke-2011 and ELSDSR for 5 EM iteration and 512 VQ clusters

Database name and Recording device		Type of speech	No. of speakers	No. of Gaussian components	Starting frequency (f_{min})	Vector dimension (D)	Testing time (sec)	Accuracy (%)
IITG-MV SR	DVR (D01)	**Noisy**	100	16	0	13	3638	96
				16	300	13	3287	95
				32	0	13	5119	96
				32	300	13	5336	**96**
	Headset (H01)	**Noisy**	100	16	0	13	4010	70
				16	300	13	4239	89
				32	0	13	5871	76
				32	300	13	5801	**93**
	Tablet PC (T01)	**Noisy**	100	16	0	13	3922	90
				16	300	13	3513	89
				32	0	13	5433	**91**
				32	300	13	5129	89
	Nokia 5130c (M01)	**Noisy**	100	16	0	13	2570	94
				16	300	13	2741	**95**
				32	0	13	4434	92
				32	300	13	4806	94
	Sony Ericsson W350i (M02)	**Noisy**	100	16	0	13	2824	86
				16	300	13	2641	87
				32	0	13	4456	86
				32	300	13	4645	**90**
Hyke-2011	Mic	**Clean**	83	16	0	13	3660	**100**
				16	300	13	3830	100
				32	0	13	6780	100
				32	300	13	6812	100
ELSDSR	Mic	**Clean**	22	16	0	13	656	**100**
				16	300	13	714	100
				32	0	13	1292	100
				32	300	13	1367	100

HMM/GMM-based SR in noisy environment performs better than only GMM-based SR. In noisy environment, the accuracy of GMM-based SR degrades more rapidly than the HMM/GMM-based SR.

6 Conclusion

SR has a very close relation with the speech recognition. Emotion extraction from speech data using corpus-based feature and sentiment orientation technique could be thought of as an extension of SR experiment [2]. In this paper, we cite SR experiment and analyze feature extraction and modeling/classification steps. It is very important to mention that number of GMM components and Mel filter bank's minimum frequency f_{min} have significant influence on the recognition accuracy. Since there are sufficient differences in accuracies between clean speech data and noisy speech data, we can infer that noise level shifts the data from its true orientation. Various normalization techniques in feature domain and modeling/classification domain could be applied to combat with the unwanted shift of data in feature space. Indeed, before transforming data into feature space various filtering techniques to reduce the effect of noise are also available.

Acknowledgements This project is partially supported by the CMATER laboratory of the Computer Science and Engineering Department, Jadavpur University, India, TEQIP-II, PURSE-II, and UPE-II projects of Government of India. Subhadip Basu is partially supported by the Research Award (F.30-31/2016(SA-II)) from UGC, Government of India. Bidhan Barai is partially supported by the RGNF Research Award (F1-17.1/2014-15/RGNF-2014-15-SC-WES-67459/(SA-III)) from UGC, Government of India.

References

1. Campbell, W.M., Sturim, D.E., Reynolds, D.A.: Support vector machines using gmm supervectors for speaker verification. IEEE Signal Process. Lett. **13**(5), 308–311 (2006)
2. Jain, V.K., Kumar, S., Fernandes, S.L.: Extraction of emotions from multilingual text using intelligent text processing and computational linguistics. J. Comput. Sci. (2017)
3. Kanagasundaram, A., Vogt, R., Dean, D.B., Sridharan, S., Mason, M.W.: I-vector based speaker recognition on short utterances. In: Proceedings of the 12th Annual Conference of the International Speech Communication Association, pp. 2341–2344. International Speech Communication Association (ISCA) (2011)
4. Madikeri, S.R., Murthy, H.A.: Mel filter bank energy-based slope feature and its application to speaker recognition. In: Communications (NCC), 2011 National Conference on, pp. 1–4. IEEE (2011)
5. Murthy, H.A., Yegnanarayana, B.: Group delay functions and its applications in speech technology. Sadhana **36**(5), 745–782 (2011)

6. Nakagawa, S., Wang, L., Ohtsuka, S.: Speaker identification and verification by combining MFCC and phase information. IEEE Trans. Audio Speech Lang. Process. **20**(4), 1085–1095 (2012)
7. Pruzansky, S.: Pattern-matching procedure for automatic talker recognition. J. Acoust. Soc. Am. **35**(3), 354–358 (1963)
8. Reynolds, D.A., Rose, R.C.: Robust text-independent speaker identification using gaussian mixture speaker models. IEEE Trans. Speech Audio Process. **3**(1), 72–83 (1995)
9. Sapijaszko, G.I., Mikhael, W.B.: An overview of recent window based feature extraction algorithms for speaker recognition. In: Circuits and Systems (MWSCAS), 2012 IEEE 55th International Midwest Symposium on, pp. 880–883. IEEE (2012)
10. Soong, F.K., Rosenberg, A.E., Juang, B.H., Rabiner, L.R.: Report: a vector quantization approach to speaker recognition. AT&T Techn. J. **66**(2), 14–26 (1987)
11. Zhao, X., Wang, D.: Analyzing noise robustness of MFCC and GFCC features in speaker identification. In: Acoustics, Speech and Signal Processing (ICASSP), 2013 IEEE International Conference on, pp. 7204–7208. IEEE (2013)

Differential Evolution-Based Sensor Allocation for Target Tracking Application in Sensor-Cloud

Sangeeta Kumari and Govind P. Gupta

Abstract In a sensor-cloud system, an optimal set of sensor nodes are generally allocated to complete the subsequent target tracking task. In this kind of system, allocation of an optimal number of sensor nodes for target tracking application is a NP-hard problem. In this paper, a meta-heuristic optimization-based scheme is used, called differential evolution-based sensor allocation scheme (DESA) for allocation of optimal sensor nodes to attain efficient target tracking. DESA uses a novel fitness function which comprises three parameters such as dwelling time, detection probability of the sensor node, and competency of the sensor. Simulation results show that proposed scheme allocates approximately 40–48% less number of sensor nodes for covering the target for its efficient tracking.

1 Introduction

Recent advancement in the wireless communication, sensing technology and cloud computing motivates the integration of the resource-constrained wireless sensor network (WSN) with the resource-rich cloud system. Such type of integration forms a sensor-cloud system to take advantages of the storage and computing power of the cloud system [1–3]. The traditional wireless sensor network applications are based on the single-user-centric approach in which user needs to deploy the whole network and program them for a single application. It also includes maintenance cost, deployment cost, and data storage cost. Due to the limitation of the WSNs in terms of memory, communication, energy consumption, and scalability, there is a need of high computing power and huge amount of data storage for real-time applications. So cloud computing is introduced to provide a flexible stack of huge computing and

S. Kumari (✉) · G. P. Gupta
Department of Information Technology, National Institute of Technology, Raipur 492010, Chhattisgarh, India
e-mail: sangeetak2606@gmail.com

G. P. Gupta
e-mail: gpgupta3@gmail.com

V. Bhateja et al. (eds.), *Intelligent Engineering Informatics*, Advances in Intelligent Systems and Computing 695, https://doi.org/10.1007/978-981-10-7566-7_34

software services at low cost, and it integrates with WSNs application which provide sensing-as-a-service on the basis of pay-per-use to the users [1–5].

In sensor-cloud environment, concept of virtualization is introduced to virtualize the physical sensor nodes and allocating them for the on-demand sensing and monitoring task [5]. Such type of the services of the sensor-cloud is called Sensors-as-a-Service (Se-aaS). Allocation of an optimal number of sensor nodes for target tracking application is a NP-hard problem. In order to schedule and allocate an optimal number of sensor nodes for target tracking application, this paper proposes a meta-heuristic optimization algorithm, called differential evolution-based sensor allocation scheme (DESA). In the proposed scheme, a novel fitness function is derived which contains three parameters such as dwelling time, detection probability of the sensor node, and competency of the sensor. Performance analysis of the proposed scheme is evaluated in terms of percentage of the sensor nodes allocated by varying the probability of target detection and accuracy.

This paper is organized as follows: Section 2 presents a brief overview of the related work on sensor cloud. In Sect. 3, a brief overview of the sensor-cloud architecture is given. Section 4 presents a detail discussion on the proposed scheme. Result analysis is presented in Sect. 5. Finally, Sect. 6 concludes the proposed work.

2 Related Work

In this section, a brief overview of the related work on the sensor-cloud is discussed. In [2], a design of sensor-cloud infrastructure was proposed in which they focus on the virtualization process of the physical sensors as a virtual sensor. Yuriyama et al. [3] focus on the new approach of sensor-cloud infrastructure for enhancing the cloud service research. Further, the work discussed in [3] was expanded the authors of [4]. In [4], a theoretical modeling of virtualization in a sensor-cloud environment has been studied. In [5], Madria et al. focus on the layered sensor-cloud architecture for connecting physical sensor networks to the virtual sensor and providing sensing-as-a-service to the end users. Also in various papers such as [6, 7], researchers provide cloud-based sensing-as-a-service to the end users.

There are very few researchers who visualize the technical issues of the sensor-cloud infrastructure. Recently, in paper [8], Chatterjee et al. have studied an optimal scheduling algorithm for sensor-cloud system. Also Chatterjee et al. [9] considered multiple targets tracking application in the presence of overlapping coverage area for maintaining the privacy and correctness of sensed information. In [10, 11], researchers focus on the problem of localization of the stationary target and also the problem of maneuvering targets with multiple sensors and analysis the performance associated with it. Wang et al. [12] proposed a method for sensor selection and sensor placement in posterior target localization and tracking with minimum entropy. There are few researchers [13, 14] who focus on energy efficiency of the sensor nodes while tracking the target. In [15], they applied suboptimal algorithm for sensor allocation to minimize the measurement error.

In [16], an information-based pruning algorithm has been proposed for minimizing the estimation error in a multi-step sensor scheduling. Zhou et al. [17] considered a blind scheduling algorithm that focuses on the fairness, simplicity, and asymptotic optimality for the mobile media cloud. Also in [18], Zhou et al. have explored blind online scheduling for mobile cloud multimedia service. Misra et al. [19] proposed a Q-SAA algorithm for sensor allocation in target tracking application. In this work, we have used meta-heuristic optimization technique to optimize the allocation of the sensor nodes for the target tracking task.

3 Overview of Sensor-Cloud

Sensor-Cloud infrastructure provides virtual sensor to the user. The user sends a request to the sensor-cloud, and according to their needs by selecting service instances (virtual sensor) from the service templates of the sensor-cloud, a set of virtual sensor is allocated. Virtual sensor collaborates the information from several sensor networks and shares it on a big scale to the multiple users at the same time [20]. There is a virtualization process between user and the sensor-cloud, it means users need not to be worry about the data maintenance, data storage, or which type

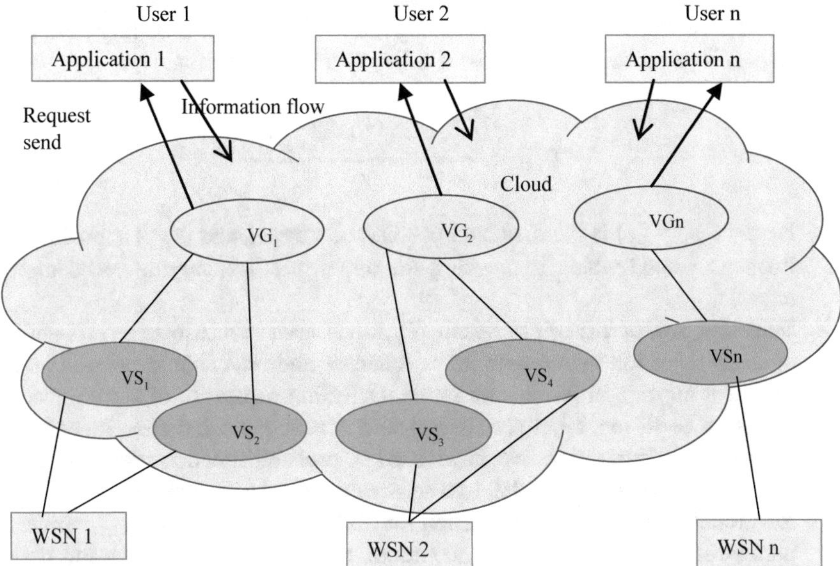

Fig. 1 Sensor-cloud architecture

of sensor are used and how to configure them. When the virtual sensor becomes useless, it can be deleted by the end users to avoid the additional utilization charges of the resources [21].

In Fig. 1, we show the architecture of the sensor-cloud, in which VS_1, VS_2, … VS_n is used to collect information according to the user demand from various WSNs. Then virtual group, i.e., VG_1, VG_2.....VG_n, is formed who serves the user application and accesses it from anywhere, at any time as pay-per-use basis.

4 Differential Evolution-Based Sensor Allocation Scheme

In this section, first, we present the derivation of the fitness function used in the differential evolution-based optimization scheme. Next, description of the proposed scheme is presented.

4.1 Description of Fitness Function (F_i)

In order to evaluate optimal allocation of sensor nodes for target tracking task, a fitness function (F_i) is derived which contains four parameters such as availability of the sensors, detection probability, accuracy, and dwelling time. Description of these parameters is described as follows:

i. ***Dwelling time of a target*** (***D_t***): It is used to determine the time a target can stay in the area which is lied in the sensing range of the node. Dwelling time with respect to a sensor can be derived using following equation.

$$D_t = \frac{\sqrt{\left(x'_{i+1} - x\right)^2 + \left(y'_{i+1} - y\right)^2}}{v} \tag{1}$$

Here, (x'_{i+1}, y'_{i+1}) is the predicted position of a target, and (x, y) is position of the sensor node which is covering the target. v is the moving speed of the target.

ii. ***Detection probability of the sensor*** (***P_{sp}***): It is used to measure the probability of target detection by a sensor node. A sensor node can able to detect a target only if it appears in its sensing range. Detection capability of a sensor node increases as distance between target and sensor node reduces. In order to evaluate this parameter, we have used a probabilistic detection model as proposed in [9]. In the model, two sensing ranges, i.e., R_1 and R_{max}, are used for modeling the sensor's detection capability. Here R_1 and R_{max} represent minimum sensing range and maximum sensing range of a sensor node, respectively. *P_{sp} can be evaluated by using following expression:*

$$P_{sp} = \begin{cases} 1 & 0 \le d \le R_1 \\ \frac{\alpha}{d^{\beta}} & R_1 \le d \le R_{max} \\ 0 & R_{max} \le d \end{cases} \quad (2)$$

Here d represents distance between target and the sensor node.

iii. ***Competency of the sensors (C_i)***: This parameter is calculated with help of the residual energy of a sensor node at a particular time.

$$C_i = \frac{E_i}{\mu} \quad (3)$$

where E_i is the residual energy of a sensor i and μ is the energy consumption rate.

Now, with help of Eqs. (1), (2), and (3), we can derive a fitness function (F_i) which will be used in the differential evolution for based optimization algorithm for allocating best sensors to achieve higher energy efficiency. F_i is a multi-objective function which can be represented by weighted sum of the three parameters as discussed above. Now, F_i can be represented by Eq. (4) as given below:

$$F_i = w_1.D_t + w_2.P_{sp} + w_3.C_i \quad (4)$$

Here, $w_1 + w_2 + w_3 = 1$ and value of each weight $w_i \in [0, 1], 1 \le i \le 3$. First, we compute the fitness value of all the n_t sensors where $n_t \epsilon N_t$, then we allocated the set of active sensors according to those sensors which have the highest fitness value with respect to their sensing range. This allows the user to select optimal sensors with highest quality of service-based tracking that provide sensing-as-a-service on the basis of pay-per-use.

4.2 Proposed Method

In this section, we describe DESA scheme. Here we assume present potion of the target at time t is (x_i, y_i), and previous location of the target was (x_{i-1}, y_{i-1}) at time t_{i-1}. In the proposed scheme, first DESA evaluate the next location of the target, (i.e., x_{i+1}, y_{i+1}) based on the knowledge of last two location coordinate of the target. In order to predict the next location of the target, speed (v_T) at which the target is moving is evaluated by Eq. (5).

$$v_T = \frac{\sqrt{(x_i - x_{i-1})^2 + (y_i - y_{i-1})^2}}{t_i - t_{i-1}} \quad (5)$$

The moving direction (δ) of the target can be evaluated using Eq. (6).

$$\delta = \cos^{-1} \frac{x_i - x_{i-1}}{\sqrt{(x_i - x_{i-1})^2 + (y_i - y_{i-1})^2}} \tag{6}$$

After evaluation of the target's velocity and its moving direction, we can easily calculate next location of the target at time t_{i+1}. Equation (7) is used for evaluation of next location of the target as follows:

$$\left.\begin{aligned} x'_{i+1} &= x_i + v_T . \cos\,\delta \\ y'_{i+1} &= y_i + v_T . \sin\,\delta \end{aligned}\right\} \tag{7}$$

After estimation of the next location of the target, i.e., (x_{i+1}, y_{i+1}), SADE selects number of sensor nodes, i.e., N_t eligible for tracking the target. After the selection of eligible sensor nodes, we apply DE-based meta-heuristic for selecting the best sensor nodes for tracking the target. Best sensor nodes (n_t) is the 30% of N_t. Flowchart of the proposed scheme is illustrated in Fig. 2.

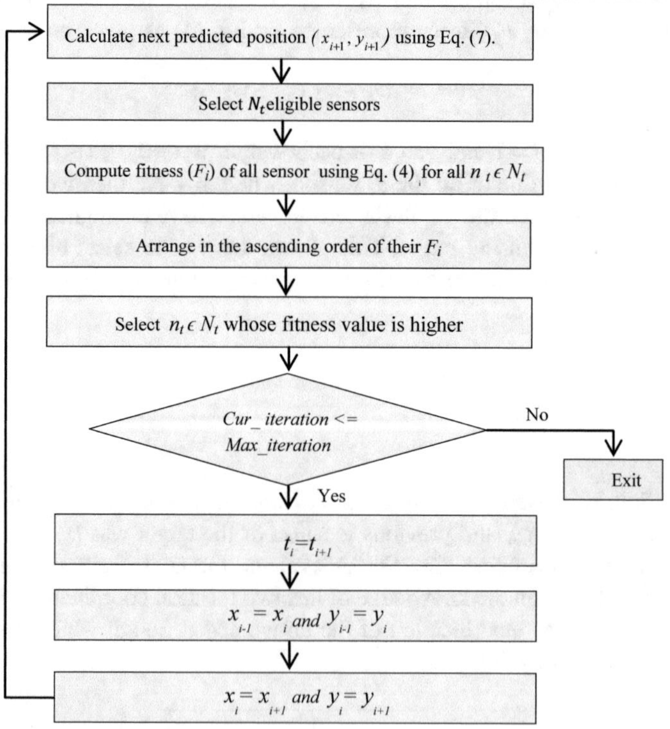

Fig. 2 Flowchart of DE-SAA algorithm

5 Results and Analysis

In this section, we analyze the performance of the proposed scheme in different scenario. For simulation of the proposed scheme, a custom WSN simulator using MATLAB R2009 is designed where network size is 200 × 200 and 200 sensor nodes are randomly deployed over these areas of interest. The result was executed by varying different parameters such as detection probability of the sensor nodes for detection of the target. Table 1 shows the simulation parameters used during experiment.

Figure 3 illustrates the prediction of the target path. It can be observed from the Fig. 3 that proposed scheme predicted the next location of the target with minimal error which can be used to allocate and scheduling of the sensor nodes for energy-efficient monitoring of the moving target over a 2D area.

Figure 4 shows the effect of the detection probably of the sensor nodes over the allocation of the sensor nodes for tracking of the target. It can be observed from the

Table 1 Target location and its best three covering sensor nodes

Node ID	Target X	Position Y	Optimal SNs allocation for covering the target	Fitness value
1	56.70	66.70	[110, 185, 133]	2.7546
2	71.00	81.00	[136, 93, 159]	1.5492
3	74.57	84.57	[120, 175, 133]	2.2870
4	74.91	84.91	[165, 5, 159]	1.2797
5	57.79	67.79	[159, 68, 85]	2.2782
6	71.24	81.24	[93, 136, 159]	1.3204
7	56.60	66.60	[189, 13, 14]	2.3560
8	70.98	80.98	[106, 171, 194]	1.3563
9	76.36	86.36	[198, 93, 136]	2.2348
10	75.30	85.30	[110, 185, 133]	1.2595

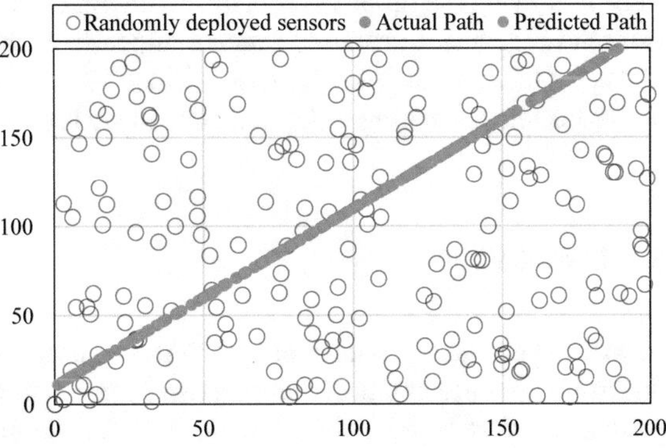

Fig. 3 Illustration of the predicted path of the target within monitoring area

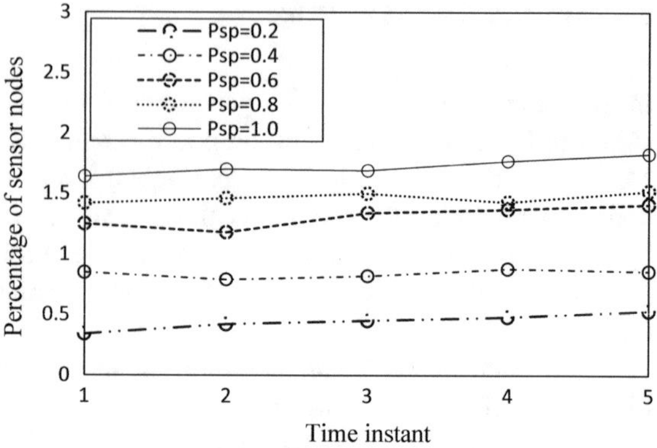

Fig. 4 Illustration of the effects of the detection probability on sensor allocation

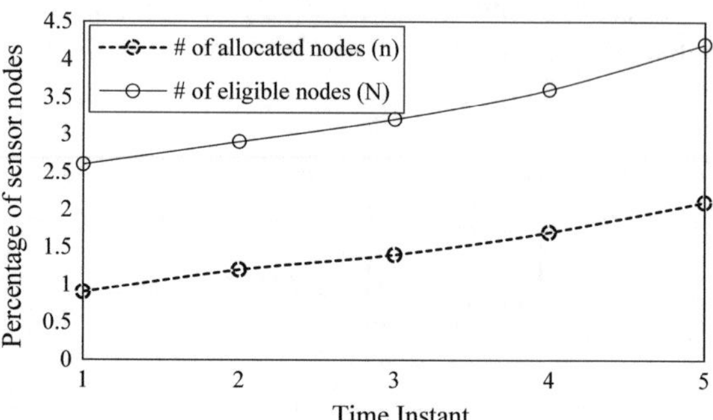

Fig. 5 Analysis of eligible and allocated nodes

Fig. 4 that as detection probability of the sensor nodes increases from 0.2 to 1.0, percentage of the sensor nodes allocated for target tracking task will also increase. This is due to fact that more sensor nodes are available for the selection and allocation of this task. Table 1 listed the location of the target and three best sensor nodes that are covering the target at instance t and their fitness value.

Figure 5 illustrates the comparative study of number of eligible nodes (N_t) for covering a given moving target against number of allocated nodes for doing the same task using the proposed scheme at different times instances. It can be observed from the Fig. 5 that proposed scheme allocates approximately 40–48% less sensor nodes for tracking the moving target. Since the proposed scheme uses 40–48% less nodes for target tracking tasks, it results in saving energy resources of the network.

6 Conclusion

In this paper, sensor allocation problem for target tracking application has been discussed. A differential evolution-based sensor allocation scheme is proposed to predict the next location of the target and for the allocation of the optimal set of active sensors for tracking the target. A novel fitness function is derived for evaluating the best sensor nodes for the tracking of the target. Performance of the proposed scheme is evaluated by varying the detection probability of the sensor nodes and also percentage of sensor nodes allocated at different time instance during the target moving path in the network. Proposed scheme will help the end user for obtaining better services with low cost. It also improves the quality of services and increases energy efficiency of the sensor nodes.

References

1. Díaz, M., Martín, C., Rubio, B.: State-of-the-art, challenges, and open issues in the integration of Internet of things and cloud computing. J. Netw. Comput. Appl. **67**, 99–117 (2016)
2. Yuriyama, M., Kushida, T.: Sensor-cloud infrastructure-physical sensor management with virtualized sensors on cloud computing. In: 13th IEEE International Conference on Network-Based Information Systems (NBiS) (2010)
3. Yuriyama, M., Kushida, T., Itakura, M.: A new model of accelerating service innovation with sensor-cloud infrastructure. In: SRII Global Conference (SRII), 2011 Annual. IEEE (2011)
4. Misra, S., Chatterjee, S., Obaidat, M.S.: On theoretical modeling of sensor cloud: a paradigm shift from wireless sensor network. IEEE Syst. J. **11**(2), 1084–1093 (2017)
5. Madria, S., Kumar, V., Dalvi, R.: Sensor cloud: a cloud of virtual sensors. IEEE Softw. **31**(2), 70–77 (2014)
6. Kim, M., et al.: Developing an on-demand cloud-based sensing-as-a-service system for internet of things. J. Comput. Netw. Commun. (2016)
7. Alamri, A., et al.: A survey on sensor-cloud: architecture, applications, and approaches. Int. J. Distrib. Sens. Netw. **9**(2), 917923 (2013)
8. Chatterjee, S., Misra, S., Khan, S.: Optimal data center scheduling for quality of service management in sensor-cloud. IEEE Trans. Cloud Comput. (2015)
9. Chatterjee, S., Misra, S.: Target tracking using sensor-cloud: sensor-target mapping in presence of overlapping coverage. IEEE Commun. Lett. **18**(8), 1435–1438 (2014)
10. Wang, H., Yao, K., Estrin, D.: Information-theoretic approaches for sensor selection and placement in sensor networks for target localization and tracking. J. Commun. Netw. **7**(4), 438–449 (2005)
11. Hamouda, Y.E.M., Phillips, C.: Adaptive sampling for energy-efficient collaborative multi-target tracking in wireless sensor networks. IET Wirel. Sens. Syst. **1**(1), 15–25 (2011)
12. Aeron, S., Saligrama, V., Castanon, David A.: Efficient sensor management policies for distributed target tracking in multihop sensor networks. IEEE Trans. Signal Process. **56**(6), 2562–2574 (2008)
13. Maheswararajah, S., Halgamuge, Saman K., Premaratne, M.: Sensor scheduling for target tracking by suboptimal algorithms. IEEE Trans. Veh. Technol. **58**(3), 1467–1479 (2009)
14. Huber, M.F.: Optimal pruning for multi-step sensor scheduling. IEEE Trans. Autom. Control **57**(5), 1338–1343 (2012)
15. Zhou, L., Wang, H.: Toward blind scheduling in mobile media cloud: Fairness, simplicity, and asymptotic optimality. IEEE Trans. Multimed. **15**(4), 735–746 (2013)
16. Zhou, L., et al.: Exploring blind online scheduling for mobile cloud multimedia services. IEEE Wirel. Commun. **20**(3), 54–61 (2013)
17. Misra, S., et al.: QoS-aware sensor allocation for target tracking in sensor-cloud. Ad Hoc Netw. **33**, 140–153 (2015)

18. Kurschl, W., Beer, W.: Combining cloud computing and wireless sensor networks. In: Proceedings of the 11th International Conference on Information Integration and Web-based Applications & Services. ACM (2009)
19. Dash, S.K., Mohapatra, S., Pattnaik, P.K.: A survey on applications of wireless sensor network using cloud computing. Int. J. Comput. Sci. Emerg. Technol. **1**(4), 50–55 (2010)
20. Iqbal, M., et al.: Multi-objective optimization in sensor networks: Optimization classification, applications and solution approaches. Comput. Netw. **99**, 134–161 (2016)
21. Fazio, M., Puliafito, A.: Cloud4sens: A cloud-based architecture for sensor controlling and monitoring. IEEE. Commun. Mag. **53**(3), 41–47 (2015)

Design and Analysis of Kite for Producing Power up to 2.6 Watts

Shabana Urooj, Urvashi Sharma and Pragati Tripathi

Abstract This paper presents a novel technique with the optimization of Airborne Wind Energy System (AWES) to generate energy that combines with experimental results. An economic and simply moving blade system has been proposed in this work as conventional wind energy methods require heavy towers, deep foundation, huge blades, and high maintenance. A simple kite with two different blade arrangements has been tested at three different locations. The adopted artificial crosswind motion has the potential to increase the power generation capacity as compared to the stationary flight. A maximum of 2.6 watts and a minimum of 0.03 watts power have been accomplished by the proposed model.

Keywords Airborne Wind Energy · Hardware design · Site specification Kite power

1 Introduction

Friction between the atmosphere and the surface of the earth produces a boundary layer effect for the low and intermittent wind. It is evident that the wind blows near to and at ground surface are persistent and stable as compared to high altitude. The relationship that governs the variation of wind at different atmospheric levels depends on the roughness of wind [1]. Most of the wind will pass effectively if the turbine rotor runs slowly. Blades act as a solid wall which obstructs the wind flow if the rotor runs fast even with the little power extraction. The goal of this work is to

S. Urooj (✉) · U. Sharma · P. Tripathi
Department of Electrical Engineering, School of Engineering,
Gautam Buddha University, Greater Noida, Uttar Pradesh, India
e-mail: shabanaurooj@ieee.org

U. Sharma
e-mail: urvashisharma.uk95@gmail.com

P. Tripathi
e-mail: pragati.knp022@gmail.com

V. Bhateja et al. (eds.), *Intelligent Engineering Informatics*, Advances in Intelligent Systems and Computing 695, https://doi.org/10.1007/978-981-10-7566-7_35

propose a system to achieve high efficiency in terms of power output. Experimental investigation has been done to create a system that can replace big pole wind turbine configuration by utilizing the wind capacity. The experiment has been conducted at three different sites.

1.1 Airborne Wind Energy (AWE)

Most of the design includes a tethered flying airfoil. These designs are classified into two categories, viz. airborne generators and mechanical linkages. Airborne generators transmit electricity to the ground via power cable and mechanical linkage connected at ground station. Here, the major focus is the modification of design. In the proposed system, the rotor blade circuitry is mounted on the kite to achieve maximum benefit of high altitude wind. This will handle fuel saving in a great deal. In this framework, a novel investigation has been done to propagate new technique in renewable energy sector, i.e., Airborne Wind Energy (AWE). Machines that harvest this kind of energy are called airborne wind energy systems (AWES) [2]. AWE is emerging rapidly in the scientific community.

2 Proposed Model of the System

2.1 Components and Specification

Two sets of blade system are used for experiment. None of the specifications regarding wind turbine size and wind speed ratio has been reported in literature. The design depends upon both the diameter of the rotor and rated capacity of the generator. Two prototypes with three and four blades are connected with rotor, respectively. The resultant system has been tested for generation of power. Four blades are mounted on chord length of the kite in series fashion (Fig. 1). It is reported that the wind turbine is faster when number of blades decreases [3]. The specification of hardware components is as mentioned in Table 1. The minimum and maximum wind speed for kite is 2 and 9 m/s. The output obtained with the help of multimeter.

2.2 Analysis

Three different sites are considered to test the system. Site-1 is a room of 35X25X15 with an average wind speed of 1440 rpm. Site-2 is a room with

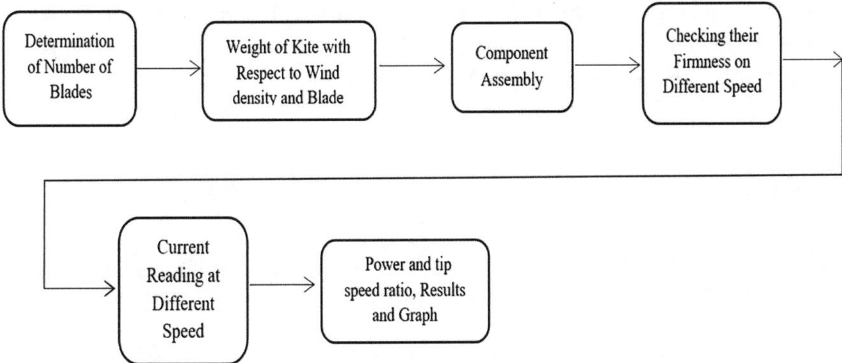

Fig. 1 Block diagram of hardware implementation

Table 1 Hardware components specification

Kite area	1080 cm^2
Kite weight	0.5 gm
Chord length	60 cm
Control lines	2

33X22X11 with an average speed of 1360 rpm. Site-3 is an open area of 4.1 m/sec speed.

Initially, the kite has been made with cotton cloth and paper. Both the used materials were not found efficient to perform the experiment. Kite made up of synthetic resin has been used for experimentation. There are two arrangements of blade system on kite. With the first arrangement, blades are mounted at the tip of kite and another setup is with the two series-connected blade systems on the chord length of kite to make it rotate with wind speed. The power output obtained by using a synthetic resin made kite is elaborated in Table 2. Site-1 with single three-blade and two four-blade systems generate 2.604 and 0.44 watt of power, respectively, with synthetic resin kite. The output obtained from Site-2 is 0.44 and

Table 2 Power rating with respect to blade

Location	Blade system	Power(watt)
Site-1	Single 3	2.604
	Two 4 blade	0.44
Site-2	Single 3	0.1848
	Two 4 blade	0.0314
	Single 3	1.248
	Two 4 blade	0.106
	Single 3	2.0
	Two 4 blade	0.3418
Site-3	No motion	0.0

0.1848 watt with paper kite, 0.0314 and 1.248 watt with cotton kite, 0.106 and 2.0 watt with synthetic resin kite. Site-2 is also examined with both blade systems, respectively. Site-3 does not support the prototype to generate power.

3 Calculation

The power generated by kinetic energy of a free-flowing wind stream and tip speed ratio can be defined by Eqs. (1) and (2) [4]:

$$P = (1/2)\rho A v^3 \tag{1}$$

$$\lambda = (2\pi r\, n)/v \tag{2}$$

The following shows the definition of various variables used in this model:

ρ = density (kg/m^3)
m = mass (kg)
A = swept area (m^2)
v = wind speed (m/s)
C_p =power coefficient
P =power obtained (W)
r = radius (m)

The swept area of the turbine can be calculated from the length of the turbine blades using the equation of the area of the circle [4]:

$$A = \pi r^2$$

where the radius is equal to the blade length as shown in figure below (Fig. 2):

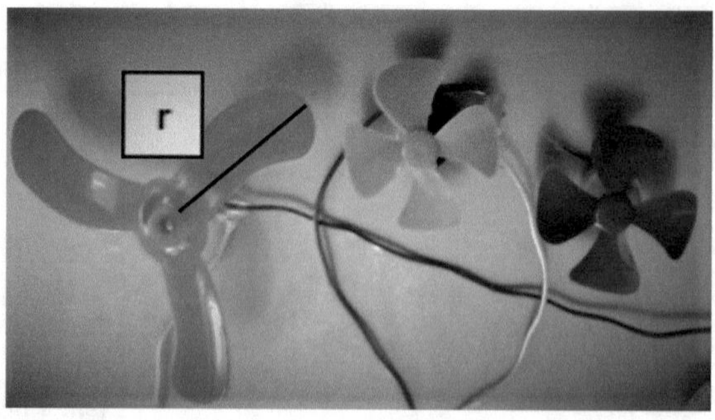

Fig. 2 Schematic radius of blade

Table 3 Tip speed ratio with respect to site

Blade system	Location	Speed of wind(m/sec)	Tip speed ratio
Single 3	Site-1	8-9	0.166
Two 4 blade			0.09
Single 3	Site-2	2-3	0.23
Two 4 blade			0.22
Single 3		3-4	0.26
Two 4 blade			0.15
Single 3		4-5	0.18
Two 4 blade			0.10
No motion	Site-3	4-5	–

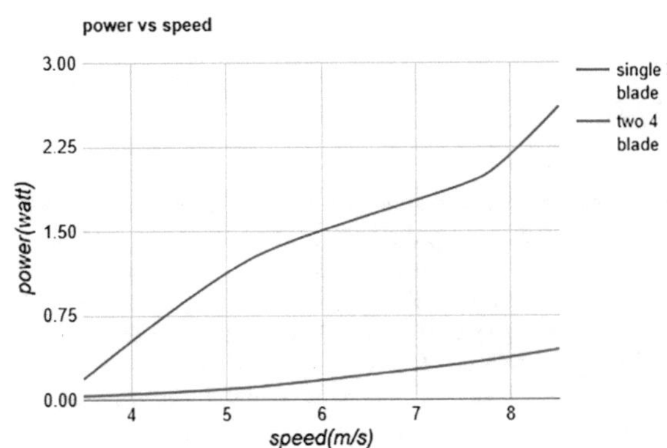

Fig. 3 Graph obtained between power and speed

Aerofoil is designed in order to obtain the optimal tip speed ratio with 25–30%. These highly efficient aerofoil's rotor blade designs increase the rotational speed of the rotor blade. With the help of power output, tip speed ratio is calculated. The maximum tip speed ratio is 0.26 on Site-2 with single three-blade system and the minimum value of 0.09 is obtained with assembly of four-blade system connected in series Site-1 as illustrated in Table 3.

The graph is plotted as shown (Fig. 3) between power and wind speed to demonstrate the efficiency of two different blade assemblies. Airborne Wind Energy power utilizes the principle that wind speed is more strong and consistent on altitudes [5].

4 Conclusion and Future Aspects

In this work, we have been proposed an experimentally combined prototype. It accessed the local terrain and become most efficient technique for mobile generation as seen from literature. The energy of 2.6 watt generated with the speed of 8–9 m/sec. The experiment carried out on two different set of blades proves the flexible characteristics of plastic kite over the cotton cloth and paper. The outcome is plotted and reported that three-blade systems are efficient with respect to configuration cost.

Future work will involve the analysis for the Site-3. It requires an arrangement that makes shaft rotate continuously with the motion of kite. This will involve drawing literature that regulates with constantly moving mechanical device. All readings must have been taken with constant wind speed.

References

1. Diwale, S.S., Lymperopoulos, I., Jones, C.N.: Optimization of an airborne wind energy system using constrained gaussian processes. In: IEEE Conference on Control Application. (2014)
2. Cherubini, A., Papini, A., Vertechy, R., Fontana, M.: Airborne wind energy system: a review of the technologies. Renew. Sustain. Energy Rev. 1461–1476 (2015)
3. Gasch, R., Twele, J.: Wind Power Plant Fundamentals Design, Construction and Operation. Solarpraxix, James &James, London (2004)
4. Sarkar, A., Behera, D.K.: Wind turbine blade efficiency and power calculation with electrical analogy. Int. J. Sci. Res. Publ. **2**(2) (2012)
5. https://spectrum.ieee.org/energywise/green-tech/wind/google-acquires-airborne-wind-power-company-makani (2013)

RSSI-Based Posture Identification for Repeated and Continuous Motions in Body Area Network

Tanmoy Maitra and Sarbani Roy

Abstract Node-wise suitable posture-based data transmission reduces the energy consumption and prolongs the network lifetime. But, the challenging task is to classify and identify the posture sequence for a repeated activity such as walk, freehand exercise, and run in body area network (BAN) with low-cost (without using motion-detecting sensors like accelerometer) solution. This study proposes a solution to identify and classify the posture-based movements in repeated activity like a freehand exercise in BAN after observing the variation of received signal strength indicator (RSSI) over time. Analysis through simulation results shows that proposed solution can achieve the goal.

1 Introduction

The network topology is rapidly altered due to the movements of body segments in body area network (BAN) when a human does some activities. These movements can be considered as postural movements. Generally, for the continuous motions like run, walk, and exercise, there is a fixed sequence in the posture-based movements which is repeated over time. As the distance between the sink and nodes is variable due to postural movements, the performance metrics (energy, throughput, delay) of the network also vary. In [1], a sender-initiated data transmission for repeated activity has been proposed where the posture-wise network performance of an exercise has been captured when all the nodes transmit their data to a sink. Here, the authors suggested that if the transmission schedule of nodes can be distributed based on their best potential posture, then the lifetime of the network increases. For

T. Maitra (✉)
School of Computer Engineering, KIIT University, Bhubaneswar 751024, India
e-mail: tanmoy.maitrafcs@kiit.ac.in

S. Roy
Department of Computer Science and Engineering, Jadavpur University, Kolkata 700032, India
e-mail: sarbani.roy@jadavpuruniversity.in

V. Bhateja et al. (eds.), *Intelligent Engineering Informatics*, Advances in Intelligent Systems and Computing 695, https://doi.org/10.1007/978-981-10-7566-7_36

this purpose, the authors assumed that the sink and all other nodes know the sequence of postures. But, in the real scenario, it is very challenging task to know the sequence of postures without knowing any pre-defined knowledge about repeated activity for continuous motions with a low-cost solution. To investigate this issue, we have found that there are some researches [2, 3] which detected the pattern of postures based on the movements of body segments. An article [2] classified different limb movements based on RSS values using K-NN algorithm, where authors suggested that RSS is a good choice to classify the posture-wise movement's pattern. Based on the training data using accelerometer sensor, a study [3] recognized the motion of a human. However, the existing studies [2, 3] have been experimentally evaluated the result using only one or two nodes, which are basically the individual sink (like smartphone) having sensing, data processing, data gathering, and data delivery capabilities with Internet facility. However, such architecture cannot be considered as BAN because, body-centric network is not formed in this case where, in BAN, different heterogeneous or homogeneous sensors form a network within the human body to deliver their data to a sink via single hop (star topology) or multihop (tree topology) transmission. Furthermore, to the best of our knowledge, the existing posture classification techniques considered the different activity as different posture. For an example, sleep (activity) is a posture, walk (activity) is a posture, and run (activity) is another posture and then the existing study classified the postures according to the pre-defined training data. But, if we observe closely, then, we can notice that an activity (repeated over time) contains some sequence of postures which is repeated over time if a person does the activity for some time. Therefore, we focus in this work on an activity (rather than several activities) which contains a posture sequence repeated over time. However, the term *posture* is used here to indicate the movements of body segment for an action like run or walk or exercise rather than *action-as-posture*. Thus, the definition of posture is different in this article over the existing works which are impossible to compare with the proposed solution.

On the other hand, posture-based data transmission increases the lifetime of BAN [1]. For this purpose, sequence of postures should be identified first so that posture-based data transmission can be achieved. Motivated by the said statement, this paper proposes a RSSI-based posture identification mechanism in BAN when a human is in a repeated and continuous motion. For this purpose, this article has chosen an exercise as a case study of repeated and continuous motion to check the validity of proposed protocol (see Fig. 1). Note that, this paper considers that "*activity is an action,*" and "*movement of body segments due to an action is a posture*.

Road map of this study is as follows: Sect. 2 introduces the model for postural movement. The proposed solution is discussed in Sect. 3. We analyze the proposed solution in Sect. 4. Finally, the conclusion is given in Sect. 5.

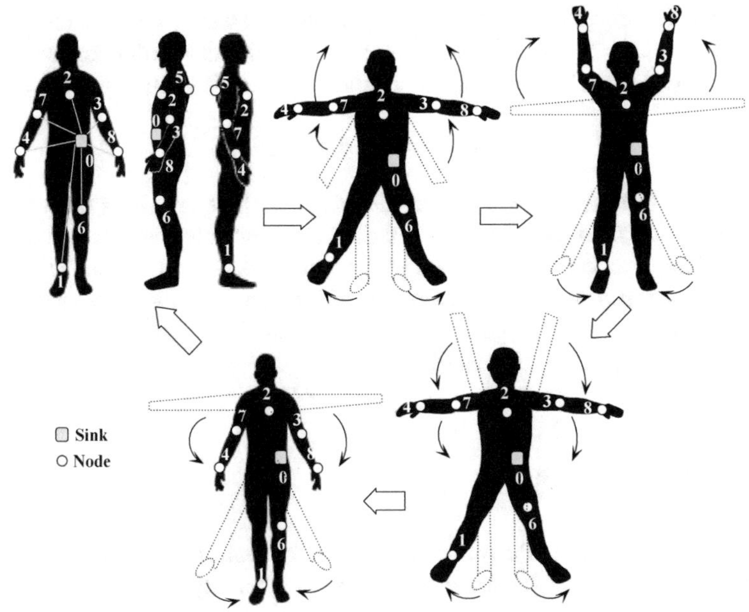

Fig. 1 Sequence of postures due to body movement in an exercise [1]

2 Model for Repeated Postural Movement

During an activity with repeated and continuous motions like freehand exercise, distance between the sink and a sensor placed on human body varies according to the movement of body segments. However, the path loss of signal proportional to the distance, therefore, the value of RSSI inversely related to the distance. In a freehand exercise as shown in Fig. 1, the distance between sensors placed on arms and sink varies according to the movement of arms, and thus, RSSI also varies. For a continuous motion, nodes change their position from one place to another over a time period. Furthermore, in a repeated and continuous motion, a posture can be stated as a displacement of nodes' position over a time span. Thus, in this study, we break the continuous motion into some equal time durations to capture the variations of RSSI for each node within each duration. Here, RSSI in each duration of time for movements denotes a posture. But, for an action like freehand exercise, posture can be repeated over time; thus, our aim is to classify and identify the occurrence of similar postures over each small time period by observing the variation of RSSI in each time span.

In this study, sink transmits the schedule packets for each time duration, and then the sensors periodically send their beacon packets to the sink within that duration in a single hop manner. However, sink calculates the small duration as follows:

$$T_{slot} = T_{sch} + n \times T_{data} \tag{1}$$

where T_{sch} is the time to transmit schedule packet; n (here, n = 8) is the number of sensors/nodes in the network; T_{data} is the time to transmit one data packet. Note that, in node-wise best posture-based data transmission mechanism [1], all the nodes may perform their best within a particular posture; therefore, sink has to select all the nodes for that posture. Thus, maximum n number of nodes may be selected for any duration of posture.

3 Proposed Solution

In this section, we discuss the proposed solution to identify the sequence of posture repeated over time for the continuous motion of an activity like freehand exercise. Sink collects RSSI for the scanning duration (denoted as *ScanTime*) which is the aggregation of some small and equal time slot $T^{j}_{slot} \forall j = [1, k]$ as calculated in Eq. (1). Then, based on RSSI for each T_{slot}, the sink will find the sequence of posture. The format of each packet of proposed solution is shown in Fig. 2.

The following steps are executed by the sink and sensors so that a sequence of postures for repeated and continuous motions can be identified and classified. Figure 3 shows the timeline of proposed solution.

Sink chooses *ScanTime* which is long enough to identify the sequence of postures. Then, the sink sends a broadcast packet containing information $\langle ScanTime, T_{slot} \rangle$.

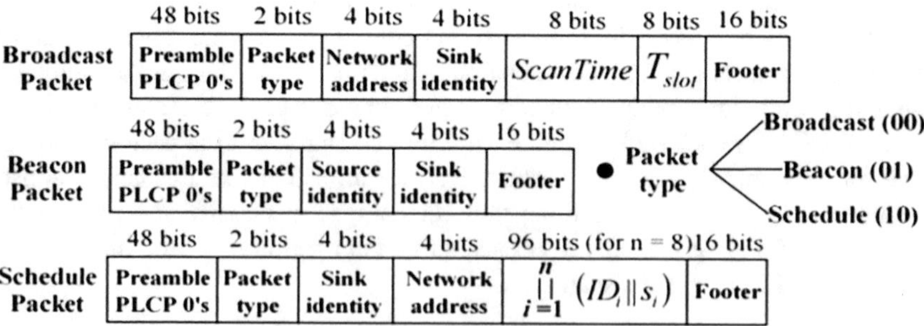

Fig. 2 Packet format of the proposed solution

After receiving the broadcast packet, all the nodes set their scanning time and T_{slot} so that they can send a beacon packet periodically within each T_{slot} throughout the scanning period. However, a sensor will wait to get its slot number to transmit its beacon for each T_{slot}.

For each T_{slot}, the sink permutes the slot numbers from initial vector $\vec{s} = [1, n]$ and assigns the slot number $s_i \forall [1, n]$ corresponding to node *i*. Then sink broadcasts the schedule packet containing node identity slot number information, i.e., $(ID_i, s_i) \forall i = [1, n]$.

After getting schedule packet for each T_{slot}, a node *i* extracts its slot number s_i and goes to sleep for the time duration $(s_i - 1) \times T_{DATA}$. After waking up, node *i* sends its beacon packet for T_{DATA} duration and again goes to sleep for the time duration $T_{slot} - \{(s_i \times T_{DATA}) + T_{sch}\}$ as shown in Fig. 3.

For each T_{slot}, after receiving a beacon packet from node *i*, sink calculates RSSI of the packet and puts the information $\langle ID_i, RSSI^i_{T^j_{slot}} \rangle \forall j = [1, k]$ and $i = [1, n]$ into a list, named as *infolist*. However, sink follows this procedure throughout the scanning duration. Table 1 shows the final *infolist* built by sink.

For each node $i = [1, n]$, sink calculates the standard deviation of RSSI σ_i^{RSSI} from *infolist* as follows: $\sigma_i^{RSSI} = \sqrt{\frac{\sum_{j=1}^{k} \left(RSSI^i_{T^j_{slot}} - RSSI^i_{avg} \right)^2}{k}}$, where, $RSSI^i_{avg}$ is the mean value of RSSI for node *i* during the scanning time as shown in Table 1. Thereafter sink follows Algorithm 1 to make another list for posture identification, named as *pinfolist* as shown in Table 2. For this purpose, sink decides the initial cluster for each node *i* to find the closest RSSI of one time slot to other time slots (see Algorithm 1) based on threshold value σ_i^{RSSI}. But in the initial cluster formation,

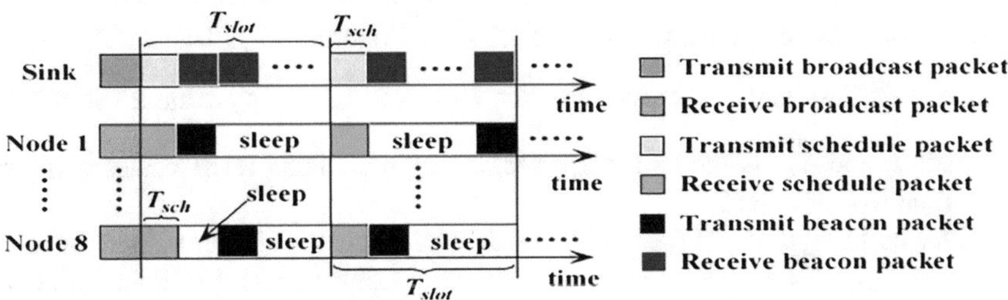

Fig. 3 Timeline of proposed solution

Table 1 *infolist*: a list of node-wise RSSI for each time slot

Node ID	T^1_{slot}	T^2_{slot}	…	T^k_{slot}
1	$RSSI^1_{T^1_{slot}}$	$RSSI^1_{T^2_{slot}}$	…	$RSSI^1_{T^k_{slot}}$ ⇒ σ_1^{RSSI}
.	.	.	.	.
.	.	.	.	.
n	$RSSI^n_{T^1_{slot}}$	$RSSI^n_{T^2_{slot}}$	…	$RSSI^n_{T^k_{slot}}$

Table 2 *pnfolist*: a list of node-wise assigned value for each time slot

Node ID	T^1_{slot}	T^2_{slot}	…	T^k_{slot}
1	v_1^1	v_1^3	.	v_1^1
.	P_1 . ⇒	.	.	P_1 .
.	.	.	.	.
n	v_n^1	v_n^6	.	v_n^1

RSSI of a time slot may close to RSSI of other multiple time slots for each node i. Thus, it is required to select one time slot which has a minimum difference of RSSI among the initial cluster of close RSSI. In Algorithm 1, $\vec{R}_i = \{RSSI^i_{T^j_{slot}}\} \forall j$ is the set of RSSI of node i for each time slot, $R^i_{j*} = RSSI^i_{T^j_{slot}} \in \vec{R}_i \forall j$ is the RSSI of T^j_{slot} for node i, and $\vec{X}_i^{j*}$ is the temporary vector which contains RSSI values close to each time slot of node i.

After building *pinfolist* (see Table 2), the sink finds the similar values for each time slot T_{slot} of all nodes $i = [1, n]$, that means sink selects the similar column vector from Table 2. For the similar column vector, the sink assigns the same posture number (e.g., P_1, P_2, P_3, $\cdots$).

Algorithm 1 *postureClassification* ()

```
∀ node i = [1,n]
                ∀ T^j_slot where j = [1,k]
                   1. Find the difference of RSSI from R^i_{j*} ∈ R⃗_i to all other RSSI in
Initial               R⃗_i | i ≠ j .
cluster            2. Select those RSSI values from R⃗_i whose difference (from R^i_{j*} )
                      is within the threshold σ_i and forms a new vector X⃗_i^{j*}.
                end for
                For each RSSI^i_j ∈ X⃗_i^{j*}
                   3. Check whether RSSI^i_j is present in other vectors X⃗_i^{m*} | m* ≠ j*
                      or not
                   if (found)
                      if (diff(RSSI^i_j, X⃗_i^{j*}) ≤ diff(RSSI^i_j, X⃗_i^{m*}))
Final                     Remove RSSI^i_j from X⃗_i^{m*}
cluster               else
                          Remove RSSI^i_j from X⃗_i^{j*}
                      end if
                   end if
                end for
                For each X⃗_i^{j*} , X⃗_i^{j*} ← v_l , where v_l is a non repeated value
end for
```

4 Result Analysis

We have used well-popular *Castalia* [4] simulator to evaluate the results of the proposed solution in a BAN environment (see Fig. 1). *Castalia* simulator (based on C ++) provides an environment by which wireless sensor networks and BAN can be simulated. Moreover, to know more details about Castalia simulator, we refer references [4–6]. The radio parameters for BAN, used in this article, are same as *Castalia* provides (see Table 3).

We considered 40 time slots (i.e., $k = 40$) for the duration of scanning time and tried to classify the repeated postures over the time slots. For this purpose, Fig. 4 is given to show the variation of RSSI for each node during the scanning time, which is divided into some small time slots (see Eq. (1)). As the distance between sink and nodes 1, 2, 5, and 6 is probably similar for each time slot; therefore, for each node, variation of RSSI is also close to each other for each time slot (see Fig. 4a). But,

Table 3 Simulation setup

Parameter	Value	Parameter	Value
Human height	5'6"	Transmission power	-25 dBm
# of slots (k)	40	# of nodes (n)	8
Modulation type	DIFFQPSK	Frequency bandwidth	20 MHz
Noise bandwidth	1000 MHz	Noise floor	-104 dBm
Sensitivity	-87 dBm	Sink	Node 0

due to huge movements of nodes placed on hands (i.e., nodes 3, 4, 7, and 8), variable distance between the sink and these nodes are highly notable. Thus, RSSI of these nodes also varies for each time slot (see Fig. 4b).

After observing the variation of RSSI for each time slot of all the nodes, the sink classifies and identifies the repeated postures over scanning time as shown in Fig. 5. By the proposed solution, nine different postures are classified which are repeated over time. For an example, the sink assigned same value for slot 1, slot 17, and slot

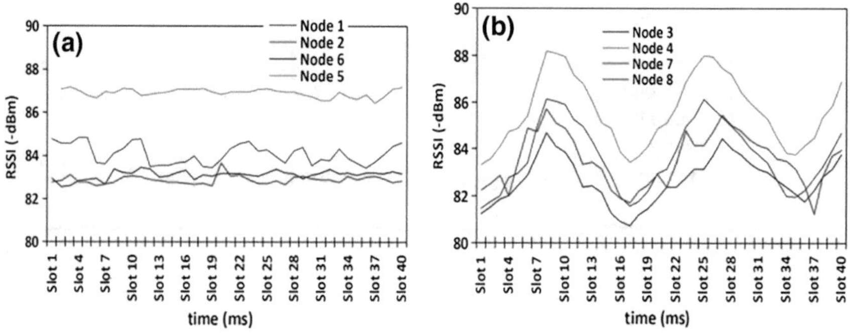

Fig. 4 Variation of RSSI over time **a** nodes with less mobility and **b** nodes with high mobility

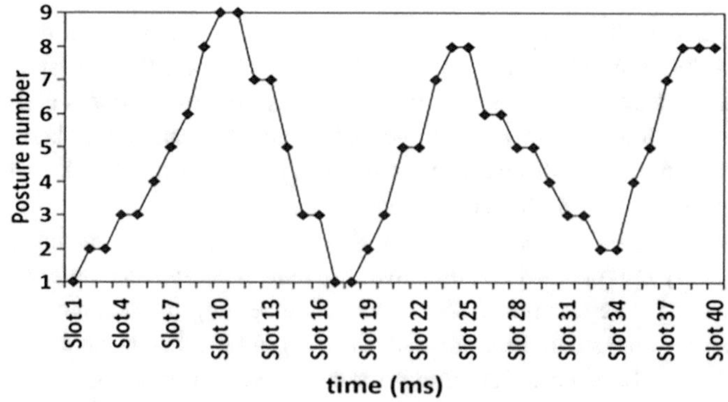

Fig. 5 Nature of posture occurrence over time (repeated manner)

18 by performing final cluster as discussed in Algorithm 1. Therefore, posture 1 (as per ascending order) is repeated on slots 17 and 18, respectively, after the slot 1 which indicates that the proposed solution can classify and identify the posture sequence repeated over time when human do an activity for some time continuously. Figure 5 shows the occurrence of posture for continuous time of the exercise.

5 Conclusion

This study proposes a solution in which any motion-detector sensors like accelerometer are not required to identify and classify the repeated postures for an action. In the proposed technique, sink does not need to know about the actions like sleep, sit, run, and walk because, in this article, our aim is to find posture sequence for an individual activity like exercise (repeated over time) rather than to identify different actions. Thus, finding posture sequence for a repeated action makes our paper different from existing works. From the initial observation of this work, it can be said that the proposed solution can classify the postures over time by getting RSSI of a signal from the nodes for different time instance. In our future work, we will focus on estimating the error rate of the proposed classification technique.

References

1. Maitra, T., Mallick, P., Roy, S.: LD-MAC: A load-distributed data transmission in body area network. In: IEEE SENSORS, Orlando, FL, pp. 1–3. (2016)
2. Guraliuc, A.R., Barsocchi, P., Potortì, F., Nepa, P.: Limb movements classification using wearable wireless transceivers. IEEE Trans. Inf. Technol. Biomed. **15**(3), 474–480 (2011)
3. Lugade, V., Fortune, E., Morrow, M., Kaufman, K.: Validity of using tri-axial accelerometers to measure human movement-Part I: Posture and movement detection. Med. Eng. Phys. **36**(2), 169–176 (2014)
4. Castalia: Wireless Sensor Networks and BAN Simulator. http://castalia.npc.nicta.com.au
5. Mukherjee, N., Neogy, S., Roy, S.: Building Wireless Sensor Networks: Theoretical and Practical Perspectives. CRC Press, London (2015)
6. Maitra, T., Roy, S.: A comparative study on popular MAC protocols for mixed wireless sensor networks: from implementation viewpoint. Comput. Sci. Rev. **22**(C), 107–134 (2016)

Old-Age Health Risk Prediction and Maintenance via IoT Devices and Artificial Neural Network

Dayashankar Prajapati and K. Bhargavi

Abstract IoT is the collaboration of physical devices, electronics, software, sensors, actuators, and network connectivity which helps in collecting and exchanging of data. It is a smart technology which cooperates with the environment by the sensors and actuators which are found very useful in monitoring the health of the people. It holds different features such as diagnosis, signal analysis, drug development, medical image analysis, and radiology. These features of IoT devices are used for monitoring the health of old-age citizen as they are highly influenced by several diseases which require continuous monitoring and treatment. In this paper, a novel IoT and neural network-based old-age health risk prediction framework is proposed, and the performance of the proposed framework is found to be good with respect to parameters like time consumption, monitoring efficiency, and cost incurred.

Keywords IoT · Health monitoring · Internet of Things · Neural network Health risk prediction

1 Introduction

Old age is the age surpassing the life expectancy, age over the retirement, and it's also almost the end of the human life cycle. Old-age citizens often are more susceptible to diseases, syndromes, and sickness such as diabetes, blood pressure, joint pain, and dental problem. According to the research by WHO [1], the life expectancy rate of people in India is 68 years for female and 67 years for the male which is very less compared to Japan. The proper health monitoring of the old-age citizens

D. Prajapati (✉) · K. Bhargavi
Department of Computer Science and Engineering, Siddaganga Institute of Technology, Tumkur 572103, Karnataka, India
e-mail: dayashankarprajapati48@gmail.com

K. Bhargavi
e-mail: bhargavi.tumkur@gmail.com

V. Bhateja et al. (eds.), *Intelligent Engineering Informatics*, Advances in Intelligent Systems and Computing 695, https://doi.org/10.1007/978-981-10-7566-7_37

can increase the life expectancy rate of them. IoT allows the health of the people to be sensed remotely across existing network infrastructure. It can help in determining the health parameters inside the patient's body such as blood pressure, pulse rate, respiratory rate, and urinary tract infections and the providers get an accurate view of their health so that they can personalize it as need.

Neural network holds the ability to solve the problems that either do not have an algorithmic solution [2]. They are the digitized model of the biological brain and can detect a complex nonlinear relationship between the health condition and the patient activities where doctors may fail to connect. IoT and neural network works on the same platform sharing the parameters to each other. For instance, consider health monitoring system which is monitored with a large number of health parameter collected through different IoT devices. Possible actions can be performed are monitoring of blood pressure, pulse, respiratory, enabling user notification, and contacting concerned department. In these cases, some cases are critical in which direct fast action needs to be taken, for example, when a patient having a low respiratory rate. In this case, the immediate response needs to be made for which a good local neural network can be crafted in order to feed the reliable and fast response.

2 Definitions

In this section, the definition of some of the medical terminologies used in the paper is provided with the normal range of corresponding terms.

- **PID**: It is a patient's unique 4-digit identification number.
  For example, 1259, 8545, 6963.
- **SPO_2**: It indicates oxygen-saturated hemoglobin level with respect to total hemoglobin in the blood of the patient.
  Normal range: 95–100%
- **Respiratory rate**: It is defined as the number of breathing per minute which is directly proportional to the health of the body.
  Normal range: 12–28 breaths per minute
- **Pulse rate**: It is defined as the ratio of heartbeat measured and the number of contraction of the heart per minute.
  Normal range: 30–100 beats per minute for adults
- **Blood pressure**: It is defined as the pressure applied by the blood against the inner wall of the arteries which is measured in terms of systolic pressure and diastolic pressure.
  Normal range: 120/80 mm Hg
- **Conscious level**: Conscious level defines the quality of being awareness within oneself or with the external world.
- **Skin disorder**: The infection and diseases on the skin which indicate the symptoms of other diseases in the patient's body.

3 Organization of Paper

The rest part of the paper is organized as follows: Sect. 4 provides some of the related work, Sect. 5 gives the proposed work and description of the framework in detail, Sect. 6 discusses detail working of different module of the architecture along with their algorithms, Sect. 7 describes the result, and finally Sect. 8 gives the conclusion.

4 Related Work

An IoT-based remote health monitoring framework for geographically distributed age citizens is discussed in [3]. The discussion is focused on the need of IoT devices, and also the network requirement and communication protocols are considered; feasibility of deploying remote health monitoring system is analyzed. However, the scope of the old-age health monitoring system is limited only to a small region.

The health monitoring system using IoT and Raspberry Pie board is discussed in [4]. It has made an attention on monitoring the patients in intensive care unit (ICU) and cardiac care unit (CCU) where the focus is limited to health disease patients.

Elderly remote health monitoring scheme using IoT from user-centered viewpoints is discussed in [5]. The primary attention focused is to collect the necessary requirement of the remote health monitoring system in order to pave the way for future system and improve the quality. However, the requirement and objectives of elderly health monitoring are briefly described from the user-centered viewpoint [6, 7].

5 Proposed Work

The proposed old-age health prediction and maintenance framework are shown in Fig. 1. It is split into four modules that have individual works and performance which include various IoT devices like UVC radiometer, Raybaby camera, and CMOS IoT devices. The IoT devices are connected together and monitor the patient within the territory and collect the health information.

The primary module in the proposed work is IoT Data Collector which collects the health parameters (HP) of the patients from all planted IoT devices. The second module is First Opinion Formulator which accepts the data from the IoT Data Collector through the network and formulates the first opinion to the patient. The third module is risk predictor which uses high-risk-appended health parameters for calculating the risk factor of the patients. The final module is Second Opinion Formulator which provides the new prescription to the patient by consulting the concerned department of the healthcare center.

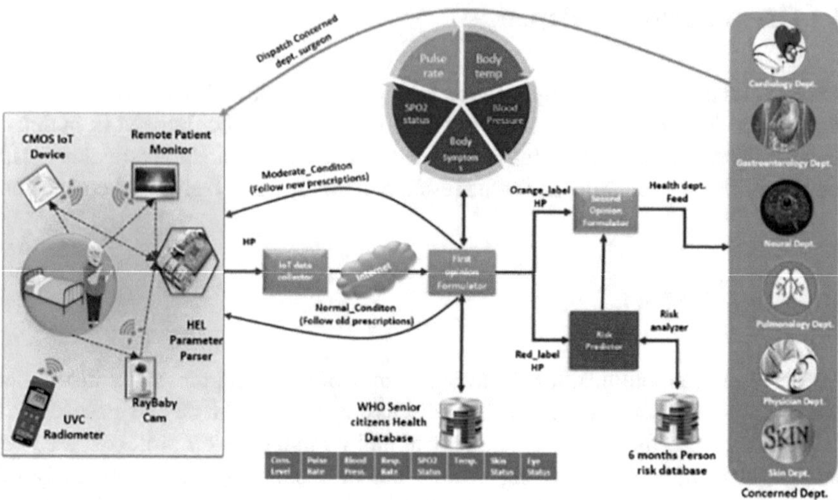

Fig. 1 Proposed old-age health prediction and maintenance framework

6 Algorithms

6.1 IoT Data Collector

The main function of this module is to aggregate the health parameters of the patients from all planted IoT devices. The working of IoT data collector is given in Algorithm 1.

Algorithm 1. IoT Data Collector

Input: Health parameters from every deployed IoT devices
Output: Aggregated health parameters from IoT devices with person id
BEGIN
  Initialize the IoT devices IoT_1, IoT_2...IoT_n
  For every IoT devices, IoT_i € IoT **do**
    Record the HP= {HP_1, HP_2...HP_K}
  End for
  For ever **do**
    If recorded HP! = NULL
      Data successfully collected from IoT devices_Send the HP to First Opinion Formulator
    Else
      Data transfer from IoT devices failed_Reinitialize the IoT devices & recollect the data
    End If
  End for
  Output aggregate HP of the patient_send to First Opinion Formulator
END

6.2 First Opinion Formulator

This module accepts the aggregated health parameter from IoT Data Collector and formulates a valid first opinion on the patient health. The working of First Opinion Formulator is given in Algorithm 2.

Algorithm 2. First Opinion Formulator

```
BEGIN
    Accept aggregated HP from IoT Data Collector
    Initialize Red_label = high risk, Orange_label= moderate, Green_label =Normal
    Read the approaching HP to HP1← (R_rate), HP2← (P_rate), HP3← (BP_rate),
    HP4←(Temp), HP5← (SPO2_rate), HP6← (E_status), HP7← (S_status) from IoT devices
    For every HP in WHO senior citizen database do
        For age group 45 to 65
          For every HPi do
            Compute sum of corresponding HPi
          End for
            Compute child_avg = sum of HPi/n
            Compute child_avg= {R_rate, P_rate, BP_rate, SPO2_rate, Temp,
            S_status, E_status}
        End for
    End for
    For age group greater than 65 Find Predecessor_avg
      For every HPi do
        Compute sum of corresponding HPi
      End for
        Compute predecessor_avg=sum of HPi/n
        Compute Predecessor_avg= {R_rate, P_rate, BP_rate, SPO2_rate, Temp,
        S_status, E_status}
    End for
    For age group with difference of 10
      For every HP and child_avg
        If HP deviation is lower to child_avg then
         Append Green_label to HPi
        Else if HP deviation is neutral to child_avg then
         Append Orange_label to HPi
        Else
         Append Red_label to HPi_ convey moderate condition to Risk Prediction
        End if
      End for
    End for
   Return HP with Orange_label to Second Opinion formulator
   Return HP with Red_label to Risk Predictor
End
```

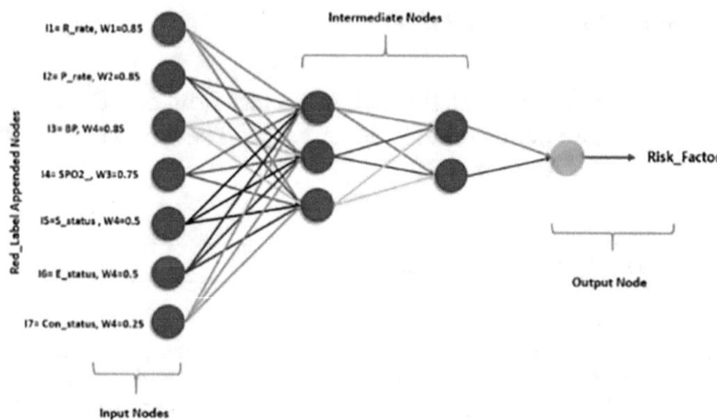

Fig. 2 Health risk prediction using neural network

7 Risk Predictor

In this module, the critical health parameters are fed as input to the neural network which assigns the risk factor of the patient shown in Fig. 2. The working of risk predictor is given in Algorithm 3.

Algorithm 3. Risk Predictor

```
BEGIN
    Initialize High_risk=0, Low_risk=0
    Initialize Old Risk_factor from person risk database
    Accept appended Red_label HP from first opinion formulator
    For every Red_label HP do
        Initialize no. of input nodes, hidden nodes and output nodes in Neural Network
        Compute Risk_factor _store to the personal risk database
    End for
    Compare Risk_factor with Old Risk_factor from Person risk database
    For ever do
        If Risk_factor is greater than old Risk_factor then
            Return High_risk=true
        Else
            Return Low_risk=true
        End if
    End for
    Output HP by appending risk factor_send to Second Opinion Formulator
End
```

8 Second Opinion Risk Formulator

The module accepts input from First Opinion Formulator and risk predictor and decides the required action to be taken on the resultant risk factor. The working of second opinion formulator is given in Algorithm 4.

Algorithm 4. Second Opinion Risk Formulator

```
BEGIN
    Accept Orange_label appended HP from the first opinion Formulator
    Accept Risk_factor from Risk Predictor
    For every HP and Risk_factor do
      If Orange_label ←(R_rate and SPO2) and High_risk then
         Connect Pulmonology department_Dispatch emergency service
      Else If Orange_label ←(R_rate and SPO2) and Low_risk then
         Connect pulmonology department_Prescribe new prescription
      Else If Orange_label ←(P_rate and BP) and High_risk then
         Connect cardiology department_Dispatch emergency service
      Else If Orange_label ←(P_rate and BP) and Low_risk then
         Connect cardiology department_Prescribe new prescription
      Else If Orange_label ←(S_status and E_status) and High_risk then
         Connect skin, physician department _Dispatch emergency service
      Else If Orange_label ←(S_status and E_status) and Low_risk then
         Connect skin, physician department_Prescribe new prescription
      Else If Orange_label ←(con_status and B_status and Temp) and High_risk then
         Connect neural department_Dispatch emergency service
      Else If Orange_label ←(con_status and B_status and Temp) and Low_risk then
         Connect neural department_Prescribe new prescription
      End if
    End for
End
```

9 Result

In this section, the performance of the proposed framework is compared with the existing work with respect to performance parameters like cost efficiency, time consumption of IoTs and estimated improved condition of old-age health [8, 9].

9.1 Health Monitoring Efficiency

The graph of the number of IoTs versus adaptability is shown in Fig. 3. The efficiency of the health monitoring scheme increases with the increasing health

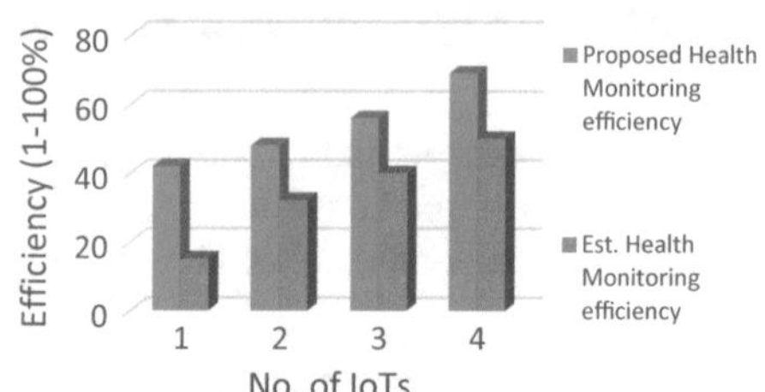

Fig. 3 Health monitoring efficiency

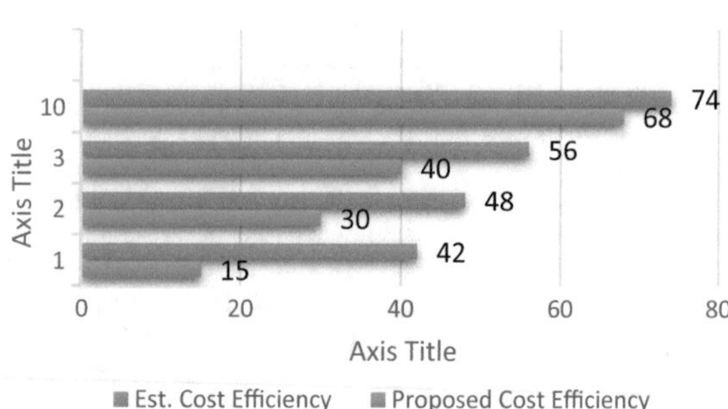

Fig. 4 Health monitoring cost efficiency

parameters [10, 11]. In the proposed work, as more IoT devices are deployed to monitor patient's health status, the efficiency of the proposed work is high.

9.2 Health Monitoring Cost Efficiency

The graph of cost efficiency versus the number of health parameter is shown in Fig. 4. The IoT which can perform multiple functions (SPO_2, body activity, respiration rate, etc.) such as CMOS IoT device and Raybaby camera is used in the proposed work which will decrease the cost efficiency compared to the multiple single-functional IoT devices [12].

10 Conclusion

This paper presents a completely unique theme for monitoring the health of old-age citizen and preventing them from the dangerous diseases like blood pressure, respiratory problem, sugar, and joint pain. The result obtained shows that the proposed framework using IoT is found to be the best method to keep the old-age citizens live healthily.

References

1. https://en.wikipedia.org/wiki/List_of_countries_by_life_expectancy/
2. Kumar, S.S., Kumar, Dr. K.A.: Neural networks in medical and healthcare. Int. J. Innovative Res. Develop. **2**(8), 241–244 (2013)
3. Khoi, N.M., Saguna, S., Mitra, K., hlund, C.: IReHMo: an efficient IoT-based remote health monitoring system for smart regions. In: IEEE Xplore, pp. 1–6 (2015)
4. Koshti, M., Ganorkar, S., Prof. Dr.: IoT based health monitoring system using Raspberry Pi and ECG signal. Int. J. Innovative Res. Sci. Eng. Technol. **5**, 8977–8985 (2016)
5. Azimi, I., Rahmani, A.M., Liljeberg, P., Tenhunen, H.: Internet of things for remote elderly monitoring: a study from user-centered perspective. J. Ambient Intell. Humanized Comput, 1–17 (2016). Springer
6. World Health Organization website (https://www.who.int/)
7. Aruna, D.S., Godfrey, W.S., Sasikumar, S.: Patient health monitoring system (PHMS) using IoT devices. Int. J. Comput. Sci. Eng. Technol. **7**, 68–73 (2016)
8. Hassanalieragh, M., Page, A., Soyata, T., Sharma, G., Aktas, M., Mateos, G., Kantarci, B., Andreescu, S.: Health monitoring and management using Internet-of-Things (IoT) sensing with cloud-based processing: opportunities and challenges. In: IEEE International Conference on Services Computing, pp. 285–292 (2015)
9. Gómez, J., Oviedo, B., Zhuma, E.: Patient monitoring system based on internet of things. In: 7th International Conference on Ambient Systems, Networks and Technologies, pp. 90–97 (2016)
10. Babu, B.S., Bhargavi, K.: CAs-based QoS scheme for remote health monitoring. Int. J. Agent Technol. Syst. **5**(4), 44–67 (2013)
11. Gautam, A., et al.: An improved mammogram classification approach using back propagation neural network. In: Data Engineering and Intelligent Computing, pp. 369–376. Springer, Singapore (2018)
12. Tiwari, A., Gautam, A.: Classification of Mammograms using Sigmoidal Transformation and SVM. In: Proceedings of International Conference on Smart Computing and Informatics (SCI-2017), pp. 1–8, March 2017

Computational Intelligence Based Chaotic Time Series Prediction Using Evolved Neural Network

N. Ashwini, M. Rajshekar Patil and Manoj Kumar Singh

Abstract A nonlinear behavior may exist intrinsically with a deterministic dynamic system which shows high sensitivity with initial condition. Such behavior is characterized as chaotic behavior. Strange attractor confines the dynamic behavior of chaotic system in a finite state space region, and the available state variables show the stochastic behavior with time. In this research work, time delay neural network has applied to predict the various chaotic time series. Adaptive social behavior optimization has applied to evolve the optimal set of weights of the time delay neural network. Comparison of learning performance has given with popularly gradient descent-based learning. Performance evaluation has defined in terms of coefficient of determination along with root-mean-square error in prediction under learning and test phase of chaotic time series. The three benchmarks of chaotic time series (logistic differential equation, Mackey–Glass, and Lorenz system) have taken for predicting purpose. We have shown with experimental results that the proposed new method of neural network learning is very efficient and has delivered the better prediction for various complex chaotic time series.

Keywords Chaotic time series · Forecasting · Time delay neural network ASBO · Gradient descent

N. Ashwini
BMS Institute of Technology and Management, Bengaluru, Karnataka, India
e-mail: ashwinilaxman@bmsit.in

M. R. Patil
TCET (UD) – College of Engineering, Hyderabad, Telangana, India
e-mail: pvsmr1@gmail.com

M. K. Singh (✉)
Manuro Tech Research Pvt. Ltd, Bengaluru, Karnataka, India
e-mail: mksingh@manuroresearch.com

V. Bhateja et al. (eds.), *Intelligent Engineering Informatics*, Advances in Intelligent Systems and Computing 695, https://doi.org/10.1007/978-981-10-7566-7_38

1 Introduction

The time series is a set of data constructed from chronological observed data. Generally speaking, time series prediction mainly includes three operation steps: observation and preprocessing of the past and present time series data samples, suitable prediction model design, and finally the use of available data samples with predicting model to predict the future data samples.

There are various useful practical applications of time series forecasting in different areas such as in the economics prediction, weather prediction, and network traffic prediction [1–3]. Time series prediction may be usually classified into the linear and nonlinear prediction methods. The typical linear time series prediction models mainly include autoregressive (AR), moving average (MA), and autoregressive moving average (ARMA) [3–5]. However, these models are mostly the statistical method-based prediction models and need some prior knowledge that is proven difficulty, so these models can hardly obtain effective forecasting results for chaotic nonlinear time series. Many practical time series data have complexity of chaotic characteristics which are difficult to approximate by linear model; hence, computational intelligence-based methods such as genetic algorithm(GA) [6, 7], neural network [8] and support vector machine(SVM) [9] have applied to meet the requirement. Volná et al. [10] has applied neural network-based time series analysis to recognize the pattern in Forex market trading system. To improve the forecasting performance of linear regression, [11] has applied the grouping-based quadratic mean loss function. Volná and Kotyrba [12] has proposed averaging value-based time series prediction. One day ahead speed of wind has predicted in [13] considering different varieties of time series models. Akpinar and Yumusak [14] has applied time series method to increase the performance accuracy in the forecasting of demand of gas tractions on midterm consumption. In long-term forecasting, self-organizing map has applied to choose the best model in ensembles [15]. Averkin et al. [16] has presented research review in the field of hybrid methods of time series forecasting. Forecasting of high-dimensional time series by dimension reduction method has been proposed in [17].

The objective of this research is to propose a new learning concept using ASBO to evolve the neural network connection weights to model nonlinear time series, in particular chaotic systems, in order to predict a system's future behavior from a sequence of time-ordered observations.

The remaining part of this paper is organized as follows. In Sect. 2, mathematical description of time series prediction has presented. Section 3 has discussed neural network and its learning details by proposed method ASBO. Sections 4 and 5 carry the information about chaotic time series and detail experimental results. Finally, a conclusion has given in Sect. 6.

2 Time Series Prediction and Dynamical Systems

Theoretically, a dynamic system can be represented as a smoothing map function as $F: R \times S \rightarrow S$, where S is an open set of a Euclidean space. Considering $F(t, x) = F_t(x)$, the map F has to satisfy the following conditions:

$$F_O(x) = x; \quad F_t(F_S(x)) = F_{s+t}(x), \forall s, t \in R$$

For any given initial condition $x_0 = F_0(x)$, a dynamical system defines a trajectory $x(t) = F_t(x_0)$ in the set S.

For various different initial conditions, analyzing the characteristics of obtained trajectories is fundamental problem in dynamic system. Our interest is in similar problem but inverse manner as stated above. With available finite part of a time series x(t), x is a component of a vector X that is evolving through an unknown dynamic system. We assume that the trajectory x(t) lies on a manifold with fractal dimension D(a 'strange attractor'), and our objective is to predict the future behavior of time series x(t). Remarkably this is possible without having the other components of vector X. Takens's embedding theorem [18] ensures that under different initial conditions almost for all τ and for some $m \leq 2D + 1$ there is a smooth map $f: \mathfrak{R}^m \rightarrow \mathfrak{R}$ such that:

$$x(n\tau) = f(x((n-1)\tau), x((n-2)\tau), \ldots\ldots\ldots\ldots x((n-m)\tau)) \tag{1}$$

The value of m used is called the embedding dimension, and the smallest value for which Eq. 1 is true is called the minimum embedding dimension m*. Therefore if the map were known, the value of x at time nτ is uniquely determined by its m values in the past. For simplicity of notation, we define the m-dimensional vector

$$\tilde{x}_{n-1} \equiv (x((n-1)\tau), x((n-2)\tau), \ldots\ldots\ldots\ldots x((n-m)\tau)) \tag{2}$$

$$x(n\tau) = f(x((n-1)\tau), x((n-2)\tau), \ldots\ldots\ldots x((n-m)\tau)) \tag{3}$$

in such a way that Eq. 1 can be written simply as $x(n\tau) = f(\tilde{x}_{n-1})$. If N observations $\{x(n\tau)\}_{n=1}^{N}$ of the time series x(t) are known, then one also knows N − m values of the function and the problem of learning the dynamical system becomes equivalent to the problem of estimating the unknown function f from a set of N − m sparse data points in $\mathfrak{R}^m$. Many regression techniques can be used to solve problems of this type. In this paper, we concentrate on the time delay neural network evolved by ASBO. Thus, given observations y(t), y(t − 1),… , y(t − m), we want to predict system behavior y(t + 1), y(t + 2),… , y(t + n). It is known that long-term dependencies, where the desired output of a network depends on inputs many steps backward in time, are difficult to learn. Modeling the statistics of a time series that changes over time represents another challenge. In order to overcome these difficulties, various network architectures have been developed.

3 Artificial Neural Networks (ANN)

A neural network is a network of processing elements which has principle of computation as parallel, distributed information processing. Processing elements also called as neurons are interconnected with unidirectional signal flow path. ANN can be characterized by describing the computational capabilities of the processing elements and the topology of the interconnection network; an interconnection strength or weight is associated with each link between two or more processing elements. Typically, an algorithm is used to set the weights such that the network performs its intended function.

3.1 Learning Algorithms in ANN

The objective of learning is to find a location in weight space such that a network performs a given learning task satisfactorily. In supervised learning schemes, the comparison between a network's output and a desired output provides the basis for defining some error measure (e.g., minimum mean-squared error). The search for a solution then is equivalent to searching for a point in weight space which has minimal error. Depending upon the way the optimal weights are obtained, the method of learning can be classified broadly as deterministic approach or stochastic approach.

3.2 Deterministic Approach: Gradient Descent

Algorithms based on gradient descent are by far the most popular neural network training methods. However, they suffer from two major disadvantages: The algorithms may converge very slowly to a minimum depending on the parameters chosen and the geometry of the error surface. Heuristics have been proposed to reduce the time needed to train a network. A more severe problem is that such algorithms are not guaranteed to find the global minimum and thus may get stuck in a local minimum; i.e., a network may fail to learn a particular mapping altogether. The initial position in the weight space, from which the search for a minimum starts, largely determines whether or not the network learns a particular mapping. Gradient descent can only be applied in networks where the discriminant functions are continuous.

3.3 Stochastic Approach: Evolving the Weights

Many real-world problems are having enumerative in nature and irregularity in their landscape. Deterministic approach may get difficulties to solve such kind of problems. In order to obtain the better solution, stochastic methods are the good alternatives. In stochastic method, there is a need to define the fitness function to estimate the fitness value of possible solutions, a mapping mechanism to encode/decode between the problem and algorithm domain.

3.4 Proposed ASBO-Based Learning of ANN

Based upon human social evolution, adaptive social behavior optimization (ASBO) has been developed in [19]. As in human social life, directly or indirectly our course of actions or behaviors is highly influenced by a number of social factors such as leader, neighbors, and self-cognition status. These influencing factors have effect in differential form, and its affect in changing the individual behavior is very complex and dynamic. Generally, it depends upon the individual cognition status as well as interaction with social environment. To model the influencing affect in ASBO, a self-adaptive mechanism based on mutation strategy has applied. Considering the direct encoding of solution, a set of solution population has defined. In Fig. 1, solution direct encoding has presented. The connection between a node 'a' and node '1' is represented by W1a. An individual, who has maximum value of fitness, has considered as leader in the current population. Neighbor group members are decided through the most closet solutions having next higher fitness value. The new

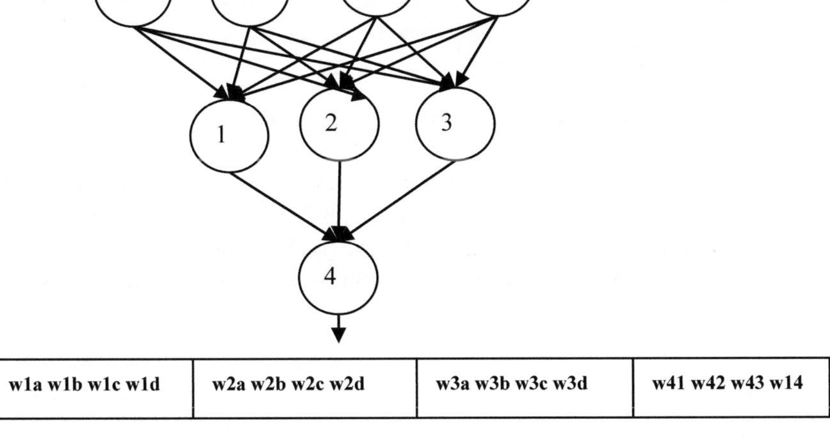

Fig. 1 A solution representation in population

status of each solution is derived through the influence environment as given in Eq. 4, and new solution is obtained by adding the status improvement with the previous existing solution as shown in Eq. 5:

$$\Delta x(i+1) = C_g R_1 (G_{bi} - X_i) + C_s R_2 (S_{bi} - X_i) + C_n R_3 (N_{ci} - X_i) \tag{4}$$

$$X(i+1) = X_i + \Delta X(i+1) \tag{5}$$

where $\Delta x(i+1)$ represents the change in the ith dimension of status; Cg, Cs, and Cn are adaptive constants and have values greater than zero; Ri, (i = 1, 2, 3) are random numbers generated through uniform distribution in the range of [0 1]; Gb is the leader; and Sb is the self best for an individual. Nc is the center position for a solution which is the mean position formed through its and neighbors. Parameters Ci(Cg, Cs, Cn) represent the influencing coefficient and estimated through self-adaptive process as defined through Eqs. 6 and 7

$$\mathrm{C'i(j)} = \mathrm{Ci(j)} + \sigma\mathrm{i(j).N(0,1)} \tag{6}$$

$$\begin{gathered}\sigma'\mathrm{i}'(\mathrm{j}) = \sigma\mathrm{i(j)}\exp(\tau'\mathrm{N}(0,1) + \tau\mathrm{N_j}(0,1)) \\ \text{for all } \mathrm{j} \in \{1,2,3\}\end{gathered} \tag{7}$$

where N(0, 1) and Nj(0, 1) are Gaussian distributed random numbers.

3.5 *Time Delay Neural Networks*

A number of important applications require neural networks to represent spatiotemporal patterns, i.e., patterns that change over time. Examples of such patterns include speech signals, share prices in financial markets, and seismic readings. Feedforward networks with tapped input delay lines can be used to represent time series with limited context. The input vector X consists now of the past samples, n * p steps backward in time:

$$X = [x(t-p), x(t-2p), \ldots\ldots\ldots, x(t-np)] \tag{8}$$

This architecture as shown in Fig. 2 can be used for nonlinear prediction of a stationary time series, i.e., a time series with statistics that remains unchanged over time. In order to learn the nonlinear characteristic of a time series, gradient descent-based learning algorithm called the error back propagation algorithm can be used.

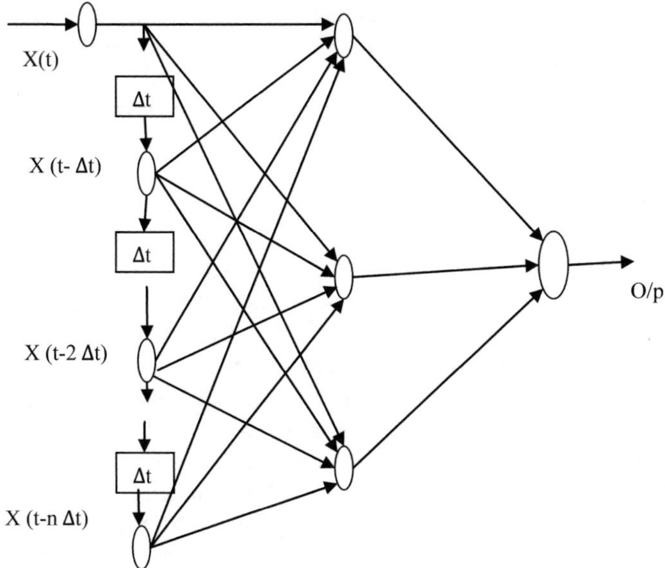

Fig. 2 Time delay neural network (TDNN) and Δt represents the delay

4 Chaotic Time Series

Chaotic system represents a very important class under the dynamic system. In practical applications, a number of systems have shown the chaotic behavior; for example, seismic signals, ECG signals, and share prices are among the few. These systems have deterministic, nonlinearity, and dynamic-like qualities. In practice, most of the time, such system behavior information is available only in terms of time series. Not only that, there is no guarantee of availability of all possible states in such time series. The prime objective of time series analysis is to develop the model for such unknown dynamic system that delivered the time series. Linear modeling is one of the most common ways to analyze the time series and carry the assumption that signal has been generated through finite-dimensional linear system that contains limited number of frequencies. But chaotic dynamic system produces the time series which contain the continuous Fourier spectrum; hence, linear prediction model does not suit in this frame. ANN is one of the finest structures which can be used to model chaotic systems efficiently. In order to compare the different network architectures, we present simulations with following different tasks:

- Prediction of logistic difference equation time series.
- Prediction of Mackey–Glass chaotic time series.
- Prediction of Lorenz system time series.

For all tasks, we present single-step ahead prediction, i.e., we predict one step into the future, given real (measured) values as input, as well as 'n' step ahead prediction.

5 Experimental Setup

Time delay neural network (TDNN) has designed with size of architecture [2 3 1], [4 3 1], and [10 5 1] for logistic time series, Mackey–Glass, and Lorenz system time series with $\Delta t = 1, 6, 1$ correspondingly. Sigmoid function is taken as a transfer function inside the hidden as well as output layer nodes. Learning constant and momentum constant for gradient descent-based learning are taken as 0.1, and bias has applied for hidden layer nodes. A number of allowed iterations are equal to 2000. For ASBO-TDNN, size of solution population is 100: Totally 10 population evolved independently in the first phase up to 150 iterations, and in second phase, 100 best of the first phase are evolved up to 500 iterations. Preprocessing of data is given by Eq. 9, and prediction performance is analyzed by the coefficient of determination R^2 and root-mean-square error (RMSE) as given in Eqs. 10 and 11 correspondingly. The higher value of R^2 equal to 1 and lower value of RMSE equal to 0 are the reference. With sample interval of 1, 1, 0.01, total 1000, 2000, and 10000 samples of chaotic time series data of logistic difference, Mackey–Glass and Lorenz systems are generated by the prediction mapping function. For training purpose, 500, 1000, and 500 samples are taken and remaining samples are taken for test purpose. Weight optimization of TDNN is given by gradient descent-based backpropagation learning and ASBO-based evolution. Performances for all the three different experiments have shown below. With the predicted results for three benchmark chaotic time series, it is very clear that performance of proposed ASBO-TDNN has given always better results compared to BP-TDNN.

5.1 Data Preprocessing

In the process of simulation experiments, this paper uses Eq. 9 to normalize original chaotic nonlinear time series, where xmax and xmin indicate respectively the maximum value and the minimum value of the time series, and the variation range of normalized time series values is between [0, 1].

$$\bar{x}(t) = [x(t) - x_{\min}]/[x_{\max} - x_{\min}] \tag{9}$$

5.2 Prediction Performance Evaluation Criteria

The predictors' accuracy was evaluated according to the coefficient of determination R^2 as given in Eq. 10 and by root-mean-square error (RMSE) as defined in Eq. 11.

$$R^2 = 1 - \left[\sum_{t=1}^{N} (x_t - \hat{x}_t)^2 / \sum_{t=1}^{N} (x_t - \bar{x})^2 \right] \tag{10}$$

where N represents the number of samples and $x(t), \hat{x}(t)$ denote the actual and the predicted process values, respectively, while $\bar{x}$ denotes the mean of the actual data. The value of coefficient of determination lies in the range of [0 1]; 0 for trivial mean predictor while 1 for perfect predictor:

$$RMSE = \sqrt{\frac{\sum_{1}^{N} E_t^2}{N}} \tag{11}$$

5.3 Experiment Case 1: Logistic Difference Equation Chaotic Time Series

Consider the logistic difference equation as given by Eq. 12

$$X[t+1] = \lambda^* X[t][1 - X[t]] \tag{12}$$

This equation shows the diverse behavior depending upon the value of λ. If $1 < \lambda < 3$, the fixed point for the equation is $X = 1 - \lambda^{-1}$. For $\lambda = 3$, system bifurcates to give a cycle of period two which is stable for $3 < \lambda < 1 + 6^{0.5}$. As λ increases beyond this value, successive bifurcation gives rise to a cascade of period doubling which leads to an apparently chaotic sequence for $3.57 < \lambda \lessdot 4$. Obtained performances have shown in Table 1 for training and test phases for obtained chaotic time series through Eq. 12.

Table 1 Performance comparison for logistic difference chaotic time series

	Coefficient of determination		Root-mean-square error	
	BP-TDNN	ASBO-TDNN	BP-TDNN	ASBO-TDNN
Tr. data	0.9859	**0.9874**	0.0429	**0.0404**
Test data	0.9854	**0.9874**	0.0418	**0.0389**

Table 2 Performance comparison for Mackey–Glass chaotic time series

	Coefficientof determination		Root–mean-square error	
	BP-TDNN	ASBO-TDNN	BP-TDNN	ASBO-TDNN
Tr. data	0.9499	**0.9734**	0.0700	**0.0510**
Test data	0.8841	**0.9042**	0.0924	**0.0840**

5.4 Experiment Case 2: Mackey–Glass Chaotic Time Series

A chaotic dynamical system is created as a model for blood flow as a delay differential equation.

$$\frac{dx(t)}{dt} = \frac{Ax(t-\tau)}{1+[x(t-\tau)]^c} - Bx(t) \tag{13}$$

where x(t) is a static variable, t is time in seconds, and A, B, C, and τ are constants, with A = 0.2, B = 0.1, C = 10, $\tau = 17$ and initial condition x(t) = 1.2 for t = 0 and equal 0 for t < 0. Fourth-order Runge–Kutta method has applied to numerical integration to generate the time series. Obtained performances have shown in Table 2, and sample values have been plotted in Figs. 3 and 4 for training and test phase.

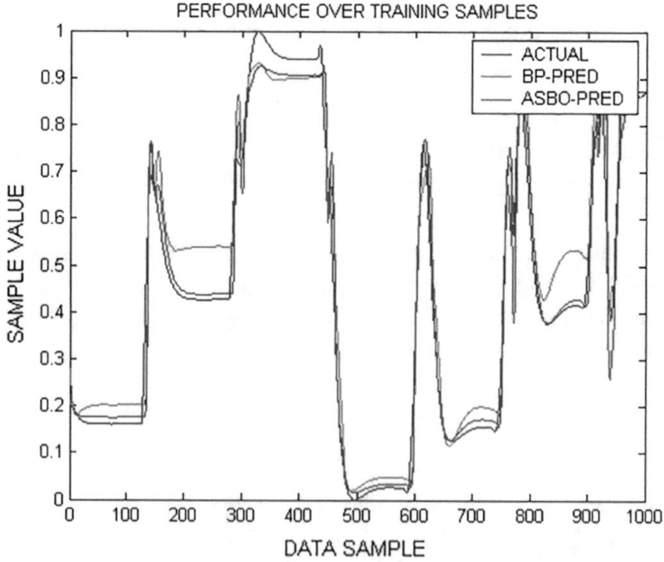

Fig. 3 Prediction over first 1000 training samples in Mackey–Glass time series

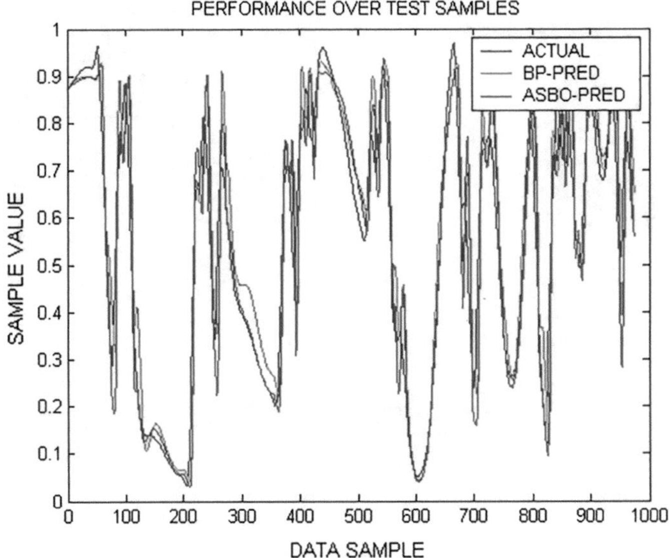

Fig. 4 Prediction over test data samples

5.5 Experiment Case 3: Lorenz System Chaotic Time Series

$$\frac{dx_1(t)}{dt} = A[x_2(t) - x_1(t)]; \; \frac{dx_2(t)}{dt} = -x_1(t)x_3(t) + Bx_1(t) - x_2(t)$$

$$\frac{dx_3(t)}{dt} = x_1(t)x_2(t) - Cx_3(t); \tag{14}$$

Initial condition $x_1(t_0) = x_{1,0}; x_2(t_0) = x_{2,0}; x_3(t_0) = x_{3,0}$.

Lorenz has shown that the behavior of the system given by above differential equation becomes non-periodic for some parameter sets. Moreover, it is unstable with respect to small modification of the initial conditions, which is the characteristic of chaotic systems. Parameters for chaotic behavior are A = 10, B = 28, and C = 8/3; with initial condition $x_1(t_0) = x_2(t_0) = x_3(t_0) = 0$; the fourth-order Runge–Kutta method has applied to generate the series; parameter $\times 3$ has selected for experiment of prediction. Obtained performances have shown in Table 3, and sample values have been plotted in Figs. 5 and 6 for training and test phases.

Table 3 Performance comparison for Lorenz system chaotic time series

	Coefficientof determination		Root-mean-square error	
	BP-TDNN	ASBO-TDNN	BP-TDNN	ASBO-TDNN
Tr. data	0.8788	**0.9934**	0.0514	**0.0357**
Test data	0.9687	**0.9949**	0.0120	**0.0114**

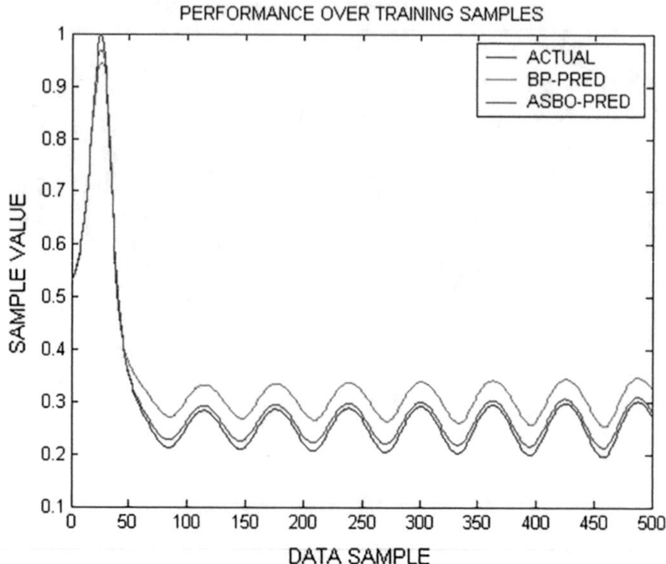

Fig. 5 Prediction outcomes at training phase

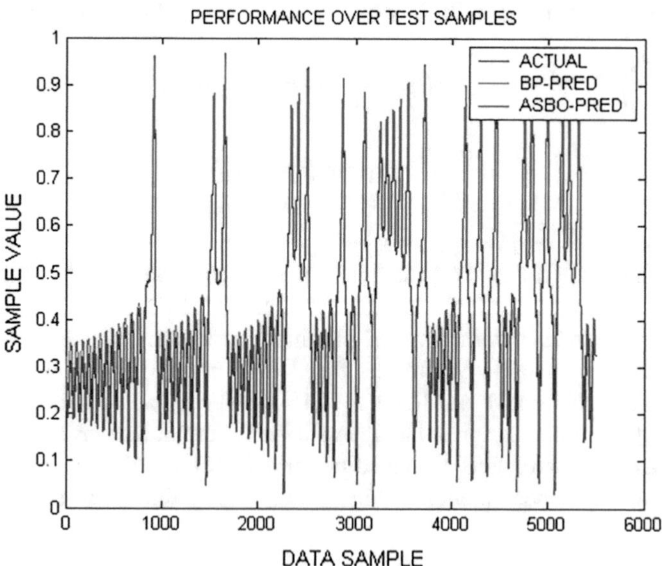

Fig. 6 Prediction outcomes at test phase

6 Conclusion

It is always a challenging task for researchers to develop the more accurate model for time series prediction. In this regard, a new approach to develop such kind of model based on social behavior-based optimization and time delay neural network has presented. It is observed that evolving the neural network with ASBO-based method not only have the better generalized capability but also performance quality is very high. Simplicity in design is the further benefit associated with the proposed method. Capability for prediction of the proposed method has verified with various benchmarks of a chaotic time series, which is very difficult to predict. Comparisons of obtained performances are made with frequently applied method of learning in neural network, namely backpropagation. It is hoped that the proposed method can be equally applied to other various different types of neural network and for real-time series data.

Acknowledgement This research has done in Manuro Tech Research Pvt.Ltd., Bengaluru, India, under Innovative solution for Future Technology program.

References

1. Bermolen, P., Rossi, D.: Support vector regression for link load prediction. Comput. Netw. **53**, 191–201 (2009)
2. Lee, C.M., Ko, C.N.: Time series prediction using RBF neural networks with a nonlinear time-varying evolutin PSO algorithm. Neurocomputing **73**, 449–460 (2009)
3. Xu, K., Xu, J.W., Ban, X.J.: Forecasting of some non-stationary time series based on wavelet decomposition. Acta Electronica Sinica **29**(4), 566–568 (2001)
4. Geva, A.B.: ScaleNet-multiscale neural-network architecture for time series prediction. IEEE Trans. Neural Netw. **9**(5), 1471–1482 (1998)
5. Chen, Y.H., Yang, B., Dong, J.W.: Time-series prediction using a local linear wavelet neural network. Neurocomputing **69**, 449–465 (2006)
6. Szpiro, G.G.: Forecasting chaotic time series with genetic algorithms. Am. Phys. Soc. 2557–2568 (1997)
7. Lamamra, K, Azar, A.T., Ben Salah, C.: Chaotic system modelling using a neural network with optimized structure. Studies in Computational Intelligence, vol. 688, pp 833–856. Springer, Cham (2017)
8. Oliveira, K.D., Vannucci, A., da Silva, E.C.: Using artificial neural networks to forecast chaotic time series. Phys. A **284**, 393–404 (2000)
9. Thissen, U., van Brakela, R., de Weijer, A.P., Melssen, W.J., Buydens, L.M.C.: Using support vector machines for time series prediction. Chemometr. Intell. Lab. Syst. **69**, 35–49 (2003)
10. Volná, E., et al.: Nonlinear time series analysis via neural networks. In: Chaos and Complex Systems, pp. 415–418. Springer, Heidelberg (2013)
11. Ristanoski, G., Liu, W., Bailey J.: Time series forecasting using distribution enhanced linear regression. In: Advances in Knowledge Discovery and Data Mining. PAKDD 2013. Lecture Notes in Computer Science, vol. 7818. Springer, Heidelberg (2013)
12. Volna, E., Kotyrba, M.: Time series prediction based on averaging values via neural networks. prediction, modeling and analysis of complex systems. In: Advances in Intelligent Systems and Computing, vol. 289. Springer, Cham (2014)

13. Prem, V., Rao, K.U.: Time series decomposition model for accurate wind speed forecast, Renewables: Wind, Water and Solar, pp. 2–18 (2015)
14. Akpinar, M., Yumusak, N.: Year ahead demand forecast of city natural gas using seasonal time series methods. J. Energies **9**(9), 727 (2016)
15. Fonseca-Delgado, R., Gómez-Gil, P.: Modeling diversity in ensembles for time-series prediction based on self-organizing maps. In: Advances in Self-Organizing Maps and Learning Vector Quantization. Advances in Intelligent Systems and Computing, vol. 428. Springer, Cham (2016)
16. Averkin, A., Yarushev, S., Dolgy, I., Sukhanov, A.: Time series forecasting based on hybrid neural networks and multiple regression. In: Advances in Intelligent Systems and Computing, vol 450. Springer, Cham (2016)
17. Gao, Y., Shang, H.L., Yang, Y.: High-dimensional functional time series forecasting. In: Functional Statistics and Related Fields. Contributions to Statistics. Springer, Cham (2017)
18. Takens, F.: Detecting strange attractors in fluid turbulence. Dynamical Systems and Turbulence. Springer, Heidelberg (1981)
19. Singh, M.K.: A new optimization method based on adaptive social behavior: ASBO. In: ASIC 174, 823–831. Springer, New Delhi (2012)

Intelligent and Integrated Book Recommendation and Best Price Identifier System Using Machine Learning

Akanksha Goel, Divanshu Khandelwal, Jayant Mundhra and Ritu Tiwari

Abstract When one wants to read something, it is tough to decide the just perfect one book that one would want to read. And, given such a system is conceived and the reader can find the book he/she wishes to read, how does he/she decide where to buy that book from? Again, there are a plethora of vendors selling the same book at different prices. Thus, this paper proposes a dynamic recommendation system by extracting the relevant data from different E-portals and applying Hybrid Filtering Approach (collaborative and content-based filtering) on the collected data to recommend the books. This system makes use of user-based collaborative filtering approach using cosine similarity rule and is optimized with bee algorithm, and the results are refined by applying natural language processing on the reviews. The paper also intends to solve cold start problem by extracting available user preferences from Facebook API.

Keywords Data analysis · Nature-inspired algorithms · Natural language processing · Data scrapping · Recommendation system

A. Goel (✉) · D. Khandelwal · J. Mundhra · R. Tiwari
Robotics and Intelligent System Design Lab, ABV-IIITM Gwalior, Gwalior, India
e-mail: akanksha.ausum05@gmail.com

D. Khandelwal
e-mail: divanshukhandelwal79@gmail.com

J. Mundhra
e-mail: jayant111294@gmail.com

R. Tiwari
e-mail: tiwariritu2@gmail.com

V. Bhateja et al. (eds.), *Intelligent Engineering Informatics*, Advances in Intelligent Systems and Computing 695, https://doi.org/10.1007/978-981-10-7566-7_39

1 Introduction

With the boom of choices in today's world, one has multiplicity of options to choose from. This multiplicity of options has received a massive fillip from the parallel boom in Internet technologies leading to an unprecedented growth in the amount of resources available to an individual for access. Users often find themselves in a dilemma while choosing the right resource. A recommender system [1, 2] can overcome this information overloaded by filtering out superfluous clutter of resources and recommending relevant resources to the user on the basis of personalized preferences.

We have amalgamated the said recommendation system into a Web portal, which gives the recommendations from scrapped Web-based dynamically fetched database. Scrapping of dynamic data from Internet resources has been a trending approach in the recent times [3]. We have implemented a hybrid approach [4] incorporating two prominent service recommendation techniques which have been proposed in the recent years, namely collaborative filtering and content-based filtering [5, 6]. Collaborative filtering is a widely used recommending technique for recommending items to a user based on similar user ratings on the items. However, this technique faces three challenges, namely sparsity, synonymy and cold start problems, where sparsity indicates when user's rating is not available, synonymy problem indicates when similar items but with different names exist in the system leading to system inability to establish similarities and cold start problem arises when new user arrives or when there is newly deployed item. To counter new item cold start problem, we have incorporated content-based filtering approach which provides recommendations based on user preferences and also on the basis of descriptive information on the books available [7].

To optimize the collaborative filtering, this paper proposes to make use of nature-inspired bee algorithm [8, 9]. Here, bee algorithm solves problems by modelling the foraging behaviour of honeybees in finding the optimum solution. On the results derived from aforementioned approaches, we apply natural language processing (NLP) on the reviews of those books in order to get top 'n' recommendations on our Web portal. NLP is a sub-domain of artificial intelligence which is aimed at assisting the computer systems to comprehend and analyse the human

Table 1 Table for computing MAE and RMSE

Accuracy metrics [13]	Equations	Proposed algorithm values
Mean absolute error (MAE)	$\mathrm{MAE} = \frac{\sum_{i=1}^{n} \lvert y_i - x_i \rvert}{n} = \frac{\sum_{i=1}^{n} \lvert e_i}{n}$ where y_i is predicted rating, xi is the actual rating, and n is the number of observations.	0.1650
Root mean square error (RMSE)	$\mathrm{RMSD} = \frac{\sum_{i=1}^{n} (\hat{y}_i - y_i)^2}{n}$ where $\hat{y}_i$ is predicted rating, yi is actual rating, and n is the number of observations.	0.1669

languages [10–13]. At the end, we have tried to evaluate the system's efficiency by making use of algorithmic approaches like mean absolute error, root mean square error, precision recall, and ROC Curve (Table 1).

2 Related Works

In this section, we have tried to share the various approaches that have been proposed by other authors in the same domain of recommendation systems.

JoeranBeel [14] researched and parsed over 200 papers, and then, information was derived about various recommendation algorithms. Heung-nam Kim [15] has given a collaborative filtering-based model to make recommendations post discovering useful and meaningful patterns and preferences of an individual user. Greg Linden [16] has proposed item-to-item collaborative filtering methodology because it responds to changes in the user's preferences and makes effective recommendations irrespective of the scale of purchases and ratings.

Chonghuan Xu [17] constructed an effective recommendation model in order to improve the current recommender system for better customer service and analysis of the collaborative filtering algorithm and propose a modified one based on different influence factors. Martin Lopez-Nores [18] has given a property-based collaborative filtering approach for making recommendations by measuring the likelihood that certain items will be appealing to the users with certain preferences. This helps to address common problems with other collaborative approaches like sparsity and latency. Ismail Sengor Altingovde [19] has proposed cluster-based collaborative filtering techniques which can be made more accurate and scalable using a specifically tailored cluster-skipping inverted index leading to a substantial reduction in neighbourhood formation time by up to 60%.

Murat Goksedef [20] discussed a hybrid recommender system which is an amalgamated result of sundry recommendation approaches based on Web usage mining. Marko Balabnovic [21] has given an implementation of a hybrid content-based, collaborative Web-page recommender system which is very effective in eliminating several handicaps of either content-based or collaborative recommendation approaches. Toon De Pessemier [22] developed a hybrid system as an amalgamation of content-based, collaborative and knowledge-based filtering based on use of rating profiles, personal interests and specific demands from the user. Nitin Pradeep Kumar [23] has highlighted how traditional collaborative filtering methods face two major shortfalls: scalability and sparsity of data. Based on that understanding, he has proposed a hybrid method based on item-based cluster filtering to achieve a personalized outcome recommendation. Further use of case-based reasoning (CBR) in tandem with average filling is proposed to handle the sparsity in data, and self-organizing map (SOM) optimized with genetic algorithm (GA) has been proposed to reduce scope for item-based collaborative filtering.

2.1 Research Gaps

Most of the papers use static datasets for giving recommendations. Many challenges like cold start problem, sparsity and synonymy mentioned in Sect. 1 are not solved by most of the researchers. Degree of preferences of similar users is not considered in traditional collaborative filtering while calculating the similarity score [17], and analysis of reviews of books for recommendation so that book with most positive reviews must be ranked in top has not been done in most of the papers. There is no prediction of ratings in most of the papers analysed which decides the goodness of books.

This paper addresses new user cold start problem with the help of fetching preferences of new users from Facebook API explained in Sect. 3.2. New item cold start problem in collaborative filtering is solved by content-based filtering, and book is automatically considered due to dynamic fetching of data rather than old static datasets.

This paper proposes the use of bee algorithm, a nature-inspired algorithmic approach to optimize the recommendations made using traditional collaborative filtering. To further optimize the outcomes and to suggest the user with top 'n' book recommendations, the paper proposes the use of review analysis using natural language processing (NLP) by awarding each book a score calculated using formula mentioned in Sect. 3.2. This paper uses modified collaborative filtering technique in which degree of preference by similar user is considered in similarity score as mentioned in Sect. 3.2.

3 Proposed System

Paper proposes a system which tends to provide the most suitable recommendation of books and their respective prices on variegated e-commerce portals. This system would thus involve four discrete stages to ensure the efficacy and delivery of results, namely data extraction, preprocessing of data, recommendation system and performance evaluation, which are further explained in brief in the passages below.

3.1 Stages

Data extraction: E-commerce platforms like Amazon.in provide an enormous database of books with preset book selection lists like trending books, which can be easily scrapped to make recommendations to even those users whose interests or preferences are not known to us. For data extraction of books and their prices from portals like Amazon [29], eBay [32] and SnapDeal [30], we intend to make use of BeautifulSoup [27] library based on Python programming language.

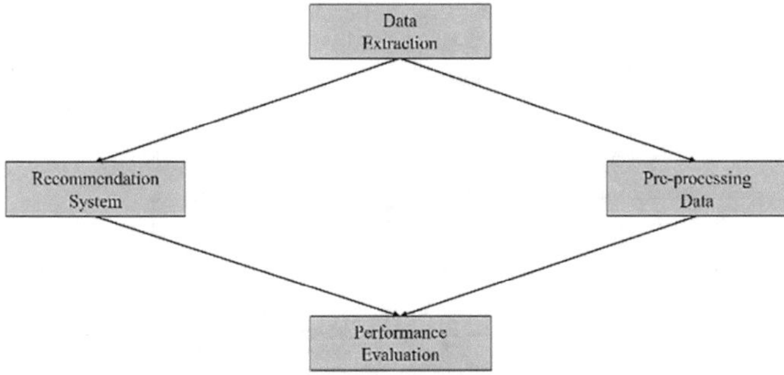

Fig. 1 Block diagram for proposed methodology

Preprocessing Data: In this stage, the previously extracted data are mined and segregated to be sub-classified with respect to sundry parameters like genre, year, author and age.

Recommendation System: Recommender systems are a representation of user choices for suggesting items to view or purchase based on user preferences. To achieve this, we have made use of hybrid approach using content-based and collaborative filtering strengthened by natured-inspired Bee algorithm.

Performance Evaluation: [26] The performance, efficiency and efficacy of the developed system have been evaluated by using sundry accuracy matrices like ROC curve and precision recall accuracy matrix and error calculation methodologies like mean absolute error and root mean square error (Fig. 1).

3.2 Working of the System

Overall functioning can be summed up in the following block diagrams (Figs. 2 and 3):

When a person is not logged into the system, then the person would only be able to access the trending books which are identified using the data scrapped from Amazon. This scrapping of selecting few lists of trending books from Amazon.in helps in making recommendations to the readers whose preferences are not known to the system, thus providing an answer to address the cold start problem. While in case when user logs into the system using the Facebook API, then the books of interest of user are saved into the database, user enters the home page. Now the user can enter the book or author name in the embedded search tab on the Web portal. This Web portal provides a workable platform even when user's ratings are not available. When a search for the books is conducted using the Amazon and relevant details are then scrapped off according to our needs and collected into our database, book link, book name, author name, book type, book image. Finally, scrapped data are displayed on the Web portal whereby the user also gets the privilege to add their own ratings for the books that they have read themselves. This new rating is then

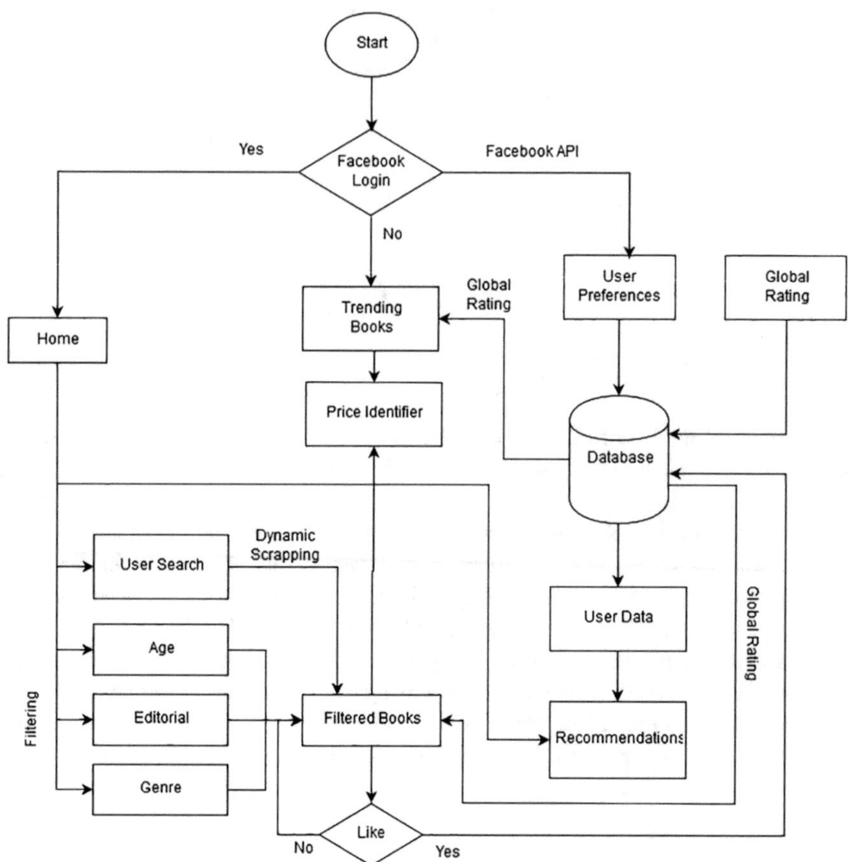

Fig. 2 Detailed block diagram of mechanism

saved in our database for further use. When user clicks on any of the books from the search on the Web portal, then the respective book's price, image, ratings, number of reviews are then fetched from multifarious sources—Amazon, eBay, Goodreads [31], SnapDeal. These fetched details are then taken to our Web page, where prices on various portals are compared and also a globalized rating is computed using ratings and number of reviews as follows:

The global rating will be measured using the average of products of number of users and overall ratings on each of the considered portals, namely Amazon, SnapDeal, eBay and Goodreads with respect to total number of users.

$$\text{Global Rating}(\text{weighted mean}) = \frac{N_a{*}Ra + Ns{*}Rs + Ng{*}Rg}{(N_a + Ns + Ng)} \quad (1)$$

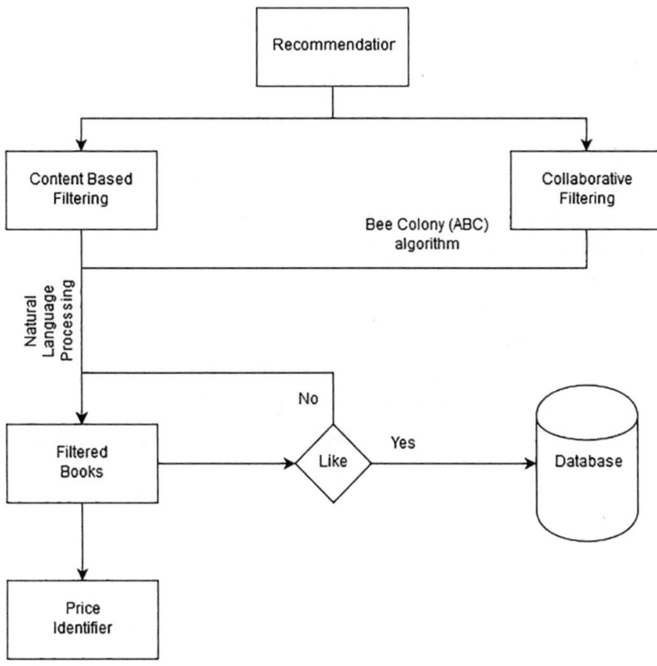

Fig. 3 Detailed flow of recommendation for the proposed system in Fig. 2

where Na refers to number of Amazon users who have given rating, Ns to number of SnapDeal users, Ng to number of Goodreads users, Ra to rating given by Amazon, Rs to rating given by SnapDeal and Rg to rating given by Goodreads.

This page again makes it possible for the user to add their own ratings for the books or update previously added rating, which is updated in our database immediately. User can get books according to genre, top trending book and age as shown in Fig. 2. The books clicked (author finds interesting) in this process are also saved into our database.

Using Facebook API and search results, our database is made which dynamically increases with time as per the searches of the user. Then, the user will be able to get recommendations on the basis user preferences which will be elaborated upon in the further passages.

To make the recommendations effectively, we have made use of content-based filtering using author name and collaborative filtering is optimized using bee algorithm (as shown in Fig. 3) and implemented using algorithm (mentioned in Sect. 3.3). Reviews of the books filtered from above approaches are scraped from Amazon.com and saved for analysis. Tokenization is used to remove unnecessary words from the reviews. Now, analysis of the reviews was done for positivity and negativity [12], and the score is formulated as:

$$Score = \frac{(Pos^*Npos - Neg^*Nneg + N^*Nn)}{Tr} \quad (2)$$

where Pos refers to positivity of review, Neg to negativity of a review, N to neutrality, Npos to number of positive reviews, Nneg to number of negative reviews, Nn to number of neutral reviews, and Tr refers to total number of reviews. This score is used to give top *'n'* recommendations.

Modified Collaborative Filtering Recommendation Method

Traditional collaborative filtering method prominently makes use of standard cosine similarity or Pearson correlation to decipher the similarity between any two users. Traditional collaborative filtering does not take into account the influence of an object's degree, leading to objects with different degrees having the same contribution to the similarity. If users ui and uj both have selected object o_l, then they have similar degree of preference for object o_l. The computation of similarity between the two users can be influenced by sundry factors, which is why we need to improve the traditional method of collaborative filtering [17].

Ideally, the similarity between two users is relative to their preference degree and trust relationship. Taking these features into account, this paper proposes a modified collaborative filtering algorithmic approach. We conclude that preference degrees are related to corresponding users' ratings. Trust relationship is derived from past ratings from the two users on corelated products by Hwang and Chen's trust computation method [24]. It is assumed that the similarity computation on user is affected by an influence degree that is proportional to $[(1-(|v_{li}-v_{li}|/M))]^{\alpha}$, where α represents a freely adjustable parameter. Accordingly, the contribution of object o_l to the similarity s_{ij} should be positively correlated with its preference degree and trust relationship. Ergo, it has been surmised that the contribution of object o_l to s_{ij} is directly proportional to $|v_{li} - v_{lj}|$.

The formulation of s_{ij} can be expressed as:

$$s_{ij} = \frac{1}{\sqrt{k(u_i)k(u_j)}} \sum_{l=1}^{n} a_{li}a_{lj}[(1-(|v_{li}-v_{li}|/M))]^{\alpha} \quad (3)$$

M here refers to the difference between the maximum and minimum rating scores. v_{li}/v_{lj} shows the preference degree from the object got from user u_i/u_j. $K(u_i)$ refers to the degree of the user, namely the count of objects selected by the user.

3.3 Proposed Algorithm

1. Gather the information about the user logging in (Let the user be denoted by 'k'). If no prior information about the user is available, then follow the steps 2–4.

2. Get a list of all other users (l).
3. Initiate a while loop with the condition for length of list l being not equal to 0. Select another user and compute the similarity score (build a path from user k) mentioned in the Sect. 3.2 and pop the user from the list.
4. Discover the minimal path from the map attained through similarity scores (most similar users).
5. If user is not new, we already have some prior knowledge of similarity scores. Check if the similarity score is changed from the most similar user.
6. If no change detected, go to step-8
   Else,
   Estimate a new map using the previous knowledge of similarity score.
7. Use the new map to discover the minimal path.
8. Glean in all the books from the most similar user.
9. Find all the books that are not read by user k and arrange them according to the ratings awarded by most similar user.
10. Get top 'k' books from the selected books.
11. Get top 'p' books from content-based algorithm.
12. Glean in the reviews for all books (k + p) and make use of natural language processing (NLP) to compute the score [Eq. 2] for the particular book.
13. Award ranks according to the score and make the top 'n' book recommendations.

4 Results and Evaluation

In the proposed Web portal, the following Web page (home page) greets the user with trending books (Fig. 4) and also provides the user with user login prompt.

Post login, the following page displays, where the user gets to either make their own personalized search for book recommendations, or choose from various filters like age and genres (Fig. 5).

Also, the way details of a book respective to a particular user are stored in the database has been shown in Fig. 9 which are updated dynamically with respect to the database scrapped from e-commerce Websites.

For each of the books, the user then gets greeted by a page as follows: displaying the prices it is being sold for at various e-commerce portals and also giving the user an option to rate a particular book if they have already read it (Fig. 6).

Post the application of NLP of the books attained from various recommendation approaches, the book recommendations are then made as shown in Fig. 7.

For the recommended books, user gives rating to the scale of [0–5] to every book as shown in Fig. 7. If user give rating in range from (0–3] then those books are considered as non-relevant and if user gives user rating from (3–5], then those books are considered as relevant. The data obtained from this process are used to make the confusion matrix shown in Tables 2, 3, 4, 5 and 6.

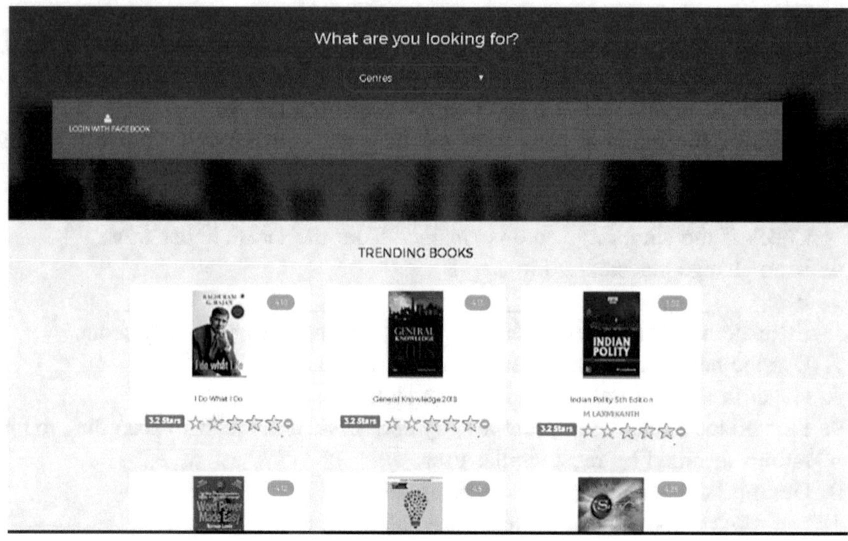

Fig. 4 Home page of the Web portal

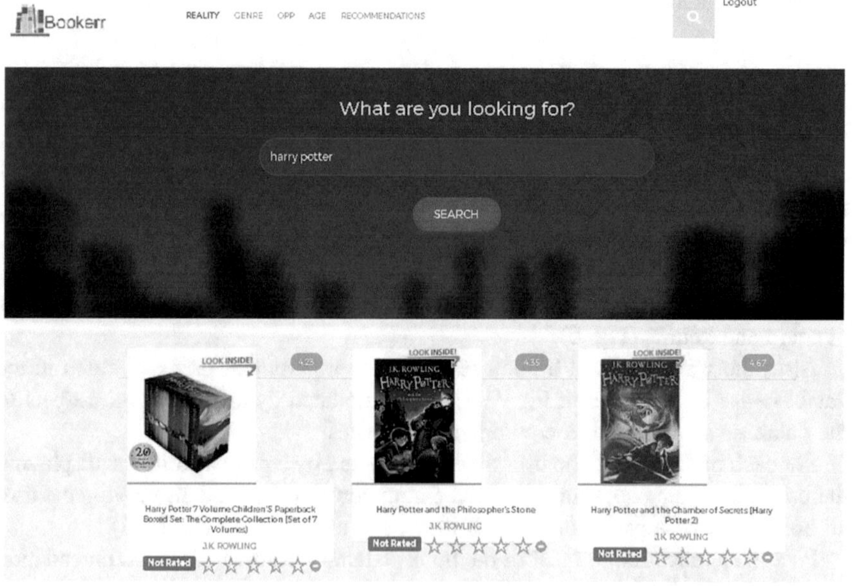

Fig. 5 Results after search

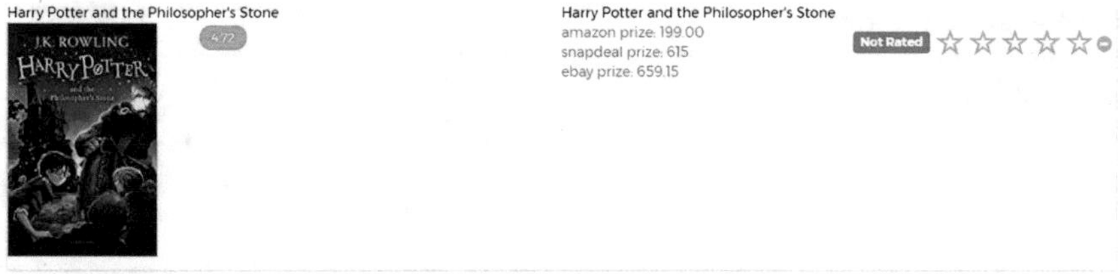

Fig. 6 Display of price and global rating of selected book

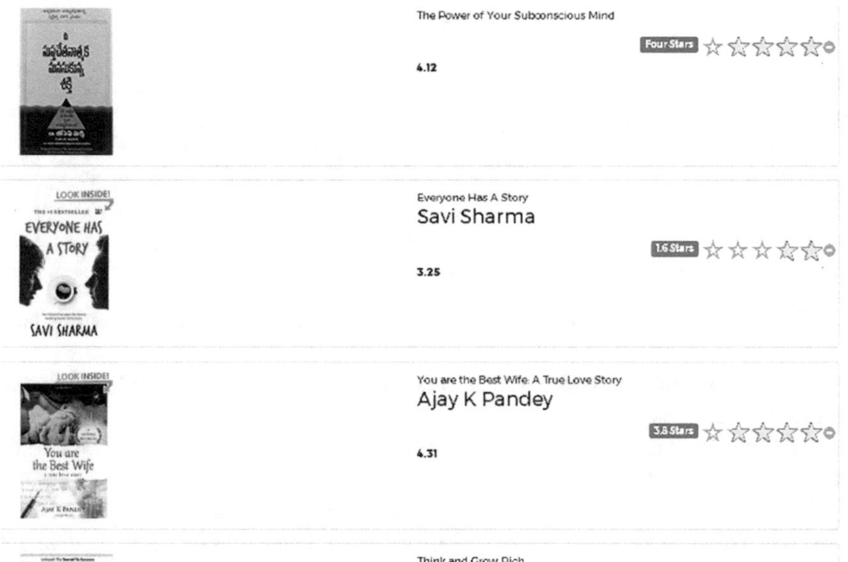

Fig. 7 Displaying recommendations in the Web portal

Table 2 Table for computing recall and precision

Books	Recommended	Not recommended
Relevant	True Positive (TP)	False Negative (FN)
Non-relevant	False Positive (FP)	True Negative (TN)

Table 3 Confusion matrix when given 5 recommendations

N = 5	Recommended	Not recommended	
Relevant	TP = 4	FN = 8	12
Non-relevant	FP = 1	TN = 7	8
	5	15	

Table 4 Confusion matrix when given 10 recommendations

N = 10	Recommended	Not recommended	
Relevant	TP = 7	FN = 6	13
Non-Relevant	FP = 3	TN = 4	**7**
	10	10	

Table 5 Confusion matrix when given 15 recommendations

N = 15	Recommended	Not recommended	
Relevant	TP = 10	FN = 3	13
Non-Relevant	FP = 5	TI\l = 2	**7**
	15	5	

Table 6 Confusion matrix when given 20 recommendations

N = 20	Recommended	Not recommended	
Relevant	TP = 12	FN = 0	12
Non-relevant	FP = B	TN = 0	8
	20	0	

There are various evaluation techniques [13] from which the following have been used:

As value of RMSD and MAE is close to 0, therefore rating predicted in this paper is well measured.

ROC curve: It is a plot of the true positive rate against the false positive rate for the various points [27].

True positive rate refers to true positives upon sum of true positive and false positive. True negative rate refers to false negative upon sum of false negative and true negative.

In ROC curve, we calculate area under the curve and accuracy is measured by it. An area of 1 represents test accuracy, and an area of .5 represents test failure [28]. Area of [.90–1] represents excellent and [.80–.90) represents good. [.70–.80) represents fair, [.60–.70) represents poor, and [.50–.60) represents fail. The area under the ROC curve is .8636; therefore, the proposed algorithm is good to function and deliver righteous recommendations (Figs. 8 and 9).

Precision [25] has been measured against different number of recommendations. Upon observing Fig. 10, it is observed that as the number of recommendations is incremented, the precision eventually falls.

$$Precision = \frac{Correctly\,recommended\,items}{Total\,recommended} = \frac{tp}{tp + fp} \quad (4)$$

1.0
0.8
0.6
0.4
0.2
0.0
true positive rate
0.00
0.02
0.04
0.06
0.08
0.10
false negative rate

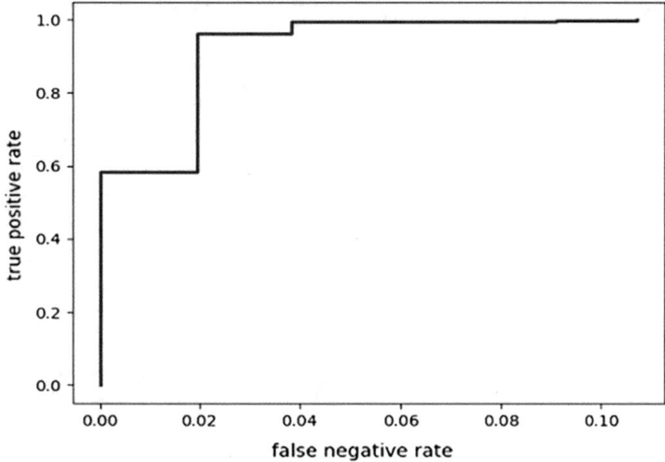

Fig. 8 ROC curve

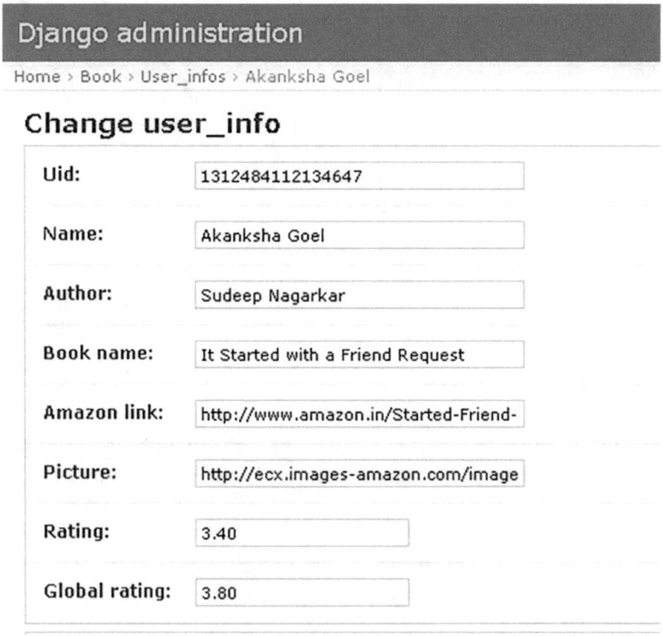

Django administration
Home › Book › User_infos › Akanksha Goel
Change user_info
Uid:
1312484112134647
Name:
Akanksha Goel
Author:
Sudeep Nagarkar
Book name:
It Started with a Friend Request
Amazon link:
http://www.amazon.in/Started-Friend-
Picture:
http://ecx.images-amazon.com/image
Rating:
3.40
Global rating:
3.80

Fig. 9 User database

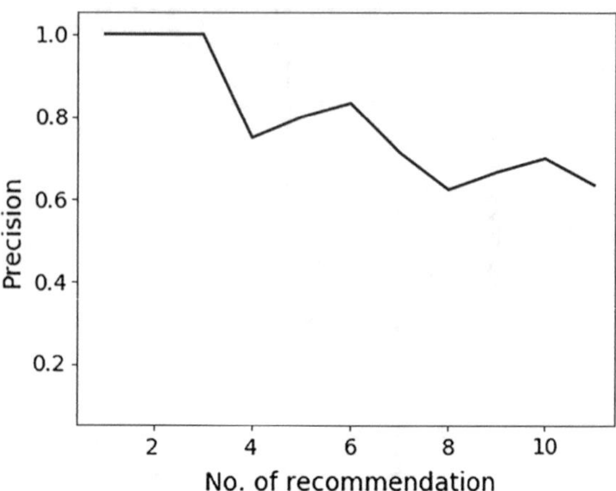

Fig. 10 Precision analysis

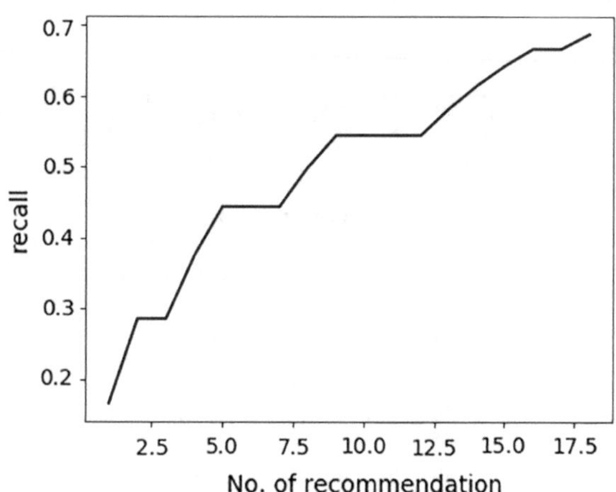

Fig. 11 Recall analysis

Recall refers to the ability of system to recommend as few non-relevant resources as possible. A resource is considered non-relevant if it is detested by the user [25].

$$Recall = \frac{Correctly\ recommended\ items}{Relevant\ items} = \frac{tp}{tp + fn} \quad (5)$$

Upon observing Fig. 11, it is observed that the count of false recommendations decreases as the number of recommendations is incremented.

5 Conclusion and Future Scope

This paper focuses on building a hybrid recommender system for books by using content-based filtering and optimized collaborative algorithm using bee algorithm, which is a nature-inspired approach. The collaborative filtering has been further modified to also take into account several other influential similarity factors to establish similarities between users like preference degree and trust relationship.

This paper also proposes the use of natural language processing (NLP) to give top 'n' book recommendations. The aforementioned system has then been amalgamated in the form of a Web portal which dynamically fetches the database of books while also providing the user with different prices from different e-commerce sources like Amazon, SnapDeal and eBay. Special features have been incorporated in the said Web portal to present books filtered on parameters like age, genre and trending books, while a search feature has also been formulated to enable the user search of books according to title or author. Furthermore, the said Web portal also provides a rating feature to using which the users can themselves judge and rate the books as well.

In future, we propose the further expansion of database from more variegated sources and apply several other algorithms as well for better recommendation. The proposed mechanism for recommendation of books can also be extended for other such commodities for which viable and robust database can be created.

References

1. Isinkaye, F.O., Folajimi, Y.O., Ojokoh, B.A.: Recommendation systems: principles, methods evaluation. Egypt. Inf. J. **16**, 261–273 (2015)
2. Adomavicius, G., Tuzhilin, A.: Toward the next generation of recommender systems: a survey of the state of-the-art and possible extensions. IEEE Trans. Knowl. Data Eng. **16**, 1041–4347 (2005)
3. Kanetkar, S., Nayak, A., Swamy, S., Bhatia, G.: Web-based personalized hybrid book recommendation system. In: IEEE International Conference on Advances in Engineering & Technology Research, Unnao, India (2014). https://doi.org/10.1109/icaetr.2014.701292
4. Chandak, M., Girase, S., Mukhopadhyay, D.: Introducing hybrid technique for optimization of book recommender system. Proc. Comput. Sci. **45**: 23–31 (2015). Elsevier
5. Kwon, K., Park, Y.: Collaborative filtering using dual information sources. IEEE Intell. Syst. **22**:1541–1672 (2007). https://doi.org/10.1109/mis.2007.48
6. Tewari, A.S., Kumar, A., Barman, A.G.: Book recommendation system based on combine features of content based filtering, collaborative filtering and association rule mining. In: Advance Computing Conference (IACC), IEEE International, Gurgaon, India (2014). https://doi.org/10.1109/iadcc.2014.6779375
7. Anne, H.H.Ngu, Segev, A., Jian, Yu., Sheng, Q.Z., Yao, L.: Unified collaborative and content based web service recommendation. IEEE Trans. Serv. Comput. **8**, 453–466 (2014)
8. Yang, X-S.: Nature-Inspired Optimization Algorithms Book. Elsevier (2014)
9. Pham, D.T., Castellani, M., Le Thi, H.A.: Nature-inspired intelligent optimisation using the Bees algorithm. In: Transactions on Computational Intelligence XIII, vol. 8342, pp. 38–69. Springer, Heidelberg (2014)

10. Lei, X., Qian, X., Zhao, G.: Rating prediction based on social sentiment from textual reviews. IEEE Trans. Multimed. **18**, 1910–1921 (2016). https://doi.org/10.1109/TMM.2016.2575738
11. Mahajan, C., Mulay, P.: E3: effective emoticon extractor for behavior analysis from social media. Proc. Comput. Sci. **50**, 610–616 (2014). Elsevier
12. Hirschberg, J., Ballard, B.W., Hindle, D.: Natural language processing. AT&T Tech. J. IEEE **67**, 41–57 (1988). https://doi.org/10.1002/j.1538-7305.1988.tb00232.x
13. Herlocker, J., Konstan, J., Terveen, L., Riedl, J.: Evaluating collaborative filtering recommender systems. ACM Trans. Inf. Syst. **22**, 5–53 (2004)
14. Beel, J., Gipp, B., Langer, S., Breitinger, C.: Research-paper recommender systems: a literature survey. Int. J. Digit. Libr. **17**, 305–338 (2015)
15. Kim, H., Ha, I., Lee, K., Jo, G., El-Saddik, A.: Collaborative user modeling for enhanced content filtering in recommender systems. Decis. Support Syst. **51**, 772–781 (2011)
16. Smith, B., Linden, G.: Two decades of recommender systems at Amazon.com. IEEE Internet Comput. **21**, 12–18 (2017)
17. Xu, C.: Personal recommendation using a novel collaborative filtering algorithm in customer relationship management. Discrete Dynamics in Nature and Society, Hindawi (2013). http://dx.doi.org/10.1155/2013/739460
18. López-Nores, M., Blanco-Fernández, Y., Pazos-Arias, J., Gil-Solla, A.: Property-based collaborative filtering for health-aware recommender systems. Expert Syst. Appl. **39**, 7451–7457 (2012)
19. Altingovde, I., Subakan, Ö., Ulusoy, Ö.: Cluster searching strategies for collaborative recommendation systems. Inf. Process. Manage. **49**, 688–697 (2013)
20. Göksedef, M., Gündüz-Öğüdücü, Ş.: Combination of web page recommender systems. Expert Syst. Appl. **37**, 2911–2922 (2010)
21. Balabanović, M., Shoham, Y.: Fab: content-based, collaborative recommendation. Commun. ACM **40**, 66–72 (1997)
22. Pessemier, T., Dhondt, J., Martens, L.: Hybrid group recommendations for a travel service. Multimed. Tools Appl. **76**, 2787–2811 (2016)
23. Kumar, N., Fan, Z.: Hybrid user-item based collaborative filtering. Proc. Comput. Sci. **60**, 1453–1461 (2015)
24. Hwang, C.S., Chen, Y.P.: Using trust in collaborative filtering recommendation. In: International Conference on Industrial, Engineering and Other Applications of Applied Intelligent Systems, Kyoto, Japan, vol. 4570, pp. 1052–1060. Springer, Heidelberg (2007). https://doi.org/10.1007/978-3-540-73325-6_105
25. Tarus, J.K., Niu, Z., Yousif, A.: A hybrid knowledge-based recommender system for e learning based on ontology and sequential pattern mining. Future Generation Computer Systems, **72**:37–48 (2017). Elsevier
26. Beautiful soup documentation. https://www.crummy.com/software/BeautifulSoup/bs4/doc/
27. https://www.medcalc.org/manual/roc-curves.php
28. http://gim.unmc.edu/dxtests/roc3.htm
29. https://www.amazon.in/
30. https://www.snapdeal.com/
31. http://www.goodreads.com/
32. https://www.ebay.in/

Online Handwritten Bangla Character Recognition Using CNN: A Deep Learning Approach

Shibaprasad Sen, Dwaipayan Shaoo, Sayantan Paul, Ram Sarkar and Kaushik Roy

Abstract In the present work, a typical Convolutional Neural Network (CNN) architecture has been used for the recognition of online handwritten isolated Bangla characters. A detailed analysis about the effects of using different kernel variations, pooling strategies, and activation functions in the CNN architecture has been performed. In this work, total 10000 character samples have been used and among the samples, 30% have been considered as test set and rest 70% have been used to train the recognition model. On test dataset, the technique has been provided 99.40% recognition accuracy. The outcome is better than some recently proposed handcrafted features used for the recognition of online handwritten Bangla characters.

Keywords Online handwriting recognition · Feature · Bangla script CNN

S. Sen (✉) · D. Shaoo
Future Institute of Engineering & Management, Kolkata, India
e-mail: shibubiet@gmail.com

D. Shaoo
e-mail: dwaipayanshaoo18@gmail.com

S. Paul · R. Sarkar
Jadavpur University, Kolkata, India
e-mail: sayantanpaul98@gmail.com

R. Sarkar
e-mail: raamsarkar@gmail.com

K. Roy
West Bengal State University, Kolkata, India
e-mail: kaushik.mrg@gmail.com

V. Bhateja et al. (eds.), *Intelligent Engineering Informatics*, Advances in Intelligent Systems and Computing 695, https://doi.org/10.1007/978-981-10-7566-7_40

1 Introduction

In pattern recognition area, online handwriting recognition (OHR) has become one of the forthcoming research topics due to the exponential growth of the usage of the devices such as Take Note, iPad, Smartphone and so on. Writing on those devices is quite similar to the writing with pen and paper. But, these devices can take input faster than a keyword and also minimize the chances of errors due to mistyping. Some good researches on online handwriting recognition for Devanagari [1–3], English [4, 5] are available in the literature. For the recognition of handwritten Bangla characters, different authors have used handcrafted features for both holistic and stroke-based approaches. Bhattacharya et al. in [6] have proposed a novel direction code-based feature for the recognition of online Bangla characters. Roy in [7, 8] has considered stroke-level sequential and dynamic information (obtained from pen movements on writing pads) as feature vector for stroke and character recognition. Sen et al. in [9, 10] and Mondal et al. in [11] have shown the effectiveness of Hausdorff distance-based feature, combination of area, mass distribution and chord length features, and direction code histogram with point-float feature for the recognition of Bangla characters. Parui et al. in [12] have used handcrafted feature for sub-stroke-level recognition and designed one hidden Markov model (HMM) for individual stroke class. After stroke recognition, characters are formed with the help of 50 lookup tables.

All the methods reported in [6–12] have used handcrafted features for the recognition of characters/strokes. Now, building of a strong handcrafted feature for the recognition of Bangla characters is a challenging job due to its inherent complex shape. Deep learning architecture has gained popularity for promising success to solve different complex computer vision and pattern recognition problems. A CNN can be viewed as a special type of feed-forward multilayer trained in supervised model. The main advantage of using CNN is that it automatically extracts the salient features from the input images which are mostly invariant to shift and distortion [13]. Another advantage of CNN is that the use of shared weight in its convolutional layers reduces the number of parameters as well as improves the performance [14].

Various methods have been proposed till date and high recognition rates are reported for the offline recognition of characters written in English [15, 16], Devanagari [17–19] using CNN. Though few researchers [20, 21] have used CNN for offline Bangla character recognition, but from best of our knowledge, no such work is found for online Bangla character recognition. Motivated by this fact, in the present work, we have used CNN for the recognition of online handwritten Bangla basic characters. The rest of the paper is organized as follows: Sect. 2 describes data collection and preprocessing, Sect. 3 deals with CNN architecture, Sect. 4 analyzes the various outcomes, and finally Sect. 5 concludes the paper.

2 Database Preparation

Here, we have used 200 unique samples for each character class written by 100 different persons from different section of society to get variability in handwriting. We have not put any restriction to the writers while collecting data from them. We have used Take Note device to store information from different users. In this experiment, firstly, we have generated the offline grayscale image from the online information and then resized it to fit into a window of size 28 × 28.

3 Proposed Work

In the present work, we have used CNN, a deep learning architecture for recognition of online handwritten Bangla characters which has two convolutional layers and two pooling layers followed by a fully connected classification layer with input and output layers (see Fig. 1).

The input layer contains 784 nodes for a 28 × 28 pixels image. Convolution layer is the first layer through which the input image passes. The functionality of this layer involves arbitrary number of learnable filters to move along the width and height of the image to produce feature map. A filter can be considered as an array of numbers where the numbers are called *weights or parameters.*

After sliding the filter over all the locations, we have achieved a 24 × 24 array of numbers, which is known as **activation map or feature map**. The first convolution layer produces 32 such feature maps. Convolution operation with kernel of spatial dimension 5 (i.e., size of 5 × 5) converts 28 spatial dimensions to 24 (i.e., 28 – 5 + 1) spatial dimensions [20]. So, size of the first-level feature maps becomes 24 × 24.

The output (32 @ 24 × 24) of first convolution layer goes to first pooling layer where each feature map has been resized. In this experiment, both Max_Pooling and Avg_Pooling schemes have been applied. As an example, when Avg_pooling has been adopted then at the first pooling layer, feature maps (produced at first

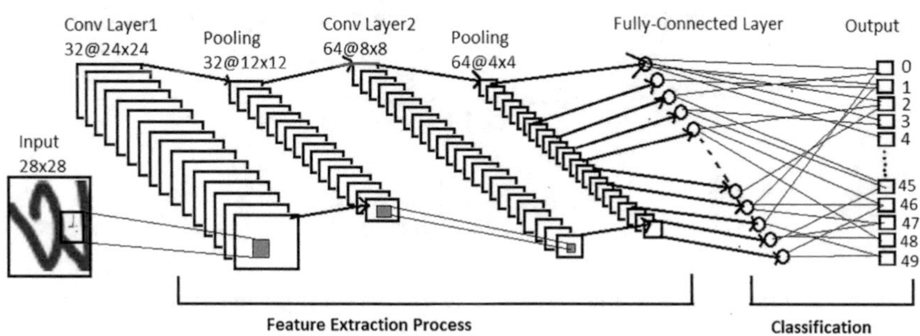

Fig. 1 Architecture of CNN used in the present work

convolution layer) have been down-sampled from 24 × 24 into 12 × 12 feature maps by applying a local averaging with 2 × 2 area, multiplying by a coefficient, adding a bias, and passing through an activation function. Formally, it can be expressed by Eq. (1) [20].

$$x_j^l = f\left(\beta_j^l \mathrm{down}\left(x_j^{l-1}\right) + b_j^l\right) \tag{1}$$

where down (.) represents a pooling function through local averaging; β and b are multiplicative coefficient and additive bias, respectively. The output image becomes twice smaller in both spatial dimensions for 2 × 2 local area averaging in down (.) function. This operation reduces the spatial resolution of the feature map, shift, and distortions sensitivity.

Following the similar fashion, second convolution and pooling operations have been performed. In this work, second convolutional operation produces 64 distinct feature maps. A 5 × 5 filter produces a feature map size of 12 × 12 into 8 × 8. Then second pooling operation resizes each feature map to size of 4 × 4. These 64 feature map values are considered as 1024 (= 64 × 4 × 4) distinct nodes which are fully connected to 50 output (corresponding to each character) class. In CNN, the error (E) has been minimized by using Eq. (2) [20].

$$E = \frac{1}{2}\frac{1}{PO}\sum_{p=1}^{P}\sum_{o=1}^{O}\left(d_o(p) - y_o(p)\right)^2 \tag{2}$$

where P is the total number of patterns (here 10000); O is the number of output nodes (i.e., 50); do and yo are the desired and actual outputs of a node, respectively, for a particular pattern p.

It is to be noted that padding of 0 valued 2 pixels has been done on all four sides of an image after passing through each convolutional layer to retain the previous size. In the current experiment, two different activation functions (Softmax and Sigmoid) have been used at the fully connected network during classification in the output layer. The Sigmoid function is generally used for the two-class logistic regression, whereas the Softmax function is used for the multiclass logistic regression. Corresponding predicted probabilities that can be computed by using Sigmoid and Softmax functions have been mentioned in Eq. (3–4) [22, 23].

$$F(X_i) = \frac{1}{(1 + \mathrm{Exp}(-X_i))} \tag{3}$$

$$F(X_i) = \frac{\mathrm{Exp}\,(X_i)}{\sum_{j=0}^{k} \mathrm{Exp}\,(X_j)} \tag{4}$$

Both Softmax and Sigmoid function use cross-entropy loss function for error calculation and can be described by Eq. (5) [24].

$$E(t, o) = -\sum_{j} t_j \log o_j \tag{5}$$

where t and o act as the target and output at neuron j, respectively. The sum is carried on over each neuron in the output layer. oj denotes the result of the Softmax function F(Xi).

4 Results and Discussion

In this experiment, we have considered 50 different character classes, where each class contains 200 different samples. We have analyzed the impact of different learnable parameters of CNN such as kernel size (3 × 3, 5 × 5 and 7 × 7), pooling strategies (Max_Pooling and Avg_Pooling), and activation functions (Softmax and Sigmoid) on the overall recognition model. Here, 30% of the entire characters have been considered as test set and rest 70% have been used to train the recognition model. Table 1 summarizes the outcomes for all possible combinations of those variations on test set. From this table, we have observed that recognition accuracy has reached to maximum (shown in bold font in Table 1) when filter size is 5 × 5; Max_Pooling is the pooling scheme and Softmax is the activation function. As the current experiment has been conducted on image of size 28 × 28, so, if filter size becomes larger (i.e., 7 × 7) then it fails to capture minute detail of the structurally similar character patterns. Whereas, if the filter size becomes very small (i.e., 3 × 3), then it may generate some redundant information which, in turn, would reduce the recognition ability of the model. Hence, in our case, 5 × 5 is proven to be optimal filter dimension. When using a Softmax classifier, we get a

Table 1 Achieved classification accuracies by CNN with varying kernel size, pooling scheme, and activation functions for classification

Kernel size	Pooling scheme	Activation function	Test accuracy (in %)
3 × 3	Max_Pooling	Softmax	99.27
	Avg_Pooling		99.13
	Max_Pooling	Sigmoid	99.07
	Avg_Pooling		99.03
5 × 5	**Max_Pooling**	**Softmax**	**99.40**
	Avg_Pooling		99.27
	Max_Pooling	Sigmoid	99.33
	Avg_Pooling		99.00
7 × 7	Max_Pooling	Softmax	99.23
	Avg_Pooling		99.13
	Max_Pooling	Sigmoid	99.20
	Avg_Pooling		99.03

Table 2 Brief comparison of the present method with some past methods on 10000 samples

Reference	Method	Recognition rate (%)
Sen et al. [9]	Feature: Haudorff distance Classifier: MLP	95.57
Roy et al. [8]	Structural + Point Based Classifier: MLP	91.03
Sen et al. [10]	Feature: Area, mass distribution and chord length Classifier: SMO (Sequential Minimal Optimization)	98.50
Bhattacharya et al. [6]	Directional code feature Classifier: MLP	91.61
Current work	**CNN**	**99.40**

probability of each class, having joint distribution and a multinomial likelihood whose sum is bound to be one. On use of Sigmoid activation function for multiclass classification, it would be having a marginal distribution and a Bernoulli likelihood. As the classes here are not completely independent of each other, the Softmax activation function becomes more effective than Sigmoid. Generally, pooling is used to reduce variance, computation complexity, and extract low-level features from neighborhood. Although both 2×2 Max_Pooling/Avg_Pooling reduces half of the image size by dividing it into four equal quadrants, Max_Pooling considers maximum value presented in each quadrant and hence can extract the most important features like edges. But the Avg_Pooling may overlook some important information due to averaging out all the values. Hence, in general, Max_Pooling has the better potential than Avg_Pooling.

We have compared the outcome of the present work with some handcrafted feature-based character recognition methods. The works performed by Sen et al. [9, 10] have used same database, whereas the features mentioned by Roy et al. [8] and Bhattacharya et al. [6] have been implemented on the present database. The comparison details are reported in Table 2. From the table, it is clear that recognition using CNN outperforms the other methods.

5 Conclusion

In the present work, a popularly used CNN architecture is investigated in terms of kernel size, pooling scheme, and activation function for online handwritten Bangla character recognition purpose. The outcome of this work is quite impressive and also it outperforms the existing handcrafted feature extraction methods. In future, we plan to study the more recent CNN architectures to improve recognition accuracy on a larger database.

References

1. Kubatur, S., Sid-Ahmed, M., Ahmadi, M.: A neural network approach to online Devanagari handwritten character recognition. In: International Conference on High Performance Computing and Simulation, pp. 209–214 (2012)
2. Swethalakshmi, H., Sekhar, C.C., Chakravarthy, V.S.: Spatiostructural features for recognition of online handwritten characters in Devanagari and Tamil scripts. In: International Conference on Artificial Neural Networks, pp. 230–239 (2007)
3. Swethalakshmi, H., Jayaraman, A., Chakravarthy, V.S., Sekhar, C.C.: On-line handwritten character recognition for Devanagari and Telugu scripts using support vector machines. In: International Workshop on Frontiers in Handwriting Recognition, pp. 367–372 (2006)
4. Tappert, C.C., Suen, C.Y., Wakahara, T.: The state of online handwriting recognition. IEEE Trans. Pattern Anal. Mach. Intell. 787–807 (1990)
5. Vescovo, G.D., Rizzi, A.: Online handwriting recognition by the symbolic histograms approach. In: International Conference on Granular Computing, pp. 686–690 (2007)
6. Bhattacharya, U., Gupta, B.K., Parui, S.K.: Direction code based features for recognition of online handwritten characters of Bangla. In: International Conference on Document Analysis and Recognition, pp. 58–62 (2007)
7. Roy, K.: Stroke-database design for online handwriting recognition in Bangla. Int. J. Mod. Eng. Res. 2534–2540 (2012)
8. Roy, K., Sharma, N., Pal, U.: Online Bangla handwriting recognition system. In: International Conference on Advances in Pattern Recognition, pp. 117–122 (2007)
9. Sen, S.P., Sarkar, R., Roy, K., Hori, N.: Recognize online handwritten Bangla characters using hausdorff distance based feature. In: International Conference on Frontiers in Intelligent Computing: Theory and Application, pp. 541–549 (2017)
10. Sen, S.P., Mitra, M., Chowdhury, S., Sarkar, R., Roy, K.: Quad-tree based image segmentation and feature extraction to recognize online handwritten Bangla characters. In: 7th IAPR TC3 Workshop on Artificial Neural Networks in Pattern Recognition, Ulm, Germany, pp. 246–256 (2016)
11. Mondal, T., Bhattacharya, U., Parui, S.K., Das, K., Mandalapu, D.: On-line handwriting recognition of Indian scripts—The first benchmark. In: International Conference on Frontiers in Handwriting Recognition, pp. 200–205 (2010)
12. Parui, S.K., Guin, K., Bhattacharya, U., Chaudhuri, B.B.: Online handwritten Bangla character recognition using HMM. In: International Conference on Pattern Recognition, pp. 1–4 (2008)
13. Shin, H.C., Roth, H.R., Gao, M., Lu, L., Xu, Z., Nogues, I., Yao, J., Mollura, D., Summers, R. M.: Deep convolutional neural networks for computer-aided detection: CNN architectures, dataset characteristics and transfer learning. IEEE Trans. Med. Imaging **35**, 1285–1298 (2016)
14. Bai, J., Chen, Z., Feng, B., Xu, B.: Image character recognition using deep convolutional neural network learned from different languages. In: IEEE International Conference on Image Processing, pp. 2560–2564 (2014)
15. Yuan, A., Bai, G., Jiao, L., Liu, Y.: Offline handwritten English character recognition based on convolutional neural network. In: International Workshop on Document Analysis Systems, pp. 125–129 (2012)
16. Bouchain, D.: Character recognition using convolutional neural networks. Inst. Neural Inf. Process. (2007)
17. Acharya, S., Pant, A.K., Gyawali, P.K.: Deep learning based large scale handwritten Devanagari character recognition. In: International Conference on Software, Knowledge, Information Management and Applications (2015)
18. Singh, P., Verma, A., Chaudhari, N.S.: Deep convolutional neural network classifier for handwritten Devanagari character recognition. Inf. Syst. Des. Intell. Appl. 551–561 (2016)

19. Mehrotra, K., Jetley, S., Deshmukh, A., Belhe, S.: Unconstrained handwritten Devanagari character recognition using convolutional neural networks. In: Proceedings of 4th International Workshop on Multilingual OCR (2013)
20. Rahman, M.M., Akhand, M.A.H., Islam, S., Shill, P.C., Rahman, M.H.: Bangla handwritten character recognition using convolutional neural network. Int. J. Image Graph. Signal Process. **8**(3), 42–49 (2015)
21. Maitra, D.S., Bhattacharya, U., Parui, S.K.: CNN based common approach to handwritten character recognition of multiple scripts. In: International Conference on Document Analysis and Recognition, pp. 1021–1025 (2015)
22. Janocha, K., Czarnecki, W.M.: On Loss Functions for Deep Neural Networks in Classification, pp. 1–10 (2017)
23. Oland, A., Bansal, A., Dannenberg, R.B., Raj, B.: Be Careful What You Backpropagate: A Case For Linear Output Activations & Gradient Boosting (2017). https://arxiv.org/abs/1707.04199
24. Jeong, J.M., Choi, S.J., Kim, C.: Correcting method of numerical errors of soft-max and cross entropy according to CNN's output value. Neural Comput. Appl. **1**(1), 15–20 (2017)

Survey on Clustering Algorithms for Unstructured Data

R. S. M. Lakshmi Patibandla and N. Veeranjaneyulu

Abstract In modern applications, clustering algorithms have been emerged learning aid to generate and analyze the huge volumes of data. The foremost clustering objective is to classify same type of data has been grouped with in the same Cluster while they are similar according to precise metrics. For various applications, clustering is one of the techniques to classify and analyze the large amount of data. On the other hand, the main issues of applying clustering algorithms for big data that causes uncertainty among the practitioners require consent in the definition of their properties in addition to be deficient in proper classification. In this paper, we studied various existing clustering methods which are suitable for large, semi-structured, and unstructured data and how we can apply same algorithms in distributed environment/hadoop.

Keywords Clustering · Data algorithms · Semi-structured · Unstructured

1 Introduction

According to substances behavior, characteristics and similarities to group some set of substances as a Cluster. Clustering is one of the techniques which can be used in data mining to classify the objects. Two types of partitioning methods are available in clustering—hard and soft. Hard partitioning allocates an object instance not only exist in part of the Cluster it stringently belongs to that Cluster. Soft partitioning means each and every object belongs to Cluster with resolute degree. Some of the objects participate in single Cluster or multiple Clusters and to construct

R. S. M. L. Patibandla (✉)
Department of CSE, Vignan's Foundation for Science, Technology & Research, Vadlamudi, Andhra Pradesh, India
e-mail: patibandla.lakshmi@gmail.com

N. Veeranjaneyulu
Department of IT, VFSTR University, Vadlamudi, Andhra Pradesh, India
e-mail: veeru2006n@gmail.com

V. Bhateja et al. (eds.), *Intelligent Engineering Informatics*, Advances in Intelligent Systems and Computing 695, https://doi.org/10.1007/978-981-10-7566-7_41

hierarchical trees on Cluster associations. There are various approaches to apply this partitioning based on different models. Several algorithms are used for every model to differentiate its properties and results. Some of the important models are distributed, centralized, connectivity, group, graph, density and so on.
Distributed methods have been related to predefined statistical models; it can group the objects which are in the same distribution. This method desires a precise and complex model to communicate in a better approach with real data. These techniques have been achieved an optimal solution, analyze dependencies, and correlations. While comparing to other Clusters, each object is a part of the Cluster with minimal value differences in centralized method. In this technique, all the elements of Cluster can be corresponding to vector values. The major problem in centralized method is the number of Clusters should be predefined. It is mostly relevant to the classification methods and used for optimization problems. Each object is associated with neighbors depends on the degree of relationship and the distance between them in connectivity methods. Among the members of a dataset with high density in a determined location, density-based algorithms have to create Clusters. These kinds of methods have less performance on detecting the limited areas of the group. There are so many methods are available in clustering, may provide different datasets.

The priority may be given for particular technique depends on the form of desired output; software and hardware facilities presented on size of the dataset and recognized performance of the technique with particular types of data. Generally, these methods are categorized into two ways found on the structure of the generated Cluster. Non-hierarchical methods segregate a datasets of X substances into Y Clusters overlie among them.

The single link technique is almost certainly the greatest notorious one of the hierarchical methods and operates by combining the similar type of objects from various Clusters. Single Link is the combining a group of Clusters through direct association.

The complete link technique has been related to the single link technique apart from that determines the inter-cluster similarity. This technique has been illustrated by minute, tightly bound Clusters.

Clustering is a very precious data analysis approach; it has several different applications in the scientific world. So, Cluster analysis provides good results on various types of data for every big dataset. This method is most efficient in distinctive objects, biological research, image processing, identifying patterns, and the classifications of medical tests. Clustering can also be used in market research and personal information combined with all the details such as shopping, location, member indications [1]. On the other hand, it has been used in robotics, mathematical, and statistical analysis for providing a broad spectrum of utilization.

Datasets contain several items in earlier. At present, new methods have been used to store and process the large amount of data. Big data refers to collection of complex and large datasets for processing and it is very difficult to process with customary tools. Big data is usually associated with volume (data with huge quantity), variety (data with various kinds), and velocity (continuously accruing novel data) [2]. Large data extends the capabilities of systems for storing, analyzing, and processing, includes veracity and value [3, 4]. Big data can be used in several applications: health care, e-commerce, finance, and social networks, etc. Not only collected the data from the computers, but also from different sensors installed in vehicles, posts by social media, through mobile phones and too many sources using big data.

Big data contains two types of data—structured and unstructured. Structured data is simple to analyze the data, because the data exists in databases as rows and columns. In unstructured data, data is not organized in a predefined way. So, the developers develop the tools for switch the data into unstructured to structured [5]. Suppose a structured dataset 'A' contains data items A1, A2,..., Ax by the features a1, a2,...., ay, where total number of data items represented by x and y represents the total number of features. So, $A = \{A1, A2,..., Ax\} = \{a_{ij}, i = 1,..., x, j = 1,..., y\}$, where a_{ij} is the jth feature value of the ith object. In high-dimensional data, the total number of features y is high. It is useful to solve the visualization problems with dimensionality reduction [6, 7]. Large datasets contain several Clusters and essential to classify Clusters.

The proper data model structures have been related to relational databases and data tables in semi-structured data. Extensible Markup Language and JSON languages are used for handling semi-structured data. Natural language processing techniques or traditional mining techniques are used to switch semi-structured data into structured data.

2 Clustering Techniques Suitable for Big Data

Given dataset A = {A1, A2,..., Ax} and m is constant integer value, clustering is to classify a plotting f:A → {1,...., m}, here all items $A_{l,}$ l € $\{1,..., n\}$ is allocated to Cluster M_jj = 1,...., m. A Cluster Mj includes the items plot to it: $Mj = \{A_l|f(A_l) = Mj, 1,..., m,$ and A_l €$A\}$. A data item inside a Cluster can be mostly connected just before the items in that Cluster than associated to data items exterior to that Cluster [8]. One of the similarity measures of Cluster is Minkowski distance; it is also called as Euclidean distance. Several clustering techniques have been implemented like K-means [9], EM, ISODATA, Fuzzy c-means, Leader, SOM [10] and so on.

2.1 K-Means Method

K-means algorithm is most important technique in clustering. In this technique, firstly the k number of preferred Clusters is chosen and to assign the initial values for Cluster centers. After that all the data items are allocated to Cluster amongst nearby centers and novel centers of all Clusters have been evaluated. These steps are frequently repeated until it satisfies the convergence criteria or stop. This convergence criterion supports squared error, i.e., the average variation among the Cluster centers and the allocated data items toward the Cluster. The stop criteria have highest number of iterations. Diverse standard K-means algorithms have been

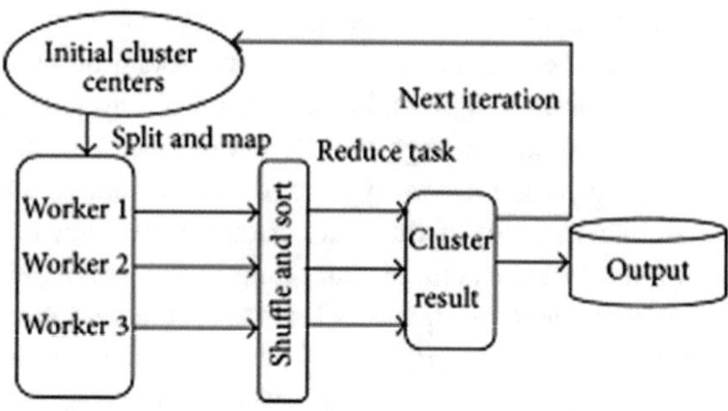

Fig. 1 Process of K-means

implemented earlier like Kernel K-Means [11], Spherical K-Means [12], Minkowski Metric Weighted K-Means [13], and Fuzzy C-Means [14] and so on. By using this customized algorithms, improve the speed up of calculations for particular tasks. The standard K-Means clustering algorithm is highly capable on big datasets according to its low cost, but for big data analytics this algorithm initiated some very particular tools. X-means technique [15] is used to expand K-means through effective assessment of total Clusters. The standard K-means clustering algorithm has been considered to solve single-view clustering data problems. Heterogeneous features for clustering multi-view K-Means clustering have anticipated. Streaming K-means algorithm [16, 17, 18] is designed for well-cluster data. The major problem of K-means algorithm is on large data that is accumulated in main memory. Various techniques of Euclidean K-means [19] have been proposed for stream data. Some other algorithms to determine the novel algorithms are mostly suitable for large datasets (Fig. 1).

2.2 *Expectation Maximization Method*

Expectation minimization algorithm is also used for clustering data [20]. Where the data can be treated as incomplete, this iterative algorithm is used to solve problems. Expectation maximization is a suitable optimization algorithm to develop the statistical models. This algorithm has broadly applied on various purposes like pattern recognition, speech processing, and computer vision. It has Expectation and Maximization steps. These steps are repeatedly executed till meets the criteria. In Expectation step, estimate the possibility of every data item belongs to each Cluster. In Maximization step, the possibility of parameter vector distribution of each Cluster is estimated. EM algorithm finishes after reaching the large amount of iterations [21].

2.3 Other Clustering Methods

BIRCH Algorithm [22] has initiated in support of capturing desired data. Agglomerative and divisive two varieties of clustering are available in hierarchical clustering.

Agglomerative clustering is firstly apiece of data item inside the Cluster, and followed by couple of Clusters have been combined. In Divisive Clustering, entire data items have allocated first to single Cluster and followed by divide that Cluster subsequently. Dissimilarity measure between the required data items has to choose what Cluster must be merged in agglomerative as well as divide in divisive. Finally, the results of this clustering have been visualized in dendrogram.

DBSCAN algorithm [23] is anticipated data items of an entire Cluster; the region radius has to include lowest amount of data items, solidity inside region have extended various Thresholds which are predefined. It requires least amount of data items within same Cluster and the Threshold rate of distance to facilitate the meticulous region of a data Item. DBSCAN and K-means methods have developed mining tools like RapidMiner and WEKA.

To overcome the disadvantages of DBSCAN algorithm, OPTICS algorithm has been introduced. This algorithm is used to identify significant Clusters in unreliable density of data items [24]. CURE algorithm is used for a fixed number of data items to specify a Cluster [23]. Clusters among the nearest pair of specified data items have chosen to be merged in this algorithm. A distance is the measure for nearest data items from various Clusters. These data items have shrunk to the center of the Cluster.

A CB-SVM has intended for managing very massive datasets. This method has been applied on hierarchical micro-clustering algorithm which checks only once to entire data and to give SVM in favor of high-value data items of statistical data summary and maximize the benefit of knowledge on SVM. Piecemeal Principal Direction Divisive Partitioning is a scalable process in the track of large datasets of Cluster to exist in memory [25]. Original data items can be divided into Clusters and keep it into memory.

The Self-Organizing Map (SOM) is excellent application for neural networks in an unsupervised learning [10]. This is used as a tool for clustering and visualizing the data [26]. Self-Organizing Map comprised earlier: Merge Self-Organizing Map (MSOM) [27], Recursive Self-Organizing Map (RecSOM) [28], WEBSOM [29] and so on. Generally, all the above extensions are formed to perform specific tasks very efficient way. The Self-Organizing Maps are commonly worn for analyzing high-dimensional data as well as visualization tool [30].

Table 1 RDBMS versus HADOOP/MapReduce

	RDBMS	Hadoop/MapReduce
Data size	Gigabytes (Terabytes)	Petabytes (Hexabytes)
Access	Interactive and batch	Batch
Updates	Read/Write many times	Write once/read many times
Structure	Static schema	Dynamic schema
Integrity	High	Low
Query response time	Can be nearly immediate	Has latency (due to batch processing)

3 Algorithms Used on HADOOP

The total number of data items is not large in the dataset; it can be analyzed in data mining systems like Weka, RapidMiner, Orange, and KNIME by using various clustering methods. If the total number of data items is huge in the dataset, then the dataset has been analyzed on distributed systems. While handling the big data, Hadoop technologies are most supportive (Table 1).

HADOOP is feasible to develop cost effective for huge amount of data storage on distributed systems with the use of less cost networking servers. Hadoop File System allocates the data into Hadoop, after that start working on all the servers exists in the Cluster. HDFS allows terabytes of data with thousands of nodes in processing the big data. If there are any node failures, HDFS makes us continue operation. Only one node failure never affects the system failure. Hadoop provides a novel approach for multiple computers in the Cluster on distributed environment is MapReduce. MapReduce is an effective programming model to handle enormous datasets through a Parallel Distributed Algorithm for dividing, processing, and aggregation of huge datasets [31, 32]. Open-source MapReduce has implemented for big Clusters in HADOOP. This contains only one master node called as Job-Tracker and many other nodes called TaskTrackers. Suppose the dataset acquired from social networks like Facebook or Twitter must be processed arbitrarily in the data must be removed.

Twitter datasets contain the data like Username, department, area, expressions, and tweets. Datasets MapReduce may apply on twitter datasets has given input to MapReduce process for the enhancement and structuring. It applies for any size of dataset and supports for valuable execution of datasets, through which unstructured data of any size must be structured effectively.

After completing the execution of MapReduce program for unstructured data, the resultant dataset has been structured in specific order as per the requirements of user.

HADOOP also have HIVE, Sqoop, Hbase, etc., like applications for implementing and exporting from other traditional and non-traditional databases. It is used within diverse applications like video/audio file processing, text analytics, and file processing on images [33].

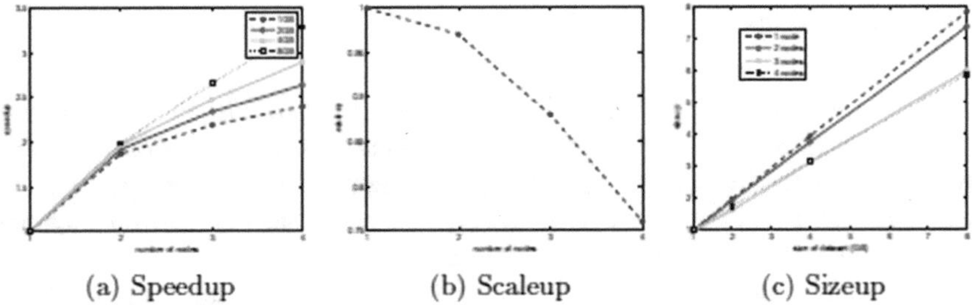

(a) Speedup (b) Scaleup (c) Sizeup

Fig. 2 Evaluation results by Zhao

For example, CCTV video data systems of an organization presently humans can monitor all these videos. By this, we can detect some incidents to be noticed from multiple videos and be alert for every time. Imagine this monitoring process has to be automated, the large volumes of data (terabytes) in few hours. Then, it will detect the incidents at correct time. HADOOP has all these capabilities and used such type of scenarios efficiently.

In the paper [34], Zhao proposed a parallel K-means clustering algorithm based on MapReduce, which have been extensively used in both academic world and engineering. Analyze the proposed algorithms performance through scaleup, speedup, and sizeup of the Clusters. The effective solutions of proposed algorithm know how to process huge datasets. While testing this algorithm, the speed up is not linear. If the dataset size increases, then the communication overhead also increases (Fig. 2).

In the paper [35], Jimmy Lin and Chris Dyer provide detailed explanation of used Expectation and Maximization algorithms on Text Processing and appropriate those algorithms into the MapReduce Programming model. Expectation and Maximization obviously fit into this model and make all the iterations of EM of one MapReduce job. Mappers map more individual instances and evaluate the summary of statistics. Reducers sum collectively the requisite instruction statistics and solve the Maximization step optimization problems. Here, in experimental work, we observe that universal data is required for synchronization of HADOOP tasks and it was most difficult in this platform.

In the paper [36], Zhenhua applied K-means algorithm for remote sensing images in Hadoop. While doing this research, HADOOP works on text and images have to be identified as text and processed, transparency in illustration and executing it on bigger or smaller images.

In the paper [37], Zganquan sun explored the use of SVM on MapReduce platform. He concludes MapReduce is capable to reduce the instruction time and execution time for SVM. There is no relationship between the portioning method and the derived performance. EM algorithm approximates the parameters for unseen variables maximize the probability.

4 Conclusions

Clustering technologies and approaches have described in this study. At present, social networking sites, CCTV devices, and other resources produce the large amount of data. Clustering algorithms which are useful for unstructured and semi-structured data are discussed in this paper. Parallel and distributed computing systems have been used for complex problems.

This study summarizes the clustering algorithms which are used on large, unstructured, and semi-structured data. How these algorithms have applied to distributed and Hadoop platforms.

References

1. Madhuri, R., RamakrishnaMurty, M., Murthy, J.V.R., Prasad Reddy, P.V.G.D., et al.: Cluster analysis on different data sets using K-modes and K-prototype algorithms. In: International Conference and Published The Proceeding in AISC and Computing, pp. 137–144. Springer (2014)
2. Schmidt, S.: Data is exploding: the 3 V's of big data. Business Computing World (2012)
3. RamakrishnaMurty, M., Murthy, J.V.R., Prasad Reddy, P.V.G.D., Sapathy, S.C.: A survey of Cross-Domain text categorization techniques. In: International Conference on Recent Advances in Information Technology RAIT-2012 IEEE Xplorer Proceedings (2012), 978-1-4577-0697-4/12
4. RamakrishnaMurty, M., Murthy, J.V.R., Prasad Reddy, P.V.G.D., et al.: Homogeneity separateness: a new validity measure for clustering problems. In: International Conference and Published The Proceedings in AISC and Computing, pp. 1–10. Springer (2014)
5. Zhai, Y., Ong, Y.-S., Tsang, I.W.: The emerging big dimensionality. In: Proceedings of the 22nd International Conference on World Wide Web Companion, Computational Intelligence Magazine, pp. 14–26. IEEE (2014)
6. Medvedev, V., Dzemyda, G., Kurasova, O., Marcinkeviˇcius, V.: Efficient data projection for visual analysis of large data sets using neural networks. Informatica, 507–520 (2011)
7. Dzemyda, G., Kurasova, O., Medvedev, V.: Dimension reduction and data visualization using neural networks. In: Maglogiannis, I., Karpouzis, K., Wallace, M., Soldatos, J. (eds.): Emerging Artificial Intelligence Applications in Computer Engineering, pp. 25–49 (2007)
8. Dunham, M.H.: Data Mining: Introductory and Advanced Topics. Prentice Hall PTR, Upper Saddle River, USA (2002)
9. MacQueen, J.: Some methods for classification and analysis of multivariate observations. In: Le Cam, L.M., Neyman, J. (eds.): Proceedings of the Fifth Berkeley Symposium on Mathematical Statistics and Probability, University of California Press, Berkeley and Los Angeles, USA, pp. 281–297 (1967)
10. Kohonen, T.: Overture. In: Self-Organizing Neural Networks: Recent Advances and Applications, pp. 1–12. Springer, New York, USA (2002)
11. Dhillon, I., Guan, Y., Kulis, B.: Kernel k-means: spectral clustering and normalized cuts. In: Proceeding KDD 2004 Proceedings of the tenth ACM SIGKDD International Conference on Knowledge Discovery and Data Mining, pp. 551–556 (2004)
12. Dhillon, I., Modha, D.: Concept decompositions for large sparse text data using clustering. Mach. Learn. 143–175 (2001)
13. de Amorim, R.C., Mirkin, B.: Minkowski metric, feature weighting and anomalous cluster initializing in k-means clustering. Pattern Recognit. 1061–1075, (2012)

14. Bezdek, J.C.: Pattern Recognition with Fuzzy Objective Function Algorithms, Kluwer Academic Publishers (1981)
15. Pelleg, D., Moore, A.W.: X-means: extending k-means with efficient estimation of the number of clusters. In: Proceedings of the 17th International Conference on Machine Learning, Morgan Kaufmann, pp. 727–734 (2000)
16. Cai, X., Nie, F., Huang, H.: Multi-view k-means clustering on big data. In: Rossi, F. (ed.): Proceedings of the 23rd International Joint Conference on Artificial Intelligence, IJCAI 2013, IJCAI/AAAI (2013)
17. Ailon, N., Jaiswal, R., Monteleoni, C.: Streaming k-means approximation. In: Proceedings of 23rd Annual Conference on Neural Information Processing Systems, NIPS, pp. 10–18 (2009)
18. Braverman, V., Meyerson, A., Ostrovsky, R., Roytman, A., Shindler, M., Tagiku, B.: Streaming k-means on well-clusterable data. In: Randall, D. (ed.): Proceedings of the Twenty-Second Annual ACM-SIAM SODA, pp. 26–40 (2011)
19. Shindler, M., Wong, A., Meyerson, A.: Fast and accurate k-means for large datasets. In: Shawe-Taylor, J., Zemel, R.S., Bartlett, P.L., Pereira, F.C.N., Weinberger, K.Q. (eds.): Proceedings of 25th Annual Conference on Neural Information Processing Systems pp. 2375–2383 (2011)
20. McLachlan, G., Krishnan, T.: The EM Algorithm and Extensions, 2nd edn, Wiley series in probability and statistics (2008)
21. Abimbola, A.A., Omidiora, E.O., Olabiyisi, S.O.: An exploratory study of k-means and expectation maximization algorithms. Br. J. Math. Comput. Sci. 62–71 (2012)
22. Zhang, T., Ramakrishnan, R., Livny, M.: BIRCH: an efficient data clustering method for very large databases. In: Proceedings of the 1996 ACM SIGMOD International Conference on Management of Data. SIGMOD 1996, pp. 103–114. ACM, New York, USA (1996)
23. Guha, S., Rastogi, R., Shim, K.: CURE: an efficient clustering algorithm for large databases. Inf. Syst. 35–58 (2001)
24. Ankerst, M., Breunig, M.M., Kriegel, H.P., Sander, J.: OPTICS: ordering points to identify the clustering structure. In: Proceedings of the 1999 ACM SIGMOD International Conference on Management of Data, pp. 49–60. ACM (1999)
25. David, L., Daniel, B.: Clustering very large datasets using a low memory matrix factored representation. Comput. Intell. 114–135 (2009)
26. Dzemyda, G., Kurasova, O., Zilinskas, J.: Multidimensional Data Visualization: Methods and Applications, Springer Optimization and Its Applications. Springer (2013)
27. Hammer, B., Micheli, A., Sperduti, A., Strickert, M.: A general framework for unsupervised processing of structured data. Neurocomputing, 3–35 (2004)
28. Voegtlin, T.: Recursive self-organizing maps. Neural Netw. 979–991 (2002)
29. Lagus, K., Kaski, S., Kohonen, T.: Mining massive document collections by the WEBSOM method. Inf. Sci. 135–156 (2004)
30. Stefanoviˇc, P., Kurasova, O.: Visual analysis of self-organizing maps. In: Nonlinear Analysis: Modelling and Control, pp. 488–504 (2011)
31. Kurasova, O., Marcinkeviˇcius, V., Medvedev, V., Rapeˇcka, A., Stefanoviˇc, P.: Strategies for big data clustering. In: IEEE 26th International Conference on Tools with Artificial Intelligence, pp. 740–747 (2014)
32. Nandakumar, A.N., Yambem, N.: A survey on data mining algorithms on apache hadoop platform. IJETAE, 563–565 (2014)
33. Veeranjaneyulu, N., NirupamaBhat, M., Raghunadh, A.: Approaches for managing and analyzing unstructured data. IJCSE, 19–24 (2014)
34. Jaatun, M.G., Zhao, G., Rong, C. (eds.): Parallel K-Means clustering based on MapReduce. In: CloudCom 2009, LNCS 5931, pp. 674–679 (2009)
35. Lin, J., Dyer, C.: Data-Intensive Text Processing with MapReduce (2010)
36. Wang, F.L., et al., (eds.): Parallel K-Means clustering of remote sensing images based on MapReduce. In: WISM 2010, LNCS 6318, pp. 162–170 (2010)
37. Sun, Z.: Study on Parallel SVM Based on MapReduce. In: Conference on WorldComp (2012)

A Chaotic Steganography Method Using Ant Colony Optimization

Shahzad Alam, Tanvir Ahmad and M. N. Doja

Abstract Since ancient times, there is a need to save data from third parties that spy on the data, leads to the practice of steganography to ensure data integrity and privacy. The advent in multimedia technologies and greater need for security has raised the popularity of image steganography. The traditional algorithm used for image steganography is the Least Significant Bit Method (LSB). A better strategy to overcome the LSB method is to hide data only in edge pixel of image. In this paper, first we apply the ant colony optimization (ACO) method to find edges of an image given by Xiaochen Liu et al. then will hide sensitive message in edges of image randomly to provide more security level. The edge detection technique and chaotic scheme-based steganography is introduced based on the ACO. This strategy efficiently finds the edges and does not attract much attraction compared to other traditional methods. Our result indicates the higher value of PSNR that shows the performance of the proposed algorithm.

Keywords Steganography · Data hiding · Edge detection · Least Significant Bit (LSB) · Pixel · Ant colony optimization

1 Introduction

The advancement in technology increased the computational powers of computers enormously. The passwords used now days are not sufficient to protect information. Hackers can now decrypt the weakly encrypted message in a very short time.

S. Alam (✉) · T. Ahmad · M. N. Doja
Faculty of Engineering and Technology, Department of Computer Engineering, Jamia Millia Islamia, Delhi 110025, New Delhi, India
e-mail: shahzad5alam@gmail.com

T. Ahmad
e-mail: tahmad2@jmi.ac.in

M. N. Doja
e-mail: ndoja@yahoo.com

V. Bhateja et al. (eds.), *Intelligent Engineering Informatics*, Advances in Intelligent Systems and Computing 695, https://doi.org/10.1007/978-981-10-7566-7_42

Transmitting password encrypted information over network is no more secure. Researchers are focused on increasing the level of encryption by hiding a secret message within a non-valuable asset such as audio, image, or text. This practice is known as steganography. In general, the main aim of steganography is to communicate under the cover being undetected. This does not mean that information is hidden from others; steganography conceals the fact that there is a transfer of information going on at first place. In steganography, the user data is hidden inside a particular shaped container but in actual it does not store the original data itself. Containers can be in form of digital images, executable files, and also sound clips. The data is hidden in cover media. The secret message is embedded in carrier medium, for example, on an image file or a video file. The main role of steganography is invisible communication which paves the way for secret undetectable message communication. It alters the characteristics of the carrier.

Steganography is composed of three elements; they are the carrier, the message, and key. This object will be responsible to "carry" the secret message. The decoding of the secret message is done with the help of key. A scheme presented by Chen et al. [1] showed the embedding of data at LSB of individual pixel cannot be detected by human visual system (HVS). The resulting image form after embedding is called stego image. If we change the LSB, the quality of image does not change from human view but this method is responsive to various image processing attacks, namely compression, cropping. The issues with LSB technique were thoroughly researched: Several techniques were put forward like data hiding in higher bits of LSB. In all the stego methods, the bit position of each pixel remains the same. For instance in the LSB method, least significant bit is exclusively used for data hiding. This process is overall weak. And if we consider using higher layers, the quality of the image will be greatly affected. The secret message is to be positioned at different pixels at completely different positions. Another non-homogeneous algorithm named GSA, i.e., Genetic Simulated Annealing, uses long chromosomes and very large generations to find the optimal solution [2]. With experimental results hands, this technique is effective while implementing but is not efficient due to large generations that are to be analyzed. Image edge detection is done by detecting pixels in image where sharp changes in intensity are found. This is an important aspect in machine vision where edge detection is used to identify and differentiate object from each other. Edge detection aims to localize an image boundary. Edge detection [3, 4] is used in various other applications that use machine vision such as a robot. Conventional algorithms of image edge detection are computationally expensive as they take large amount of time to detect edges in image because the entire algorithm has to run for each pixel. So, the time increases with the size of image [5]. The edges are sharp contrast area of a digital image or where brightness changes abruptly in a digital image. There are different types of methods to identify the edges of input digital image such as Prewitt, Sobel, Roberts, Kirsh, Canny edge detection algorithms [6, 7]. The Canny method for edge detection gives better result. Liu and Fang [8] used ACO and Canny [9, 10] as hybrid approach to find out the edge pixel of image for chaotic steganography. In this work, we implemented the ant colony optimization proposed by Xiaochen Liu et al. to find the edge pixels of the input image and then apply embedding method for data hiding techniques on it and will be discussed in the subsequent section of this paper.

This paper is framed as follows: Sect. 2 introduces the ant colony optimization technique and proposed algorithm for chaotic steganography, and Sect. 3 describes the result of proposed algorithm. Finally, Sect. 4 concludes this paper.

2 Proposed Algorithm

In this paper, a chaos-based edge steganography method is discussed. The embedding procedure has three phases. Firstly, we find the edge pixel of source image using ant colony optimization techniques. After that we apply chaotic algorithm named as Random-SelectPixel (RSP) which is based on Fisher—Yates shuffle techniques for selecting the random pixel for hiding the secret bits of encrypted message based on the input key [11, 12]. The block diagram for embedding is illustrated in Fig. 1. The ant colony optimization and 1D chaotic logistic map are used in the proposed design algorithm.

2.1 Ant Colony Optimization

Ant colony optimization is an algorithm which follows the behavior of ants for finding solutions for optimization problems. It is a meta-heuristic algorithm that replicates searching behavior of ants in order to find the optimal path of combinatorial optimization problems, weighted construction graphs are used. This optimization technique is taken up from the natural behavior of ants living together. In an ant colony, the ant at the first position drops pheromone on the path which it walks to mark the way for the other ant members following behind. With the

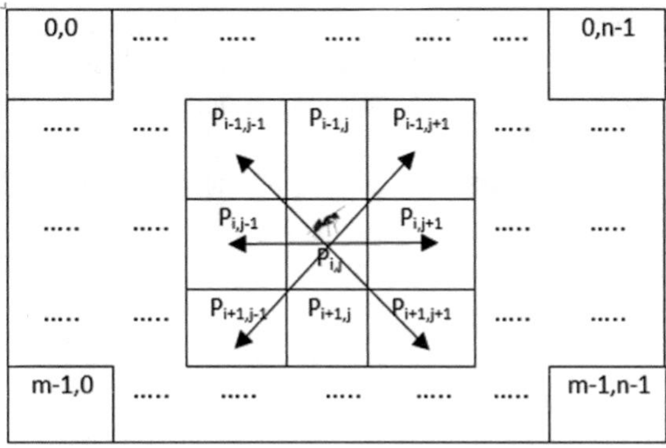

Fig. 1 A local configuration of MxN size image for calculating the intensity of pixel at $P_{i,j}$

passing time, the pheromone trail vaporizes. If the ant will take longer duration to travel the path back and forth, the pheromone will also take longer time to vaporize. The shorter duration is a favorable as the density of pheromone is higher on shorter paths. Henceforth, a feedback mechanism helps the ants to lead behind shorter paths. In this approach, the message is hidden in the pixels that cause the least suspicion in the image. We use the edges of the image to hide data. The change in edge pixels is least noticeable to the human eye. This causes a slight increase in the brightness of pixel which gives an impression of gradient fill rather than raising suspicion of a hidden message. To increase the efficiency of the algorithm, chaotic and RSP [12] algorithms are used that distribute a payload over edge pixel which makes decoding the message in them almost impossible even if one detects a hidden message in them.

Image edge detection can be percept as a problem of finding the edge pixels in a digital image. An m × n two-dimensional digital image can be illustrated as a 2D matrix having image pixels as its elements as shown in Fig. 1.

A pixel of image is connected to adjacent pixel as depicted in Fig. 1. Each Pixel of image is consider as a node of graph. Ants visit the graph by moving from one pixel to another adjacent pixel by their connections. An ant cannot jump to a non-neighboring pixel from the current position. Artificial ants are dispersed over the entire image and proceed from one pixel to adjacent pixel which belongs to edge pixel. The movement of the ants is determining by the intensity values of the pixel. The pheromone matrix is updated by Eq. (1). The objective of the ants' operation is to build a final pheromone matrix that shows the edge pixel information of image [13] that shows the edge pixel information of image.

$$p_{ij}^{n} = \begin{cases} \dfrac{[\tau_{ij}^{n-1}]^{\alpha}[\eta_{ij}]^{\beta}}{\sum_{j \in N_i^k} [\tau_{ij}^{n-1}]^{\alpha}[\eta_{ij}]^{\beta}} & if \quad j \in N_i^k \end{cases} \tag{1}$$

where τ_{ij}^{n-1}, η_{ij}, and N_i^k are pheromone intensity of the pixel $P_{i,j}$, the heuristic information of pixel $P_{i,j}$ and neighborhood pixel of the pixel $P_{i,j.}$ α and β control the relative value of the pheromone matrix. The following are the steps to obtain the edge image as below:

Step 1: k ants are taken initially and placed on the randomly selected pixels on an input digital image of size M × N.

Step 2: Apply the ACO techniques to obtain pheromone matrix governed by Eq. (1).

Step 3: Update the global pheromone matrix by following equations

$$\tau_{i,j}^{(n)} = (i - \rho) \cdot \tau_{i,j}^{(n-1)} + \sum_{1}^{K} \Delta\tau_{i,j}^{(k)} \tag{2}$$

$$\Delta\tau_{i,j}^{(n)} = \begin{cases} \frac{1}{L_k}, & \text{if ant k used } P_{i,j} \\ 0, & otherwise \end{cases} \tag{3}$$

where ρ is pheromone evaporation rate, k is total number of ants, $\Delta\tau_{i,j}^{(k)}$ amount of pheromone deposited by ant, L_k tour length of kth ant.

Step 4: Finally pheromone matrix is obtained which indicate image pixel either as an edge pixel or non-edge pixel.

2.2 Chaotic 1D Logistic Map

In this work, we used 1D Logistic map [14] for generating the chaotic sequence based on the initial key. It is useful for selecting random edge pixel of the input image. It is nonlinear chaotic discrete systems that exhibit chaotic nature. It has following equation.

$$X_{n+1} = \alpha X_n(1 - X_n) \tag{iv}$$

where X_n is initial condition lies between zero and one, α is the initial parameter, and n is the total number of iterations for system. It possesses strong chaotic characteristics for $3.57 < \alpha < 4$ and $X_{n+1} \in (0, 1)$ for all n.

2.3 Embedding Algorithm

In the embedding process, firstly, we obtained edge image of input image by ant colony optimization (ACO) method. We calculate the hash value of secret message and append it to original secret message and finally encrypt it [15]. After that, bits of encrypted message are taken for embedding process. The random pixel is selected by RSP algorithm which generates random pixel based on the secret key. At the receiver end, extracting method is used to obtain the encrypted bits of secret message. The process of proposed embedding algorithm is shown in Fig. 2. The steps of the algorithm are as follows:

A.1: First we obtained the edge image by ACO algorithm.
A.2: Secret message is appended by its hash value. Then it is encrypted.
A.3: Encrypted message bits are hidden in edge pixel of image. The RSP algorithm governs the selection of random pixels.
A.4: Finally, we get the stego image.

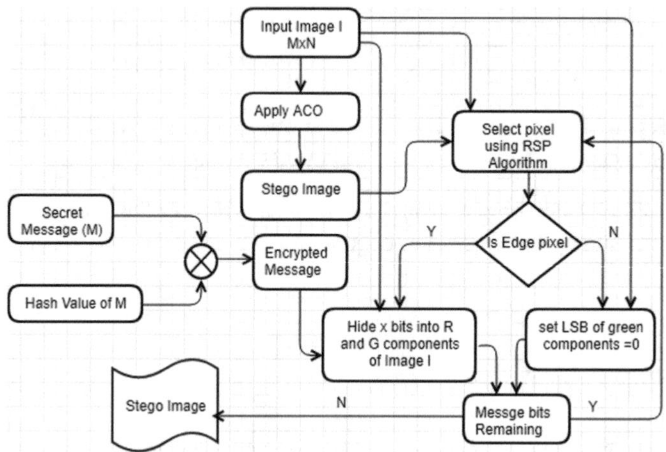

Fig. 2 Diagram of propose embedding algorithm

The random pixel $P_{i,j}$ is chosen to dispersed the data. RSP gives the pixel number randomly based on the secret key. The two bits of LSB of green and red components of selected pixel of source image are changed by the bits of encrypted message, whereas blue components indicate the whether pixel is edge or not. If selected pixel by RSP algorithm is edge pixel, then LSB value of blue component of that selected pixel is set as 1. The entire process occurs again in the same way until all the bits of secret encrypted message are obtained. The process of embedding of secret encrypted message is described in following Fig. 2.

3 Experimental Results and Analysis

In this paper, work we have used the 256 × 256 grayscale images of "Lena" and "Camera" to evaluate the performance of proposed algorithm as shown in Fig. 3. We have taken chaotic 1D logistic map initial parameter $\alpha = 3.89$ and key

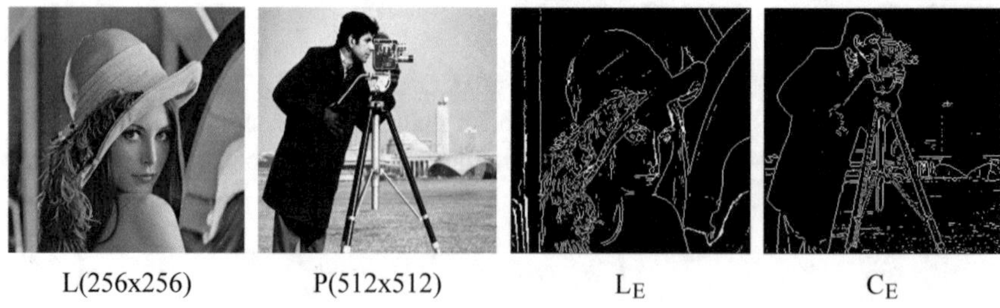

Fig. 3 L: lena, P: pepper, images and L_E and C_E corresponding edge images

Table 1 Result of lena image L (size 256 × 256) M: secret bits, $L_{i=1,2,3,4}$ = Stego image containing different secret bits, d_L = difference of edge pixel ($L–L_E$)

Image L	Total pixels	Edge pixels	Edge pixels (%)	M (bits)	Edge pixels of stego image	d_L	Diff. (%)	MSE	PSNR
L_2	65536	5032	7.67	2048	5027	5	0.0994	0.0280	62.4505
L_3	65536	5032	7.67	4096	5025	7	0.1393	0.0554	60.1078
L_4	65536	5032	7.67	8192	5021	11	0.2190	0.1129	57.6597
L_5	65536	5032	7.67	16384	5019	13	0.2590	0.2253	54.7464

Table 2 Result of camera image C (size 256 × 256), $C_{i=1,2,3,4}$ = Stego image, and d_C = difference in edge pixel ($C–C_{Ei}$)

Image C	Total pixels	Edge pixels	Edge pixels (%)	M (bits)	Edge pixels of stego image	d_C	Diff. (%)	MSE	PSNR
C_1	65536	4171	6.36	2048	4164	7	0.1678	0.0343	62.7815
C_2	65536	4171	6.36	4096	4162	9	0.215	0.0648	60.0127
C_3	65536	4171	6.36	8192	4160	11	0.263	0.1296	57.0053
C_4	65536	4171	6.36	16384	4159	12	0.2877	0.2564	54.0414

X_n = 0.67635635. We have hidden the variable size of bits of secret message in images. We have shown the results in tabular form Tables 1 and 2. In addition, we used mean square error (MSE) and peak to signal noise ratio (PSNR) values to measure the quality of the stego images. Tables 1 and 2 show the result of the experiments.

After embedding the secret bits of data, the stego image edge pixel leads to change as compare to the original image edges pixel [5, 16]. The difference of edge pixel of original image and stego image is d as shown in Figs. 4 and 5. The difference is 5, 7, and 11; as a result, it is very difficult to extract the encrypted data from the stego image. To overcome this problem, we have used blue components as

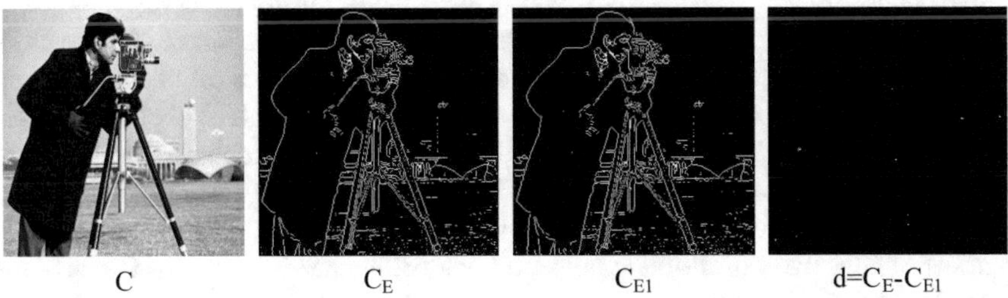

Fig. 4 C original image, C_E edge pixel image of C, C_{E1} image of stego image and d contain pixel which are not in original image

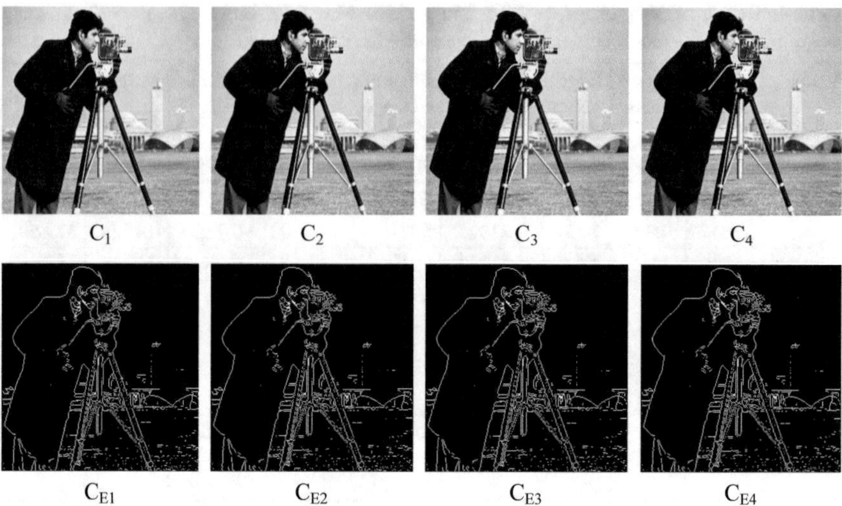

Fig. 5 C_1 original image, C_2 and C_3 are stego image with different size of message bits, C_{E1}, C_{E2}, and C_{E3} are the edge images, respectively

indicator which store a 1 bit for edge pixel and 0 bit if non-edge pixel in LSB. The PSNR value is greater than 54 dB which prove that the performance of algorithm based on chaotic steganography using ACO.

4 Conclusion

In this paper, we presented a novel chaotic edge-based steganography method using ACO technique has been proposed. In our experiment, we measured the value of PSNR and MSE on variable size of secret bits hidden in source image. The performance of proposed method is measured by the PSNR and MSE. For better result, PSNR standard value should be greater than 40 dB. Our work results indicate PSNR values greater than 54 dB which is acceptable. For more security, we have added hash value of message bits and encrypted message in the image. A secret key is used to disperse the bits of encrypted message to the edge pixel of input image. The extraction of secret bits from stego image is not possible without the information of secret key. At the receiver end, message is extracted and hash value is compared. If hash matches, then message is not altered by any unauthorized person. The result from the work shows that the total edge pixel of source image and total number of edge pixel of resultant or stego image difference is very less. Our scheme provides authenticity of received message by checking the hash value of secret message. Further, it becomes very burdensome for the HVS to concede the changes made in pixel of the input image.

References

1. Chen, W.J., Chang, C.C., Le, T.H.N.: High payload steganography mechanism using hybrid edge detector. Expert Syst. Appl. **37**(4), 3292–3301 (2010)
2. Gu, J., Pan, Y., Wang, H.: Research on the improvement of image edge detection algorithm based on artificial neural network. Optik **126**, 2974–2978 (2015)
3. Alam, S., Kumar, V., Siddiqui, W.A., Ahmad, M.: Key-Dependent image steganography using edge detection. In: International Conference on Advanced Computing and Communication Technologies, pp. 85–88 (2014)
4. Verma, O.P., Sharma, R.: An optimal edge detection using universal law of gravity and ant colony algorithm. Inf. Commun. Technol (WICT) (2011)
5. Islam, S., Gupta, P.: Revisiting least two significant bits steganography. In: Paper presented a the 8th International Conference on Intelligent Information Processing (ICIIP), Seoul, Republic of Korea, pp. 90–93 (2013)
6. Modi, M.R., Islam, S., Gupta, P.: Huang, D.-S. Bevilacqua, V., Figueroa, J.C., Premaratne, P.: Edge Based Steganography on Colored Images. Lecture Notes in Computer Science: 9th International Conference on Intelligent Computing (ICIC), vol. 7995, pp. 593–600. Springer, Berlin (2013)
7. Canny, J.: A computational approach to edge detection. IEEE Trans. Pattern Anal. Mach. Intell. PAMI **8**(6), 679–698 (1986)
8. Liu, X., Fang, S.: A convenient and robust edge detection method based on ant colony optimization
9. Ahmad, S.A,T., Doja, M.N.: A novel edge based chaotic steganography method using neural network. In: Proceedings of the 5th International Conference on Frontiers in Intelligent Computing: Theory and Applications, pp. 467–475
10. Tian, J., Yu, W., Xie, S.: An ant colony optimization algorithm for image edge detection. In: Evolutionary Computation, CEC 2008. (IEEE World Congress on Computational Intelligence) (2008). https://doi.org/10.1109/cec.2008.4630880
11. Alam, S., et al.: Analysis of modified LSB approaches of hiding information in digital images. In: International Conference on Computational Intelligence and Communication Systems, pp. 280–285. IEEE (2013). ISBN: 978-1-4799-7633-1/14/2014 (HIS)
12. Alam, S., Zakariya, S.M., Akhtar, N.: Analysis of modified Triple-A steganography technique using Fisher Yates algorithm. In: International Conference on Hybrid Intelligent Systems. ISBN: 978-1-4799-7633-1/14/2014
13. Rafsanjani, M.K., et al.: Edge detection in digital images using Ant Colony Optimization. Comput. Sci. J. Moldova, **23**(3), p. 69 (2015)
14. May, R.M.: Simple mathematical model with very complicated dynamics. Nature **261**, 459–467 (1967)
15. Alam, S., et al.: Digital Image Authentication and Encryption using Digital Signature. In: International Conference on Advances in Computer Engineering and Applications (ICACEA), pp. 332–336 (2015). Print ISBN: 978-1-4673-6911-4/15
16. Islam, S., Modi, M.R. Gupta, P.: Edge-based image steganography. EURASIP J. Inf. Secur. (2014)

Gene Selection for Diagnosis of Cancer in Microarray Data Using Memetic Algorithm

Shemim Begum, Souravi Chakraborty, Abakash Banerjee, Soumen Das, Ram Sarkar and Debasis Chakraborty

Abstract Selecting a small subset of genes that helps to build a good classification model for prediction of disease on the microarray data is a very demanding optimization problem. Genetic algorithm (GA) is a population-based optimization algorithm, which has a lot of applications in the field of molecular biology. But the premature convergence is one of the limitations of GA. Memetic algorithm (MA), an extension of GA, diminishes the possibility of such premature convergence. Microarray technology enables to measure the expression level of thousands of genes to recognize the changes in expression level among different biological states. In this paper, superiority of MA is established over GA, simulated annealing (SA), and tabu search (TS), while selecting the genes in microarray data. Experiments on three well-known data sets, namely DLBCL, leukemia, and prostate cancer, exhibit that MA yields more promising results than classical GA, SA, and TS.

Keywords Memetic algorithm · Symmetrical uncertainty · Microarray data Gene selection

S. Begum · S. Chakraborty · A. Banerjee · S. Das
Government College of Engineering & Textile Technology, Berhampore, Murshidabad, India
e-mail: shemim_begum@yahoo.com

S. Chakraborty
e-mail: souravichakraborty11@gmail.com

A. Banerjee
e-mail: abakash481@gmail.com

S. Das
e-mail: das.suvo1234@gmail.com

R. Sarkar (✉)
Jadavpur University, Kolkata, India
e-mail: ramjucse@gmail.com

D. Chakraborty
Murshidabad College of Engineering and Technology, Berhampore, Murshidabad, India
e-mail: debasismcet@yahoo.in

V. Bhateja et al. (eds.), *Intelligent Engineering Informatics*, Advances in Intelligent Systems and Computing 695, https://doi.org/10.1007/978-981-10-7566-7_43

1 Introduction

The project related to human genome generates huge amount of data. As a consequence, the job to determine the functional relationship among genes seems to be one of the major challenges to the scientist. To alleviate the difficulties, DNA microarray technologies [1] have a great achievement. Microarray data is characterized with small samples but thousands of genes. A noteworthy discovery has been done, and new informative genes have been diagnosed with microarray data. But this data contains non-relational and redundant data, which influence mostly the accuracy of the recognizer. Also microarray data suffers from the curse of dimensionality problem [2]. Moreover, microarray data need a good data mining technique to derive information from the same. Hence, researchers have shown interest to solve this problem by some traditional or non-traditional optimization techniques. The traditional optimization method provides assurance of optimal solution. But the traditional techniques like simplex method, brute force method are very unskillful for solving such problem due to its restrictive complexity (memory and time complexity). Non-traditional optimization method involves in finding an adequate solution to the problem. Examples of such methods are GA, MA, SA, and TS that yield desired outcome in solving the problems. These non-traditional optimization techniques exhibit its acceptability in the field of microarray technology as reported in the literature [3, 4]. GA is highly preferable in producing high quality solution even on complicated problem [5]. But the main drawback of GA is its inherent nature, which causes a large time to come across a convergence, even sometimes fails to determine the optimum solution with sufficient precision. On the other hand, instead of genetic parameters other issues such as selection of genetic encoding, selection of fitness function provide a great aspect in getting a good efficiency of the system. But MA, which is an extension of GA, is proficient to improve the result generated by GA more precisely and efficiently [6, 7]. In addition, MA has a great impact in the field of multiobjective optimization problem. The main cause for the enhancement of MA is its "global convexity," which can be exploited by meta-heuristic technique. One such implementation of MA technology in cancer classification is reported in [8, 9]. The need of selecting optimal genes in microarray data has motivated us to conduct the experiments on three data sets, viz. DLBCL, leukemia, and prostate cancer. Here, MA is used as feature selection algorithm and support vector machine (SVM) is used as a classifier to measure the fitness of the chromosomes. In [10], back-propagation artificial neural network (BP-ANN) has been proposed as a classification tool to distinguish mammograms into normal or abnormal. In our literature, we have compared our results with GA, SA, and TS, and from the result it is evident that MA outperforms the others.

2 Methodology

MA starts with an initial population (randomly generated), P, composed of subsets of features. After that each selective pair of chromosome meet with crossover and each chromosome goes through mutation [11] operations successively. The above procedure continues to a number of iterations (generations). The remarkable feature of MA is its local search, which is used to optimize the individual solution. After certain number of generations, it is supposed to produce the optimal solution.

2.1 *Memetic Algorithm*

An initial population of gene subset is generated randomly. Each chromosome is represented by a binary string, a bit of "1" ("0") represents the specific gene is selected (excluded). The fitness of each chromosome is measured by an objective function

$$Fitness(cr) = J(Gcr) \tag{1}$$

where Gcr denotes the gene subset, which is encoded in a chromosome cr, and $J(Gcr)$ calculates the classification accuracy for the subset.

	Algorithm
1	Input an initial population of feature subset, encoded with binary string, which is generated randomly.
2	Repeat step 2 to step 6 until convergence or maximum number of iteration is reached.
3	Compute fitness of each population using Eq. (1)
4	Each chromosome ***cr*** undergoes a local search.
5	Chromosome ***cr*** is replaced with the improved Chromosome ***cr''***, which is having higher fitness value otherwise it remains same.
6	Perform crossover on a pair of selected chromosomes and mutation on each of the selected chromosomes.
7	Output is the chromosome with good fitness value and with minimum number of features.

Local search: It deals with two operators, namely *addition* and *deletion*. Let A and B are the two subsets of included and eliminated genes encoded in any *cr*.

During the *addition* operation, genes from subset B are inserted into the subset A. While *deletion* operator takes of gene from subset A and inserted into subset B. The genes, to be added/deleted to/from a specific chromosome, expect to produce efficient chromosome.

Addition: All the genes in the subset B are ranked in descending order with C-correlation [12] measure in *B*. Select the gene *Bi* in *B* with the least rank and move it into *A*.

i.e. $A = AUBi$, where i is a randomly generated number, $i = 1 \ldots p$, where p is the size of subset B.

Deletion: All the genes in the subset *A* are ranked in ascending order with C-correlation measure in *A*. Select the gene *Aj* in B with the largest rank and move it into *B*.

$B = BUAj$, j is a randomly generated number, $i = 1 \ldots\ldots.q$, where q is the size of the subset *A*.

The computational cost of the local search can be computed as $p \times q$, where p and q are the number of addition and number of deletion operations on the two subsets *A* and *B*, respectively. The fitness function of each chromosome, *cr*, in the population is evaluated by the classifier SVM.

Crossover: It is a genetic operation which allows for the creation of child chromosomes so as to pass information combinations from parent to child. There are numerous available methods like one-point crossover, two-point crossover, simplex crossover, uniform crossover. We perform one-point crossover [13] in our method due to its simplicity and non-requirement of a probability for the crossover whose value in selection process is important.

Mutation: It allows to us explore the unexplored parts of the feature space. In the mutation operation, we have selected the probability of mutation as a random variable so as to allow for variable degrees of mutation at various levels.

3 Results and Discussion

3.1 Data Sets

In this paper, we have done the performance analysis of MA along with GA, SA, and TS over three popular gene expression data sets, viz. prostate cancer, leukemia, and DLBCL. Details of the data sets are available in [14].

Prostate cancer: It contains 102 samples with 12533 numbers of genes. Here, 52 samples belong to prostate tumor subtype and rest 50 belong to non-tumor prostate subtype.

Leukemia: Acute lymphoblastic lymphoma (ALL) and acute myeloid lymphoma (AML) are two subtypes of leukemia. This data set contains 72 samples along with 5147 genes. Among these 47 samples belong to ALL and 25 samples belong to AML.

DLBCL: Diffuse large B-cell lymphomas (DLBCL) and follicular lymphomas (FL) are two B-cell malignancies in DLBCL data set. These two malignancies have different medical appearances and therefore different therapies would be provided during the treatment. DLBCL data set contains 77 samples blended with 7070 genes. There are 58 samples from DLBCL and 19 samples from FL subtypes.

3.2 Evaluation Measures

We have conducted an experiment taking all the features from three data sets and passed it through five classifiers, namely naïve Bayes, random forest, decision tree, k-nearest neighbor, and SVM (with RBF kernel). We have reported the results obtained using different classifiers using tenfold crossvalidation scheme in Table 1. From the table, it can be shown that SVM produces best result for DLBCL and prostate cancer data sets, whereas naïve Bayes outperforms the other in case of leukemia data set. It is to be noted that SVM deals with regularization parameter, which helps in avoiding overfitting problem and it uses the kernel function, which helps to build an expert knowledge regarding the problem. SVM is entitled with four kernel functions, namely RBF, linear, polynomial, and sigmoid. SVM with RBF kernel provides a superior classification accuracy in gene expressed-based data set [15]. Keeping these facts in mind, we have conducted the experiment using SVM on said data sets.

By applying symmetrical uncertainty (SU) [16], a filter feature selection method, we have identified the top 200 genes from each of the data sets. Then 100 chromosomes are generated randomly with each chromosome is of size 200. A random number (n) is generated in the range [1, 200]. After that top n number genes are selected from the subset of genes that are ranked with C-correlation measures. Then a population with randomly selected gene subset is generated. The fitness of each chromosome is measured by the classifier SVM.

Now each chromosome undergoes a local search, which attempts to find out the local optimum in an organized way. The local search is a combination of addition and deletion operations. Addition and deletion operations are continued up to m

Table 1 Performances (% of accuracy) of five classifiers on prostate cancer, leukemia, and DLBCL data sets considering all the features (genes) of the data sets

Data set	Naïve Bayes	Random forest	Decision forest	SVM	K-NN
Prostate	62	87.25	85.29	**88.23**	49.01
Leukemia	**92**	91.66	79.16	73.61	65.27
DLBCL	80.51	81.81	72.72	**85.71**	75.32

times, where the range of m is restricted upon the size of the gene subsets A and B individually. After local search, the fitness of each chromosome is measured using the SVM and the task of pair mating is performed. In this case, consecutive two chromosomes are considered as parents and each pair meet up with crossover and each chromosome goes through with mutation procedure, respectively. The crossover and mutation processes certify that MA can investigate new gene subsets that are not there in the previous population yet. The above procedure continues for the n times or until the satisfactory result is achieved. In this literature, efficiency of MA is compared with GA, SA, and TS in terms of accuracies and number of features. The chromosomes in the tenth generation with top five fitness values with respect to number of features are depicted in Tables 2, 3, 4, 5, 6, and 7 respectively. Moreover, the best results in terms of fitness value of chromosome along with the

Table 2 Overall accuracies and number of features at tenth generation of the MA and GA on prostate cancer data set

Accuracy in MA (%)	Number of genes	Accuracy in GA (%)	Number of genes
96.34	18	97.56	189
96.34	18	97.56	19
96.34	5	96.34	132
97.56	**7**	96.34	178
97.56	7	96.34	106

Table 3 Overall accuracies and number of features at tenth generation of the MA and GA on leukemia data set

Accuracy in MA (%)	Number of genes	Accuracy in GA (%)	Number of genes
100	17	100	63
100	35	100	54
100	9	100	59
100	25	100	151
100	**5**	100	175

Table 4 Overall accuracies and number of features at tenth generation of the MA and GA on DLBCL data set

Accuracy in MA (%)	Number of genes	Accuracy in GA (%)	Number of genes
100	31	100	120
100	46	100	150
100	52	100	83
100	**4**	100	115
100	24	100	175

number of features obtained at iterations (generations) second, fourth, sixth, eighth, and tenth are depicted in Fig. 1(a–c), respectively.

We have compared the performances of MA over three data sets, viz. Prostate, Leukemia, and DLBCL. It can be observed from these tables that results of MA over three data sets outperform the results of GA, SA, and TS. The accuracy of the chromosome is considered as the prevailing objective, while number of genes are considered as the secondary objective. From the results shown on Tables 2, 3, 4, 5, 6, and 7, it can be concluded that MA outperforms other feature selection algorithm such as GA, SA, and TS.

Table 5 Overall accuracies and number of features at tenth generation of the SA and TS on prostate cancer data set

Accuracy in SA (%)	Number of genes	Accuracy in TS (%)	Number of genes
92	70	90	37
94	38	86	46
86	54	77	69
85	87	85	50
90	73	94	59

Table 6 Overall accuracies and number of features at tenth generation of the SA and TS on leukemia data set

Accuracy in SA (%)	Number of genes	Accuracy in TS (%)	Number of genes
87.5	70	90	83
92	61	86	44
89	62	77	160
94	54	85	66
92	73	94	59

Table 7 Overall accuracies and number of features at tenth generation of the SA and TS on DLBCL data set

Accuracy in SA (%)	Number of genes	Accuracy in TS (%)	Number of genes
91	35	93	80
92	51	86	34
89	32	91	70
93	24	95	66
90.5	43	89	39

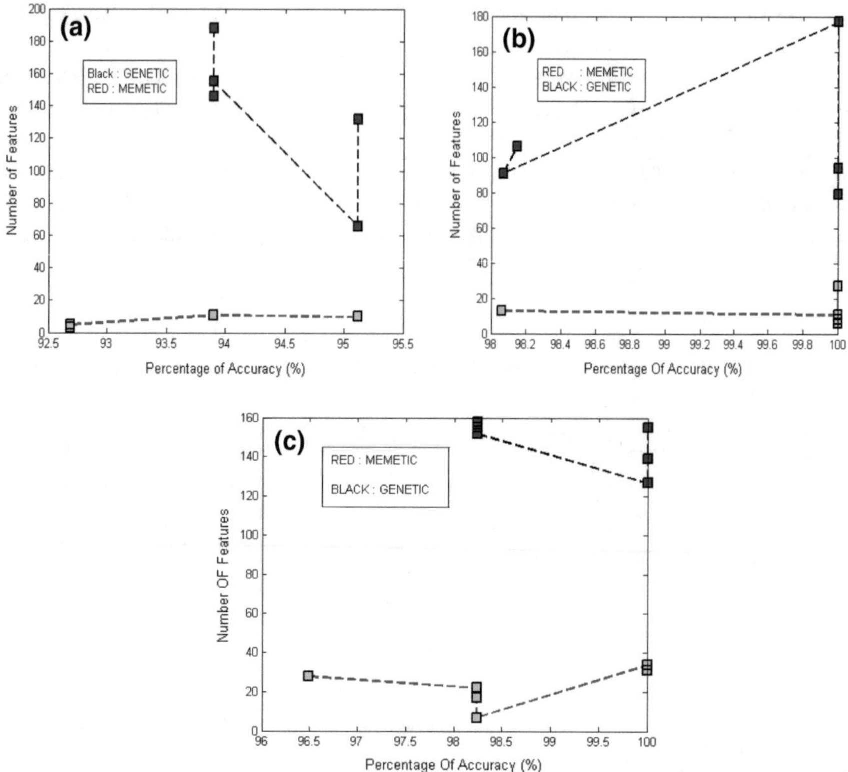

Fig. 1 **a–c** Accuracy versus number of features for prostate cancer, leukemia, and DLBCL data set in second, fourth, sixth, eighth, and tenth generations of MA, respectively

4 Conclusion

In this paper, we have applied MA for gene selection for the classification of microarray data. Performance of MA is assessed on three microarray data sets. It can be said that removing local search from MA ensures its weakness in efficiency. This explores the significance of memetic approach, which blends local search with GA. From the experiments, it can be concluded that MA has been converged to satisfactory result with minimum number of features. In clinical diagnosis, smallest number of genes with good predictive accuracy is highly preferable. As a further scope, we can try to get variation of the proposed method by applying different local search methods. Also the accuracy of this method can be compared with other statistical models to propose a competent model with proficient accuracy.

References

1. Ruskin, J.: Computational modeling and analysis of microarray data. Microarrays (Basel) **5** (4), 26 (2016)
2. Keogh, E., Mueen, A: Curse of dimensionality. Encycl. Mach. Learn., 257–258 (2010)
3. Garg, P., et al.: A comparison between memetic algorithm and genetic algorithm for the cryptanalysis of simplified data encryption standard algorithm. In: Inter. J. Net. Sec. & Its App. (IJNSA) **1**(1), (2009)
4. Duval, B., Hao, J.-K., Hernandez Hernandez, J.C.: A memetic algorithm for gene selection and molecular classification of cancer. In: GECCO'09, July 8–12, Montréal Québec, Canada (2009)
5. Dash, R., Misra, B.: Gene selection and classification of microarray data: a Pareto DE approach. Intell. Decis. Technol. **11**(1), 1–15 (2016)
6. Ayadi, W., Hao, J.K.: A memetic algorithm for discovering negative correlation biclusters of DNA microarray data. Neurocomputing **145**, 14–22 (2014)
7. Cotta, C., Moscato, P., Garcia, V., Frana, P., Mendes, A.: Gene ordering in microarray data using parallel memetic algorithms. In: 2012 41st International Conference on Parallel Processing Workshops Oslo, Norway, June (2005)
8. Duval, B., Hao, J.K.: Advances in metaheuristics for gene selection and classification of microarray data. Brief. Bioinform. **11**(1), 127–141 (2010)
9. Sekhara Rao, A.C., Dara, S., Haider, B.: Cancer microarray data feature selection using multi-objective binary particle swarm optimization algorithm. EXCLI J., 460–473 (2016)
10. Gautam, A., et al.: An improved mammogram classification approach using back propagation neural network. In: Proceedings of the 3rd International Conference on Computer and Communication Technology (IC3T-2016). Springer (2016)
11. Luke, S., Spector, L.: A comparison of crossover and mutation in genetic programming. In: Koza, J., et al. (eds.) Proceedings of the Second Annual Conference on Genetic Programming (GP-97). Morgan Kaufmann (1997)
12. Huang, J., Huang, N.: A method for feature selection based on the correlation. In: International Conference on Measurement, Information and Control (MIC) (2012)
13. Umbarkar, A.J., Sheth, P.D.: Crossover operator in genetic algorithm: a review. ICTACT J. Soft Comput. **6**(01) (2015)
14. http://www.biolab.si/sup/bi-cancer/projections/
15. Byun, H., Lee, S.W.: Applications of support vector machines for pattern recognition: a survey. In: SVM 2002. LNCS, vol. 2388, pp. 213–236. Springer, Berlin (2002)
16. Sarhrouni, E., Hammouch, A., Aboutajdine, D.: Application of symmetric uncertainty and mutual information to dimensionality reduction and classification of hyperspectral images. Int. J. Eng. Technol. (IJET) **4**(5) (2012)

Handwritten Bangla City Name Recognition Using Shape-Context Feature

Samanway Sahoo, Subham Kumar Nandi, Sourav Barua, Pallavi, Samir Malakar and Ram Sarkar

Abstract A segmentation-free approach is proposed to recognize the handwritten city names written in Bangla script. Initially, all the word images are converted into virtually single connected component following the refraction properties of light in order to design a unique shape-context of the same. Then a 64-dimensional feature vector is estimated from the said shape-context of each word image. A database containing 150 samples of 50 most popular city names of West Bengal, a state of India is prepared for evaluating the present method. Performance of this feature vector is also compared with some recently published feature vectors, and it is observed that the newly designed feature vector has outperformed the others.

Keywords Shape-context feature • Handwritten word recognition
City name recognition • Bangla script

S. Sahoo (✉) · S. K. Nandi · S. Barua · Pallavi
Department of Computer Science and Technology, Indian Institute of Engineering Science and Technology, Shibpur, Shibpur, India
e-mail: samanwaysahoo2@gmail.com

S. K. Nandi
e-mail: subhamadmnclg@gmail.com

S. Barua
e-mail: souravbarua934@gmail.com

Pallavi
e-mail: pallavi071996@gmail.com

S. Malakar
Department of Computer Science, Asutosh College, Kolkata, India
e-mail: malakarsamir@gmail.com

R. Sarkar
Department of Computer Science and Engineering, Jadavpur University, Kolkata, India
e-mail: raamsarkar@gmail.com

V. Bhateja et al. (eds.), *Intelligent Engineering Informatics*, Advances in Intelligent Systems and Computing 695, https://doi.org/10.1007/978-981-10-7566-7_44

1 Introduction

This technologically advanced era motivates the present research to automate every sphere of the human life by reducing manual labor needed in any system. One such example is the digitization and interpretation of the document images, either printed or handwritten, through computerized system. Relevance of such systems can be found in many real-life applications like industrial robots, automated teller machine, automated highway system, and many more. But the system which relies on handwriting recognition, for example, automation of postal system, bank cheque/ money transfer receipts, forensic verification suffers from lack of improvement in comparison with others. Possible reasons for this are multifold such as layout analysis, text/non-text separation, character, or word recognition. Along with these, when the research considers the handwritten text images, some add-on complexities like skew, slant, shape, and size variations of the characters/word are also found.

Irrespective of script, considerable progress in research has been found in the literature for handwritten character or digit recognition than handwritten word recognition (HWR). A similar statement can also be drawn when comparison is made for the research advancement in printed text images and its handwritten counterpart. Complexities in the handwriting, said earlier, are reasons for the slow progress of the research in this domain. But all these facts make the HWR a challenging and interesting research problem.

Moreover, due to the exponential expansion of the Internet usage over the past decade, document image analysis research on different local/regional language has observed a significant growth. Bangla, an Indo-Aryan script, is the sixth most popular language in terms of number of native users (around 205 million). It is also national language of Bangladesh. This script contains 11 vowels, 39 consonants [1]. In addition to this, presence of compound characters, a complex shape formed by taking vowel and consonant character together, and modified shape makes this script more complex. Apart from Bangla language, presently this script is also used to write languages like Assamese, Manipuri, and Sylheti with slight modification in some cases.

Two different approaches of HWR, in general, are found in the literature, namely analytic [1] and holistic [2]. In analytic approach, the words are segmented into characters and/or character sub-parts, and then each of the components is separately recognized in order to obtain machine editable form of the whole words. This method is ideal for application when the large number of vocabularies is to be considered to develop a recognition system. But unfortunately this approach mostly suffers from segmentation ambiguities [3] due to huge variation in shape and size of the basic units within different words which, in turn, reduces its practical usage significantly.

Therefore, researchers have looked for an alternative solution for this which is called holistic approach. This approach mainly works better on limited and pre-defined word vocabulary. In holistic approach, machine learning algorithms are used to recognize a word at one go instead of segmenting the word into the characters present therein. Keeping the said facts in mind, in this work, an attempt has been made to develop a HWR system for recognizing 50 major city names of West Bengal, a state of India.

2 Related Work

A number of research attempts found in literature that have dealt with Bangla HWR in holistic approach [2, 4–6]. The work reported in [2] has extracted convex hull and concentric rectangle-based features for classifying a 54 common word. In [4, 5], the authors have used basic HOG-based feature descriptor and elliptical feature respectively, to perform handwritten city name recognition considering 20 different city names belong to West Bengal, India. A non-symmetric half plane-hidden Markov model (NSHP-HMM) has been introduced by the authors in [6] for recognizing Bangla city names collected from postal documents.

Among the other Indic scripts, authors in [7, 8] have described methods for recognizing words written in Devanagari script. A set of topological features has been extracted from handwritten words of all capital names of Indian states and territories [7]. Authors in [8] have used curvelet transform to do the same job. Other than the Indic scripts, for recognition of Arabic words, a holistic approach using classifier ensemble mechanism is found in [9]. In this work, statistical and contour-based features (SCF) and moment-based features are used. In [10], authors have employed holistic approach for recognizing words from historical document. Longest run features are computed from hypothetically segmented sub-images in [11]. It uses the most frequent words from a standard database called CMATERdb 1.2.1 [12].

3 Present Work

An HWR system using a holistic approach is introduced here for recognizing 50 most popular city names [13] of West Bengal, written in Bangla script. The word patterns are classified in supervised way. The detail description of different stages of this technique is described hereafter.

3.1 Database Preparation

Database is primary pre-requisite for any pattern recognition research. Therefore, in the present work, a handwritten Bangla city name database comprising 150 samples for each of the 50 different city names is prepared. Handwritten samples are collected on request from different individuals varying in age, sex, and profession to incorporate the variations in writing styles. These sample words are collected in the pre-formatted A4 sheets using black/blue ink pen. Then, these filled-up datasheets are scanned using flat-bedded scanner with resolution of 300 dpi and stored in 24-bit bitmap file format.

Word images are cropped out programmatically from theses scanned datasheets with minimal rectangular bounding box enclosing all the data pixels therein and stored as grayscale image. These grayscale word images are then preprocessed using the algorithm described in the work [14]. Any of these preprocessed word images contain only data and non-data pixels that are represented hereafter by '1' and '0' respectively, and is formally defined as $I = \{f(x, y): (x, y) \in [1, H] \times [1, W] \wedge f(x, y) \in \{'0', '1'\}\}$, where H and W are height and width of a word image respectively.

3.2 Shape Formation

In this stage, a word image I is converted into *virtually Connected Component* (VCC) by using refraction property of light [15] for accurate estimation of the shape topology of word images. In this regard, the following assumptions are made:

(i) An Image I is composed of two mediums that are separated along line segment $y = d\,(= \frac{H}{2})$. The refractive indices of upper and lower mediums are δ_U and δ_L, respectively.
(ii) A light ray (hypothetical) is incident at an angle θ on top boundary of I. The light ray traverses a path within the image. Some paths traversed by light rays incident at angle θ are shown in Fig. 1b considering $\delta_U = 1$ and $\delta_L = -1$.

Now, it is assumed that successive light rays incident from left to right direction of the image at the same angle θ that are one-pixel distance apart. A maximum of n, where $n = \lfloor W - d^* \tan\theta \rfloor$, number of such light rays can be placed there and therefore n such paths can be found. On each of these paths all the non-data pixels lie between first and last data pixels are labeled as '#'. Therefore, a newly labeled image (say I_L) is generated which can formally be defined as

$$I_L = \left\{f(x, y): (x, y) \in [1, H] \times [1, W] \wedge f(x, y) \in \{'0', '1', '\#'\}\right\}$$

However, these light rays fail to cover the some region of I (shaded in Fig. 1c), where few portions of I may lie. Therefore, another set of light rays with incident angle θ are made to fall along left boundary of I (see Fig. 1d) to label previously uncover region. Let this newly generated region is defined as below

$$I'_L = \left\{f(x, y): (x, y) \in H \times W \wedge f(x, y) \in \{'0', '1', '\#'\}\right\}.$$

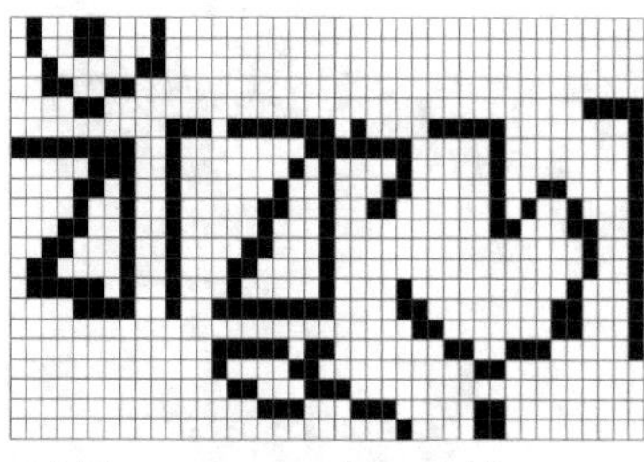

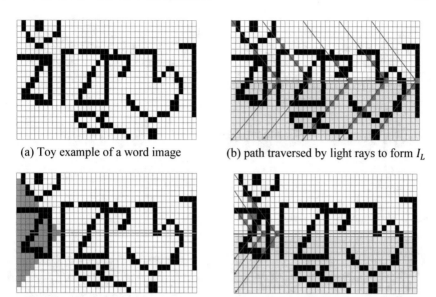

(a) Toy example of a word image

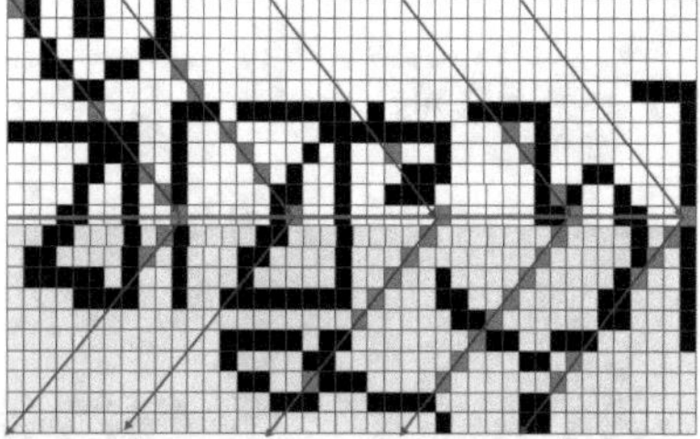

(b) path traversed by light rays to form I_L

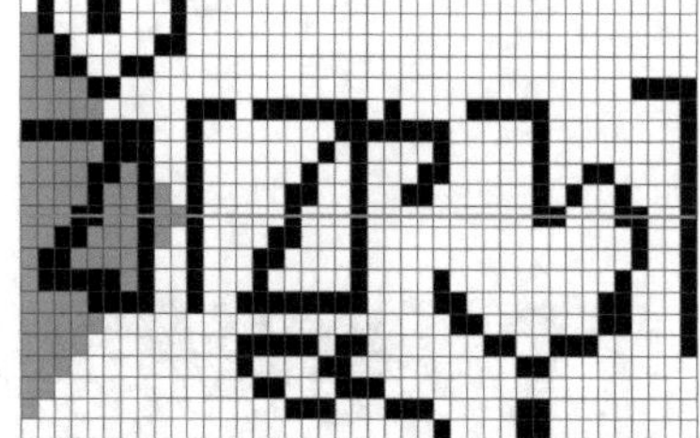

(c) Uncover region by light rays in path I_L

(d) path traversed by light rays to form $I_L{}'$

Fig. 1 Generation of I_θ, where $\theta = \frac{\pi}{4}$, $\delta_U = 1$ and $\delta_L = -1$ (greenish pixels are shaded pixels (i.e., $f(x,y) = '\#'$) and red arrows are the path traversed by the light rays)

Finally, a shape structure (let, I_θ) of I can be formed by taking union of I_L and I'_L. Different values of θ would provide different shape structures keeping δ_U and δ_L fixed. Finally, the shape of a word image (W_S) is obtained by

$$W_S = \bigcap_\theta I_\theta = \left\{ f(x,y) \colon (x,y) \in [1,H] \times [1,W] \wedge f(x,y) \in \{'0', '1', '\#'\} \right\}.$$

This W_S is hereafter considered for calculating feature values. The generation of W_S on real data considering $\theta = \frac{\pi}{4}$ and $-\frac{\pi}{4}$ is illustrated in Fig. 2 (*where* $\delta_U = 1$ and $\delta_L = -1$).

3.3 Feature Extraction

The unique shape structure, described in the previous section, is used here to estimate a 64-dimensional feature vector, from each of the word images. The features are described hereafter.

Histogram of Chain Codes: Histogram of Chain Codes (HCC) is a popular shape coding technique. In this method, first boundary of W_S is extracted considering data and shaded pixels (i.e., $f(x,y) = '\#'$) as content pixels (see Fig. 3). Next, the contour is visited following Freeman eight directional chain code representation

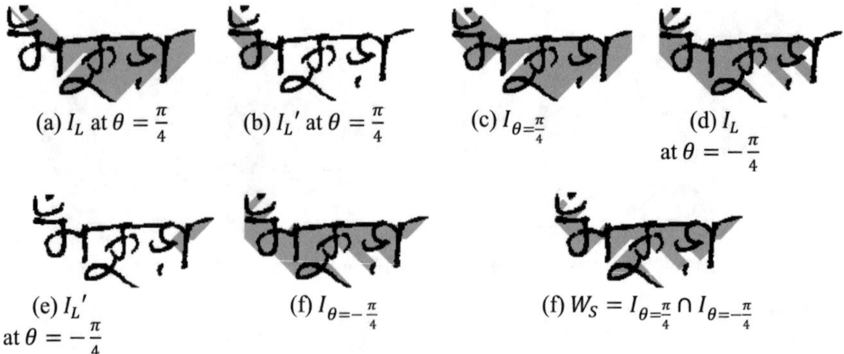

Fig. 2 Generation of W_S, where $\delta_U = 1$ and $\delta_L = -1$

Fig. 3 Illustration of boundary extraction of a word image (Purulia)

technique [16]. Therefore, the contour of the word image is represented by eight directional values, i.e., each of the boundary pixels possessess directional value. Counts of such directional values in each direction are considered here as feature value.

Number of Dominant Shaded Region: Regions formed by shaded pixels have been generated by subtracting the data pixels from W_S (see Fig. 4b). Next, these regions are labeled using connected component labeling algorithm. A region having more shaded pixels than a pre-defined threshold value (here, 200) is considered here as dominant shaded region. Such consideration of threshold value is made to avoid the regions that are falsely generated due to presence of noise components or word boundary pixels (or less informative may be). The count of such dominant shaded regions (generation of dominant component is shown in Fig. 4), varies with word shapes, is considered here as a feature value.

Pixel Ratio: Pixel ratio [7] is a global feature which is, in general, calculated as the ratio of number of data pixels to number of non-data pixels. But it is defined here as ratio of number of data pixels to number of shaded pixels that are contained inside any shape defined with different values of θ and W_S.

Density: Density, extracted locally, is measured as number of data pixels and shaded pixels in unit area. This has been done to consider all the pixels that help in forming the shape structure of a word image. To get local information, W_S is hypothetically segmented in different ways that are described in Fig. 5a–e, and from each segment this feature value is estimated.

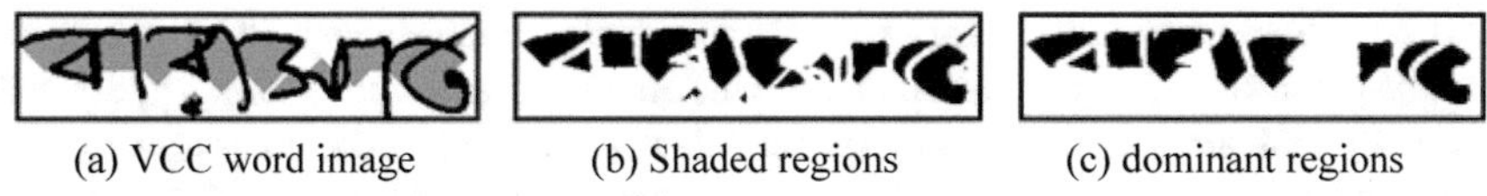

(a) VCC word image (b) Shaded regions (c) dominant regions

Fig. 4 Generation of dominant shaded region

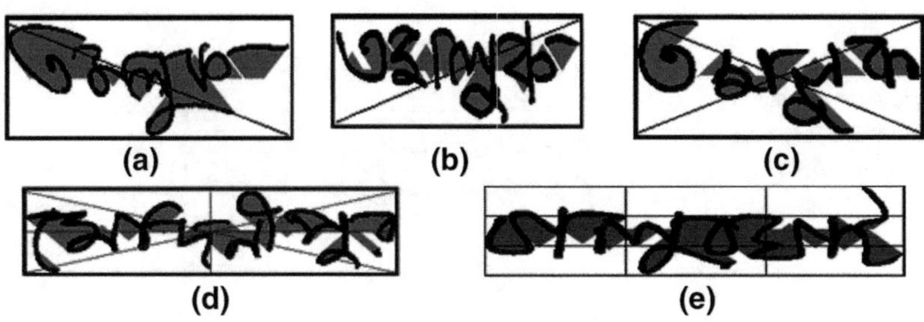

Fig. 5 Different hypothetical word image segment for generating density feature values

Centroid: Centroid feature [1] is the positional information based on center of gravity of image pixels. This feature is mostly extracted from local regions of a pattern to differentiate one pattern from others. It is worth to mention that during calculation of this feature values W_S is divided into nine parts and both the data and shaded pixels are considered there as content pixels.

Distance of Centroid from Origin: Along with the positional information of centroid [1], described above, Euclidian distance of the same from top leftmost point (i.e., origin) of a word image is considered here as a feature value. To calculate this, centroid of entire shape W_S considering shaded and data pixels as content pixels is first calculated and then distance from the origin is calculated.

4 Result and Discussion

As in the present work, word samples for 50 most popular city names of West Bengal are collected from 150 different individuals; hence, number of total word samples becomes 7500 that are used for assessing the performance of the newly designed feature vector. These samples are divided into train and test sets to perform nine set of experiments. A well-known data mining tool, WEKA [17], is used here to perform the recognition. Some popularly used classifiers, *namely* multilayer perceptron (MLP), sequential minimal optimization (SMO), multiclass classifier (MC), and CVParameter embedding with SMO (CVP-SMO) are used for comparing recognition accuracies. The experimental result is provided in Fig. 6. From the experimental result, it is found that SMO and CVP-SMO provide very close recognition results for all the experiments.

Four different feature vectors published recently, namely basic HOG [4], elliptical [5], and SCF [9] (two different versions are considered called here Type 1 and Type 2) that are having feature length 80, 65, 116, and 148, respectively, are extracted from present dataset. Next, same set of experiments as mentioned above are performed using SMO. Comparative results are depicted in Fig. 7a. In addition to this, a voting schema is designed using majority rule by embedding the four

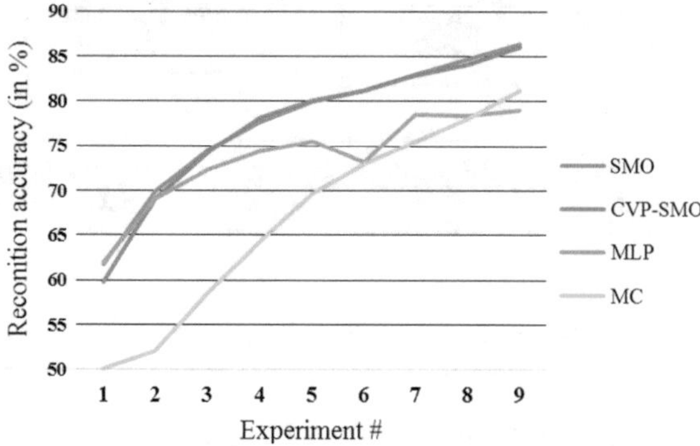

Fig. 6 Recognition accuracies for different experiment using four different classifier

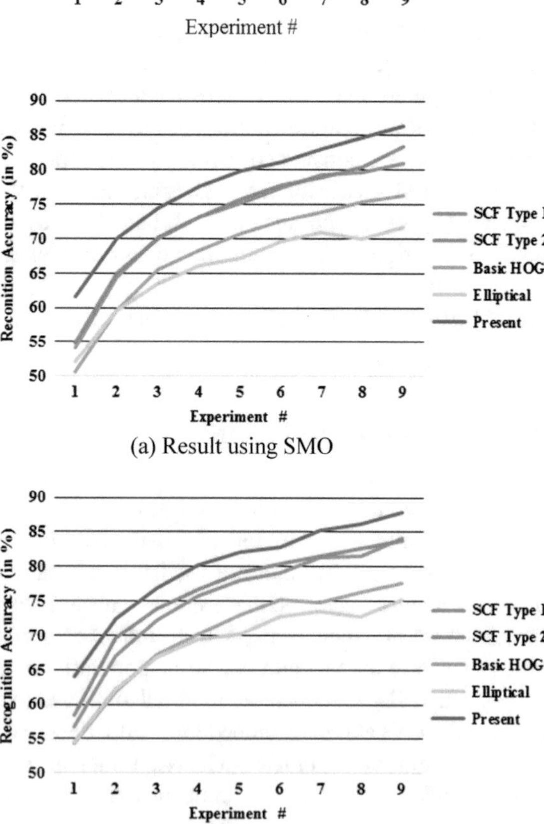

Fig. 7 Comparison of the present feature vector with state-of-the-art feature vectors

classifiers namely, MLP, SMO, MC, and CVP-SMO therein. This schema is applied for all the said features and the results are shown in Fig. 7b. From Fig. 7a, b, it is clear that the present feature vector provides better result than the others.

5 Conclusion

In this work, a holistic HWR approach is proposed to recognize the handwritten city names written in Bangla script. In doing so, a novel shape-context of the word images is designed based on the refraction properties of light. After that, from each of word images, a 64-dimensional feature vector is extracted. The system is evaluated on database comprising 50 popular city names of West Bengal, and the word samples are collected from more than 150 native writers. The experimental outcome shows that the newly designed feature vector provides better recognition accuracy than some of the state-of-the-art feature vectors. In addition to this, the dimension of this proposed feature vector is less than these state-of-the-art feature vectors.

In spite of this success, there are still some rooms for improvement. To be specific, by considering variations in the refraction index, medium position and incident angle, more generic and accurate shape structure of the word images could be formed and that may provide even better performance. Combining topological and texture-based features with the present one can be thought of in future. Last but not the least, adding more sample variations to each word class with increased number of word classes would be a good idea to prove the scalability of the present method.

References

1. Basu, S., Das, N., Sarkar, R., Kundu, M., Nasipuri, M., Basu, D.K.: A hierarchical approach to recognition of handwritten Bangla characters. Pattern Recogn. **42**(7), 1467–1484 (2009)
2. Bhowmik, S., Polley, S., Roushan, M.G., Malakar, S., Sarkar, R., Nasipuri, M.: A holistic word recognition technique for handwritten Bangla words. Int. J. Appl. Pattern Recogn. **2**(2), 142–159 (2015)
3. Sarkar, R., Malakar, S., Das, N., Basu, S., Kundu, M., Nasipuri, M.: Word extraction and character segmentation from text lines of unconstrained handwritten Bangla document images. J. Intell. Syst. **20**(3), 227–260 (2011)
4. Bhowmik, S., Roushan, M.G., Polley, S., Malakar, S., Sarkar, R., Nasipuri, M.: Handwritten Bangla word recognition using HOG descriptor. In: Fourth International Conference of Emerging Applications of Information Technology (EAIT), pp. 193–197. IEEE (2014)
5. Bhowmik, S., Malakar, S., Sarkar, R., Nasipuri, M.: Handwritten Bangla word recognition using elliptical features. In: 2014 International Conference on Computational Intelligence and Communication Networks (CICN), pp. 257–261. IEEE (2014)
6. Vajda, S., Roy, K., Pal, U., Chaudhuri, B.B., Belaid, A.: Automation of Indian postal documents written in Bangla and English. Int. J. Pattern Recogn. Artif. Intell. **23**(08), 1599–1632 (2009)

7. Malakar, S., Sharma, P., Singh, P.K., Das, M., Sarkar, R., Nasipuri, M.: A holistic approach for handwritten Hindi word recognition. Int. J. Comput. Vis. Image Process. **7**(1), 59–78 (2017)
8. Singh, B., Mittal, A., Ansari, M.A., Ghosh, D.: Handwritten Devanagari word recognition: a curvelet transform based approach. Int. J. Comput. Sci. Eng. **3**(4), 1658–1665 (2011)
9. Tamen, Z., Drias, H., Boughaci, D.: An efficient multiple classifier system for Arabic handwritten words recognition. Pattern Recogn. Lett. **93**, 123–132 (2017)
10. Lavrenko, V., Rath, T.M., Manmatha, R.: Holistic word recognition for handwritten historical documents. In: Proceedings of First International Workshop on Document Image Analysis for Libraries, pp. 278–287. IEEE (2004)
11. Acharyya, A., Rakshit, S., Sarkar, R., Basu, S., Nasipuri, M.: Handwritten word recognition using MLP based classifier: a holistic approach. Int. J. Comput. Sci. Issues **10**(2), 422–427 (2013)
12. Sarkar, R., Das, N., Basu, S., Kundu, M., Nasipuri, M., Basu, D.K.: CMATERdb1: a database of unconstrained handwritten Bangla and Bangla-English mixed script document image. Int. J. Doc. Anal. Recogn. (IJDAR) **15**(1), 71–83 (2012)
13. Languages with at least 50 million first-language speakers. https://www.ethnologue.com/statistics/size. Accessed from Summary by language size Ethnologue
14. Das, B., Bhowmik, S., Saha, A., Sarkar, R.: An adaptive foreground-background separation method for effective binarization of document images. In: proceedings on 8th International Conference on Soft Computing and Pattern Recognition (2016)
15. Refraction. https://en.wikipedia.org/wiki/Refraction.
16. Freeman, H.: On the encoding of arbitrary geometric configurations. IRE Trans. Electron. Comput. **10**, 260–268 (1961)
17. Hall, M., Frank, E., Holmes, G., Pfahringer, B., Reutemann, P., Witten, I.H.: The WEKA data mining software: an update. SIGKDD Explor. **11**(1), 10–18 (2009)

Driver Safety Approach Using Efficient Image Processing Algorithms for Driver Distraction Detection and Alerting

Omar Wathiq and Bhavna D. Ambudkar

Abstract Currently, due to different reasons, the road accidents are increasing. Road accidents are prone to number human deaths. There are different reasons which lead to road accidents, but drivers fatigue or distraction is main threat in major accidental cases. Therefore, recently various methods are explained by many authors for untimely identification of driver sleepiness in the manner of prohibiting mischance on road. In this paper, we are presenting the novel approach called hybrid method in which automatic care of driver safety and hospitality management services. Our approach aims at determining first if a driver is distracted or not based yawing, eye position, head position, mouth position etc., second if driver is detected as distracted instance alarming will perform on both driver side and near hospital services in order to be available in case of accident happen. Based on computer vision techniques, we propose four different modules for features extraction, focusing on arm position, face orientation, facial expression and eye behaviour, and then, the outputs of all these phases combined together and feed to the classifier feed-forward neural network (FFNN) for alarming the distraction detection and type of distraction. The outcome of this paper is efficient driver safety approach by considering the RGB-D sensor and image processing algorithms.

Keywords Feature extraction · Driver safety · Driver distraction Fatigue · Face detection · Eyes detection · Yawing · SVM

O. Wathiq (✉) · B. D. Ambudkar
(E&TC), Dr. D.Y. Patil Institute of Engineering & Technology,
Pimpri, Pune 411018, India
e-mail: projectspune.app@gmail.com

B. D. Ambudkar
e-mail: expert.herani@gmail.com

V. Bhateja et al. (eds.), *Intelligent Engineering Informatics*, Advances in Intelligent Systems and Computing 695, https://doi.org/10.1007/978-981-10-7566-7_45

1 Introduction

To minimize the road accidents, nowadays intelligent transportation systems (ITSs) have been designed for human safety. Driver disturbance is a main reason in many accident cases on rural roads. Drowsiness or fatigue breaks the driver concentration while driving which resulted in loss of decision-making functionality for controlling car. From fundamental research and surveys, it is viewed as that for keeps driving case, the driver is exhausted following 2–3 h and henceforth, the execution of guiding is disintegrated. The driver drowsiness is more at midnight, after lunch, afternoon hours as compared to other times in day. Also, the alcohol and drugs are the reasons for loss of driver concentration. All over world, many countries presented their own different statistics over the mischance's which occurred because of driver sleepiness as well as interruption. Normally, around 20–30% accidents happen because of the driver drowsiness and distraction. The fatigue, drowsiness and distraction this term is explained view of physiology as well as psychology. In viewpoint of physiology as well as psychology, session of enervation as well as sleepiness is dissimilar; however, surveys discovered in scope of ITS, enervation as well as sleepiness are compatible sessions.

Another term called Hyper vigilance means that shortage of consciousness as well as have to be involve sleepiness, interruption else each. A selective as well as research-based statement for enervation has not been granted till; hence, there is no some numerical standard for be alive it. Enervation occurs almost in three kinds of sensory enervation, muscle enervation and cognitive enervation. Sensory enervation as well as muscular enervation is just quantifiable as well as there is no some systems or method for compute subjective enervation [1, 2]. The specific statement for enervation is not granted still; however, there is association within enervation, sleepiness as well as body inversion, electrical skin opposition, eye motion, respiratory rate, pulse rate as well as brain actions [3]. But, greatest successful technique else method for compute enervation as well as sleepiness is brain actions convention, but in that resemble, brain signals have to accepted from electrodes which related to driving constraint head which generate it as an obtrusive resemble. Examination of brain actions is a leading significant indication of enervation which is emerged in the eye. Regular in the company of investigations, intermission within optical incentive as well as that's acknowledgement is single between almost all calculate to develop responsiveness. This intermission is comprehended through a PVT. It is also known as psychomotor vigilance task as well as this present's acknowledgement speed of an independent to her optical incentive. Investigates distinguish the it has truly close connection amongst PVT as well as accordingly stage of close eyelids between a calculate of time. Stage of eyelid termination after some time is called as PERCLOS [4]. Eventually, there is nearby connection in the middle of weakness and level of eye conclusion. Driver confront perception frameworks recognize this significance appraise driver weariness or tiredness.

Monotony in a particular task will scale back the centralization of person as well as have reason of interruption. Monotony is due to three important causes

(1) shortage of personal attentiveness, (2) performing a repeated work for well-established and (3) exterior components. Monotony in driving normally occurs through the second as well as third causes. Consistent driving on highways in the company of congestion consists of dismissive influence on driver centralization. In this instance, driver is not exhausted because of the monotony of operating; his or her centralization can step by step be reduced and also, the driver would not have a cautious administration on vehicle. Driver distraction may be caused by rebuke individuals or mobile and paying attention to music [1, 2]. Driver obstruction is measureable through head as well as eyes towards resolution. Main disadvantage for obstruction identification is which if head is onward as well as eyes look in the direction of road, the driver does not necessarily centralize to road. In other expressions, trying in the direction of road is not become perceptive to it [2].

As it is vital to identify driver interruption untimely and sleepiness in order to prevent the accidents, several researches were done on this subject within the past decade. The investigation on techniques for driver obstruction identification is become over but is low established than techniques of driver enervation identification. Enervation as well as obstruction is generally thought as two separate ideas and each of those factors scale back driver alertness, all classes are investigated in studies. This is divided into foremost vital approaches for fatigue and distraction detection in three categories: (1) resembles assist bioelectric signals (e.g. EEG and ECG), (2) resembles depend on guiding movement and (3) resembles depend on driver face analysis. These resembles are normally discovered by utilizing the whole dissimilar point of views cooperatively in the company of capacity of enervation disgust, capacity of obstructive identification, accuracy, facility as well as identification speed. In this paper, we are presenting new algorithms for image processing to efficiently detect the driver unsafe position-based facial movements. In Sect. 2, related works are described. In Sect. 3, proposed methodology and algorithms are presented. In Sect. 4, simulation results are discussed, finally in Sect. 5 conclusion and future work.

2 Related Work

Writer Wang et al. proposed methods. It is Identify which stage of face determined through upright estimate of grey-level picture in [5]. In this paper, the driver situation was supposed non-confused while face has brighter pixels than background. On opposite hand, learning basis on face detection techniques depends on range of training sample. This strategy improves the face detection accuracy and having additional process complexness.

Writer Viola et al. granted an algorithm for article recognition, and it utilizes very simple characteristics specified Haar in [6]. In that algorithm, various numeral of Haar-like characteristics are extricated from picture as well as various effectual attributes are divided to utilize in AdaBoost algorithm, after that this attribute is prepared into a hierarchical data structure same as the decision tree. This algorithm is comparatively quick and robust because of the easy extricated attributes as well as option of easiest attributes. This technique is uncommonly fast as well as proficient than several techniques.

In [3], author Hamada et al. utilized neural network to face identification. In that system, fringe identification system is used on picture as well as after that the outcome picture caught via window as well as is measured via neural network to identify face.

Author Zhu et al. used imaging in IR spectrum-based technique for eye identification, excluding which after primary eye identification, utilized support vector machine (SVM) for developing corrections of eye identification in [7]. In that investigation, several SVM kernels are discovered as well as this is presented that Gaussian kernel has better corrections.

Author Zhao et al. utilized IR source of illumination for eye detection that is set with visual axis of television camera in [8]. Regular in the company of level of bright supply accompanied by about to driver, pupil is normally noticed intelligently in picture. So that, eye position were identified o\in the company of utilization of incepting morphological performance as well as deducting out coming picture from primary picture. After that, few applicant spots are extricated via a threshold technique on inequality picture as well as utilization of related component observation. In addition, SVM as well as generalized symmetry transformation (GST) is used to enlarge the correction of eye identification. Eyes positions are decided through the integrated outcomes of SVM as well as GST.

In [9], author shows approaches supposition which eye is black spots on face, eye position catches. By this motivation, binaries of face picture as well as enormous outlines are identified. Initially, middle point of two highest outcomes of face picture is recognized as eye middle.

In [1, 4], author Smith et al. obtain binary picture of face area after face identification is based on skin tone. Binaries of face region are worked depending on skin tone as well as reason of eyes displays black as well as other regions of face are white. After that, associated element observation is utilized for maximizing corrections of eye identification.

In [2, 10], author invented an eye recognition approach. That is used to recognize eye in scheme of HSV colour space and also invented other method that depends on the approach of innovation for the particular eye recognize. In that approach, a linear transform is enforced on image in YCbCr colour space and transformed image was change to binary image. These technologies have a very better accuracy for eye recognition in the image, but these are not better when the lighting of surrounding is less.

In [11], author proposed project-based method in which prediction was not applied straight on image (this image is binary image or grey-level image), but prediction of the edge recognition image is used. These technologies are less careful for face colour, but these are not good in recognizing eyes with eyeglass.

In [12], author Batista recognizes eye area using a face model concept. For this motive, the evaluated monobrow region is acquired from face image and it is performed with. Then, the eyebrows area, eyes region recognize. It concluded that the pupil is the black area in eye. With this premise, after strength improvement, image binarization accomplished and after edge recognition, student is recognition.

In [13], author used two various technologies for eye recognition in day and darkness. The used method for eye recognition in day is based on intent on economical grey-level model in the top-half region of face. In the darknest time, an IR lighting and imaging systems are used, and it recognizes student directly. It looks that the integration of these two technologies is created for our proposed system robust and efficient.

3 Methodologies

We proposed computer vision and RGB-D sensor-based approach for driver safety and medical services support using alarming. Figure 1 shows the system block diagram for proposed approach.

Algorithm: Driver Safety Algorithm

Input: Input video frame from RGB-D Sensor
Output: Distracted or Normal
Step 1: Input Image Acquisition from Video Sequence
Step 2: Image Pre-processing
Step 3: F1 = Face Detection and Extraction using Viola-Jones Algorithm
Step 4: F2 = Eyes Detection and Extraction using Viola-Jones Algorithm
Step 5: F3 = Mouth Detection and Extraction using Viola-Jones Algorithm
Step 6: HO = Head Orientation Features Extraction on Input F1
Step 7: G1 = GLCM (F1)
Step 8: G2 = GLCM (F2)
Step 9: G3 = GLCM (F3)
Step 10: Features Fusion (HO + G1 + G2 + G3)
Step 11: Load Training Data
Step 12: Perform Classification
Apply Classification as:
Detect = ffnn (*test, train*)
If Detect == 1
Alert ("Driver is in danger");
Else
Alert ("Driver is in safe");
Step 13: Detection Result and Alarm
Step 14: Stop

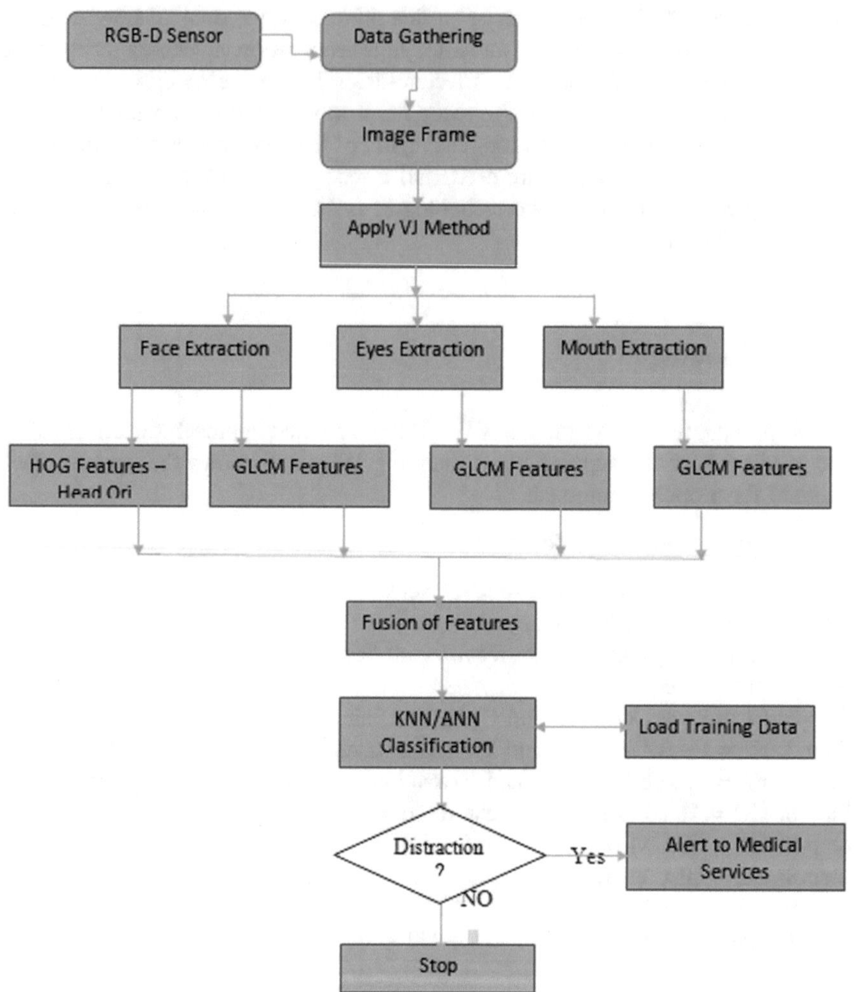

Fig. 1 Proposed system block diagram

4 Simulation Results and Discussion

This algorithm is applied on publically available driver fatigue data set which containing different drivers with different fatigues in video format. After pre-processing on input frame, face detection and extraction are performed. From input image, we have detected the important part of image for further analysis which is face. This is done by using VJ method. After the face detection, we further applied VJ method for eyes detection and extraction. VJ method is most accurate method for detection of eyes. We are using four different components for detecting the driver fatigue such as face, eyes, mouth and head rotation. For head rotation, we

Table 1 Comparative analysis for accuracy and time

Method name	Alert accuracy (%)	Recognition time (s)
AdaBoost method [14]	89.36	1.09
HMM method [15]	84	0.835
Detection method [16]	87.44	0.83
Proposed method	95.62	0.782

Table 2 Comparative analysis for precision rate

Method name	Precision (%)
AdaBoost method [14]	78.2
HMM method [15]	81.45
Detection method [16]	86.78
Proposed method	95.62

Table 3 Comparative analysis for recall rate

Method name	Recall (%)
AdaBoost method [14]	84.73
HMM method [15]	88.42
Detection method [16]	89.32
Proposed method	97.81

have used HOG method which gives us the face orientation features for detecting the rotation of head. Figure 1 shows the HOG features extraction for input face. After extracting all the important parts of face, we further applied GLCM method for feature extraction of all the facial components. Features are combined, and input is given for classifier (KNN) in order to detect whether input frame of drive shows any fatigue or not. At last for system evaluation, we measured the accuracy, precision and recognition time parameters as shown in Tables 1, 2 and 3, respectively.

The performance of precision, recall and accuracy in percentage is computed and compared in Tables 1, 2 and 3 with recent methods. The overall efficiency is improved in proposed approach as compared to all existing methods.

5 Conclusion and Future Work

In this paper, our motive was to present the efficient approach for driver safety. We proposed the efficiently automated framework for driver drowsiness detection by considering the various factors of driver such as eyes, eye blinking, mouth, head orientation. This framework is based on VJ algorithm, GLCM feature extraction and HOG feature extraction methods. For recognition, we used KNN and/or FFNN for accurate recognition and performance evaluation. The results demonstrate efficiency of proposed approach against recent methods of driver drowsiness detection. The overall accuracy for correct alarm generation is approximately 95–96%. For future work, we suggest to work on real-time application development of this framework and testing.

References

1. Smith, P., Shah, M., Lobo, N.V.: Monitoring head/eye motion for driver alertness with one camera. In: Proceeding of 15th IEEE International Conference on Pattern Recognition, Barcelona, Spain (2000)
2. Tabrizi, P.R., Zoroofi, R.A.: Open/closed eye analysis for drowsiness detection. In: Proceeding of 1st Workshops on Image Processing Theory, Tools and Applications, Sousse, Tunisia, Nov 2008
3. Hamada, T., Ito, T., Adachi, K., Nakano, T., Yamamoto, S.: Detecting method for drivers' drowsiness applicable to individual features. In: Proceeding of IEEE Intelligent Transportation Systems, Shanghai, China, Oct 2003
4. Smith, P., Shah, M., Lobo, N.V.: Determining driver visual attention with one camera. IEEE Trans. Intell. Transp. Syst. **4**(4) (2003)
5. Wang, F., Qin, H.: A FPGA based driver drowsiness detecting system. In: Proceedings of IEEE International Conference on Vehicular Electronics and Safety, Xian, China, Oct 2005
6. Viola, P., Jones, M.: Rapid object detection using a boosted cascade of simple features. In: Proceeding of International Conference on Computer Vision and Pattern Recognition (CVPR), Kauai, HI, USA (2001)
7. Zhu, Z., Fujimura, K., Ji, Q.: Real-time eye detection and tracking under various light conditions. In: ACM Eye Tracking Research and Application symposium, New Odeans, LA, USA (2002)
8. Zhao, S., Grigat, R.R.: Robust eye detection under active infrared illumination. In: Proceeding of 18th IEEE International Conference on Pattern Recognition (ICPR), Hong Kong, China, Sept 2006
9. Brandt, T., Stemmer, R., Mertsching, B., Rakotonirainy, A.: Affordable visual driver monitoring system for fatigue and monotony. In: Proceedings of IEEE International Conference on Systems, Man and Cybernetics, Hague, Netherlands, Oct 2004
10. Tabrizi, P.R., Zoroofi, R.A.: Drowsiness detection based on brightness and numeral features of eye image. In: Proceeding of 5th International Conference on Intelligent Information Hiding and Multimedia Signal Processing, Kyoto, Japan, Sept 2009
11. Horng, W.B., Chen, C.Y., Chang, Y., Fan, C.H.: Driver fatigue detection based on eye tracking and dynamic template matching. In: Proceeding of IEEE International Conference on Networking, Sensing & Control, Taipei, Taiwan, Mar 2004
12. Batista, J.: A drowsiness and point of attention monitoring system for driver vigilance. In: Proceeding of IEEE Intelligent Transportation Systems Conference, Seattle, USA, Oct 2007
13. Flores, M.J., Armingol, J.M., Escalera, A.: Driver drowsiness warning system using visual information for both diurnal and nocturnal illumination conditions. EURASIP J. Adv. Signal Process. (2010)
14. Hariri, B., Abtahi, S., Shirmohammadi, S., Martel, L.: A yawning measurement method to detect driver drowsiness
15. Singh, I., Banga, V.K.: Development of a drowsiness warning system using neural network. Int. J. Adv. Res. Electr. Electron. Instrum. Eng. **2**(8) (2013). (An ISO 3297: 2007 Certified Organization)
16. Craye, C., Karray, F.: Driver distraction detection and recognition using RGB-D sensor. arXiv:1502.00250v1 [cs.CV] 1 Feb 2015

17. Jimenez-Pinto, J., Torres-Torriti, M.: Face salient points and eyes tracking for robust drowsiness detection. Robotica **30**(5) (2012)
18. Grace, R., Byme, V.E., Bierman, D.M., Legrand, J.M., Gricourt, D., Davis, R.K., Staszewski, J.J., Carnahan, B.: A drowsy driver detection system for heavy vehicles. In: Proceedings of 17th AIAA/IEEE/SAE Digital Avionics Systems Conference (DASC), Washington, USA, Nov 1998
19. Rang-Ben, W., Ke-You, G., Shu-ming, S., Jiang-wei, C.: A monitoring method of driver fatigue behavior based on machine vision. In: Proceeding of IEEE Intelligent Vehicles Symposium, Columbus, Ohio, USA, June 2003
20. Veeraraghavan, H., Papanikolopoulos, N.: Detecting driver fatigue through the use of advanced face monitoring techniques. In: Intelligent Transportation System Institute, Department of Computer Science and Engineering, University of Minnesota (2001)
21. Dong, W., Wu, X.: Driver fatigue detection based on the distance of eyelid. In: Proceeding of IEEE International Workshop VLSI Design & Video Technology, Suzhou, China, May 2005

Feature Selection Using Histogram-Based Multi-objective GA for Handwritten *Devanagari* Numeral Recognition

Manosij Ghosh, Ritam Guha, Riktim Mondal, Pawan Kumar Singh, Ram Sarkar and Mita Nasipuri

Abstract In this paper, we propose an efficient feature selection method, called Histogram-Based Multi-objective Genetic Algorithm (HMOGA), for finding informative features from high-dimensional data which also improves the classification accuracy. This approach is applied on two previously proposed feature sets for handwritten *Devanagari* numeral recognition problem. With the feature set selected by HMOGA, final recognition is performed using the Multi-layer Perceptron (MLP)-based classifier. The rise in classification accuracy using only 50% of the original feature vector portrays the applicability of the developed idea for multi-objective optimization.

Keywords Handwritten numeral recognition · *Devanagari* script
Feature selection · Histogram-Based Multi-objective Genetic Algorithm

1 Introduction

Handwritten numeral recognition has become one of the most important research areas in pattern recognition since decades because of its various applications such as mail sorting using pin code, processing of bank checks, form processing of different government job applications [1]. Recognition of handwritten numerals is a complicated task due to the variation of size, shape, thickness, and style of writing of the individuals. Many techniques have been developed so far to solve the problem of digit recognition. However, the use of feature selection methods for this problem has not been explored much by the researchers. Feature selection is a process of selecting the relevant attributes and removing the irrelevant and redundant attributes which enhances the model interpretability, shortens the training times, and improves the overall performance of the system.

M. Ghosh · R. Guha · R. Mondal · P. K. Singh (✉) · R. Sarkar · M. Nasipuri
Department of Computer Science and Engineering, Jadavpur University, Kolkata, India
e-mail: pawansingh.ju@gmail.com

V. Bhateja et al. (eds.), *Intelligent Engineering Informatics*, Advances in Intelligent Systems and Computing 695, https://doi.org/10.1007/978-981-10-7566-7_46

Genetic Algorithm (GA) is a population-based meta-heuristic wrapper method for feature selection based on natural genetics as well as selection. The operations involved in GA are iterative in nature that manipulates initial population of chromosomes to produce a new and improved population through genetic operations such as crossover and mutation.

This paper proposes a GA-based feature selection method called the Histogram-Based Multi-objective Genetic Algorithm (HMOGA) where MLP is used as a learning algorithm. This concept is applied on handwritten *Devanagari* numeral recognition. The main reason of choosing this script is that *Devanagari* script is one of the oldest and most widely used Indian scripts since ancient times and used by around 500 million people [2]. It is also used to write some popular languages such as *Nepali*, *Pali*, *Marathi*, *Konkani*, *Sanskrit*.

Few important and recently published methods, described in [3–8], are applied for the recognition of handwritten *Devanagari* numerals. Being a classic algorithm, many works have been proposed till date based on GA for feature selection. Moreover, GA has been applied for selection of features in different domains such as classification of schizophrenia using functional magnetic resonance imaging data [9], classification of ovarian cancer [10], text clustering and text classification [11], Persian font recognition [12], breast cancer diagnosis [13], detection of premature ventricular contractions [14]. Based on generalness nature of the GA in feature selection procedure, in this paper, we have introduced a method called HMOGA which is used for selection of features in handwritten *Devanagari* numeral recognition. The proposed approach has been applied on two previously proposed feature sets for numeral recognition, *namely* Mojette Transform [15] and Regional Weighted Run Length Features [16].

2 Feature Extraction

Two feature sets previously proposed by Singh et al. [15, 16] for handwritten *Devanagari* numeral recognition are chosen to conduct the feature selection process. These feature sets are briefly described below.

2.1 Mojette Transform

Mojette Transform [17] is a mathematically defined transformation for feature extraction from images which is similar to “projection histograms features.” It is defined as a discrete version of “Radon transform,” which is a set of vectors, where each element is also known as a “bin.” This bin is calculated as the sum of the value of pixels centered on a projection line defined by m and θ_i, where θ_i is the projection angle defined as below:

$$\theta_i = \tan^{-1}\left(\frac{q_i}{p_i}\right) \tag{1}$$

m is an integer that determines which of the various projection lines at the angle θ_i are being represented by the element. For each of the values of m and θ_i, corresponding element of the Mojette Transform vector [18] is given by:

$$Proj_{p_i, q_i}(m) = \sum_{y=-\infty}^{\infty} \sum_{y=-\infty}^{\infty} f(x, y)\delta(m + q_i x - p_i y) \tag{2}$$

where (x, y) defines the position of the pixel, $f(x, y)$ represents pixel value of the image, p_i and q_i define the projection angles, m determines the particular line of projection at the angle θ_i, and δ is the Kronecker delta function defined as below:

$$\begin{aligned} \delta(w) &= 1 \; if \; w = 0 \\ &= 0 \; otherwise \end{aligned} \tag{3}$$

For the present work, each of the digit images is transformed for $\theta_i = 0^\circ, 45^\circ, 90^\circ$ and 135°. Given the size of the digit image as 32×32 pixels, the number of bins for each of the above-mentioned projection angles is calculated as 32, 63, 32, and 63, respectively. Hence, a set of 190 features is extracted after the application of Mojette transformation [15].

2.2 Regional Weighted Run Length Features

This procedure [16] consists of three key steps: (a) Contour Extraction, (b) Mask Orientation, and (c) Feature Extraction. The contour of the digit images is estimated, and the feature vector is extracted from the pixels of contour in the first step which is used to minimize the overall computation time of the system. Then, in the second step, four different types of masks are designed depending on four different orientations of the data pixels such as *vertical*, *horizontal*, and the two oblique lines slanted at $\pm 45^\circ$ to each black pixel. Here, eight-connectivity neighborhood analysis is used to determine the direction of a black pixel. In the final step, the input image of size 32×32 is divided into imaginary grids of size 8×8. Then, a mask of size 16×16 is made to slide over the image in such a manner that each mask overlaps the preceding 8 pixels (both in horizontal and vertical directions) of the adjacent grid. After that, four concentric regions (squares) are formed considering each such mask, *namely* R_i, i = 1, 2, 3, 4. R_1 covers a region of size 4×4 in the center of the corresponding sub-image. R_2 comprises a region of size 8×8 excluding region R_1. R_3 is selected as a region of size 12×12 which excludes the preceding two regions.

Finally, R_4 is the region of 16×16 dimensions excluding all three prior regions. The binary transition count (from foreground to background pixels and vice versa) is calculated from each of these regions. As a result, from each of the regions, in a total of 16 features are extracted. Finally, a feature vector of size 144 (i.e., 16 * 9) is designed using *Regional Weighted Run Length* Features from each of the handwritten *Devanagari* numerals.

3 Histogram-Based Multi-objective Genetic Algorithm

The very nature of the MOGA [19] is randomness. This is a well-accepted fact that this random nature gives MOGA a strong foothold in the meta-heuristic optimization research world. But, due to this volatility of this algorithm, at times, it ignores certain regions of the search space. As a consequence, some important feature attributes might be overlooked if we consider only the best case of the MOGA. To address this issue, we have formed the histogram of the feature attributes from *i* runs of MOGA. As the outcomes of the different runs are considered in the formation of histogram, therefore, the much-required exploration of the search space can be expected by this approach. After appropriate thresholding, only those feature values are included in the final list that appears in almost every run of MOGA.

In MOGA, feature sets are represented in the form of chromosomes. The child chromosomes are generated using the genetic operations of uniform crossover and uniform mutation. The chromosomes are encoded into binary strings where '0' represents that the feature is not selected and '1' otherwise.

At first, the population is created randomly from the entire feature space and then crossover and mutation operations are done in order to generate the new and improvised set of chromosomes. Poorly performing chromosomes of the current generations are substituted by the fitter children of the next generation, thereby allowing for the gradual improvement of the population. So, we follow elitism rule [20] while producing new generations. We have used uniform crossover [21] which involves part-wise aggregation of features from both the parents to create children chromosomes. This operation allows us to exploit the search space by checking new combinations. The use of mutation allows for the exploration of search space through inclusion and removal of random features. It involves making a '0' to a '1' and vice versa with a probability of q. The value of q is chosen to be very less to allow for only small changes in each mutation operation.

Here, the feature selection process is conceptualized as multi-objective optimization problem, where the objectives are set so as to reduce the number of features and the increment in accuracy of the recognition model. However, recognition accuracy is given higher priority than number of features selected. This is because lower dimensional feature vector cannot be accepted if it does not provide reasonable recognition accuracy. The fitness function is given in Eq. (4).

$$fitness = accuracy*w1 + (1 - c)*w2 \tag{4}$$

where c is the ratio of features selected to the total number of features, and w1 and w2 are the weights assigned to accuracy and feature set dimension, respectively.

The accuracy is calculated using MLP classifier [22]. The variation of the data (using only β of the data each time for MOGA) used to generate the i populations allows the final result to have more generalization than the feature set found using MOGA on the whole data due to the data-driven nature of MOGA.

Instead of considering the best chromosomes of 1 MOGA run only, we combine the i populations by forming a histogram of the features, to find out the importance (or weight) of each feature. The new feature subset is found from the histogram by fixing a threshold value. Considering we apply MOGA on the selected data subset i times, therefore, i population will be generated containing n chromosomes in each. We form a histogram of the features from these $i*n$ chromosomes. However, the use of a plain histogram does not take into consideration the quality of the feature sets during formation of the histogram. So, we use the following formula to account for the quality (recognition accuracy) of the feature sets.

$$value\ of\ jth\ feature = \sum_{l=1}^{i} \sum_{k=1}^{n} f_{jk}\ of\ lth\ population * a_k\ of\ lth\ population \tag{5}$$

Here, f_{jk} is the value of jth bit in kth chromosome (either 0 or 1) of a population and a_k is the accuracy of the kth chromosome of a population. The histogram of the frequency of each feature is modified using the recognition accuracy to differentiate between two features with the same frequency and give preference to the feature which is present in feature sets with better recognition accuracies. The average of the values is taken as our threshold, and then only those features with value above the average are taken in the final set.

4 Dataset Description

We have prepared in-house database consisting of 10,000 handwritten numerals. A large number of people are involved in the data collection process belonging to varying age, sex, profession etc. Mostly, they used a black or blue ink pen. All data sheets are scanned in a flatbed scanner with 300 dpi resolution and stored in .bmp file format. Finally, the numeral images are cropped automatically from the scanned sheets to prepare the isolated samples for the experiment. Noise present in the numeral images are removed using well-known Gaussian filter [23]. Then, binarization (for converting the numeral images into two-tone images '0' and '1') is accomplished using an adaptive global threshold value computed as the average of minimum and maximum intensities in that image. Figure 1 shows some samples of handwritten *Devanagari* numerals. A set of 6000 training samples is chosen where 600 samples are selected from each numeral class. A testing set of 4000 remaining samples are used by considering equal number of numeral samples from each class.

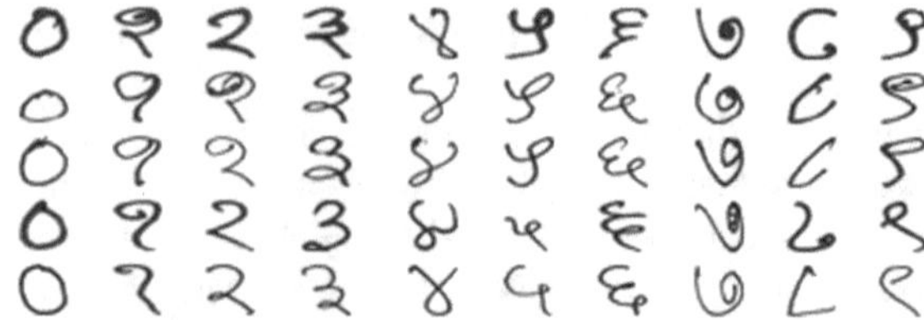

Fig. 1 Samples of handwritten *Devanagari* numerals

5 Experimental Results

An analysis of the performance of Mojette Transform and *Regional Weighted Run Length* feature sets is done by varying the number of iterations, population size, and number of features to be used in the HMOGA model. The values of the various parameters used in the present work are as follows: (a) $p = 0.5$, (b) $q = 0.01$, (c) $w_1 = 100$, (d) $w_2 = 1$, (e) $i = 5$, and (f) $l = 140$. Firstly, the population size is varied keeping number of iterations constant (say, 15) which is graphically illustrated in Fig. 2a. Then, the number of iterations is varied keeping population size constant (in our case, 30) which is depicted in Fig. 2b. After that, the values of population size are set to 30 and 20 for Mojette Transform and *Regional Weighted Run Length* feature sets, respectively. On the other hand, the number of iterations is taken as 15 for both the feature sets. Finally, the overall performances of the present feature selection technique are recorded for two different sets of features (shown in Fig. 3a, b). It can be seen from Fig. 3a that using MOGA i times and their subsequent combination, the optimum number of features is chosen to be 97 for Mojette Transform where the overall recognition accuracy is found to be 92.25%. Similarly, the optimum number of features is found to be 79 for *Regional Weighted Run Length* Features (refer to Fig. 3b) where the highest accuracy of 96.075% is achieved.

In the present work, a comparison of our method with MOGA is done for both the feature sets which is shown in Table 1. The accuracy of the original feature set

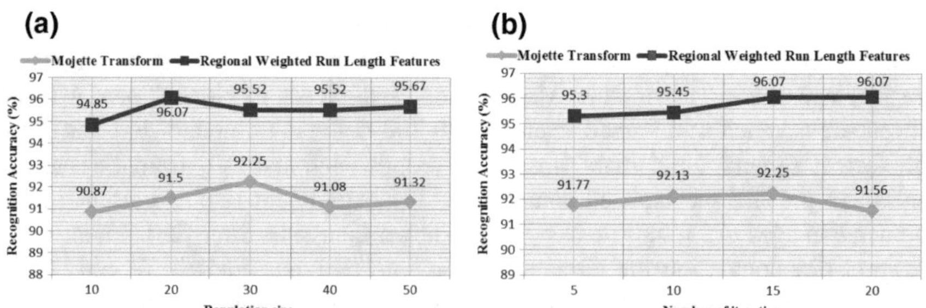

Fig. 2 Graphical comparison showing the recognition accuracies with varying: **a** population size for fixed number of iterations and **b** number of iterations having fixed population size of HMOGA model

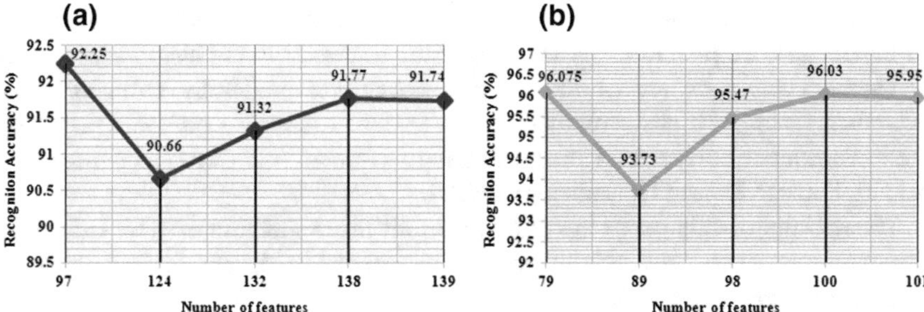

Fig. 3 Graph showing the overall recognition accuracies with variation in the number of optimal features selected using MOGA *i* times for: **a** Mojette Transform and **b** *Regional Weighted Run Length* Feature sets

Table 1 Comparison of the present HMOGA feature selection technique for two feature sets with some previous methods (best case is shaded in gray)

Feature set	Method	Number of features	Accuracy (%)
Mojette Transform [15]	Without using FS	190	91.75
	MOGA [19]	125	92.02
	HMOGA	**97**	**92.40**
Regional Weighted Run Length Features [16]	Without using FS	144	95.53
	MOGA [19]	98	95.88
	HMOGA	**79**	**96.075**

without using any feature selection scheme is also tabulated. For Mojette Transform as well as *Regional Weighted Run Length* Features, the proposed methodology makes a significant improvement of about 0.65% and 0.545% in the overall recognition accuracies, respectively, by using only about 50% of the original features. Similarly, using the basic MOGA [19] implementation, we have also attained slightly higher recognition accuracies using comparatively lesser number of selected features. This confirms the suitability of our proposed HMOGA methodology in order to select optimal subset of features for handwritten *Devanagari* numeral recognition problem.

6 Conclusion

In this paper, we develop an intelligent feature selection method, called HMOGA, in order to choose the optimal feature subset which can improve the performance of any pattern recognition problem. Here, we apply this feature selection method on two recently designed feature vectors used for handwritten *Devanagari* numeral recognition problem. We observe that this feature selection model acquires enhanced recognition accuracy using only about 50% of entire feature set in both cases. The developed method can also be used as a feature ranking mechanism. As a future scope, we plan to apply some heuristic to threshold the histogram in order to augment the accuracy of the overall recognition model by reducing more redundant features, if any. Another plan is to extend the idea on some popularly used evolutionary algorithms in the literature such as Ant Colony Optimization (ACO), Particle Swarm Optimization (PSO).

The authors of this paper are thankful to all those individuals who had given appropriate consents and contributed wholeheartedly in developing the database used in the current research.

References

1. Gharde, S.S., Ramteke, R.J., Kotkar, V.A., Bage, D.D.: Handwritten devanagari numeral and vowel recognition using invariant moments. In: Proceedings of IEEE International Conference of Global Trends in Signal Processing, Information Computing and Communication (ICGTSPICC), pp. 255–260 (2016)
2. http://web.archive.org/web/20071203134724/, http://encarta.msn.com/media_701500404/Languages_Spoken_by_More_Than_10_Million_People.html. Accessed 3 Aug 2017
3. Chaudhary, M., HasnineMirja, M., Mittal, N.K.: hindi numeral recognition using neural network. Int. J. Sci. Eng. Res. **5**(6), 260–268 (2014)
4. Hanmandlu, M., Ramana Murthy, O.V.: Fuzzy model based recognition of handwritten hindi numerals. Pattern Recogn. **40**(6), 1840–1854 (2006)
5. Bajaj, R., Dey, L., Chaudhury, S.: Devnagari numerals recognition by combining decision of multiple connectionist classifiers. Sadhana **27**(part I), 59–72 (2002)
6. Pandey, A., Kumar, A., Kumar, R., Tiwari, A.: Handwritten devanagri number recognition using majority voting scheme. Int. J. Comput. Sci. Inf. Technol. Secur. (IJCSITS) **2**(3), 631–636 (2012)
7. Arora, S., Bhattacharjee, D., Nasipuri, M., Kundu, M., Basu, D.K., Malik, L.: Handwritten devnagari numeral recognition using SVM and ANN. Int. J. Comput. Sci. Emerg. Technol. (IJCSET) **1**(2), 40–46 (2010)
8. Dongre, V.J., Mankar, V.H.: Devnagari handwritten numeral recognition using geometric features and statistical combination classifier. Int. J. Comput. Sci. En. (IJCSE) **5**(10) (2013)
9. Shahamat, H., Pouyan, A.A.: Feature selection using genetic algorithm for classification of schizophrenia using fMRI data. J. AI Data Min. **3**(1), 30–37 (2015)
10. Khare, P., Burse, K.: Feature selection using genetic algorithm and classification using weka for ovarian cancer. Int. J. Comput. Sci. Inf. Technol. (IJCSIT) **7**(1), 194–196 (2016)
11. Hong, S.S., Lee, W., Han, M.M.: The feature selection method based on genetic algorithm for efficient of text clustering and text classification. Int. J. Adv. Soft Comput. Andits Appl. **7**(1), 22–40 (2015)
12. Imani, M.B., Pourhabibi, T., Keyvanpour, M.R., Azmi, R.: A new feature selection method based on ant colony and genetic algorithm on persian font recognition. Int. J. Mach. Learn. Comput. **2**(3), 278–282 (2012)
13. Aalaei, S., Shahraki, H., Rowhanimanesh, A., Eslami, S.: Feature selection using genetic algorithm for breast cancer diagnosis: experiment on three different datasets. Iranian J. Basic Med. Sci. **19**(5), 476–482 (2016)
14. Kaya, Y., Pehlivan, H.: Feature selection using genetic algorithms for premature ventricular contraction classification. In: Proceedings of 9th IEEE International Conference on Electrical and Electronics Engineering (ELECO), pp. 1229–1232 (2015)

15. Singh, P.K., Das, S., Sarkar, R., Nasipuri, M.: Recognition of handwritten indic script numerals using mojette transform. In: Proceedings of 1st International Conference on Intelligent Computing and Communication (ICIC2), AISC 458, pp. 459–466 (2016)
16. Singh, P.K., Das, S., Sarkar, R., Nasipuri, M.: Recognition of offline handwritten devanagari numerals using regional weighted run length features. In: Proceedings of 1st IEEE International Conference on Computer, Electrical and Communication Engineering (ICCECE) (2017). ISBN (Print)-978-1-4799-4446-0. https://doi.org/10.1109/c3it.2015.7060113
17. Guedon, J.P., Normand, N.: The Mojette transform: the first ten years. In: Discrete Geometry for Computer Imagery, LNCS 3429, pp. 79–91, Poitier, France (2005)
18. Vásárhelyi, J., Serfözö, P.: Analysis of mojette transform implementation on reconfigurable hardware. In: Dagstuhl Seminar Proceedings 06141, Dynamically Reconfigurable Architectures (2006). http://drops.dagstuhl.de/opus/volltexte/2006/746
19. Murata, T., Ishibuchi, H.: MOGA: multi-objective genetic algorithms. In: Proceedings of IEEE International Conference on Evolutionary Computation (1995)
20. Dehuri, S., Paitnaik, S., Ghosh, A., Mall, R.: Application of elitist multi-objective genetic algorithm for classification rule generation. Appl. Soft Comput. **8**(1), 477–487 (2008)
21. Spears, W.M., De Jong, K.A.: On the virtues of parameterized uniform crossover. In: Proceedings of 4th International Conference on Genetic Algorithms, pp. 230–236 (1991)
22. Hornik, K., Stinchcombe, M., White, H.: Multilayer feedforward networks are universal approximators. Neural Netw. **2**, 359–366 (1989)
23. Gonzalez, R.C., Woods, R.E.: Digital Image Processing, vol. I. Prentice-Hall, India (1992)

Extended Kalman Filter for GPS Receiver Position Estimation

N. Ashok Kumar, Ch. Suresh and G. Sasibhushana Rao

Abstract Navigation, tracking, and positioning of an object are a customary problem in many fields. Global Positioning System (GPS) is the best solution to this problem. Being GPS is wireless communication through space, the received ephemeris data are erroneous. Hence, the extraction of original ephemeris data from this erroneous data is the main hurdle. Adaptive algorithms provide better results to overcome this hurdle and also for the nonlinear type of system processing. In this paper, GPS receiver position is estimated by extended Kalman filter. A dual-frequency GPS receiver is used for input data, which is located at the Department of ECE, Andhra University, Visakhapatnam (17.73° N/83.31° E). The estimated GPS receiver position is compared with the original position coordinates to check the accuracy. Receiver clock error is also estimated. The result shows that the extended Kalman filter provides a good accuracy of the estimated results for GPS receiver positioning.

Keywords GPS · Kalman filter · Dual-frequency receiver

N. Ashok Kumar (✉) · G. Sasibhushana Rao
Department of Electronics and Communication Engineering,
Andhra University College of Engineering (A), Andhra University,
Visakhapatnam, India
e-mail: ashok0709@gmail.com

G. Sasibhushana Rao
e-mail: sasigps@gmail.com

Ch. Suresh
Department of Information Technology, Anil Neerukonda Institute of Technology
and Science, Visakhapatnam, India
e-mail: sureshchittineni@gmail.com

V. Bhateja et al. (eds.), *Intelligent Engineering Informatics*, Advances in Intelligent
Systems and Computing 695, https://doi.org/10.1007/978-981-10-7566-7_47

1 Introduction

In the present world, Global Positioning System (GPS) [1, 2] is widely used for positioning, navigation, and timing. GPS receiver location can be obtained as two-dimensional coordinates and also as three-dimensional coordinates by trilateration principle. To determine the position of the receiver, one must have to know the GPS satellite's orbital position in space and distance between satellites to the receiver. The GPS receiver position can be calculated with the help of four or more satellites. In this paper, more than four satellites are considered in each epoch, and then, extended Kalman filter [2–4] is applied to estimate GPS receiver position [5] along with the receiver clock error. Section 2 covers GPS and its basic equations for finding receiver position. Section 3 covers all of extended Kalman filter. In Sect. 4, extended Kalman filter is applied to find the GPS receiver position. The paper ends with conclusions in the last section.

2 Global Positioning System

The Global Positioning System (GPS) is a real-time radio navigation system, which provides, 24 h a day, accurate GPS receiver position. GPS positioning tasks are satellite signal acquisition, tracking, and positioning. Satellite signal acquires approximated estimates of signal parameters. Tracking is to keep track of these parameters as they change over time. After tracking is done, GPS navigation data will be obtained and pseudoranges are calculated, and then, the receiver position can be computed. The GPS receiver position estimated with the help of four satellites is as shown in Fig. 1.

More than four satellites can also be used to find a receiver position. The basic observed pseudorange equation is given by

$$P_r^{obs} = \rho^{Gr} + c(\Delta t^{sat} - \Delta t) + \Delta_{ion} + \Delta_{trop} + \xi_{mr} \tag{1}$$

where Δt is the receiver clock offset, Δt^{sat} is the satellite clock offset, Δ_{ion} is the delay imparted by ionosphere, Δ_{trop} is the delay imparted by the troposphere, ξ_{mr} is the effects of multipath and receiver measurement noise, c is the velocity of light, and ρ^{Gr} is the geometric range between a particular satellite and receiver. From ephemeris data, position of the satellite $(x^{sat}, y^{sat}, z^{sat})$ can be computed. ρ^{Gr} is the geometric range between satellite and receiver, given by

$$\rho^{Gr} = \sqrt{(x^{sat} - x_r)^2 + (y^{sat} - y_r)^2 + (z^{sat} - z_r)^2} \tag{2}$$

where (x_r, y_r, z_r) is the position of the receiver. Except the receiver clock error, remaining error terms in Eq. (1) are ignored for simplification. Therefore, Eq. (1) will be given as,

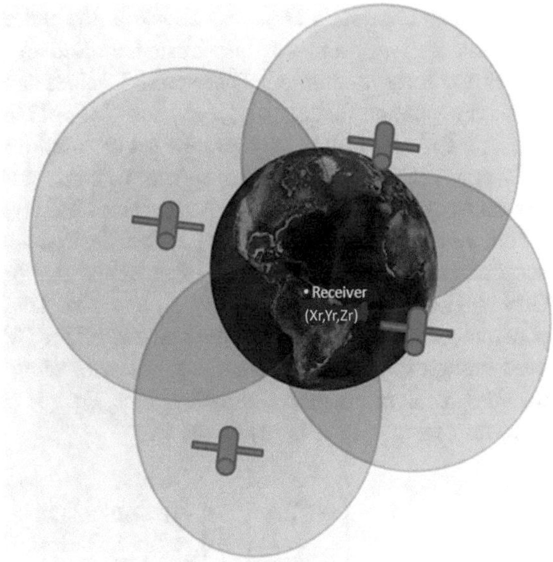

Fig. 1 GPS receiver 3D-positioning using four satellites

$$P_r^{obs} = \sqrt{(x^{sat} - x_r)^2 + (y^{sat} - y_r)^2 + (z^{sat} - z_r)^2} + clk_r \tag{3}$$

where clk_r is the receiver clock error expressed in the distance (equal to $-c.\Delta t$). In Eq. (3), there are four unknown parameters x_r, y_r, z_r, and clk_r. To determine these unknown parameters, we need four range equations from four different satellites. Therefore,

$$\begin{aligned}
P_{r\ 1}^{obs} &= \left(\left(x^{sat1} - x_r\right)^2 + \left(y^{sat1} - y_r\right)^2 + \left(z^{sat1} - z_r\right)^2 \right)^{1/2} + clk_r \\
P_{r\ 2}^{obs} &= \left(\left(x^{sat2} - x_r\right)^2 + \left(y^{sat2} - y_r\right)^2 + \left(z^{sat2} - z_r\right)^2 \right)^{1/2} + clk_r \\
P_{r\ 3}^{obs} &= \left(\left(x^{sat3} - x_r\right)^2 + \left(y^{sat3} - y_r\right)^2 + \left(z^{sat3} - z_r\right)^2 \right)^{1/2} + clk_r \\
P_{r\ 4}^{obs} &= \left(\left(x^{sat4} - x_r\right)^2 + \left(y^{sat4} - y_r\right)^2 + \left(z^{sat4} - z_r\right)^2 \right)^{1/2} + clk_r
\end{aligned} \tag{4}$$

Above equations are nonlinear simultaneous equations. To get four unknown parameters, first linearize the nonlinear Eq. (3) and then solve. Differentiating Eq. (3) results

$$\Omega P_r^{obs} = \frac{-(x^{sat} - x^r)\Omega x_r - (y^{sat} - y^r)\Omega y_r - (z^{sat} - z^r)\Omega z_r}{\sqrt{(x^{sat} - x_r)^2 + (y^{sat} - y_r)^2 + (z^{sat} - z_r)^2}} + \Omega clk_r \tag{5}$$

Consider Ωx_r, Ωy_r, Ωz_r, and Ωclk_r as the unknown parameters in the above equation. x_r, y_r, z_r, and clk_r are treated as known values by taking some initial values for these quantities. These initial values are substituted in Eq. (5) which yields the solution for Ωx_r, Ωy_r, Ωz_r, and Ωclk_r. These are used to update the initial x_r, y_r, z_r, and clk_r to find another new set of solutions for Ωx_r, Ωy_r, Ωz_r, and Ωclk_r. This new set is used again to update x_r, y_r, z_r, and clk_r. This recursive process continues until the values of the parameters Ωx_r, Ωy_r, Ωz_r, and Ωclk_r are small and within a certain predefined limit. The desired solution is obtained when the parameters Ωx_r, Ωy_r, Ωz_r, and Ωclk_r reached this predefined limit. Then, the final GPS receiver coordinates are x_r, y_r, z_r, and clk_r. This is an iteration method. In this paper, more than four satellites are considered for GPS receiver position calculation even though four satellites are enough. For example, if we consider nine satellites for GPS receiver position estimation, then Eq. (5) can be composed as a

Matrix form, which is expressed as

$$\begin{bmatrix} \Omega P_{r\ 1}^{obs} \\ \vdots \\ \Omega P_{r\ 9}^{obs} \end{bmatrix} = \begin{bmatrix} \xi_{11} & \xi_{12} & \xi_{13} & 1 \\ \vdots & \vdots & \vdots & \vdots \\ \xi_{91} & \xi_{92} & \xi_{93} & 1 \end{bmatrix} \begin{bmatrix} \Omega x_r \\ \Omega y_r \\ \Omega z_r \\ \Omega clk_r \end{bmatrix} \tag{6}$$

where the elements ξ_{i1}, ξ_{i2}, and ξ_{i3} are given by

$$\xi_{i1} = -\frac{(x^{sat} - x_r)}{P_r^{obs} - clk_r}, \xi_{i2} = -\frac{(y^{sat} - y_r)}{P_r^{obs} - clk_r}, \xi_{i3} = -\frac{(z^{sat} - z_r)}{P_r^{obs} - clk_r} \tag{7}$$

3 Extended Kalman Filter

The extended Kalman filter (EKF) [2–4, 6–8] is the nonlinear genre of the Kalman filter [9, 10] which linearizes about an estimate of the current mean and covariance. The state transition and observation models for the extended Kalman filter are taken as

$$\chi_t = f(\chi_{t-1}) + \omega_{t-1} \tag{8}$$

$$z_t = h(\chi_t) + \upsilon_t \tag{9}$$

where ω_{t-1} is the process noise with zero mean and covariance Q_t, and υ_t is the observation noise with zero mean and covariance R_t.

The functions $f(\chi_{t-1})$ and $h(\chi_t)$ are used to compute the predicted state from the previous estimate and predicted measurement from the predicted state, respectively. Instead of applying $f(\chi_{t-1})$ and $h(\chi_t)$ to the covariance directly, a Jacobian matrix is applied which is evaluated with current predicted states at each time step. This process essentially linearizes the nonlinear function around the current estimate.

Discrete-time extended Kalman filter's prediction (time update) and correction (measurement update) equations are given by,

Prediction (time update):

i. Predicted state estimate

$$\hat{\chi}_{t|t-1} = f\left(\hat{\chi}_{t-1|t-1}\right) \tag{10}$$

ii. Predicted covariance estimate

$$P_{t|t-1} = F_{t-1} P_{t-1|t-1} F_{t-1}^{T} + Q_{t-1} \tag{11}$$

Correction (measurement update):

i. Kalman gain

$$K_t = P_{t|t-1} H_t^T \left(H_t P_{t|t-1} H_t^T + R_t\right)^{-1} \tag{12}$$

ii. Updated state estimate

$$\hat{\chi}_{t|t} = \hat{\chi}_{t|t-1} + K_t\left(z_t - h\left(\hat{\chi}_{t|t-1}\right)\right) \tag{13}$$

iii. Updated covariance estimate

$$P_{t|t} = (I - K_t H_t) P_{t|t-1} \tag{14}$$

Where the Jacobian for state transition and observation matrices are defined as

$$F_{t-1} = \left.\frac{\partial f}{\partial \chi}\right|_{\hat{\chi}_{t-1|t-1}} \tag{15}$$

$$H_t = \left.\frac{\partial h}{\partial \chi}\right|_{\hat{\chi}_{t|t-1}} \tag{16}$$

The Jacobian H_t is taken from Eqs. (6) and (7) as

$$H_t = \begin{bmatrix} \xi_{11} & \xi_{12} & \xi_{13} & 1 \\ \vdots & \vdots & \vdots & \vdots \\ \xi_{91} & \xi_{92} & \xi_{93} & 1 \end{bmatrix} \tag{17}$$

The above method is extended Kalman filter (EKF) of first order. Higher-order EKFs can be composed by retaining more terms of the Taylor series expansions.

4 Results and Discussion

Real-time GPS data, recorded for 24 h, has been taken, at the Department of Electronics and Communication Engineering, Andhra University, Visakhapatnam (latitude 17.73° N/longitude 83.31° E), from a dual-frequency GPS receiver. In each epoch, data include more than four different satellites' position and the pseudorange between satellites to GPS receiver. In this paper, extended Kalman filter is applied to this data. Assume that initial receiver position coordinates are (0 0 0). The resultant GPS receiver position coordinates obtained, from the filtering operation, are compared with the original GPS receiver position coordinates to show how much they are close to the original position. This shows the accuracy provided by the application of extended Kalman filter. These results are presented in Table 1.

In Table 1, x-position error (Xe), y-position error (Ye), z-position error (Ze) in meters, and receiver clock error (Rclk) in nanosecond (10^{-9}) are presented. From the results in Table 1, it is observed that error in coordinates and receiver clock error are slightly varied for every 2 h due to neglecting other error sources in GPS pseudorange expression. The mean values of x-position error, y-position error, and z-position error for 24 h data are 48.47 m, 100.63 m, and 34.6 m, respectively. These mean values are the error in three-dimensional coordinates while comparing with the original GPS receiver position coordinates. The mean value of receiver clock error estimated is 100 ns, due to neglecting ionosphere, troposphere, and multipath errors.

In Table 2, x-position variance, y-position variance, z-position variance, and receiver clock variance in distance, i.e.,σ_x^2, σ_y^2, σ_z^2, and σ_{clk}^2 respectively, are presented. From Table 2, variances obtained are very low which ensures the deviations

Table 1 Position error estimation

Extended Kalman filter				
GPS time in hours of the day	Xe (m)	Ye (m)	Ze (m)	Rclk (ns)
00–02	50.98	105.89	39.93	58
02–04	49.59	108.63	33.37	74
04–06	42.20	98.29	30.79	118
06–08	41.83	83.59	32.06	150
08–10	45.82	83.37	34.02	152
10–12	51.16	98.10	39.71	120
12–14	50.90	103.07	35.50	105
14–16	51.37	107.50	34.08	95
16–18	47.39	106.48	35.72	90
18–20	51.69	107.76	34.51	83
20–22	48.84	101.60	31.83	89
22–24	50.35	117.94	43.60	33

Table 2 Variance estimation

Extended Kalman filter				
GPS time in hours of the day	σ_x^2	σ_y^2	σ_z^2	σ_{clk}^2
00–02	0.130	0.188	0.126	0.127
02–04	0.128	0.211	0.133	0.139
04–06	0.127	0.204	0.130	0.136
06–08	0.128	0.198	0.139	0.138
08–10	0.124	0.192	0.134	0.129
10–12	0.133	0.204	0.132	0.132
12–14	0.131	0.193	0.129	0.132
14–16	0.133	0.186	0.127	0.131
16–18	0.139	0.193	0.130	0.136
18–20	0.135	0.185	0.127	0.128
20–22	0.134	0.203	0.133	0.143
22–24	0.131	0.189	0.130	0.135

are low from its mean values. From Tables 1 and 2 results, it is observed that extended Kalman filter provides good accuracy for estimating GPS receiver position.

5 Conclusions

Extended Kalman filter approximates nonlinear function. Here, linearizing the nonlinear functions helps to use linear Kalman filter for the estimation of GPS receiver position with good accuracy. For GPS receiver position estimation, extended Kalman filter provides less variance and good accuracy in position estimation and also provides receiver clock error of 100 ns; it is because of neglecting other sources which influences the clock error. However, the mean values for the error in GPS receiver position while comparing with the original GPS receiver position coordinates are 48.47 m, 100.63 m, and 34.6 m, respectively. Therefore, extended Kalman filter is better algorithm for GPS receiver position estimation and also for nonlinear type of estimation problems.

References

1. Rao, G.S.: Global Navigation Satellite Systems. McGraw Hill Education Private limited (2010). ISBN (13): 978-0-07-070029-1
2. Grewal, M.S., Andrews, A.P., Bartone, C.G.: Global Navigation Satellite Systems, Inertial Navigation, and Integration, 3rd edn. (2013)

3. Gordan, N., Pitt, M.: A comparison of sample based filters and the Extended Kalman Filter for the bearings only tracking problem. In: Signal Processing Conference (EUSIPO) (1998)
4. Burl, J.: A reduced order Extended Kalman Filter for sequential images containing a moving object. IEEE Trans. Image Process. **2**(3), 285–295 (1993)
5. Wang, Y.: Position estimation using Extended Kalman Filter and RTS-smoother in a GPS receiver. In: IEEE Conference, 5th International Congress on Image and Signal Processing, (CISP 2012), pp. 1718–1721
6. Simon, D.: Optimal State Estimation. John Wiley & Sons, Inc. (2006)
7. Song, Y., Grizzle, J.W.: The Extended Kalman Filter as a local asymptotic observer for discrete-time nonlinear systems. J. Math. Syst. Estim. Control **5**(1), 59–78 (1995)
8. Yin, Jianjun, Ming, Gu: The expanded state space Kalman filter for GPS navigation. Inf. Technol. J. **10**, 2091–2097 (2010)
9. Welch, G., Bishop, G.: An Introduction to the Kalman Filter. University of North Carolina, 5 April 2004
10. Tolman, B.W.: GPS precise absolute positioning via Kalman filtering. In: 21st International Technical Meeting of the Satellite Division of the Institute of Navigation, pp. 1640–1650 (2008)

Energy Detection Based Spectrum Sensing for AWGN and MIMO Wireless Channel

Babji Prasad Chapa, Suresh Dannana and Sasibhushana Rao Gottapu

Abstract The spectrum-sensing aspect has gained importance with the advent of cognitive radio and 4G cellular system with multi-input multi-output (MIMO). It is one of the most demanding issues in both cognitive radio and 4G. Different facets of spectrum-sensing methods are studied from a cognitive radio and MIMO systems perspective. The important task of 4G system is to provide user demanding bandwidth by detecting the existence of the user in order to allocate the unused subcarrier channels to the required users. Hence, spectrum sensing is a crucial part of the 4G and 5G cellular systems. In this paper, a new spectrum-sensing method based on energy detection is proposed for Additive White Gaussian Noise (AWGN) and MIMO wireless channels. Implementation of the proposed probability of detecting a channel in use indicates that for the same values of false alarm (0.096) and SNR (5 dB), the probability of detection observed to be better in case of the MIMO (0.9763) wireless channel when compared to non-fading AWGN (0.9215) channel.

1 Introduction

Spectrum is a most valuable commodity, and the authorized spectrum is accessed only by those who own the spectrum. Cognitive radio is an advanced approach of reutilizing the allocated spectrum in an unauthorized manner [1]. It is the radio that can continuously monitor and modify the transmission specifications based on synergy with the geographical surroundings in which it operates. One main aspect

B. P. Chapa (✉) · S. Dannana · S. R. Gottapu
Department of Electronics and Communication Engineering, A.U.C.E (A), Andhra University, Visakhapatnam 530003, Andhra Pradesh, India
e-mail: babjiprasad.ch@gmail.com

S. Dannana
e-mail: suress445@gmail.com

S. R. Gottapu
e-mail: sasigps@gmail.com

V. Bhateja et al. (eds.), *Intelligent Engineering Informatics*, Advances in Intelligent Systems and Computing 695, https://doi.org/10.1007/978-981-10-7566-7_48

of cognitive radio is self-determining manipulation of the unused spectrum. Cognitive radio does the spectrum sensing, managing, sharing, and mobility [2]. In these, spectrum sensing is one of the necessary features of cognitive radio. Another important application of spectrum sensing is the Orthogonal Frequency Division Multiplexing (OFDM)-based 4G cellular system where it provides user demanding bandwidth on move from the available subcarriers which are separated by a 15 kHz carrier spacing in 20 MHz spectrum. OFDM signal transmission comprises different subcarriers; each subcarrier is modulated with the different data stream [3]. It is important to detect unused spectrum, thereby providing that unused spectrum to the users who require high demanding data rates. A new energy detection-based spectrum-sensing method for AWGN and MIMO wireless channels is proposed.

2 System Model

Spectrum sensing has two essential parameters called detection probability and false alarm probability. The model has been set up for detecting the signal. y is assumed to be a received signal vector of size N that consists of a signal and noise. That is

$$\mathbf{y} = \mathbf{x} + \mathbf{q}$$

where x is the signal and q is the noise.

The signal x is unknown and noise q is assumed to be iid gaussian random variables of zero mean with variances σ_x^2 and σ^2, respectively. If characteristics of the primary user signal are not known, energy detection technique is opted. The main intent is to find whether the signal is present or not. If the signal is not present, it is treated as H_0 and if the signal is present, it is treated as H_1 [4, 5].

2.1 Spectrum Sensing for AWGN Channel

Spectrum sensing can be done based on received signal energy; if the energy of the measured signal is greater than certain threshold, then spectrum is not vacant. In order investigate spectrum sensing in an AWGN channel, consider two hypothesis called null hypothesis H_0 and alternate hypothesis H_1.

$$\text{Under } H_0: \bar{y} = \bar{q} \tag{1}$$

$$\text{Under } H_1: \bar{y} = \bar{x} + \bar{q} \tag{2}$$

where $\bar{x}$, $\bar{q}$, and $\bar{y}$ are vectors of x, q, and y, respectively. Assume that signal and noise samples are uncorrelated and independent.

The expressions for the detection probability and the false alarm probability can be obtained as follows.

Received signal energy is $\|y\|^2 \geq \mathrm{E}^1$, where E^1 is the threshold of the detector.

The false alarm probability is given as,

$$P_{FA} = Q_{\chi_N^2}\left(\frac{\mathrm{E}^1}{\sigma^2}\right) \tag{3}$$

The detection probability is given as,

$$P_D = Q_{\chi_N^2}\left(\frac{\mathrm{E}^1}{\sigma^2 + \sigma_x^2}\right) \tag{4}$$

2.2 Spectrum Sensing for MIMO Wireless Channel

The binary hypothesis testing problem for spectrum sensing in the MIMO wireless channel is given as,

$$H_0 : y = q \tag{5}$$

$$H_1 : y = Kx + q \tag{6}$$

where K is the channel matrix. If the channel matrix K is unknown, one can employ energy detection. Now compare the output $\|y\|^2$ with a suitable threshold E to yield the detector. The hypothesis H_1 is chosen when $\|y\|^2 \geq E$; otherwise, null hypothesis is chosen.

The detection probability and false alarm probability can be calculated as follows,

The false alarm probability is given as,

$$P_{FA} = Q_{\chi_{2r}^2}\left(\frac{E}{\sigma^2/2}\right) \tag{7}$$

The detection probability is given as,

$$P_D = Q_{\chi_{2r}^2}\left(\frac{E}{\frac{1}{2}(P + \sigma^2)}\right) \tag{8}$$

3 Simulation Results

The efficiency of an energy detector for spectrum sensing is evaluated by simulating using MATLAB version R2016b. Monte Carlo (MC) simulations are used to assess the energy detector performance. The receiver performance can be determined by illustrating the receiver operating characteristics. Plotting of receiver operating characteristics is done by keeping one parameter constant and varying the other parameter. The detection probability versus false alarm probability curves are plotted for various signal-to-noise ratios.

Signal-to-noise ratio affects the detection performance and is measured by using an energy detector. Figure 1 shows the detection performance of the detector simulating over an AWGN channel. Here, the false alarm probability is set at different values like 10^{-8}, 10^{-7}, 10^{-6}, and 10^{-4} number of MC sample points set to 10000, and results are plotted between signal-to-noise ratio and detection probability. Analytical and simulated results show that the detection probability increases with increasing the signal-to-noise ratio (Table 1).

In Fig. 2, for different values of the signal-to-noise ratios like −10, −5, 0, and 5 dB, the receiver operating characteristics are plotted. Analytical and simulated results show that the detection probability is high (0.9215) for high SNR (5 dB) at a given probability of error (Table 2).

In Fig. 3 for different values of the signal-to-noise ratios like −10, −5, 0, and 5 dB, the receiver operating characteristics curves are plotted for the MIMO wireless channel. It shows the plot of probability of false alarm versus probability of detection. Analytical and simulated results show that the detection probability is high (0.9763) for the signal-to-noise ratio of 5 dB at a given probability of false alarm (Table 3).

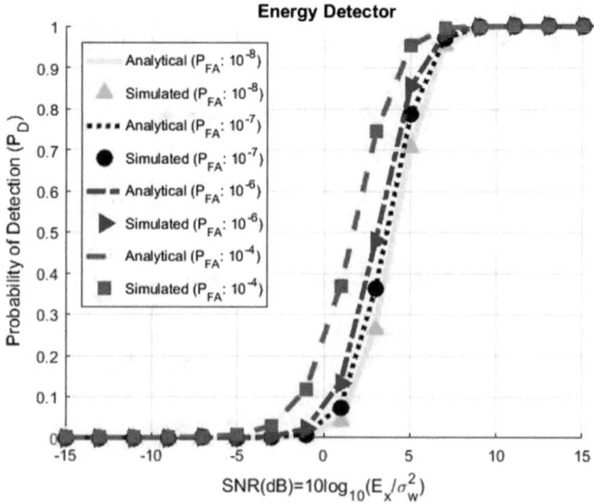

Fig. 1 Effect of SNR on the detection probability for the false alarm probabilities of 10^{-8}, 10^{-7}, 10^{-6}, and 10^{-4}

Table 1 Detection probability at a given SNR

False alarm probability	Signal-to-noise ratio (SNR in dB)	Probability of detection
10^{-4}	5	0.9498 (Analytical)
10^{-4}	5	0.9498 (Simulated)
10^{-6}	5	0.8506 (Analytical)
10^{-6}	5	0.8506 (Simulated)
10^{-7}	5	0.777 (Analytical)
10^{-7}	5	0.777 (Simulated)
10^{-8}	5	0.7016 (Analytical)
10^{-8}	5	0.7016 (Simulated)

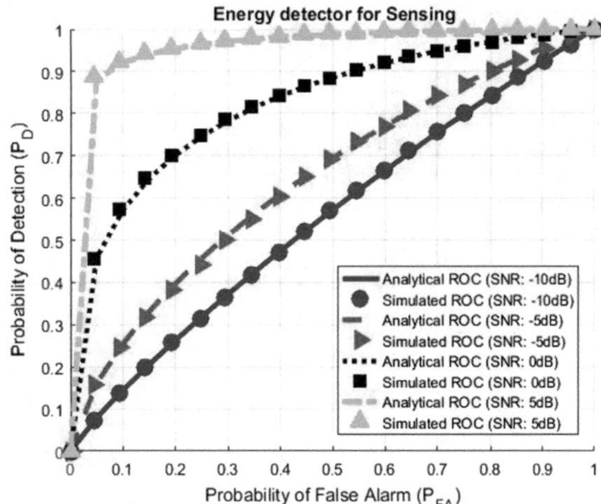

Fig. 2 Receiver operating characteristics for energy detection over AWGN channel

Table 2 Detection probability for a given false alarm probability over AWGN channel

Signal-to-noise ratio (SNR in dB)	False alarm probability	Detection probability
5	0.096	0.9215 (Analytical)
5	0.096	0.9215 (Simulated)
0	0.096	0.5718 (Analytical)
0	0.096	0.5718 (Simulated)
−5	0.096	0.2517 (Analytical)
−5	0.096	0.2517 (Simulated)
−10	0.096	0.1425 (Analytical)
−10	0.096	0.1425 (Simulated)

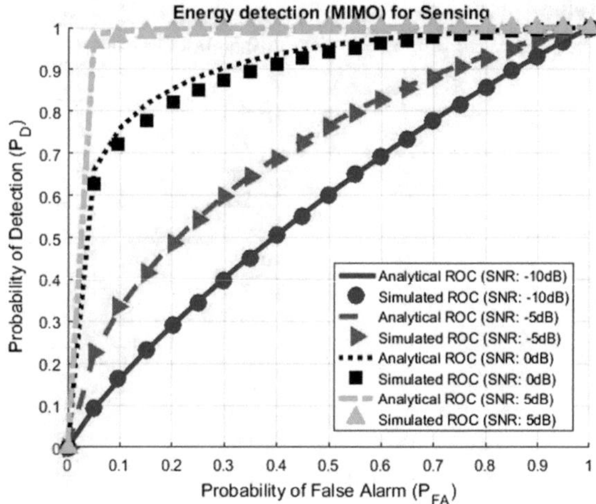

Fig. 3 Receiver operating characteristics for energy detection over MIMO wireless channel

Table 3 Detection probability for a given false alarm probability over MIMO wireless channel

Signal-to-noise ratio (SNR in dB)	False alarm probability	Detection probability
5	0.096	0.9763 (Analytical)
5	0.096	0.9763 (Simulated)
0	0.096	0.7272 (Analytical)
0	0.096	0.7272 (Simulated)
−5	0.096	0.3280 (Analytical)
−5	0.096	0.3280 (Simulated)
−10	0.096	0.1585 (Analytical)
−10	0.096	0.1585 (Simulated)

4 Conclusions

This study imparts the behavior of the energy detection technique for the AWGN and MIMO wireless channels. The energy detector performance in finding an unoccupied spectrum was measured. Simulations are carried out to measure the performance of the primary user signal detector over AWGN and MIMO wireless channels. The detection performance of this particular detector can be enhanced by proper selection of threshold value, and detection probability can be enhanced by increasing the SNR. In case of non-fading channel (AWGN), the detection probability at a given false alarm is better for SNR of 5 dB among selected values. On simulating the system over the MIMO wireless channel, it is observed that the detection probability is high for the signal-to-noise ratio of 5 dB. Finally, it is observed that the detection probability is better in case of MIMO (0.9763) wireless channel when compared to AWGN (0.9215).

References

1. Federal Communication Commission, Spectrum Policy Task Force Report, ET Docket No. 02–155, Nov 2002
2. Wang, J., Ghosh, M., Challapali, K.: Emerging cognitive radio applications: a survey. IEEE Commun. Mag. **49**(3), 74–81 (2011)
3. Amich, A., Imran, M.A., Tafazolli, R., Cheraghi, P.: Accurate and efficient algorithms for cognitive radio modeling applications under the i.n.i.d. paradigm. IEEE Trans. Veh. Technol. **64**(5), 1750–1765 (2015)
4. Digham, F.F., Alouini, M.S., Simon, M.K.: On the energy detection of unknown signals over fading channels. IEEE Trans. Commun. **55**(1), 21–24 (2007)
5. Yucek, T., Arslan, H.: A survey of spectrum sensing algorithms for cognitive radio applications. IEEE Commun. Surv. Tutor. **11**, 116–130 (2009)

Spectrum Sensing Using Matched Filter Detection

Suresh Dannana, Babji Prasad Chapa and Gottapu Sasibhushana Rao

Abstract Increasing use of wireless applications is putting a pressure on licensed spectrum which is insufficient and expensive. Indeed, because of allocation of fixed spectrum, more portion of spectrum is underutilized. Spectrum sensing can be used for efficient use of the radio spectrum. It detects the unused spectrum channels in cognitive radio network. In cognitive radio, spectrum sensing techniques such as energy detection, cyclostationary feature-based spectrum sensing technique, matched filter detection, etc., have been used. When user information is available, matched filter-based sensing gives better performance. In this paper, the probability of detection (PD) and probability of false alarm (PFA) at different SNR levels are observed. Matched filter detection performance depends on threshold value to detect the primary user. At 25 dB SNR, better probability of detection is observed for a given PFAs.

1 Introduction

At present, wireless users have enormously increased. Allocation of spectrum to the users is cumbersome. Cognitive radio networks have been introduced to meet the demand for spectrum. Since radio spectrum is limited, some users' needs spectrum continuously while others underutilize its allocated spectrum, it causes spectrum scarcity. Research on cognitive radio [1–4] has developed the interest on spectrum sensing and detection of spectrum users. Maximizing the probability of detection is without sacrificing much probability of false alarm while minimizing

S. Dannana (✉) · B. P. Chapa · G. S. Rao
Department of Electronics and Communication Engineering, A.U.C.E (A),
Andhra University, Visakhapatnam 530003, Andhra Pradesh, India
e-mail: suress445@gmail.com

B. P. Chapa
e-mail: babjiprasad.ch@gmail.com

G. S. Rao
e-mail: sasigps@gmail.com

V. Bhateja et al. (eds.), *Intelligent Engineering Informatics*, Advances in Intelligent Systems and Computing 695, https://doi.org/10.1007/978-981-10-7566-7_49

the hardware and the time to sense the spectrum is the main intention of the spectrum sensing. The radio spectrum can be sensed by various detection techniques, i.e., energy detection, [5] matched filter [6–8], cyclization feature detection, [9] and stochastic process techniques to detect PU signal in the channel. When the secondary user has complete information (such as type of modulation, frequency of carrier, bit rate, shape of the pulse, etc.) of primary user, matched filter detection is the better choice than other techniques.

2 Spectrum Sensing

In this paper, to sense primary user, matched filter-based likelihood ratio test introduced. Consider a signal model, having binary hypotheses H_0 and H_1 to represent the presence of radio signal.

Under binary hypothesis H_0, signal model is

$$x(n) = w(n) \tag{1}$$

Under hypothesis H_1, signal model is

$$x(n) = s(n) + w(n) \tag{2}$$

where x(n) is received signal, s(n) is primary user signal, typically binary phase shift keyed signal follows the Gaussian distribution. w (n) is the Gaussian noise signal with a zero mean, and variance σ^2.

2.1 Likelihood Ratio Test

Consider the signal s(i) and Gaussian noise w(i), having zero mean, variance $\boldsymbol{\sigma}^2$. Likelihood ratio test [7, 8, 10] is used to detect primary user. Here all N signals are independent, so that the joint probability density function (PDF) under $H_{0,}$ i.e., primary user is absent in,

$$\prod_{i=1}^{N} \frac{1}{\sqrt{2\pi\sigma^2}} e^{-\frac{[x(i)]^2}{2\sigma^2}} \tag{3}$$

where x(i) is *i*th received signal.

The joint PDF under $H_{1,}$ i.e., primary user is present in,

$$\prod_{i=1}^{N} \frac{1}{\sqrt{2\pi\sigma^2}} e^{-\frac{[x(i)-s(i)]^2}{2\sigma^2}} \tag{4}$$

Likelihood ratio is the ratio of $p(H_1)$ and $p(H_0)$, which is used to detect PD and PFA. This ratio is compared with threshold value to detect primary user under two hypothesis conditions.

$$\gamma(x) = \frac{p(H_1)}{p((H_0)} = \frac{\left(\frac{1}{\sqrt{2\pi\sigma^2}}\right)^N e^{-\frac{1}{2\sigma^2}\|\bar{x}-\bar{s}\|^2}}{\left(\frac{1}{\sqrt{2\pi\sigma^2}}\right)^N e^{-\frac{1\|\bar{x}\|^2}{2\sigma^2}}} \tag{5}$$

where p(H$_0$) and p(H$_1$) are priori probabilities. The decision can be taken accordingly

$$\gamma(x) = \begin{cases} \geq \gamma & \text{choose } \mathrm{H}_1 \\ < \gamma & \textit{choose } H_0 \end{cases}$$

where γ is the threshold value used to detect primary user. Likelihood ratio test is tedious one to detect primary user.

2.2 *Matched Filter*

Since likelihood ratio test is tedious, so that log-likelihood ratio test is used to simplify Eq. (5).

Under H$_1$ is

$$\begin{aligned} \ln\gamma(x) &= -\frac{1}{2\sigma^2}\left[\|\bar{x}-\bar{s}\|^2 - \|\bar{x}\|^2\right] \geq \ln\gamma \\ &= \|x\|^2 - \|\bar{x}-\bar{s}\|^2 \geq (\ln\gamma)2\sigma^2 \\ &= \bar{s}^{\mathrm{T}}\bar{x} \geq \frac{2\sigma^2\ln\gamma + \|\bar{s}\|^2}{2} \\ \gamma' &= \frac{2\sigma^2\ln\gamma + \|\bar{s}\|^2}{2} \end{aligned}$$

where γ' represents a new threshold value.

Where $\bar{s}^{\mathrm{T}}\bar{x}$ represents matched filter, which is having maximum value when the received signal matched with the transmitting signal.

$$\begin{array}{ll} \text{If } \bar{s}^{\mathrm{T}}\bar{x} \geq \gamma' & \textit{choose } H_1 \\ \text{Otherwise} & \textit{choose } H_0 \end{array}$$

$$\begin{array}{lll} H_1 & \textit{if} & \bar{s}^{\mathrm{T}}\bar{x} \geq \gamma' \\ H_0 & \textit{if} & \bar{s}^{\mathrm{T}}\bar{x} < \gamma' \end{array}$$

Under hypothesis H_0, $\bar{s}^T\bar{x} = s(1)w(1) + s(2)w(2) + \ldots s(N)w(N)$
Signal $\bar{s}^T\bar{x}$ is Gaussian with mean which is zero, variance is $\|\bar{s}\|^2\sigma^2$
variance of $\bar{s}^T\bar{x}$ is $\tilde{\sigma} = \|s\|^2\sigma^2$
Probability of false alarm (PFA) is

$$P_{FA} = P_r(\bar{s}^T\bar{x} \geq \gamma'/H_0)$$
$$= P_r(\frac{\bar{s}T\bar{x}}{\sigma\|s\|} \geq \frac{\gamma'}{\sigma\|s\|}/H_0) \tag{6}$$
$$P_{FA} = Q\left(\frac{\gamma'}{\sigma\|s\|}\right)$$

so that threshold

$$\gamma' = \sigma\|s\|Q^{-1}(PFA) \tag{7}$$

Under H_1
Probability of detection PD

$$P_D = P_r(\bar{s}^T\bar{x} \geq \gamma'/H_1)$$
$$\text{Here}\bar{x} = \bar{s} + \bar{w}$$

Signal $\bar{s}^T\bar{x}$ is Gaussian with mean $\|s\|^2$ and Variance $= \sigma^2\|s\|^2$

$$P_D = P_r(\bar{S}^T\bar{x} \geq \gamma^{|}/H_1)$$
$$P_r(\frac{\bar{S}^T\bar{X} - \overline{\|S\|}^2}{\sigma\|S\|} \geq \frac{\gamma^{|} - \|\bar{S}\|^2}{\sigma\|S\|}/H_1)$$

$$P_D = [Q\ (\frac{\gamma^{|} - \|\bar{S}\|^2}{\sigma\|S\|})] \tag{8}$$

From Eqs. (7) and (8), probability of detection is

$$P_D = Q(Q^{-1}(PFA) - \sqrt{\frac{\varepsilon}{\sigma^2}}) \tag{9}$$

where $\sqrt{\frac{\varepsilon}{\sigma^2}}$ is deflection coefficient. $\frac{\varepsilon}{\sigma^2}$ is signal to noise ratio.

Probability of detection increases by either increasing SNR or by increasing the PFA.

3 Results

The matched filter performance is observed for various SNR values as well as for a given probability of false alarms (PFAs).

The sensing performance of the proposed scheme, in terms of its receiver operating characteristics (ROC) curve, is evaluated using Monte Carlo simulations. The performance of matched filter detection technique mainly depends on signal to noise ratio (SNR) gained by secondary users. Figure 1 shows analysis between probability of detection (PD) and probability of false alarms (PFAs) for different SNR values. In this paper, four SNR values (−5, 0, 5 and 10 dB) are considered. From Fig. 1, it is observed that at 10 dB SNR, more PD is achieved. As the SNR increases, matched filter performs well. Simulation results are compared with analytical values.

Figure 2 shows plot between probability detection and SNR for a fixed value of PFA. Here four values of PFA are considered. They are 10^{-4}, 10^{-3}, 10^{-2}, and 10^{-1}. From Fig. 2, it is observed that as SNR increases, the probability of detection is improved. It can be verified from Eqs. (8) and (9). Simulation results are also compared with analytical values.

Table 1 shows threshold values for different SNR for a fixed PFA. Here four values of PFA are considered. They are 10^{-4}, 10^{-3}, 10^{-2}, and 10^{-1}. From Table 1, it is observed that at SNR = 0 dB, the threshold values are 119.008, 98.8874, 74.4431, and 41.0097 obtained for different values of PFA such as 10^{-4}, 10^{-3}, 10^{-2}, 10^{-1}, respectively. That is, as PFA increases, threshold value is decreased

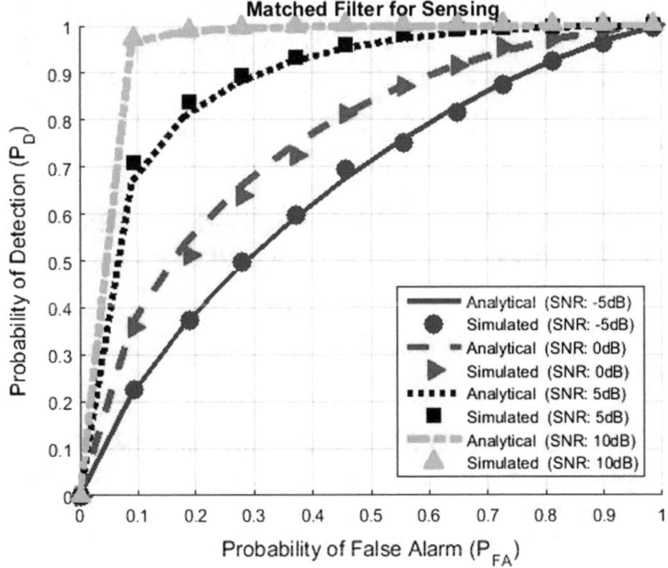

Fig. 1 Receiver operating characteristics

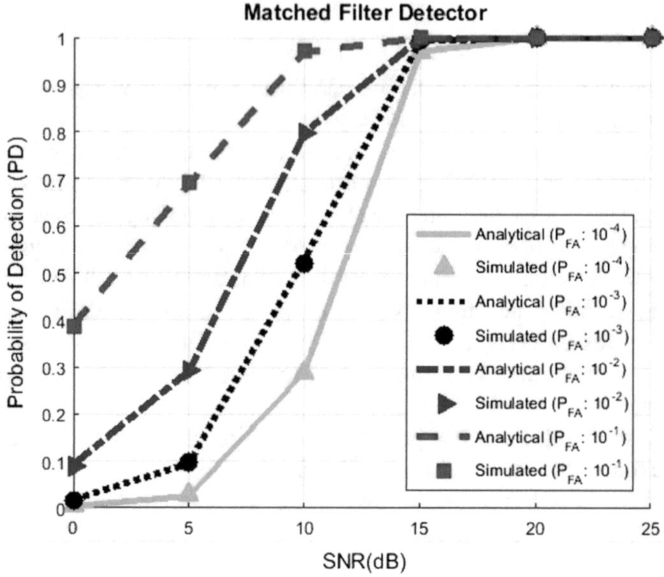

Fig. 2 Probability of detection versus SNR

Table 1 Threshold values for probability of detection

SNR in dB	PFA = $\mathbf{10^{-4}}$	PFA = $\mathbf{10^{-3}}$	PFA = $\mathbf{10^{-2}}$	PFA = $\mathbf{10^{-1}}$
0	119.008	98.8874	74.4431	41.0097
5	66.9234	55.6085	41.8624	23.0614
10	37.6338	31.2710	23.5410	12.9684
15	21.1630	17.5849	13.2381	7.2927
20	11.9009	9.8887	7.4443	4.1010
25	6.6923	5.5608	4.1862	2.3061

which leads to better probability of detection. Similarly, as SNR increases, the corresponding threshold values are decreased for a fixed PFA.

4 Conclusion

In order to sense the spectrum holes consistently and resourcefully, this paper proposed a spectrum sensing based on matched filter detection in CR networks. It needs complete information (such as type of modulation, frequency of carrier, bit rate, shape of the pulse, etc.) of primary user in advance. Monte Carlo simulation is used, which shows good performance. Simulations are carried out to measure the performance of matched filter to detect primary user over AWGN channel.

For 25 dB SNR, better probability of detection is achieved for a given PFA. As PFA increases, threshold values are decreased. This can be observed from Table 1. For a 0 dB SNR, lower threshold is achieved for PFA 0.1. If the signal characteristics are known, matched filter gives better probability of detection.

References

1. Kay, S.M.: Fundamentals of Statistical Signal Processing, Estimation Theory, vol. I. Prentice Hall (1993)
2. Kay, S.M.: Fundamentals of Statistical Signal Processing, Detection Theory, vol. I. Prentice Hall (1998)
3. Van Trees, H.L.: Detection, Estimation, and Modulation Theory. Wiley (2004)
4. Subhedar, M., Birajdar, G.: Spectrum sensing techniques in cognitive radio networks: a survey. Int. J. Next-Gener. Netw. **3**(2), 37–51 (2011)
5. Plata, D.M.M., Reátiga, Á.G.A.: Evaluation of energy detection for spectrum sensing based on the dynamic selection of detection-threshold. Procedia Eng. **35**, 135–143 (2012)
6. Salahdine, F., Ghazi, H.El., Kaabouch, N., Fihri, W.F.: Matched filter detection with dynamic threshold for cognitive radio networks. In: 2015 International Conference on Wireless Networks and Mobile Communications (WINCOM), pp. 1–6. IEEE (2015)
7. Alvi, S.A., Younis, M.S., Imran, M., Amin, F.: A log-likelihood based cooperative spectrum sensing scheme for cognitive radio networks. Procedia Comput. Sci. **37**, 196–202 (2014)
8. Vadivelu, R., Sankaranarayanan, K., Vijayakumari, V.: Matched filter based spectrum sensing for cognitive radio at low signal to noise ratio. J. Theor. Appl. Inf. Technol. **62**(1) (2014)
9. Derakhshani, M., Le-Ngoc, T., Nasiri-Kenari, M.: Efficient cooperative cyclostationary spectrum sensing in cognitive radios at low SNR regimes. IEEE Trans. Wireless Commun. **10** (11), 3754–3764 (2011)
10. Axell, E., Leus, G., Larsson, E.G.: Overview of spectrum sensing for cognitive radio. In: 2010 2nd International Workshop on Cognitive Information Processing (CIP), pp. 322–327. IEEE (2010)

Multi-antenna System Performance Under ICSIT and ICSIR Channel Conditions

Vinodh Kumar Minchula, G. Sasibhushana Rao and Ch. Vijay

Abstract In 5G cellular systems, MIMO plays an important role in the access technology and radio access network topology. The MIMO uses an additional space dimension beyond time and frequency to enhance the QoS by providing diversity gain, data rate (capacity) by multiplexing gain, coverage and outage by array gain. In this paper, different multi-antenna system performances are compared using a capacity performance metric. The capacity from performance metric is expressed in terms of spectral efficiencies versus SNR in dB. Results indicate that by increasing the number of transmitting and receiving antennas, there is an improvement in the capacities of each SIMO, MISO and MIMO system configurations without boosting the transmitted power and additional spectral requirement. Results also indicate that the capacity of MIMO wireless system is more when compared to the other diversity techniques (SIMO and MISO).

Keywords SISO · SIMO · MISO · MIMO · ICSIT · ICSIR
Singular value decomposition (SVD)

1 Introduction

Wireless communication systems have become ubiquitous. With the demand for multimedia services, the ability to provide spectral efficiency and data rates is one of the vital considerations for next-gen wireless systems [1, 2]. Wideband wireless communication is a rapidly growing technology, with the requirement to offer an access to broad range of data packet services, which supports low- and

V. K. Minchula (✉) · G. S. Rao · Ch. Vijay
Department of ECE, AUCE(A) Andhra University, Visakhapatnam 530003, India
e-mail: vinodh.edu@gmail.com

G. S. Rao
e-mail: sasigps@gmail.com

Ch. Vijay
e-mail: vchanamala@gmail.com

V. Bhateja et al. (eds.), *Intelligent Engineering Informatics*, Advances in Intelligent Systems and Computing 695, https://doi.org/10.1007/978-981-10-7566-7_50

high-mobility applications with high-quality capabilities. 4G wireless systems such as 3GPP LTE (long-term evolution) and LTE-A are developed for achieving global broadband mobile communications. Its primary goals are to have high data rate radio access services, improved coverage and capacity systems, significant improvement in spectral efficiency, low latency, diversity support and backward compatible with the existing cellular systems.

The maximum achievable data rate is a primary requirement of LTE standards. According to Shannon's channel capacity, the available received signal-to-noise ratio or received signal power limits the data rate. Another factor limiting the data rate is the transmission bandwidth.

(a) If data rates are lower than the available bandwidth (low bandwidth utilization), then increase in received signal power will give high data rates.
(b) If data rates are higher than or equal the available bandwidth (high bandwidth utilization), then much larger relative increase in received signal power will give high data rates.

Wider transmission bandwidth supports higher data rate with a challenge of multipath fading on channel. Besides the direct path (LoS), there are other paths because of diffraction, scattering, reflection or other propagation scenarios (NLoS).

The other way to increase the receiver power for achieving higher data rates is to use multiple antenna system such as diversity techniques like SIMO, MISO and MIMO systems [3–6].

In this paper, the capacities of multi-antenna systems under the knowledge of instantaneous channel conditions known at both transmitter (ICSIT) and receiver (ICSIR) are determined.

2 MIMO System Model

MIMO system model can be discussed by considering a communication channel with n_T transmitters and n_R receivers (Fig. 1).

The mathematical model of MIMO channel can be represented as

$$r_k = Hs_k + z_k \tag{1}$$

where $r_k = \left[r_k^1, r_k^2, \ldots, r_k^{n_R}\right]^T$ is the signal received at k time instant, $s_k = \left[s_k^1, s_k^2, \ldots, s_k^{n_T}\right]^T$ is the transmitted signal, and z_k is AWGN with variance 1 and spatially uncorrelated among the n_r receiver antennas. The antenna j receives a superposition of messages transmitted from transmitter antenna i, multiplied by the channel response, and Gaussian noise is added [7, 8].

The $n_R \times n_T$ channel matrix with elements h_{ji} is represented as follows

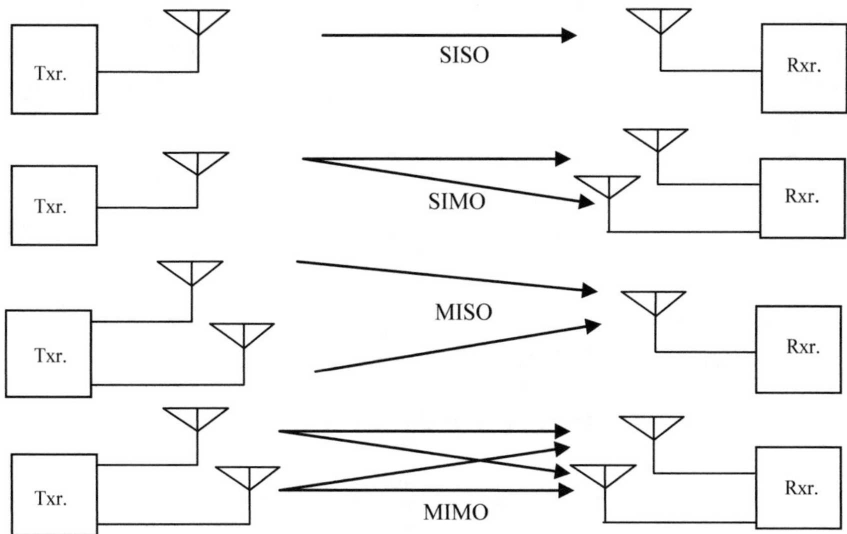

Fig. 1 Schematic block diagram of SISO, SIMO, MISO and MIMO systems

$$\text{channel matrix } (\mathrm{H}) = \begin{bmatrix} h_{11} & \cdots & h_{1n_T} \\ \cdots & \cdots & \cdots \\ h_{n_R 1} & \cdots & h_{n_R n_T} \end{bmatrix}_{n_R \times n_T} \tag{2}$$

where h_{ji} is the complex channel coefficient between the ith antenna at transmitter and the jth antenna at receiver side.

3 Performance Characteristics of Multi-antenna Systems

In this section, the channel capacities are determined and performance improvement in systems by multiplexing gain with increase in number of transmitting (n_T) and receiving antennas (n_R) is discussed.

Capacity is a performance measure and explains about the maximum transmission rate achieved for a practical communication. If this rate is larger than the capacity, then high probability of error occurs at the receiver side [9, 10]. So it is important parameter to characterize the performance of multi-antenna systems.

Wireless channels exhibit different performance characteristics according to its propagation environment. The channels are time-varying, and having the information about channel is difficult at both the receiver and transmitter. The obtained channel information is not the same as the instantaneous channel information. It is assumed in this paper that channel is deterministic and noise is circular symmetric and complex Gaussian (ZMCSCG) random variable with zero mean [3] denoted as $z_k \sim N(0, \sigma_n^2 I)$, instantaneous channel state information is available at the receiver

(ICSIR), and also instantaneous channel knowledge is known to transmitter (ICSIT). So for this scenario, the single-input single-output (SISO), multiple-input single-output (MISO), single-input multiple-output (SIMO) and multiple-input multiple-output (MIMO) capacities are built.

- SISO channel capacity C_{SISO} is given by

$$C_{SISO} = \log_2(1 + \frac{\bar{P}}{\sigma_n^2}|h|^2) \qquad \text{bits/s/Hz} \tag{3}$$

  where the received signal is expressed as $r_k = hs_k + z_k$, $\frac{\bar{P}}{\sigma_n^2}$ is the SNR at a receiver antenna, and h is the normalized channel gain.

- SIMO channel capacity C_{SIMO} is given by

$$C_{SIMO} = \log_2(1 + \frac{\bar{P}}{\sigma_n^2}\|h\|^2) \qquad \text{bits/s/Hz} \tag{4}$$

  where the received signal is expressed as $r_{j,k} = h_{j,k}s_k + z_{j,k}$

  $h = (h_{1,k}, h_{2,k}, \ldots, h_{n_R,k})^T$ is the channel coefficients at time instant k and can be expressed as

$$\|h^2\| = |h_1|^2 + |h_2|^2 + \cdots + |h_{n_R}|^2 \tag{5}$$

- MISO channel capacity C_{MISO} is given by

$$C_{MISO} = \log_2(1 + \frac{\bar{P}}{n_T\sigma_n^2}\|h\|^2) \qquad \text{bits/s/Hz} \tag{6}$$

  where the received signal is expressed as $r_k = h_{i,k}^T s_{i,k} + z_k$

  $h = (h_{1,k}, h_{2,k}, \ldots, h_{n_T,k})^T$ is the channel coefficients at time instant k for the input preprocessing signal representation $s_{i,k} = \frac{h_i^*}{\|h\|}s_k$

- MIMO channel capacity C_{MIMO} is given by

$$C_{MIMO} = \log_2(\det(I + \frac{\bar{P} \cdot HH^*}{\sigma_n^2})) \tag{7}$$

$$C_{MIMO} = \sum_{k=1}^{n} \log_2(1 + \frac{\bar{P}}{\sigma_n^2}|\lambda_k|^2) \quad \text{bits/s/Hz} \qquad n = \min(n_R, n_T) \tag{8}$$

where SVD of the channel matrix $H = TSK^H$

T is a $n_R \times n_R$ unitary matrix, K is also a $n_T \times n_T$ unitary matrix, and S is a $n_R \times n_T$ diagonal matrix with all other elements equal to zero. (H) is a complex conjugate transpose. λ_k is also the eigenmodes of the channel [7], where $\lambda_1 > \lambda_2 > \cdots > \lambda_n$

$$\mathrm{tr}(\mathrm{HH^H}) = \sum_{i=1}^{n_T} \sum_{j=1}^{n_R} \left|h_{ji}\right|^2 = \sum_{k=1}^{n} |\lambda_k|^2 \tag{9}$$

4 Results and Discussions

This section presents the results to illustrate the performance improvement in terms of capacity under ICSIT and ICSIR conditions. Figure 2 illustrates the spectral efficiencies of SISO, SIMO, MISO and MIMO systems for different SNR value ranges from 0 to 40 dB.

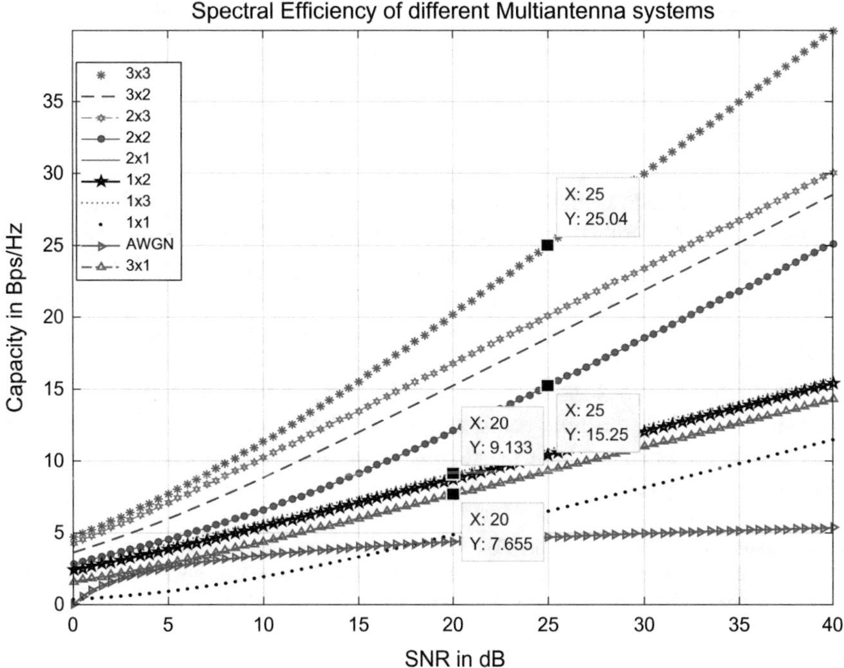

Fig. 2 Spectral efficiency of SISO, SIMO (1 × 2, 1 × 3), MISO (2 × 1, 3 × 1), MIMO (2 × 2, 3 × 2, 2 × 3, 3 × 3) systems versus SNR in dB

Table 1 Spectral efficiencies versus SNR in dB

S. no.	SNR	1 × 1	1 × 2	1 × 3	2 × 1	2 × 2	2 × 3	3 × 1	3 × 2	3 × 3
1.	1	0.44	2.68	3.01	1.79	3.18	4.81	1.82	4.06	5.23
2.	5	0.93	3.86	4.23	2.85	4.56	7.10	2.88	5.99	7.69
3.	10	1.94	5.45	5.83	4.36	6.61	10.24	4.40	8.84	11.33
4.	15	3.32	7.09	7.48	5.98	9.13	13.50	6.01	11.99	15.54
5.	20	4.88	8.75	9.13	7.62	12.07	16.80	7.65	15.25	20.18
6.	25	6.51	10.40	10.79	9.28	15.25	20.11	9.31	18.55	25.04
7.	30	8.16	12.06	12.45	10.94	18.52	23.43	10.97	21.87	29.98
8.	35	9.82	13.73	14.11	12.60	21.83	26.75	12.63	25.19	34.95
9.	40	11.47	15.39	15.77	14.26	25.15	30.08	14.29	28.51	39.93

By considering equations given in Sect. 3 (Eqs. (3), (4), (6), (7)) for multi-antenna systems respective capacities under the noise considered in the channel study (ICSIT, ICSIR) is obtained. All simulations are done in MATLAB.

Table 1 gives the capacities of SIMO, MISO and MIMO systems, where 3 × 3 MIMO at 25 dB of SNR is giving the spectral efficiency of 25.04 Bps/Hz and 2 × 2 MIMO is giving the spectral efficiency of 15.25 Bps/Hz. The MISO at 20 dB of SNR is giving the spectral efficiency of 7.655 Bps/Hz which is less when compared to SIMO spectral efficiency of 9.133 Bps/Hz because of its array gain.

5 Conclusions

The capacities of various multi-antenna systems are estimated, and performances are compared by considering ICSIT and ICSIR channel conditions. Spectral efficiency of SIMO is observed to be better than that of MISO system in case of the Rayleigh fading channel, whereas the performance is same in case of AWGN condition for both SIMO and MISO systems. MIMO system performance improves because of considering multiple antennas at receiver and transmitter. This is one of the important features in 4G systems. It is found that for a given SNR value of 25 dB, in case of 3 × 3 multi-input multi-output (MIMO) performance (25.04 bps/Hz) is quadruple the Shannon limit of a single-input single-output (SISO) system performance (6.51 bps/Hz).

Acknowledgements The work undertaken in this paper is supported by Ministry of Social Justice and Empowerment, Govt. of India, New Delhi, under UGC NFOBC Fellowship Vide Sanction letter no. F./2016-17/NFO-2016-17-OBC-AND-26194 / (SA-III/Website).

References

1. Sasibhushana Rao, G.: Mobile Cellular Communication. New Delhi, Pearson Education (2013)
2. Telatar, E.: Capacity of multi-antenna Gaussian channels. Eur. Trans. Telecommun. **10**(6), 585–595 (1999)
3. Brown, T., DeCarvalho, E., Kyritsi, P.: Practical Guide to the MIMO Radio Channel with MATLAB Examples. Wiley (2012)
4. Goldsmith, A.: Wireless Communications, pp. 299–310. Cambridge University Press (2005)
5. Winters, J.: On the capacity of radio communication systems with diversity in a rayleigh fading environment. IEEE J. Sel. Areas Commun. **5**, 871–878 (1987)
6. Molisch, A., Win, M., Winters, J.H.: Reduced-complexity transmit/receive-diversity systems. IEEE Trans. Signal Proc. **51**, 2729–2738 (2003)
7. Foschini, G.J., Gans, M.J.: On limits of wireless communications in a fading environment when using multiple antennas. Wirel. Pers. Comm. **6**(3), 311–335 (1998)
8. Vaughan, R.G., Anderson, J.B.: Antenna diversity in mobile communications. IEEE Trans. Antennas Propag. **49,** 954–960 (1987)
9. Goldsmith, A., Jafar, S., Jindal, N., Vishwanath, S.: Capacity limits of MIMO channels. IEEE J. Select. Areas Comm. **21**, 684–701 (2003)
10. Paulraj, A., Nabar, R., Gore, D.: Introduction to Space-Time Wireless Communition, pp. 66–153. Cambridge Uniersity Press (2003)

Cloud Adoption: A Future Road Map for Indian SMEs

Biswajit Biswas, Sajal Bhadra, Manas Kumar Sanyal and Sourav Das

Abstract The Cloud Environment has taken an important role in present Indian organizations mainly for small and medium-scaled enterprises (SME). The Software as Service (SAAS) model of cloud implementation has enabled them pay-per-use strategy to leverage their business needs with minimum investments. The various initiatives, like moving toward cashless transactions, taken by Government of India in recent past, forced them to adapt software system for their daily business requirements. It has created a huge business opportunity for the cloud solution providers. In this paper, authors have surveyed and outlined the current scenario of cloud adoption by Indian enterprises. The authors also tried to focus on various facets of cloud adoption along with the factors that are blockers for cloud adoption for Indian enterprises.

Keywords Cloud adoption · Dashboard · Indian SMEs

The original version of this chapter was revised: Author names in reference 9 has been corrected. The correction to this chapter is available at https://doi.org/10.1007/978-981-10-7566-7_67

B. Biswas (✉) · S. Bhadra · M. K. Sanyal · S. Das
Department of Business Administration, University of Kalyani,
Kalyani 741235, West Bengal, India
e-mail: biswajit.biswas0012@gmail.com

S. Bhadra
e-mail: sajal.bhadra@gmail.com

M. K. Sanyal
e-mail: manassanyal123@gmail.com

S. Das
e-mail: Sourav_das@live.in

V. Bhateja et al. (eds.), *Intelligent Engineering Informatics*, Advances in Intelligent Systems and Computing 695, https://doi.org/10.1007/978-981-10-7566-7_51

1 Introduction

Digitization is the main thrust for Government of India (GoI) and already taken few initiatives to achieve this. "Digital India" initiatives in one of them which started from the year 2015. Cloud computing could play a pivotal role to make it a successful one. Cloud-based ERP system is also getting popularity among the Indian organization to perform the internal business process smoothly and extract value from the transactional data to chalk out the plan and activities for future strategies of the enterprises. They orchestrated various departmental processes and take pivotal role to utilize all its available resources aiming at more outcome as an enterprise. Cloud ERP solutions, instead of On-Premises ERP solutions, are more cost effective. Any organization can plan for using Cloud ERP solution for their business needs only. Instead of purchasing license of various modules, organizations can only be opted for selected services depending on their business needs. Cloud ERP solution is a game changer in current business situation that can allow enterprise to go for pay-per-usage model [1]. Thus, with paying minimal cost enterprises can avail the software services to perform all of it business process [2]. There is no need for them to invest huge cost to maintain whole infrastructure along jumbo ERP modules. Authors have performed survey using well-defined questionnaires among the Indian small and medium-sized enterprises (SMEs) and tried to identify the status of Cloud ERP adoption status for Indian SMEs.

2 Related Works

The main characteristics of business, today's, are dynamism of business process and gather a huge number of information from various sources [3]. The top management formulates their strategy and opted for the software solutions that could help them to minimize the implementation as well as recurring cost of the business. Cloud ERP solutions, available in the market, offer significant computing and analytic power and capacity to help top management to devise strategies for future demand and supply. The Cloud ERP solutions are allowing the features to manage real-time disaster and crisis management [4]. The Cloud ERP solution is used for analyzing, predicting of objective, and controlling the process [5]. According to Gartner report [6], "it was pointed out that Cloud ERP with analytics will remain top focus through the year 2017 and 2018 for CFOs (chief information officer), and it was also reported that cloud-based ERP business is projected to grow from $0.75 billion in 2014 to $2.94 billion in 2018" [7]. The Cloud ERP solutions should be adopted in a business to make sure that the profit margin will be beneficial from the past. The built-in data analysis techniques available with Cloud EPR is used to identify the present business scenario and help to overcome today's challenges [8]. It improves the business administration quality [9]. In a survey of Xia and Gong cited that Cloud ERP Solutions are 81% accurate to make a business report. 78% for decision making in a business, 56% for customer service and 49% increasing the company revenue [10]. To improve the efficiency of business organizations, it is necessary to implement the Cloud EPR solutions depending on the enterprise's business needs. As per the Gartner report 2016, Microsoft, SAP, Oracle is the market leader as Cloud ERP vendors [11].

3 Methodology

We have done stratified survey to collect primary data. We prepared well-formatted questionnaires. The survey was conducted mainly in the selected four metropolitan cities from India and collected the detailed feedback from all the respondents. The data collection has involved 167(One hundred and sixty-seven) interviews from various organizations in duration of 6 months (from January 2017 to June 2017). The data collection was planned such a way that the feedbacks were collected from respondents working in various organization sectors like Information Technology Organizations, Banking and Insurance Institutions, Educational Institutions, Healthcare Industries, Transport and Hospitality Organizations along with Fast-Moving Consumer Industries (Fig. 1). The collected data are from four

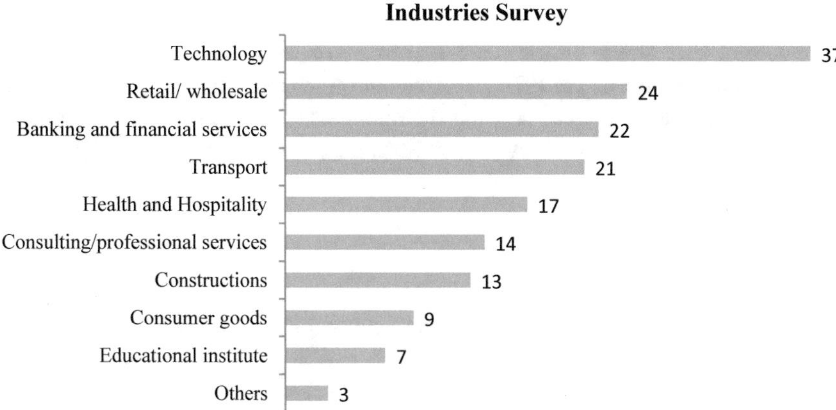

Fig. 1 Distribution of industries covered during survey

Table 1 Respondent numbers from various cities

City	Number of respondents	Active users	Potential users	No users
Delhi	34	24	7	3
Mumbai	30	23	5	2
Bangalore	40	35	14	2
Kolkata	63	31	27	5

Total respondents = 167
Active users = 113
Potential users = 42
No users = 12
[a]Primary data source: Collected by the authors

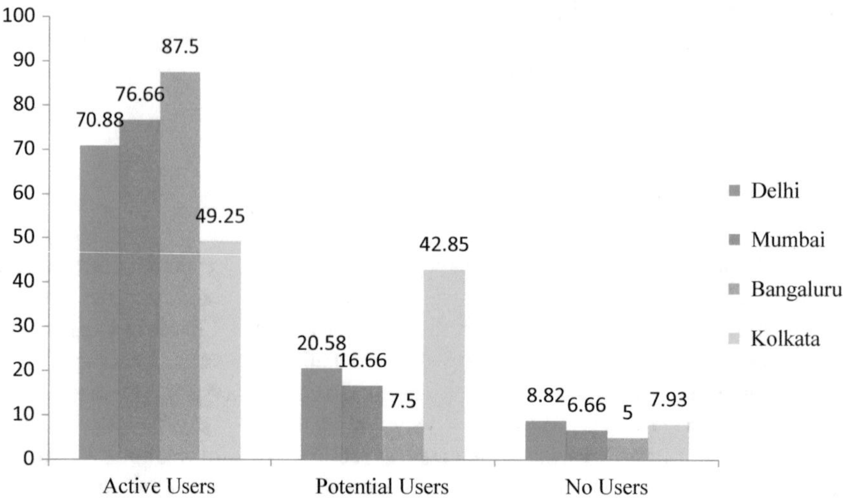

Fig. 2 Respondent graph

metropolitan cities from four parts in India, i.e., Delhi, Mumbai, Bangalore, and Kolkata (Table 1). The respondents were of three categories like those who are using Cloud ERP solutions for last few months or years, i.e., active users, those who have planned for Cloud ERP implementation, or cloud solution adoption is in progress, i.e., potential users and those respondents belonging to the enterprises who have not planned yet for cloud solutions adoption, i.e., no users (Fig. 2).

4 Result and Analysis

Depending on the survey data, authors have plotted the cloud users in a graphical view shown in Figs. 3, 4, and 5. The Indian SMEs are going toward the cloud solution.

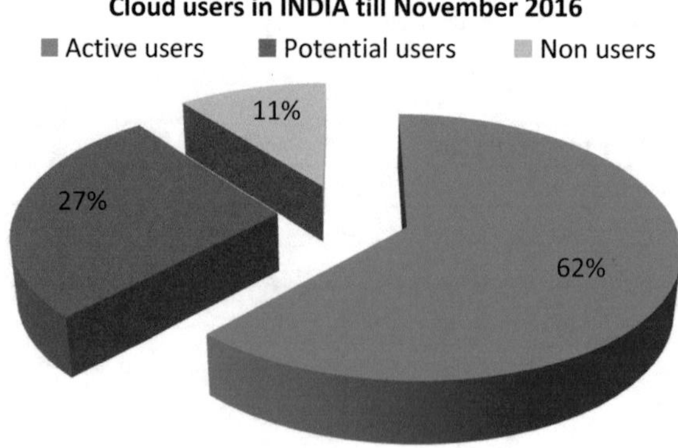

Fig. 3 Segmentation of cloud users till Nov 2016

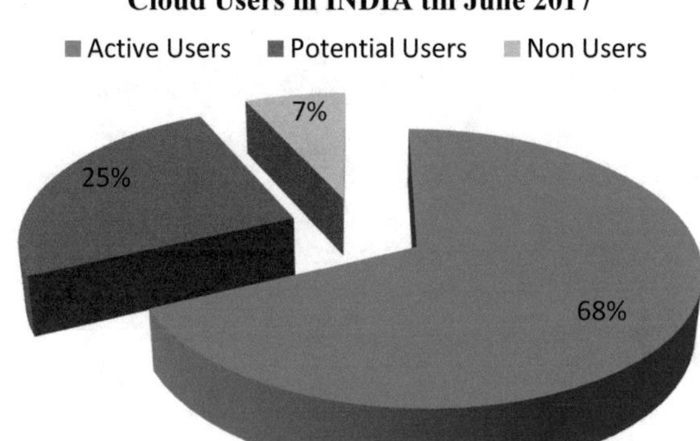

Fig. 4 Segmentation of cloud users till June 2017

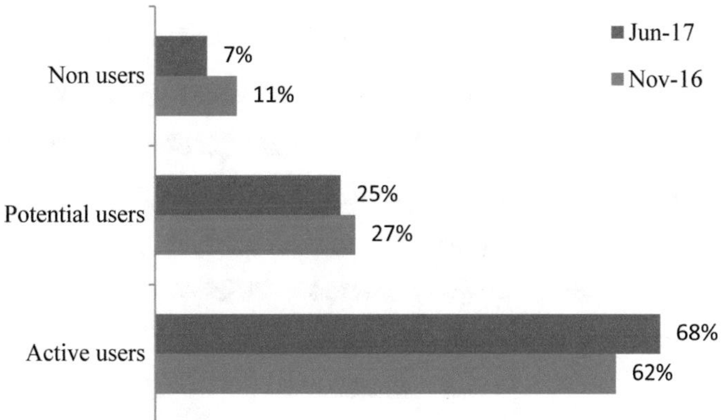

Fig. 5 Comparison of cloud adoption

As Fig. 4 shows, there are around 68% respondents from Indian enterprises who have already implemented Cloud ERP solutions for their business needs, whereas around 25% of the respondents are from the organizations that are planning or implementing Cloud ERP solutions. Only 7% of them belong to the organization who has not thought of cloud solutions.

The main aim of this research work is to identify the current scenario of Indian enterprises in the context of adoption of cloud solutions or Cloud ERPs. We have noticed from Table 1 that in each city the number of cloud users is increased daily. Most of the MNC business house has already known the cloud business intelligence, but the SMEs are not well informed. From the Figs. 4 and 5, it is very easy to understand that from month of November 2016 the small and medium-sized enterprises (SMEs) are adopting or planning for adopting cloud solutions for their

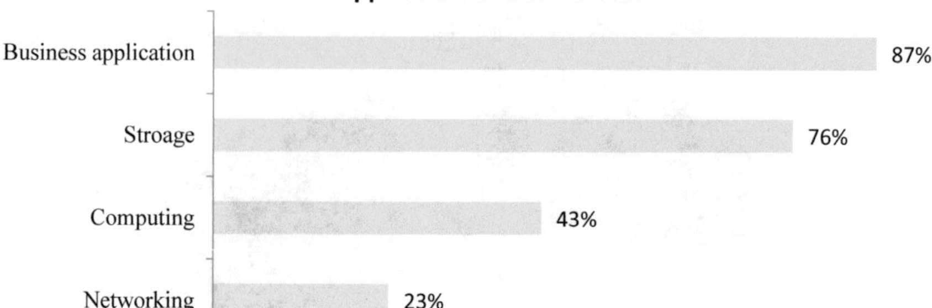

Fig. 6 Distribution of cloud adoption purposes

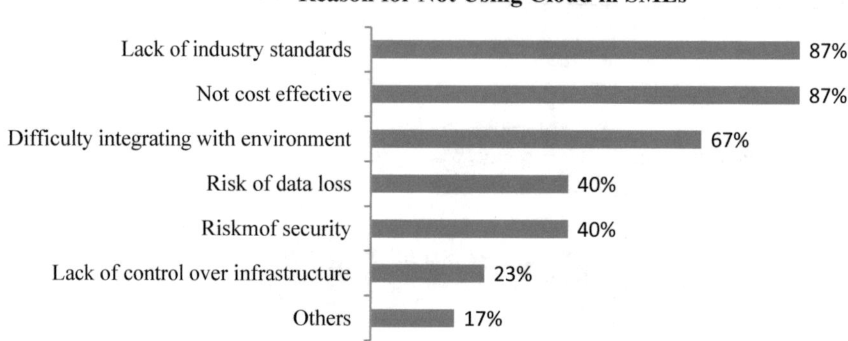

Fig. 7 Factors that are blockers for cloud adoption

business needs. Figure 5 shows clearly that the cloud adoption is gaining momentum and has increased number of active users with time in India (Fig. 6).

From January 2017 cloud Business is going to be a major role in the SMEs. It is also noticed most of the SMEs in India start to refine their business and ensured Government of India that they are prepared in digitalization (Fig. 7).

Depending on the survey data, it is identified that most of the Indian SMEs have already adopted the cloud solutions or are in process to adopt cloud solutions in near future, though there are very few enterprises that are still in doubt to adopt the cloud solution for their business.

Figure 8 shows clearly that Amazon with Amazon Web Services is market leader for providing cloud solutions for Indian enterprises followed by Microsoft, Google, IBM, Alibaba, Oracle, Netmagic, etc.

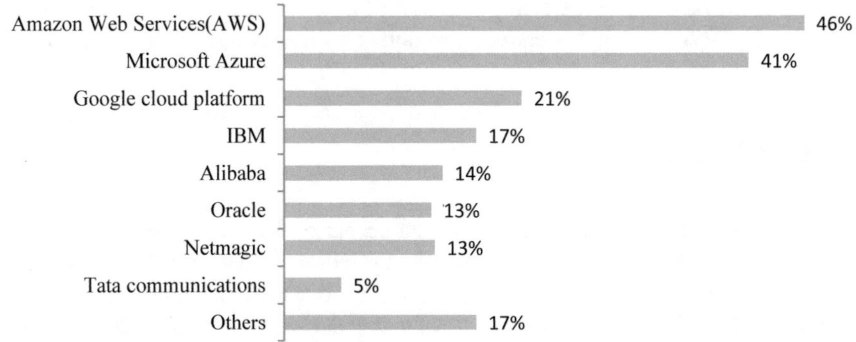

Fig. 8 Cloud service providers for Indian SMEs

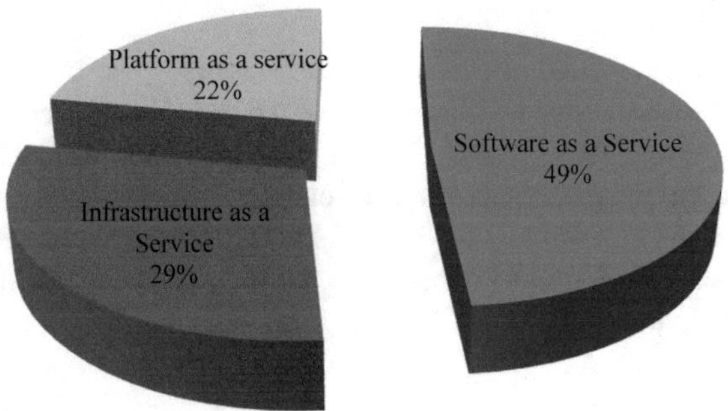

Fig. 9 Cloud model adoption for Indian SMEs

As per the survey, it is clear that 48.7% organizations adopt SaaS for its flexibility, 29.2% organization adopt IaaS for its more security, and 22.1% adopt PaaS (Fig. 9).

5 Limitations and Future Work

We have concentrated to conduct survey to gather the primary data only from four metropolitan cities from four regions in India and have represented graphically the current progress of cloud solution adoptions among Indian SMEs based on it. The survey should be executed for more enterprises located to other parts of India. It may be covered at least Tier-1 and Tier-2 cities in India along with all the metropolitan cities in India. By coving more and more enterprises from different parts of India would help us to refine result and conclusion. The authors have identified few factors that are acting as blocker for cloud adoption for Indian SMEs. No causal analysis or steps mentioned to mitigate these issues for cloud adoption.

6 Conclusion

This research paper has outlined the present scenario along with the opportunities and pitfalls of cloud implementations among Indian SMEs and has shown the progress of cloud solutions adoption for Indian business houses. This research article would help any enterprises to take the decision for planning for cloud implementation for their business needs. The sample size may be increased for tuning the results in future.

7 Declaration

We declare that we had conducted survey work for this paper after we got approval from our supervisor, Prof. Manas Kumar Sanyal. Also, we had taken proper approval from the consent authority, Department of Business Administration, University of Kalyani, Nadia, PIN-741235, West Bengal, India. For any further clarification, please feel free to contact the authors, Biswajit Biswas, Sajal Kanti Bhadra, Manas Kumar Sanyal, and Sourav Das.

References

1. Larson, D.: BI principles for agile development: keeping focused. Bus. Intell. J. **14**(4), 36–41 (2009)
2. Davenport, T.H.: Big Data @ Work: Dispelling the Myths, Uncovering the Opportunities. Harvard Business Review Press (2014)
3. Kiryakova, G., Angelova, N., Yordanova, L.: Application of cloud computing service in business. Trakia J. Sci. **13**(Suppl. 1), 392–396 (2015)
4. Al-Aqrabi, H., Liu, L., Hill, R., Antonopoulos, N.: Cloud BI: future of business intelligence in the cloud. J. Comput. Syst. Sci. **81**, 85–96 (2015)
5. Coleman, S., et al.: How can smes benefit from big data? Chall. Path Forw. Qual. Reliab. Eng. Int. **32**, 2151–2164 (2016)
6. Gartner. http://www.gartner.com/newsroom/id/2637615/
7. Amoore, L., Piotukh, V.: Life beyond big data: governing with little analytics. Econ. Soc., ISSN: 0308-5147(Print) 1469–5766 (online)
8. Thompson, W.J.J., VanderWalt, J.S.: Business intelligence in the cloud's. J. Inf. Manag. **12**(1), 445–449 (2010)

9. Ilieva, G., Yankova, T., Klisarova, S.: Cloud business intelligence: contemporary learning opportunities in MIS training. In: Proceedings of Balkan Conference on Informatics: Advances in ICT-2015
10. Xia, B.S., Gong, P.: Review of business intelligence through data analysis. Benchmarking Int. J. **21**, 300–311 (2014)
11. Sallam, R., Hostmann, B., Schlegel, K., Tapadinhas, J., Parenteau, J., Oestreich, T.: Magic quadrant for business intelligence and analytics platforms. In: Gartner (2015)

A Computational Intelligent Learning Architecture in Cloud Environment

Mridul Paul and Ajanta Das

Abstract Being a paradigm shift, cloud computing provides various utility services through virtualization of different computing resources. With fast, progressive, and simultaneous advancement of technology in distributed environment, provisioning of learning and teaching services in the cloud is a novel arena to virtually teach common individual across the globe. In order to monitor the accessing services to satisfy those individual consumer or society, negotiating and signing service level agreement (SLA) is essential. Hence, the objective of the paper is not only to provisioning e-Learning services in cloud, but also to enlist the functional parameters related to the services in SLA for monitoring purposes. This paper proposes a computational intelligent service provisioning and monitoring architecture along with the functionality and role of each component. This paper also presents the instantiation and evaluation of the proposed computational intelligent e-Learning architecture using the public cloud provider, Google App Engine, and monitors the quality of service.

Keywords Cloud computing · E-Learning services · Service provisioning
SLA management

1 Introduction

Cloud computing has created a revolution in the world of computing. Based on the concept of virtualization and service-oriented architecture (SOA), cloud computing has enabled researchers, academicians, and organizations to leverage enormous computing power for solving complex problems. The key characteristics of cloud

M. Paul · A. Das (✉)
Department of Computer Science & Engineering, Birla Institute of Technology,
Mersa, Ranchi, India
e-mail: ajantadas@bitmesra.ac.in

M. Paul
e-mail: mridulpaul2000@yahoo.com

V. Bhateja et al. (eds.), *Intelligent Engineering Informatics*, Advances in Intelligent Systems and Computing 695, https://doi.org/10.1007/978-981-10-7566-7_52

are resource provisioning on demand, access to resources from different computing devices (tablets, desktops, and mobiles), serve multiple users through shared resources, automatic commissioning (or decommissioning) of resources, and measure service for predictive planning [1].

E-Learning systems revolutionized the way training content was delivered. The concept of delivering content anywhere and anytime to the learners became evident as e-Learning systems evolved. Besides, traditional systems had several limitations such as restricted access within a local network, scalability issues when volume of content increased, and tight coupling that made it difficult to add new functionalities. Such limitations are addressed by the novel approach of provisioning e-Learning service in cloud. Cloud computing provides abundant hardware infrastructure that can address scalability challenges for services. Besides, loose coupling of services can be achieved by implementing e-Learning services based on design principles of SOA. The objective of the paper is to propose a computational intelligent e-Learning architecture that can capture both functional and nonfunctional requirements to access e-Learning services in cloud. These requirements can be mapped to the SLAs agreed upon by service consumer and provider, and appropriate metrics can be derived to monitor QoS. The contribution of this paper is toward establishment of SLAs for e-Learning services and enablement of monitoring of SLAs through proposed architecture.

The following sections of this paper are organized as follows: Sect. 2 presents proposed architecture and interactions among various subsystems and modules. An evaluation of proposed architecture is described through test case implementation in Sect. 3. The paper concludes with Sect. 4.

2 E-Learning SLAs and Architecture

While simulation of e-Learning services has been researched on grid computing [2, 3], several attempts have been made to leverage cloud infrastructure for hosting e-Learning services. Iqbal et al. [4] proposed algorithm that minimizes response time for Web applications in cloud achieving maximum resource utilization. Lodde et al. [5] proposed service response time as a key parameter to measure and control service provisioning in cloud. The framework proposed in this research work provides a mechanism to maintain user request history and response time associated with each request and derives optimal amount of computing resources (number of VM instances) required. Masud and Huang [6] highlighted the significance of institutions focusing on creating content and managing education process while cloud provider managing infrastructure in cloud for hosting e-Learning system. Similarly, Fernandez et al. [7] articulated functions of cloud service layers that can be used to develop an e-Learning system. SLA, by definition [8], is a negotiated agreement that contains details on agreed service, service level parameters to measure performance of the services and guarantees that services shall comply. It acts as a legal contract between service provider and consumer and describes obligations and actions for any violation when the service is administered. However, in order to frame SLAs for the e-Learning services, it is important to understand the functionalities of different user groups.

2.1 Functionalities of User Groups

The user groups that are going to access the e-Learning services are primarily—learner, trainer, and author. Each of these user groups have different set of requirements that are summarized in Table 1.

2.2 SLA Specifications

As the requirements from the various user groups are presented in the previous section, the SLAs specifications pertaining to these user groups are illustrated in Fig. 1 for trainer groups. The pseudo-code for learner, trainer, and author groups is represented in Fig. 2. The author SLA specification will be of similar structure as that of the trainer. The key distinction between these SLA specifications is the access levels where the access to the e-Learning courses and content is defined. For instance, the trainer and author user groups will have read, write, and update access, whereas learner user group will have read access only.

Table 1 User groups and associated requirements for e-Learning service

Entities	Requirements
Learner	• Able to view courses before enrollment • Able to select courses (enrollment) and appropriate duration • Able to undergo assessments or tests • Able to view course usage reports
Trainer	• Able to view existing courses and relevant contents • Able to add new courses and relevant contents • Able to modify (update or delete) existing courses • Able to add new tests or update existing tests
Author	• Able to view existing courses and relevant contents • Able to add new contents • Able to modify (update or delete) existing contents

```
<?xml version='1.0' encoding='UTF-8'?>
<xs:schema xmlns:xs='http://www.w3.org/2001/XMLSchema'>
     <xs:element name='e-learning_service'>
          <!-- Service Provider details -->
          <provider>
               <provider-id></provider-id>
                    <provider-name></provider-name>
                    <service-hosting>
                    <service-name></service-name>
                    <service-location></service-location>
                    </service-hosting>
               </provider>
          <!-- Service Consumer (teacher) details -->
          <consumer>
               <consumer-id></consumer-id>
               <consumer-role></consumer-role>
               <consumer-name></consumer-name>
               <agreement-date></agreement-date>
          </consumer>
          <!- trainer access details -->
          <access>
               <access-type id=''>
                    <access-name></access-name>
               </access-type>
          </access>
          <courses>
               <course>
                    <course-id></course-id>
                    <course-name></course-name>
                    <subjects>
                         <subject-id></subject-id>
                         <subject-name></subject-name>
                    </subjects>
               </course>
          </courses>
          <availability>
               <service-availability></service-availability>
                    <service-outage></service-outage>
                    <service-backup>
                          <backup-interval></backup-interval>
                          <backup-duration></backup-duration>
                    </service-backup>
          </availability>
          <performance>
               <response-time></response-time>
          </performance>
      </xs:element>
</xs:schema>
```

Fig. 1 SLA specification for e-Learning services for trainer role

(a)

```
Learner{

    Login();
    View_course();
    Negotiate_SLA();
    If (Enroll) then {
        select_course();
        download_material();
    }
    If (select_test) then {
        appear_test();
    }
    Else {
        do_common_functions() {
            view_usage_report();
            view_service_availability_report();
    }
    }
}
```

(b)

```
Trainer{

Login();
View_course();
Negotiate_SLA();
If (new_course) then {
    add_course();
    add_content();
}
Else
If (existing_course) then {
    update_course();
    update_content();
}
If (new_test) then {
    add_tests();
}
Else
If (existing_test) then {
    update_tests();
}
}
```

(c)

```
Author {

Login();
View_course();
Negotiate_SLA();
If (new_content) then {
    add_content();
}
Else
If (existing_content) then {
    update_content();
}
```

Fig. 2 Pseudo-code for learner, trainer, and author groups

2.3 *Proposed Architecture*

This paper proposes e-Learning architecture that can be virtually extended to every e-Learning service in cloud. This computational intelligent architecture can form the base for e-Learning services that can provide components for providing functional services and enable services to measure and monitor SLAs. The proposed e-Learning architecture in Fig. 3 comprises two subsystems—Learning Management and SLA Management. The interactions among these subsystems are briefly described as follows:

(a) Learning Management—This subsystem predominantly contains e-Learning services that expose learning courses to the learners. Learning courses are created by administrators that are stored in course management layer managed by trainers through content management layer. The content management layer can store content in the form of learning objects that encapsulate raw learning materials such as documents (files and videos), while course management layer provides template structure for trainers to define courses using content stored as

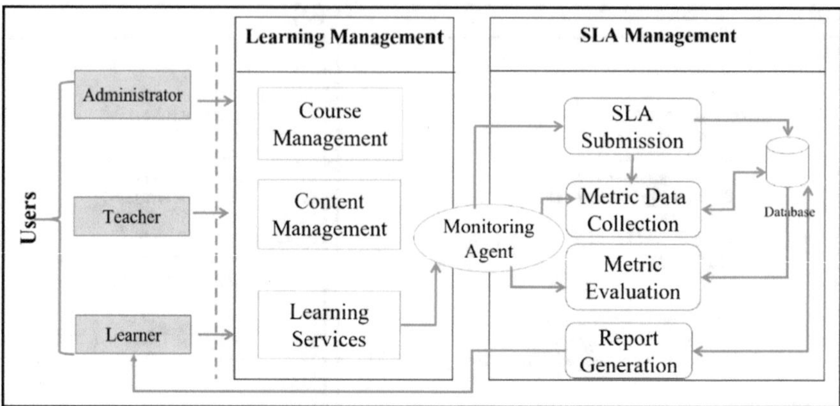

Fig. 3 Proposed e-Learning architecture in cloud

learning objects. The users (trainers, learners, etc.) can interact with this subsystem through a portal. The user management layer authenticates users to access learning services. The administrator is responsible for maintaining this layer.

(b) SLA Management—This subsystem focuses on SLA definition and enforcement. The service consumer and provider negotiate on functional requirements such as course details, content download. This step results in creation of SLAs for each of these requirements. *Monitoring Agent* plays a key role in communication between the two subsystems using dynamic generation of XML format of SLA parameters. It also ensures that statistics are collected on a regular basis through *Metric Data Collection* module. This module stores the data in the relational database. The SLAs negotiated during the start of the service through the *SLA Submission* module is also stored in the database along with the formulas. The *Metric Evaluation* module calculates the metric values based on the formulas from a database at regular intervals. These results are captured by the *Report Generation* module that generates reports.

The proposed architecture in cloud comes with some distinct benefits. First, the underlying cloud infrastructure handles the scalability for provisioning e-Learning services. This is contrary to grid architecture proposed by Roy et al. [3] where computational grid is explored for managing e-Learning services. Second, the interaction among the *Learning Management* and *SLA Management* modules are managed by *Monitoring Agent* that provides robust integration benefits such as real-time tracking of user activities. This is different from implementation approach [9] provided by service-oriented architecture (SOA) where service prototypes are responsible for designing, organizing, and utilizing distributed functionalities. Service prototypes are not robust in nature when it comes to real-time monitoring of e-Learning services.

3 Evaluation of Proposed Architecture

This section explains the evaluation of the proposed architecture using Google App Engine. Nonfunctional requirements are explained in [10]. Functional requirements are considered during evaluation. Functional requirements are the key requirements that pertain to the activity or functionality of the specific services. In context to this case study, the following functional requirements are considered:

(a) *Course details and duration*: This is a fundamental requirement from both consumer and provider perspective for provisioning of the specific service. On selecting a course, the duration of those particular courses needs to be agreed between consumer and provider. This test case implementation provides an interface where a consumer can submit course details such as choice of course and duration of course (in months or days).

(b) *Content download*: While consumers have access to any material pertaining to courses registered, the providers first verify the payment terms and then provides access to the e-Learning system. The consumer is required to submit content size (in GB) and type (document extensions such as pdf, doc) in the test case implementation. The Monitoring Agent tracks content downloads on behalf of the consumer during the session and sends this information for further processing for Metric Evaluation.

(c) *Usage reporting*: When the consumer starts accessing e-Learning services, the tracking of consumer activities in the e-Learning system becomes critical. Reports pertaining to these activities provide views to consumers and providers for any action that may be required for upholding the service agreement. Figure 4 presents a screenshot of usage report for a particular consumer with

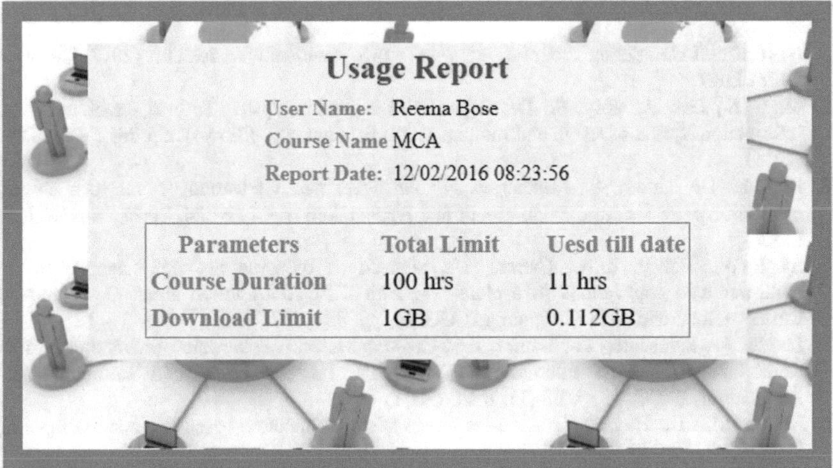

Usage Report
User Name: Reema Bose
Course Name MCA
Report Date: 12/02/2016 08:23:56
Parameters
Total Limit
Uesd till date
Course Duration
100 hrs
11 hrs
Download Limit
1GB
0.112GB

Fig. 4 Usage report for e-Learning services in cloud

agreed-upon parameters as well as current statistics on duration of access to e-Learning service and content downloaded per day. This usage report is generated on a daily basis.

The evaluation of this test case was monitored for seven (7) days for a user and for a particular course for 100 h with download limit of 1 GB. The SLA Management subsystem maintains these SLA parameters and provides usage information that gets displayed in the usage report. In this test case, after 7 days, the course duration was 11 h and content download was 0.112 GB.

4 Conclusion

Provisioning of cloud service has become a real challenge with virtualization of physical resources, infrastructure, and applications. This paper provides scope to access e-Learning services in cloud through a proposed e-Learning architecture. This paper also identifies SLA parameters for functional requirements to provision e-Learning service in cloud. Consumers need to go through the SLA parameters and need to be agreed with the specific values for measuring each parameter before accessing service. The proposed architecture limits itself to the e-Learning services that deliver textual and video content. The future scope of this work can be to evaluate this work for real-time lectures. The proposed e-Learning architecture in this paper leverages Google App Engine architecture to provide scalable and flexible services to the end users.

References

1. NIST Cloud Computing Program. http://www.nist.gov/itl/cloud/index.cfm (2017). Accessed 20 Jul 2017
2. Mitra, S., Das, A., Roy, S.: Development of e-learning system in grid environment. In: Information Systems Design and Intelligent Applications, pp. 133–141. Springer, New Delhi (2015)
3. Roy, S., De Sarkar, A., Mukherjee, N.: An agent based e-learning framework for grid environment. In: E-Learning Paradigms and Applications, pp. 121–144. Springer, Heidelberg (2014)
4. Iqbal, W., Dailey, M.N., Carrera, D.: SLA-driven dynamic resource management for multi-tier web applications in a cloud. In: 10th IEEE/ACM International Conference on Cluster, Cloud and Grid Computing (CCGrid), pp. 832–837. IEEE (2010)
5. Lodde, A., Schlechter, A., Bauler, P., Feltz, F.: SLA-driven resource provisioning in the cloud. In: 2011 First International Symposium on Network Cloud Computing and Applications (NCCA), pp. 28–35. IEEE (2011)
6. Masud, M.A.H., Huang, X.: An e-learning system architecture based on cloud computing. System **10**(11), 255–259 (2012)

7. Fernandez, A., Peralta, D., Herrera, F., Benítez, J.M.: An overview of e-learning in cloud computing. In: Workshop on Learning Technology for Education in Cloud (LTEC'12), pp. 35–46. Springer, Berlin, Heidelberg (2012)
8. Paul, M., Das, A.: A review on provisioning of services in cloud computing. Int. J. Sci. Res. **3**(11), 2692–2698 (2014)
9. Srinivasan, L., Treadwell, J.: An overview of service-oriented architecture, web services and grid computing. HP Software Global Business Unit 2 (2005)
10. Paul, M., Das, A.: SLA Based e-learning service provisioning in cloud. In: Information Systems Design and Intelligent Applications, pp. 49–57. Springer, New Delhi (2016)

Performance Scrutiny of Source and Drain-Engineered Dual-Material Double-Gate (DMDG) SOI MOSFET with Various High-K

Himanshu Yadav and R. K. Chauhan

Abstract The source and drain engineering on dual-metal dual-gate (DMDG)-based 50 nm SOI MOSFET for various high-k gate oxides has been investigated in this work to improve its electrical performance. The proposed structure is designed by modifying source and drain (MSMD) side, and its performance is evaluated on ATLAS device simulator. The effect of this heterogeneous doping on source and drain side of the DMDG transistor led to the reduction of leakage current, decreases DIBL effectively, and improves ION current, thereby enabling the proposed device appropriate for low-power digital applications.

1 Introduction

Continuous scaling of the conventional MOSFETs has given us much higher speed, improved performance as well as low-operational power devices. But as come the advantages, there are some disadvantages too. When devices are scaled down below 100 nm gate length, short-channel effects (SCEs) started to show its major impact on the device performance. Higher leakage current, low device current (ION), DIBL, and reduced ION to IOFF ratio came into picture due to these effects.

Researchers had proposed numerous solutions to overcome the drawbacks of the scaled devices, especially below 100 nm gate length. The structural solutions proposed by them are multi-gate MOSFETs, FINFETs, SOI-based MOSFETs, junction-less MOSFETs, etc. [1–10]. Dual gate (DG), dual material dual gate (DMDG), high-k material technology, and other technologies like source engineering or gate engineering are also being proposed in the recent past to enhance the performance, as well as to counter the various issues related to SCE.

H. Yadav (✉) · R. K. Chauhan
Department of Electronics and Communication, Madan Mohan Malviya University of Technology, Gorakhpur, India
e-mail: himanshuyadav1504@gmail.com

R. K. Chauhan
e-mail: rkchauhan27@gmail.com

V. Bhateja et al. (eds.), *Intelligent Engineering Informatics*, Advances in Intelligent Systems and Computing 695, https://doi.org/10.1007/978-981-10-7566-7_53

DMDG [1] and DMG [4] technologies show better control over gate and threshold voltage of the device and reduce leakage currents. The practice of using different metals also improves the carrier mobility in the channel, thereby improving the device performance also. Modified source SOI [2–4] technology improves the overall I_{ON} to I_{OFF} ratio of the device. As a consequence, better results are expected from such device compared to the conventional SOI devices. Various other technologies, such as the use of high-k materials as gate oxide, have also been reported to improve the performance of transistor at sub-100 nm level. Through which one cope with the leakage problem at such dimensions. So, the device structure modification tends to get more intricate and perilous task. The reliability of such devices becomes dicey with too much multifaceted structure modification.

The technological modification not only improves the device performance, but also it improves reliability. There are still lots of ideas and scope for improving the device performance through various methods and modifications at structure level.

In this work, various technologies like the source and drain engineering and use of high-k material on DMDG SOI MOSFET have been done to enhance its performance. Main aim of the work was twofold: (a) to improve ION to IOFF ratio and (b) keep the threshold voltage under check, so that the device can work as a low-power device in digital circuit applications. All the results have been obtained using ATLAS device simulator, and it has been validated with earlier reported results.

2 Model Description

The device design of the proposed source-engineered DMDG and source/drain-engineered DMDG SOI MOSFET is shown in Fig. 1a, b, respectively. These models have been designed with the help of ATLAS simulation tool. The schematic diagram of the DMDG MSMD SOI MOSFET has been shown in Fig. 2. The gate length of the proposed model is 50 nm in comparison to the DMDG SOI MOSFET [1] which has a gate length of 100 nm.

The source region has been engineered with two different levels of doping regions. A higher doped source region of 2 nm is present above a lightly doped region. The drain region has a similar kind of symmetric dual doping regions same as source. A higher doped drain region is present at the top side, with a lightly doped drain region present just below it. The device description has been shown in Table 1.

The DMDG technology has been employed in the proposed model. With the gate structure having two different workfunctions, material $M_{1(4.7\ eV)}$ and $M_{2\ (4.4\ eV)}$ and length of first M_1 gate $L_1 = 25$ nm and second metal M_2 gate length $L_2 = 25$ nm. Similar analogy has been followed with the bottom dual material gate also. While designing NMOS, the metal M_1 closer to the source has higher

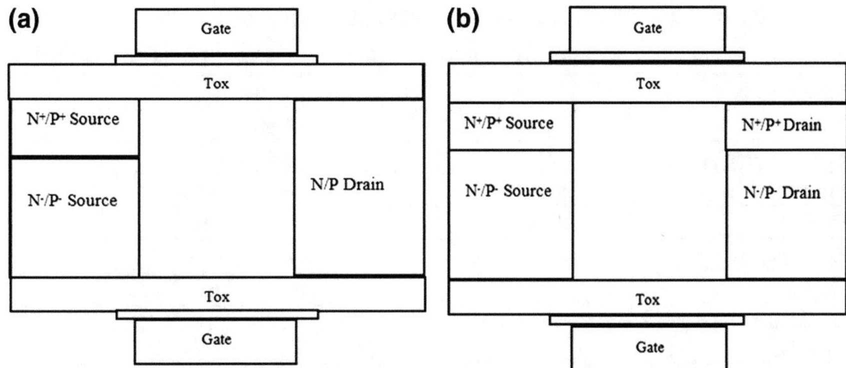

Fig. 1 Device design of source-engineered DMDG SOI (**a**) and DMDG source and drain-engineered SOI MOSFET (**b**) for a channel length of L = 50 nm

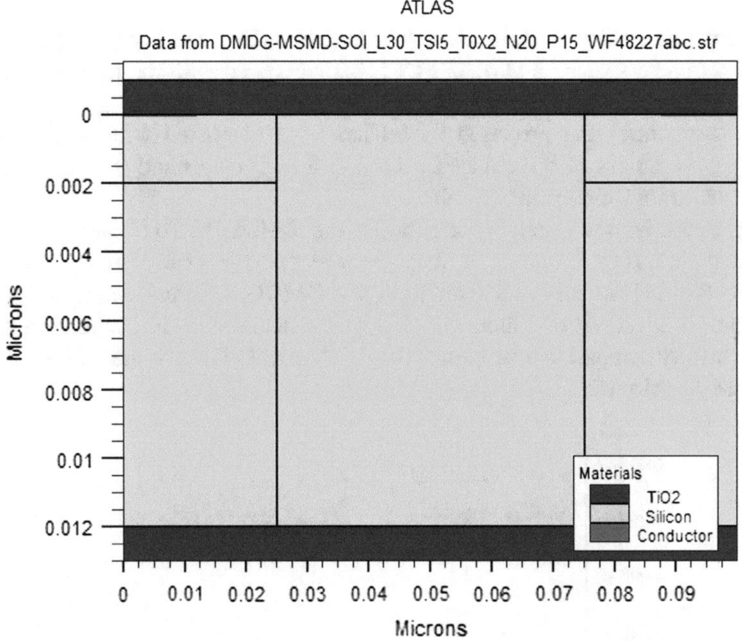

Fig. 2 DMDG SOI MOSFET with source and drain engineering structure of the proposed model for a channel length of L = 50 nm ($L_1 = L_2 = 25$ nm)

workfunction compared to the M_2 material workfunction. While designing PMOS, M_2 metal workfunction would be higher than the M_1 material [4]. The various high-k materials such as HfO_2 ($\varepsilon = 30$), ZrO_2 ($\varepsilon = 50$) and TiO_2 ($\varepsilon = 85$) have been employed on the proposed DMDG MSMD SOI MOSFET.

Table 1 Parameters of DMDG MS SOI and DMDG MSMD SOI (Proposed)

Parameters	DMDG MS SOI MOSFET	DMDG MSMD SOI MOSFET
Gate length	50 nm	50 nm
Tox	1 nm	1 nm
Substrate doping	1e18 cm^{-3}	1e18 cm^{-3}
Source region	n + 3e20 cm^{-3}	n + 3e20 cm^{-3}
Doping	n−1e17 cm^{-3}	n−1e17 cm^{-3}
Drain region	n−1e18 cm^{-3}	n + 3e20 cm^{-3}
Doping		n−1e17 cm^{-3}
Workfunction of gate metal (M_1–M_2)	4.7–4.4 eV	4.7–4.4 eV
Silicon film thickness	14 nm	14 nm

3 Result and Discussion

The proposed model of DMDG MSMD SOI MOSFET has been varied with different parameters such as DMDG SOI [1] is compared with the proposed DMDG source engineered as well as with DMDG source and drain-engineered model. Along with these, the proposed model has been compared with various high-k materials to scrutinize the device to the fullest and understand the device performance at various circumstances also.

Fig. 3. shows the comparison between the DMDG SOI [1] and DMDG SOI designed on ATLAS simulator. We can see that the characteristics shown by DMDG SOI [1] are very much similar to the DMDG SOI designed in the ATLAS simulator. As a result of which we came to a conclusion with a validation that we can rightfully compare our proposed DMDG MSMD SOI with the DMDG SOI [1] and DMDG MS SOI.

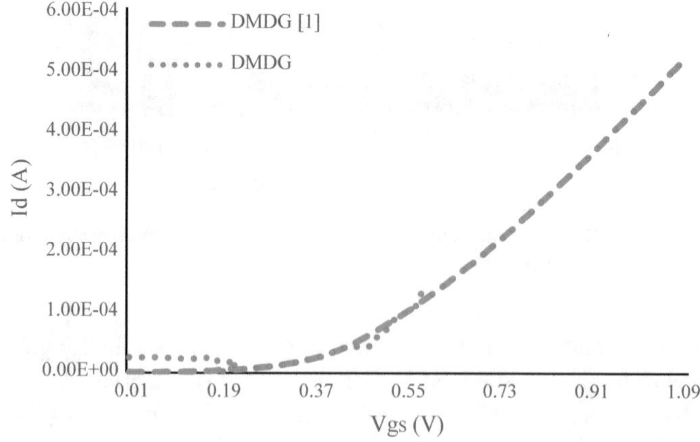

Fig. 3 Id-Vgs comparison of DMDG SOI [1] with DMDG SOI proposed model for V_{ds} = 1.1 V

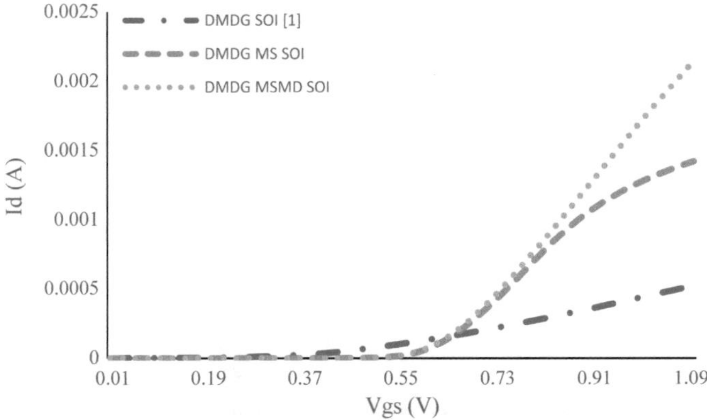

Fig. 4 Id-Vgs comparison of DMDG SOI [1], source-engineered DMDG SOI and DMDG SOI source and drain-engineered MOSFET for a channel length of L = 50 nm ($L_1 = L_2$ = 25 nm) and V_{ds} = 1.1 V

Figure 4 shows the Id-Vgs comparison between the DMDG SOI [1], DMDG source engineered, and DMDG source and drain-engineered SOI MOSFET. We can see that the DMDG MSMD SOI performs much better as compared to the other two devices. The DMDG MSMD SOI has better gate control, and the carrier velocity has been taken care of by the source engineering [2–4]. As a result of which we can see DMDG MSMD SOI performance is much superior with the other two in the respective graph.

Figure 5 further justifies the aim and objective of this work. The figure shows that the ION to IOFF ratio of the proposed DMDG MSMD SOI is much more

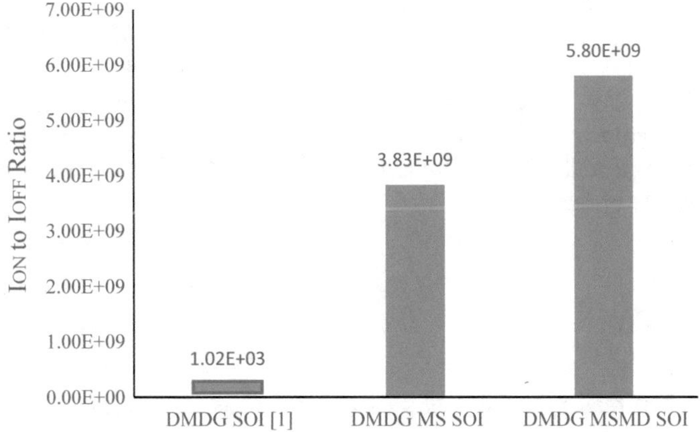

Fig. 5 I_{ON} to I_{OFF} ratio comparison of DMDG SOI [1], source-engineered DMDG SOI and DMDG SOI source and drain engineered for a channel length of L = 50 nm ($L_1 = L_2$ = 25 nm) and V_{ds} = 1.1 V

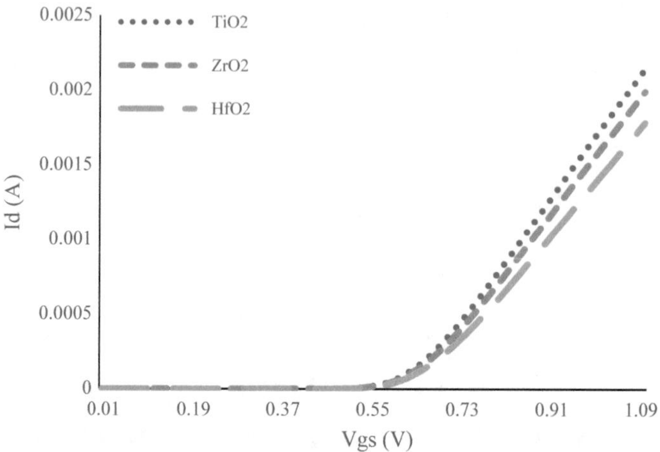

Fig. 6 Id-Vgs comparison of various high-k materials on DMDG SOI source and drain-engineered MOSFET for a channel length of $L = 50$ nm ($L_1 = L_2 = 25$ nm) and $V_{ds} = 1.1$ V

superior to the DMDG SOI [1] and DMDG MS SOI. Higher ION to IOFF ratio results in a much better switching characteristics of the device as well as better possibilities of digital applications also.

With the help of the results, we can see that the device current performance of DMDG MSMD SOI is much higher as compared to the other two. The higher ION is ensured by the DMDG MSMD SOI structure as the dual material dual gate ensures better gate control, and higher average velocity of the carriers is maintained by the source engineering. As we know that higher the drain doping, lesser will be the depletion width. So higher drain region doping ensures reduced DIBL effect compared to the other two preceding structures.

Figure 6 compares the I_D-V_{GS} curve of proposed DMDG MSMD SOI structure for varying high-k material. After comparing the proposed device, we concluded that for higher-k value, the device current is higher. The impact of the high-k material is mainly on reduction of leakage current through gate and controls the threshold voltage of the device. It has been observed that ION current increases for higher-k values material.

4 Conclusion

The concept of dual material dual gate (DMDG) has been encapsulated with the source and drain engineering in order to exhibit desirable characteristics in the proposed structure. We further analyzed DMDG MSMD SOI on various high-k materials and reached a conclusion that its effect is majorly on the ION current. The dual gate material not only reduces IOFF current, but it also gives better gate control which leads to better device performance as the drain conductance and DIBL are reduced. The dual material gate with the source engineering further adds on to the feature of the proposed structure. It augments the average carrier velocity in the channel which leads much better device current response [4]. The drain engineering has been done mainly to remove SCEs and DIBL effect. The proposed DMDG MSMD SOI (2.15 mA) has 51% more ION current than DMDG MS SOI (1.42 mA), while 378% more current than DMDG SOI (0.5 mA) [1] at Vds = 1.1 V. At the same Vds, the ION to IOFF ratio of proposed structure compared to the DMDG MS SOI has 51.4% higher ratio. The only point to ponder would be the manufacturability of the device but that also has been rectified by Zhou [8, 9]. He suggested that the DMG formation process could be made with one additional masking step. As we are reaching the verge of scaling, the high-performance need would still be there. DMDG, source or drain or gate engineering could be the possible solution to our need of improved device performance even at the smaller gate length levels.

References

1. Venkateshwar Reddy, G., Jagadesh Kumar, M.: A new dual-material double-gate (DMDG) nanoscale SOI MOSFET-two-dimensional analytical modeling and simulation. IEEE Trans. Nanotechnol. **4**(2), 260–268 (2005)
2. Mishra, V.K., Chauhan, R.K.: Performance analysis of modified source and TDBC based fully-depleted SOI MOSFET for low power digital applications. J. Nanoelectron. Am. Sci. Publ. **12**(1), 59–66 (2017)
3. Mishra, V.K., Chauhan, R.K.: Performance analysis of modified source and TDBC based fully-depleted SOI MOSFET for low power digital applications. J. Nanoelectron. Optoelectron. Am. Sci. Publ. **12**(1), 59–66 (2017)
4. Tripathi, S., Mishra, V.K., Chauhan, R.K.: Performance analysis of dual metal gate modified source fully depleted SOI MOSFET in i-manager's. J. Embed. Syst. **5**(2), 07–12 (2016)
5. Colinge, J.P., Gao, M.H., Rodriguez, A.R., Claeys, C.: Silicon-on insulator gate-all around device. In: Int. Electron Devices Meeting Tech. Dig., pp. 595–598 (1990)
6. Long, W., Chin, K.K.: Dual material gate field effect transistor (DMGFET). IEDM Tech. Dig., pp. 549–552 (1997)
7. Long, W., Ou, H., Kuo, J.-M., Chin, K.K.: Dual material gate (DMG) field effect transistor. IEEE Trans. Electron Devices **46**, 865–870 (1999)
8. Zhou, X., Long, W.: A novel hetero-material gate (HMG) MOSFET for deep-submicron ULSI technology. IEEE Trans. Electron Devices **45**, 2546–2548 (1998)
9. Zhou, X.: Exploring the novel characteristics of hetero-material gate field-effect transistors (HMGFETs) with gate-material engineering. IEEE Trans. Electron Devices **47**(1), 113–120 (2000)
10. Cheng, K.G., Khakifirooz, A.: Fully depleted SOI (FD SOI) technology. Sci China Inf Sci **59**, 1–15 (2016)

Analysis of N^+N^- Epi-Source Asymmetric Double Gate FD-SOI MOSFET

Narendra Yadava, Vimal K. Mishra and R. K. Chauhan

Abstract The conventional scaling of device dimension uses high doping which leads to degraded mobility and large undesirable junction capacitances. This paper provides a new idea to overcome the short-channel effects (SCEs) in the MOSFETs whose gate length lies in the deep submicrometer range. Here asymmetric DG-FD-SOI MOSFET having N^+N^- epi-source is proposed whose performance is analyzed. The I_{ON}/I_{OFF} ratio of the proposed device is nearly 10^{10} which is good enough for fast switching applications.

1 Introduction

To improve the performance of the electronic devices which are diminishing drastically day by day [1], it is very necessary to study and analyze its behavior and characteristics. Scaling of device leads to rise of various parasitic components, whose effect must be studied to ensure that diminished device is working properly. Taking about the MOSFET, study of SCE which arises due to its diminished geometry is very important. Engineers and scientist are making very hard effort to make the device (MOSFET) more immune to SCE which degrades the device performance by making the charge field division in the body due to gate and drain potentials.

To get proper device scaling, i.e., to minimize SCEs and to achieve good sub-threshold behavior, Brew has given an empirical formula [2]:

$$1.\quad L_{eff} > L_{min} \tag{1}$$

$$2.\quad L_{min} = A\left[T_j T_{ox}(W_S + W_D)^2\right]^{1/3} \tag{2}$$

N. Yadava (✉) · V. K. Mishra · R. K. Chauhan
Department of Electronics and Communication Engineering MMMUT, Gorakhpur, India
e-mail: narendrayadava5@gmail.com

V. Bhateja et al. (eds.), *Intelligent Engineering Informatics*, Advances in Intelligent Systems and Computing 695, https://doi.org/10.1007/978-981-10-7566-7_54

which implies that effective channel length of the MOSFET must be greater than its minimum channel length. The minimum channel length is the minimum length up to which the device performs well.

Yan [3] has further modified the above formula to get the direct relationship between its channel doping and to its intrinsic parameters:

$$N_A \geq 1.8 \times 10^{17} \text{cm}^{-3} \cdot \frac{(V_{bi})^{1/2} + (V_{bi} + V_{DD})^{1/2}}{VOLTS} \cdot \frac{T_j}{500\,\AA} \cdot \left[\frac{0.1\,\mu\text{m}}{L_{eff}} \cdot \frac{T_{ox}}{40\,\AA}\right]^3 \quad (3)$$

where $A = (0.4\,\AA)^{-1/3}$

The above two equations implies that for scaled device having $N_A \geq 10^{17}$ cm^{-3} leads to increase in the substrate capacitance which results in severe limitation in the circuit speed, degraded surface mobility, and high vertical electric field [4, 5]. To solve the above problem, SOI technology has been introduced which uses an insulator region under the channel.

This insulator is used to suppress the leakage current from drain to source which arises due to short geometry of the device. The advantage of introducing insulator is reduced junction capacitance, no need of high doping, good subthreshold characteristics which in short depicts that the SOI technology is more immune to SCE in comparison to the bulk MOS devices [6]. But due to continuous diminishing of the electronic devices, the SCE immunity of SOI technology also decreases and needs furthermore improvement. Mishra and Chauhan [7] have given the concept of source modification in which lightly doped source region is used underneath the heavily doped source. The use of lightly doped source region helps in reducing the effect of parasitic components, while the heavily doped source region is used to maintain the device performance as in the conventional MOS devices. But the design uses a single gate electrode to control the field region between source and the drain regions; hence, the control over the field is weaker.

In this paper, asymmetric double gate FD-SOI MOSFET having N^+N^- epi-source is simulated and analyzed which provides a good concept in improving the device performance by reducing the parasitic components, increasing mobility, and also the use of double gate improves the control over the field of the device.

2 Device Overview

The detailed overview of the conventional as well as proposed DG-FD-SOI MOSFET structures is explained below.

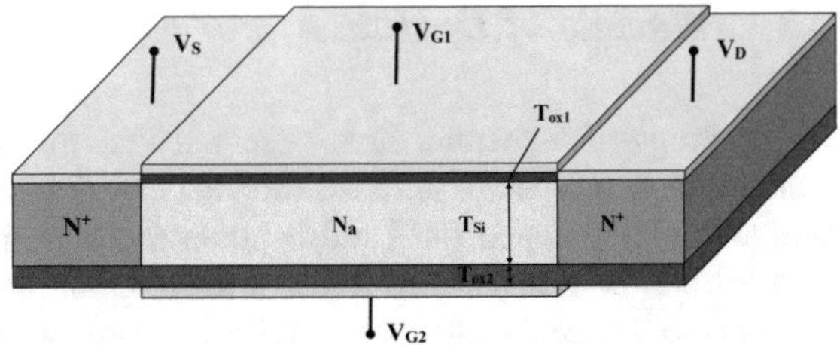

Fig. 1 Structure of conventional asymmetric DG-FD-SOI MOSFET

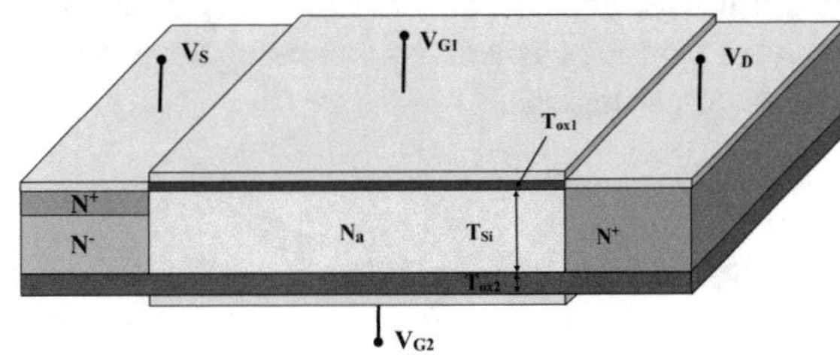

Fig. 2 Structure of proposed N^+N^- epi-source asymmetric DG-FD-SOI MOSFET

2.1 Structure

See (Figs. 1 and 2).

2.2 Parameters

The device parameters used for both conventional as well as proposed MOSFETs are taken almost similar except that the proposed structure of this paper has modified source doping region. Table 1 represents the parameters used in the modeling of the device.

Table 1 Various device parameters used for device modeling

Symbolic representation	Parameters	Value
L	Gate length	50 nm
T_{ox1}	Front gate oxide thickness	1 nm
T_{ox2}	Back gate oxide thickness	5 nm
T_{Si}	Silicon substrate thickness	12 nm
W	Device width	100 nm
L	Gate length	50 nm
N^+	Upper source/drain doping	1e20 cm^{-3}
N^-	Lower source doping	1e17 cm^{-3}
N_A	Channel doping	1e16 cm^{-3}

2.3 Principle of Operation

In the proposed structure, it is observed that the overall source doping profile consists of two types of uniform (donor type) doping profiles in which the upper profile is highly doped (N^+) and its underneath profile is lightly doped (N^-). The idea behind the using of low doped source region under the heavily doped source region is to reduce the effect of parasitic elements without compromising the device performance. The upper highly doped (N^+) region helps in maintaining the device performance, while the underneath lightly doped (N^-) region is used to decrease the effect of parasitic elements by increasing the potential difference between the source and the drain regions.

3 Results and Discussion

In this section results of the proposed device are analyzed and compared with the conventional FD-SOI MOSFET.

3.1 Parasitic Capacitances

The parasitic capacitance of both conventional and the proposed devices is compared correspondingly. Figures 3 and 4 represent the junction capacitances in both the designs.

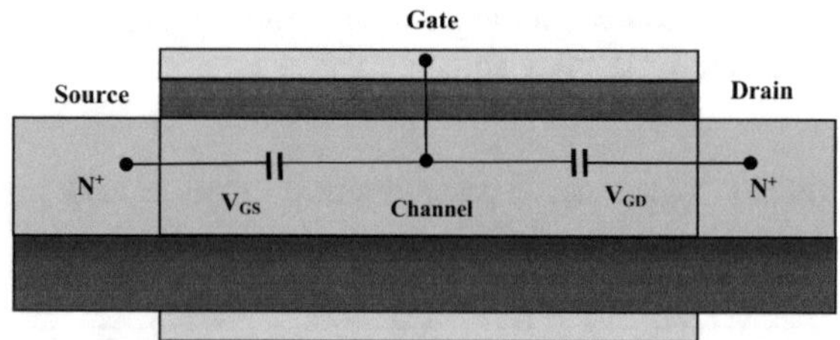

Fig. 3 Junction capacitances in conventional asymmetric DG-FD-SOI MOSFET

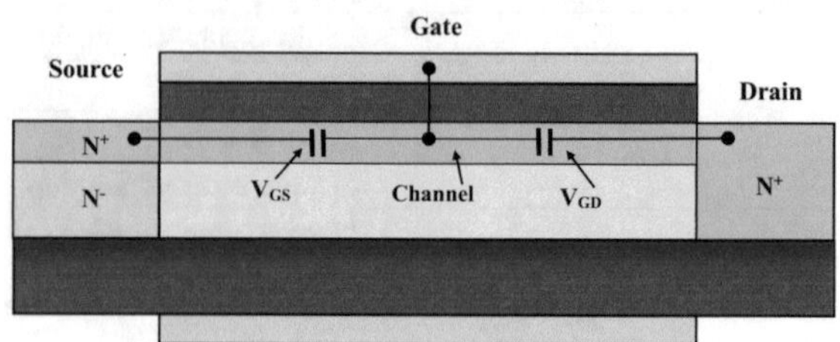

Fig. 4 Junction capacitances in proposed asymmetric DG-FD-SOI MOSFET

In both the above diagram, it is observed that the value of junction capacitances formed in the proposed structure is lesser in comparison to the conventional SOI structure. The value capacitances under the condition are evaluated by assuming that the channel is just formed and linear in nature. The value capacitances are as follows:

$$C_{GS} = C_{GD} = (C_{OX}.W.\ L)/2 \quad (4)$$

$$C'_{GS} = C'_{GD} = \left(C_{OX}.W'.\ L\right)/2 \quad (5)$$

$$\text{Also: } C = \varepsilon.A/d;\ A = W \times L \rightarrow W = A/L \quad (6)$$

$$\text{And: } A' = (A)/3$$

$$\text{Therefore: } C'_{GS} = C'_{GD} = \left(C_{ox}.W'.L\right)/2 = \left(C_{ox}.\left(A'/L\right).L\right)/2 \quad (7)$$

$$\text{Which gives: } C'_{GS} = C'_{GD} = C_{GS}/3 = C_{GD}/3 \quad (8)$$

where; C_{GS} and C_{GD} are the gate to source and gate to drain junction capacitances of the conventional SOI-MOS device, respectively. C'_{GS} and C'_{GD} are the gate to source and gate to drain junction capacitances of the proposed SOI-MOS device, respectively. A, W, and L are the channel area, width and length of conventional device, respectively. W' and L' are the channel width and length of proposed device, respectively.

Equation (5) represents that the parasitic capacitances of the proposed SOI-MOS device are nearly three times lesser than that of the conventional SOI-MOS device which results in improved control gate field of the device. Here the channel between the N^+ source and N^+ drain is only considered, due to very lightly doped N^- source region, the channel between N^- source and N^+ drain is not considered because the channel formed between this region contains very few conducting carriers because the upper channel pushes more holes into its lower channel area hence almost negligible inverted conducting path created between these two regions.

3.2 Built-in Potentials

From Fig. 2, it is observed that the overall doping profile of the source region consist of two types of doping profile in which above one third region is heavily doped in order to maintain the device performance as in conventional MOS device, and the remaining source region is lightly doped in order to reduce the effect of the parasitic capacitances. The change in built-in potential of the designed device with that of the conventional device is disused below:

In conventional SOI MOSFET:

$$V_{Sbi} = V_{Dbi} = \frac{kT}{q}.\ln\left(\frac{N_a.N^+}{n_i^2}\right) \tag{9}$$

In the proposed SOI MOSFET:

$$V_{Dbi} = \frac{kT}{q}.\ln\left(\frac{N_a.N^+}{n_i^2}\right)$$

$$V_{Sbi} = \frac{1}{3}\frac{kT}{q}.\ln\left(\frac{N_a.N^+}{n_i^2}\right) + \frac{2}{3}\frac{kT}{q}.\ln\left(\frac{N_a.N^-}{n_i^2}\right) \tag{10}$$

Where; V_{Sbi} and V_{Dbi} are the built-in potential at the source and at the drain ends, respectively.

3.3 Potential Distribution, Energy Distribution, and Mobility Distribution

In Fig. 5, potential distribution of the both conventional and proposed devices at $V_{DS} = 0$ V and at $V_{DS} = 1.5$ V is shown. From the figure, it is observed that by using N^+N^- source, the potential difference between the source and the drain region increases, which helps in minimizing the flow of leakage current from the drain end to the source end. Also, the control of gate electrode on channel field gets improved, and due to modification of the source region, the horizontal field dominates the lateral field.

Figure 6 represents the energy bands in both the devices. From the diagram, it is observed that the energy bands at the drain end of the proposed device are getting

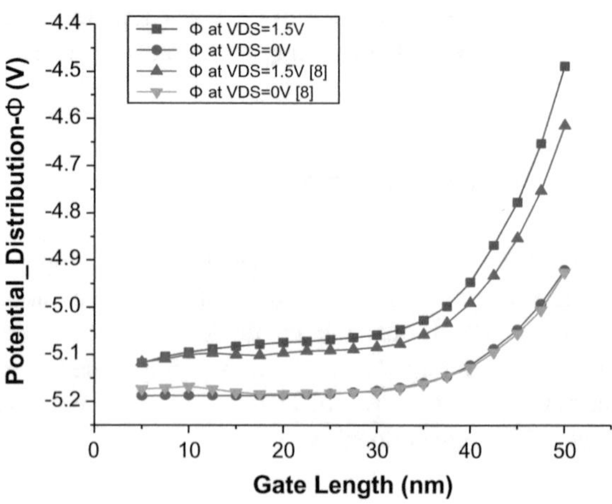

Fig. 5 Potential distribution in conventional and proposed asymmetric DG-FD-SOI MOSFET

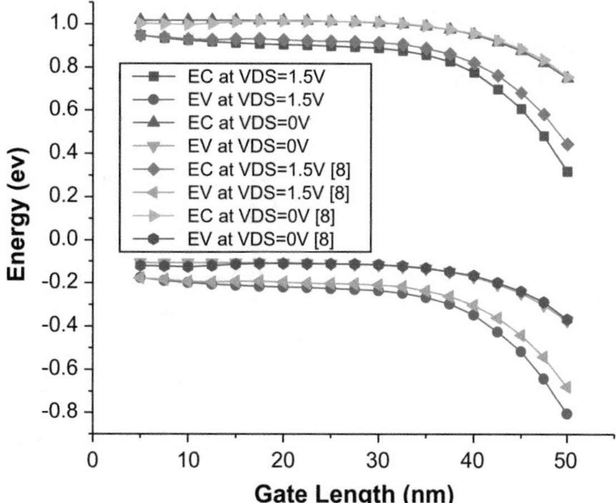

Fig. 6 Energy bands in conventional and proposed asymmetric DG-FD-SOI MOSFET

lowered in comparison to the conventional device which implies that the device can operate at lower potential while maintaining same performance in comparison to the conventional device.

Figure 7 represents the mobility of electron in the body of both the conventional and the proposed devices. From the figure, it is observed that with the use of N^+N^- source region, the mobility of electron changes, it becomes more uniform throughout the channel with respect to the conventional device. The above figure represents the mobility of electron throughout the channel length in the case when V_{DS} is applied and also when V_{DS} is not applied. The more uniform electron mobility results in improved flow of drain current through the channel.

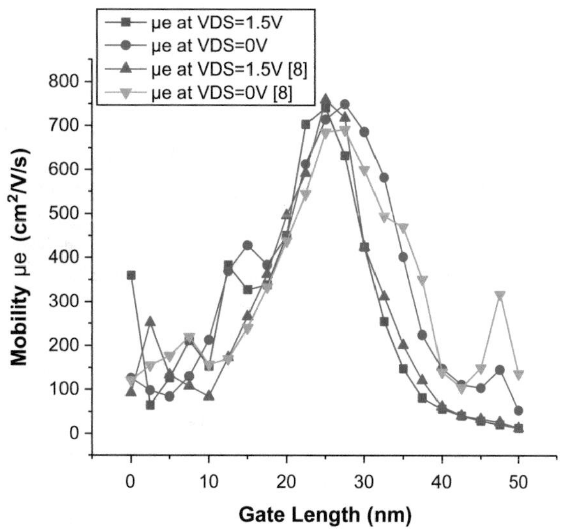

Fig. 7 Mobility of electron in conventional and proposed asymmetric DG-FD-SOI MOSFET

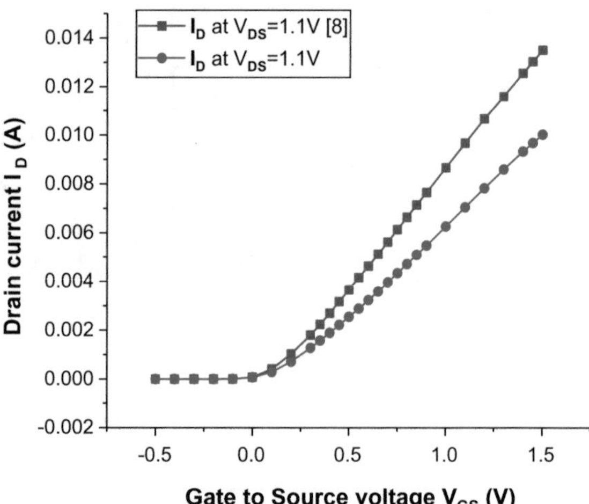

Fig. 8 I_D versus V_{GS} curve of conventional and proposed DG-FD-SOI MOSFET

3.4 *Transfer Characteristics and I_{ON}/I_{OFF} Ratio*

From Fig. 8, it is observed that with the use of N^+N^- source region, the drain current decreases will maintain the same threshold voltage, which implies that both proposed and conventional devices need almost equal potential to get turned ON, but the current of the proposed device is lower than that of the conventional device which is due to the modification in the source doping region.

For switching application, high I_{ON}/I_{OFF} ratio is required; from Fig. 9, the I_{ON}/I_{OFF} ratio of the proposed device is nearly 10^{10} which is good enough for fast switching applications.

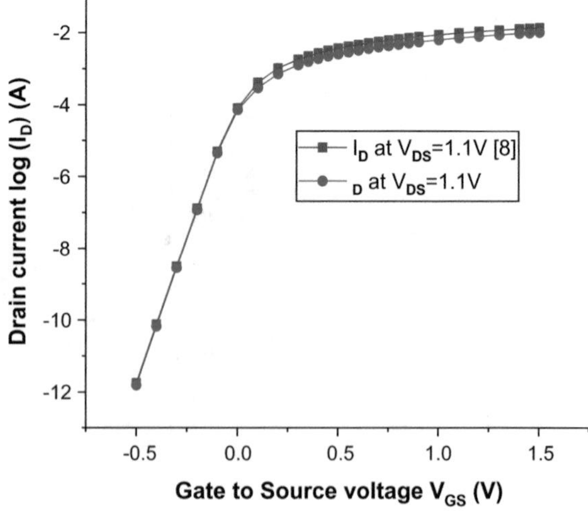

Fig. 9 log(I_D) versus V_{GS} curve of conventional and proposed DG-FD-SOI MOSFET

Therefore, from aforementioned statements, it is observed that the proposed device can be used in low power as well as fast switching devices.

4 Conclusions

Asymmetricity in gate oxide of the DG-SOI MOSFET, and source engineering is used to enhance the performance of the device. The use of N^- source region underneath the N^+ source region plays a great role in minimizing the SCE by reducing the parasitic components and provides a path divergence which reduces the electric field underneath the drain region. Also, the I_{ON}/I_{OFF} ratio of the proposed device is nearly 10^{10} which is good enough for fast switching applications.

Appendix

1. $W_D = L_B[2\beta(V_{DS} + V_{bi} + V_{BS})]^{1/2}$
2. $L_B = [\varepsilon_S/ (\beta q N_a)]^{1/2}$
3. $\beta = (kT/q)^{-1}$

References

1. International Technology Roadmap for Semiconductors. http://www.itrs2.net/itrs-reports.html (2014)
2. Brews et al.: Generalized guide for MOSFET miniaturization. IEEE Electron Device Lett. **EDL-1**, 2 (1980)
3. Yan, R.-H., Ourmazd, A., Lee, K.F.: Scaling the Si MOSFET: From Bulk to SOI to Bulk. IEEE Trans. Electron Devices **39**(7), 1704–1710 (1992)
4. Pearce, C.W., Yaney, D.S.: Short-channel effects in MOSFET's. IEEE Electron Device Lett. **EDL-6**, 326–328 (1985)
5. Xie, Q., Xu, J., Taur, y.: Review and critique of analytic models of MOSFET short-channel effects in subthreshold. IEEE Trans. Electron Devices **59**, 1569–1579 (2012)
6. Yamada, T., Nakajima, Y., Hanajiri, T., Sugano, T.: Suppression of drain-induced barrier lowering in silicon-on-insulator mosfets through source/drain engineering for low-operating-power system-on-chip applications. IEEE Trans. Electron Devices **60**, 260–267 (2013)
7. Mishra, V.K., Chauhan, R.K.: Performance analysis of modified source and tunnel diode body contact based fully-depleted silicon-on-insulator MOSFET for low power digital applications. JNO Am. Sci. Publ. **12**(1), 59–66 (2017)
8. Suzuki, K., Pidin, S.: Short channel single-gate SOI-MOSFET model. IEEE Trans. Electron Devices **50**(5), 1297–1305 (2003)

Solar Tracking for Optimizing Conversion Efficiency Using ANN

Neeraj Kumar Singh, Shilpa S. Badge and Gangadharayya F. Salimath

Abstract In order to maximize the amount of radiation collected by a solar PV panel, the tracker must follow the sun throughout the day. The tracking mechanism of sun required electric motors, light sensors, gearbox, and electronic control to accurately focus at the sun at all times which make the tracking system complex. Also to get maximum power from solar PV panel, MPPT technique must be implemented to the system. This paper deals with new approach for solar tracking and MPPT using single neural network control scheme aiming to reduce overall cost and complexity without nixing efficiency of solar photovoltaic system. The simulation model is done in the MATLAB Simulink for system analysis.

Keywords Artificial neural network (ANN) • Tracking reference neural network control (TRNNC) • Maximum power point tracking (MPPT)

1 Introduction

A lot of things are happening in the field of solar photovoltaic (PV) these days. In the year 2015 alone, the worldwide production of PV modules was more than 25 GW, accounting for an annual growth rate of more than 100%. This production is

N. K. Singh (✉)
Department of Electrical Engineering, PES College of Engineering,
Aurangabad 431004, Maharashtra, India
e-mail: neerajksssingh@gmail.com

S. S. Badge
Department of Electronics and Telecommunication Engineering,
Hi-Tech Institute of Technology, Aurangabad 431005, Maharashtra, India
e-mail: shilpasbadge123@gmail.com

G. F. Salimath
Department of Electrical Engineering, Shreeyash College of Engineering and Technology,
Aurangabad 431004, Maharashtra, India
e-mail: salimath.1984@gmail.com

V. Bhateja et al. (eds.), *Intelligent Engineering Informatics*, Advances in Intelligent Systems and Computing 695, https://doi.org/10.1007/978-981-10-7566-7_55

more than the cumulative PV modules produced in the world in the history of PV till the previous years. Wind and solar energy are the obvious paths to be followed if one wants clean and sustainable energy. The sun transmits energy in the form of electromagnetic radiation. The earth continuously receives 174 × 〖10〗^15 W of incoming irradiation at the upper atmosphere [1]. The insight into the importance of solar energy for our future energy supply has been growing vigorously with the growing concern about the price and availability of fossil fuels [2]. This solar energy can be used in various means such as lighting the street lamps [2, 3], for water pumping [3], domestic, and institutional electricity demands [4–7]. Solar sector has been growing at a breathtaking rate of more than 40% per year over the last 10 years. At the end of 2007, the size of photovoltaic market exceeded 4 GW. The solar photovoltaic device is based on solar radiation. The technical and economic performance of these devices depends on the amount of solar radiation falling at a given location and time [8]. Therefore, the measurement and estimation of solar radiation are an important aspect of work in this area [9]. The amount of radiation intercepted by the earth varies inversely with square of the distance between sun and the earth [10]. Since sun–earth distance is not constant, the amount of radiation intercepted by the earth also varies. Due to variation in solar radiation, it affects the efficiency and power output of solar panel [6, 7]. So to increase the efficiency of the solar PV system, solar tracker plays a very vital role. Presently, two types of tracking are possible: single-axis tracking and two-axis tracking [11]. In a single-axis tracking, the collector is rotated on a single axis only, while in two-axis tracking the panels are rotated across two axis. Single-axis tracking using microprocessor-based is quite complex and low efficient method for tracking sun [12]. Also to ensure the operation of PV modules for maximum power transfer, a special method called maximum power point tracking (MPPT) is employed in PV systems. In present-day scenario, various MPPT algorithms are developed, out of which most popular are as follows—constant voltage method, P&O method, inductance method, and hill climb method. Perhaps, the most popular algorithm is the hill climbing method. Major drawback of this method is we cannot define the highest possible hill. In constant voltage method, the ratio of array voltage to open-circuit voltage is nearly constant. During shadow, this method becomes weak as this method cannot find the real voltage peak. Perturb and observe method is slow to find MPP if the voltage is far away from MPP.

2 Solar Tracker Design Using ANN

It is known that the angle of incidence of sunrays on solar collector changes with the time of the day and with the day of the year. Thus, orientation of the PV module should also be changing to harness maximum sunlight. This paper proposed three-layer neural logic. Proposed three-layer neural logic consists of an input, output, and hidden layer. Hidden layer consists of 12 neurons in addition to input layer with 2 logic neurons, and output layer consists of linear neuron logic.

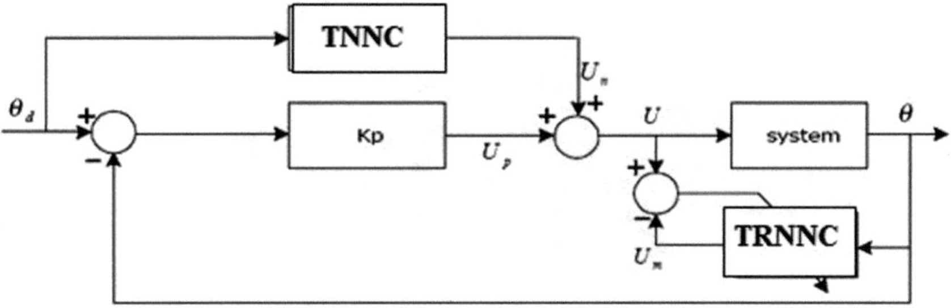

Fig. 1 TNNC network controller

2.1 Tracking Neural Network Controller (TNNC)

Figure 1 shows the basic control scheme which consists of DC servo system. LDR-type pyranometer detects the declination angle (θ), which is defined as the angle between the lines joining the center of the earth to the center of the sun given by Eq. 1.

$$\theta = 23.34 \sin \frac{360}{365} (284 + n) \tag{1}$$

Where n is the nth day of the year.

TNNC is the logic used to control the declination angle by moving the DC servo mechanism. TNNC neurons get trained and used for feedback system to locate correct position of DC servo motor which is called as tracking reference neural network control (TRNNC). The system control voltage U is changed with respect to signal given by TNNC (U_n) and TRNNC (U_p).

2.2 Proposed ANN for Solar Tracker

The three-layer structure of TNNC is shown in Fig. 2 where ith input is $I_i(k)$, x_j are logic weight connecting different layers, $O(k)$ is the output logic for neurons connecting weight different weight and hidden layer neurons. $s_j(k)$ represent the reference value used to determine the declination angle to track the exact position of θ.

The mathematical approach of the model is as follows shown in Eqs. 2–5. Suppose $w_j(k)$ is the output of jth recursion neurons, x_j is the recursion weight of jth hidden neurons, $f(\cdot)$ is sigmoid function.

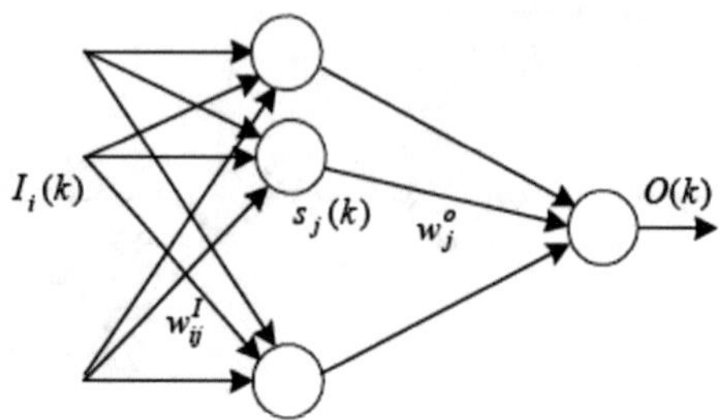

Fig. 2 Three-layer TNNC

When TNNC is used as TRNNC, $O(k) = U_n(k)$

$$s_j(k) = x_j w_j(k-1) + \sum_i I_i I w_{ij}(k) \tag{2}$$

The algorithm for the ANN logic controller is defined as:

$$\Delta w(k) = -\varepsilon \frac{\partial J}{\partial w} \tag{3}$$

$$= \varepsilon e_m(k) \frac{\partial U}{\partial w} \tag{4}$$

$$= \varepsilon e_m \frac{\partial O(k)}{\partial w} \tag{5}$$

Where Δw is any weight of TRNNC, J is the cost function of TRNNC. For the position tracking, DC servo mechanism is used. The inputs required by ANN logic controller and TNNC to control the DC servo motor voltage is given by Eq. 6.

$$U(k) = l_1\theta(m-3) + l_2\theta(m-2) + l_3\theta(m-1) \tag{6}$$

Where $U(k)$ is the control voltage for servo motor, l_i are the system parameters, and θ is position tracking. Also for different speed, the control voltage can be related by Eq. 7.

$$U(k) = z_1(k-3) + z_2 w(k-2) + z_3 w(k-1) \tag{7}$$

Where z_1 are the system parameters or speed and $w(k-i)$ is TNNC speed parameter.

3 MPPT with Artificial Neural Network

3.1 Importance of MPPT

To illustrate the importance of MPPT, consider a load and PV module. The load I-V curve and P-V curve intersect at point know as operating point shown in Fig. 3. But as solar radiation varies with time, the P-V curve also shifts accordingly. So for the same load, the operating point shifts due to P-V curve of solar PV system. To locate exact operating point such that every time I-V curve of load intersects P-V curve at it knee point to get maximum power, required tracking of P-V curve. This tracking is done through a device known as maximum power point tracker. To have the maximum power from PV panel, the operating point should be near knee of PV I-V curve. Both solar tracker and MPPT are getting logic signal from single controller shown in Fig. 4.

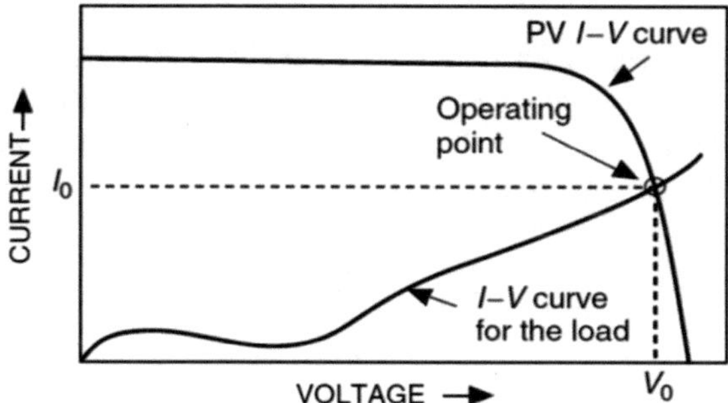

Fig. 3 Maximum power point indication

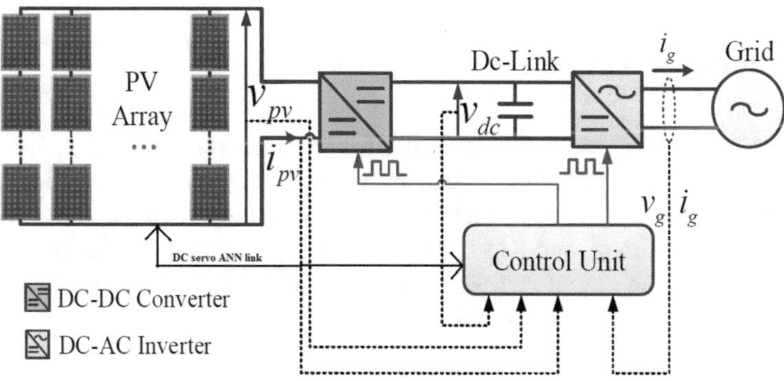

Fig. 4 Proposed ANN solar controller

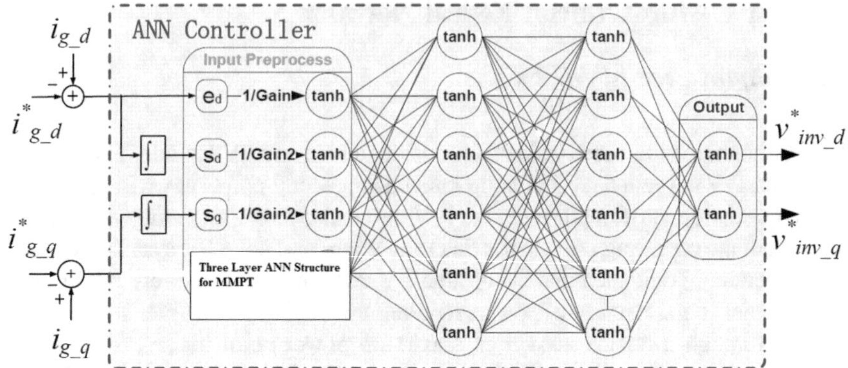

Fig. 5 ANN logic for MPPT

3.2 *Neural Network Logic for MPPT*

The design of proposed ANN controller is similar to the conventional controller. The input layer consists of three layers; out of these, one consists of error term (e_d) and other two terms are (s_{dq}), the integral of error term. The above terms are defined by Eqs. 8 and 9 as follows

$$e_d(k) = i_{g\_dq}(k) - i^*_{g\_dq}(k) \tag{8}$$

$$s_d(k) = \int_0^k e_d(k)\partial t \tag{9}$$

The ANN logic is shown in Fig. 5.

4 Simulation Results

To evaluate the proposed ANN controller for solar tracker and MPPT, MATLAB software is used. The switching frequency is 9 kHz for DC–DC converter and for AC–DC converter it is 7.5 kHz. Figure 7 shows case study of single-phase solar PV system (voltage, current, and power output). Now by varying the solar irradiation as shown in Fig. 6, the output is observed in Fig. 8 by using proposed ANN control. Comparative results are shown in Table 1.

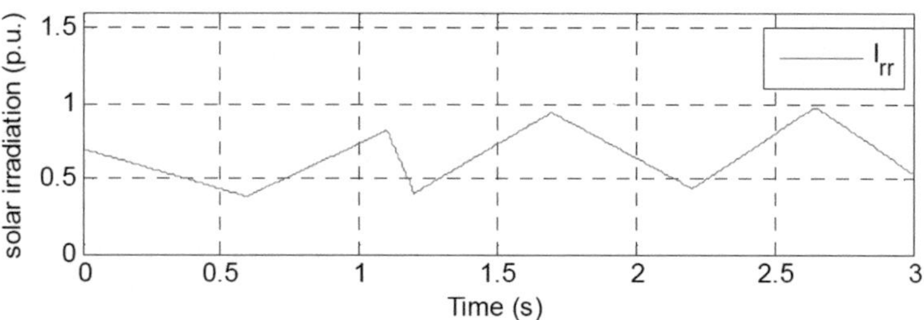

Fig. 6 Solar irradiation

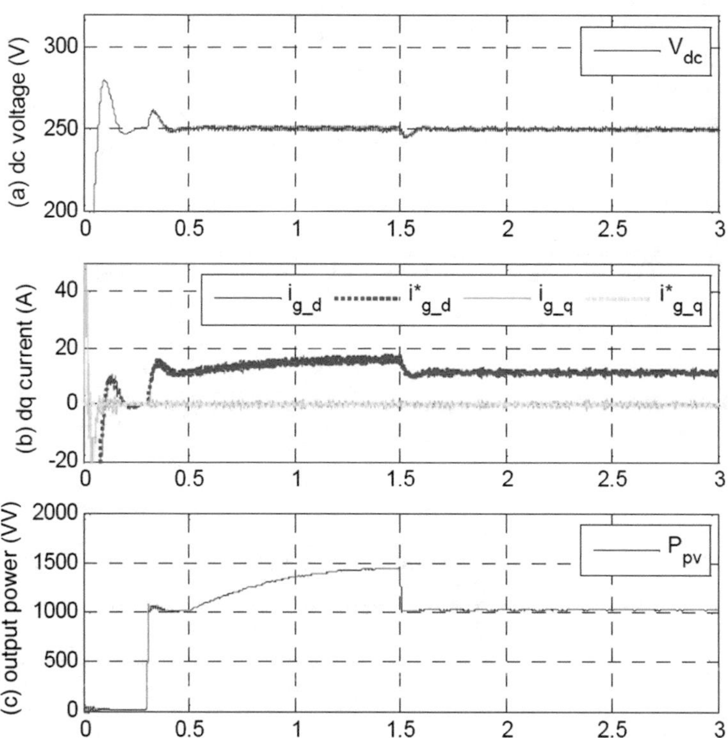

Fig. 7 Simulation result for tracker using ANN

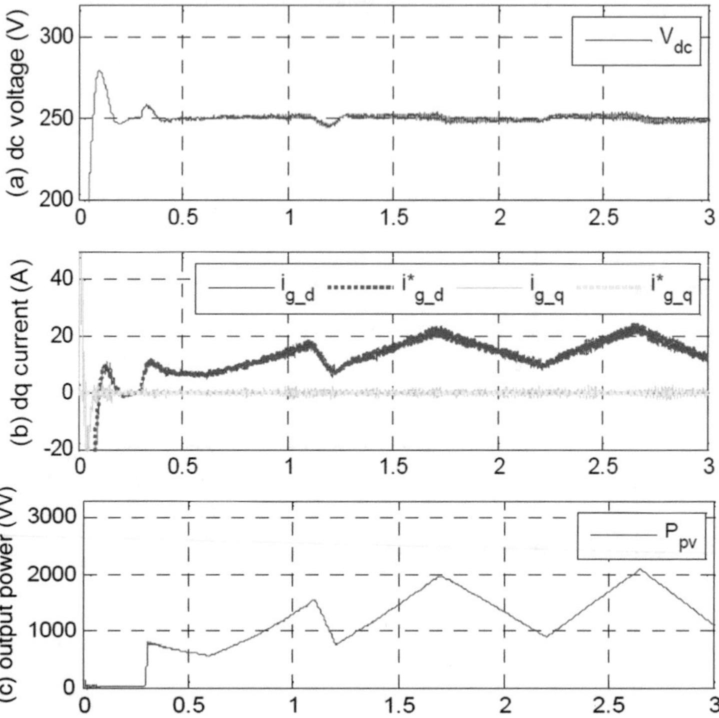

Fig. 8 Solar PV output using ANN for MPPT

Table 1 Comparison of different methods

Methods	Efficiency (%)	Tracking time (s)
Conventional P&O	94.2	0.04
Open-circuit voltage	92	0.08
Incremental conductance	96.8	0.04
Proposed ANN control	99.02	0.02

5 Conclusion

This paper presents a proposed single ANN control system which controls both solar tracker as well as MPPT of PV system. The ANN structure reduces the tracking time; also, the output efficiency of the PV system is increased. The maximum power point is tracked in 0.02 s which is very less compared with the other methods. The proposed ANN controller implements the optimal control based on approximate programming.

References

1. Renewable Energy World Editors (12 Nov 2014) Residential Solar Energy Storage Market Could Approach 1 GW by 2018. www.renewableenergy.com
2. Kumar, N.M., Singh, A.K., Reddy, K.V.K.: Fossil fuel to solar power: a sustainable technical design for street lighting in Fugar City, Nigeria. Procedia Comput. Sci. **93**, 956–966 (2016)
3. Cota, O.D., Kumar, N.M.: Solar energy: a solution for street lighting and water pumping in rural areas of Nigeria. In: Proceedings of International Conference on Modelling, Simulation and Control (ICMSC-2015), vol. 2, pp. 1073–1077. https://doi.org/10.13140/rg.2.1.4007.8486
4. Kumar, N.M., Dinniyah, F.S.: Impact of tilted PV generators on the performance ratios and energy yields of 10 kWp grid connected PV plant at Jakarta and Jayapura regions of Indonesia. In: Proceedings of 1st International Conference in Recent Advances in Mechanical Engineering (ICRAME-2017), pp. 73–75. Kingston Engineering College, Vellore, India (2017)
5. Kumar, N.M., Kumar, M.R., Rejoice, P.R., Mathew, M.: Performance analysis of 100 kWp grid connected Si-poly photovoltaic system using PVsyst simulation tool. Energy Procedia **117**, 180–189 (2017). https://doi.org/10.1016/j.egypro.2017.05.121
6. Kumar, N.M., Reddy, P.R.K., Praveen, K.: Optimal energy performance and comparison of open rack and roof mount mono c-Si photovoltaic systems. Energy Procedia **117**, 136–144 (2017). https://doi.org/10.1016/j.egypro.2017.05.116
7. Kumar, N.M., Reddy, P.R.K., Sunny, K.A., Navothana, B.: Annual energy prediction of roof mount PV system with crystalline silicon and thin film modules. In: Special Section on: Current Research Topics in Power, Nuclear and Fuel Energy, SP-CRTPNFE 2016, International Conference on Recent Trends in Engineering, Science and Technology, Hyderabad, India, vol. 1, no. 3, pp. 24–31 (2016)
8. Kumar, N.M., Navothna, B., Minz, M.: Performance comparison of building integrated multi-wattage photovoltaic generators mounted vertically and horizontally. In: Proceeding of 2017 IEEE International Conference on Smart Technology for Smart Nation (SmartTechCon), 17th–19th August 2017. REVA University, Bangalore, India (2017)
9. Kumar, N.M., Das, P., Krishna, P.R.: Estimation of grid feed in electricity from roof integrated Si-amorph PV system based on orientation, tilt and available roof surface area. In: Proceedings of 2017 IEEE International Conference on Intelligent Computing, Instrumentation and Control Technologies (ICICICT), 6th & 7th July 2017, Kerala, India (2017)
10. Franke, W.T., Kürtz, C., Fuchs, F.W.: Analysis of control strategies for a 3 phase 4 wire topology for transformerless solar inverters. In: Proceedings of IEEE International Symposium Industrial Electronics, Bari, pp. 658–663 (2010)
11. Lorenzo, E., Araujo, G., Cuevas, A., Egido, M., Miñano, J., Zilles, R.: Solar Electricity: Engineering of Photovoltaic Systems. Progensa, Sevilla, Spain (1994)
12. Hagan, M.T., Menhaj, M.B.: Training feedforward networks with the marquardt algorithm. IEEE Trans. Neural Netw. **5**(6), 989–993 (1994)

Human Gesture Recognition in Still Images Using GMM Approach

Soumya Ranjan Mishra, Tusar Kanti Mishra, Goutam Sanyal and Anirban Sarkar

Abstract Human gesture and activity recognition is an important topic, and it gains popularity in the field research in several sectors associated with computer vision. The requirements are still challenging, and researchers are proposing handful of methods to come up with those requirements. In this work, the objective is to compute and analyze native space-time features in a general experimentation for recognition of several human gestures. Particularly, we have considered four distinct feature extraction methods and six native feature representation methods. Thus, we have used a bag-of-features. As a classifier, the support vector machine (SVM) is used for classification purpose. The performance of the scheme has been analyzed using ten distinct gesture images that have been derived from the Willow 7-action dataset (Delaitre et al, Proceedings British Machine Vision Conference, 2010). Interesting experimental results are obtained that validates the efficiency of the proposed technique.

1 Introduction

There exist several video-based human action and gesture recognition schemes in the literature [2, 3]. However, few schemes are available for recognition of human gestures and actions in static images. The recognition of human gestures from static

S. R. Mishra (✉) · T. K. Mishra · G. Sanyal · A. Sarkar
Department of Computer Science and Engineering, NIT Durgapur, Durgapur, India
e-mail: soumyaranjanmishra.in@gmail.com

T. K. Mishra
e-mail: tusar.k.mishra@gmail.com

G. Sanyal
e-mail: nitgsanyal@gmail.com

A. Sarkar
e-mail: sarkar.anirban@gmail.com

S. R. Mishra · T. K. Mishra · G. Sanyal · A. Sarkar
Department of Computer Science and Engineering, ANITS, Visakhapatnam, India

V. Bhateja et al. (eds.), *Intelligent Engineering Informatics*, Advances in Intelligent Systems and Computing 695, https://doi.org/10.1007/978-981-10-7566-7_56

images plays a very important role, especially in forensic science and investigations. Thus, this problem has dragged the attention of many a researchers over the last decade. The PASCAL-VOC action recognition competition [4] has been organized for the purpose to facilitate research work in this scenario. There exists certain core difference between human action and gesture recognition in video and that in a still image. That is, in video, there are the temporal sequences of image frames available for sufficient analysis [5, 6]. However, in still image the researcher has to establish correlation among all the available objects, human body components, and the background [7].

Through the existing technique for addressing the challenge of gesture recognition in static images, the main limitation is that, they involve manual intervention for the purpose of drawing bounding boxes and labeling/annotation [8–14]. The same manual annotation is also required during all the training and testing phases. This manual intervention is cost consuming in terms of time and also there may be chances of introduction of unwanted human errors. In [15], the authors proposed a scheme that does not require human intervention for the purpose. However, here also during the testing phase, proper manual intervention is used. Similarly, in [16], automated recognition of human gestures have been performed. Still, it remains as a research challenge to classify and recognize human gestures in still images.

In this work, we have proposed a systematic and automated approach for addressing the current challenge. These are stated below:

- Candidate objects are first generated out of the input still image. The image is decomposed into different parts containing individual objects. These are suitable for detailed and individual shape analysis. Among these, the human only objects are extracted and rest are eliminated. One of the successful methods for this is reported in [17].
- An efficient product quantization scheme is applied to annotate the predictions to the morphed objects (human only). This requires no bounding box around during the input phase.

Most often, the proposed scheme accurately delineates the foreground regions of underlying human (*action masking*).

2 Proposed Scheme

The entire recognition problem has been divided into two sub-problems at par with the divide-and-conquer strategy. An overview for this has been depicted in Fig. 1. The first step involves in delineating action masks. As a solution, we have successfully used an optimization technique as given below. Corresponding sample from the experiment is shown in Fig. 2.

$$y^{N}_{p,q,r} = \max[\mathrm{Z}_{\mathrm{p}',\mathrm{q}',\mathrm{r}}, \mathrm{p} \le \mathrm{p}' < \mathrm{p} + \mathrm{N}, \mathrm{q}' < \mathrm{q} + \mathrm{N}] \quad (1)$$

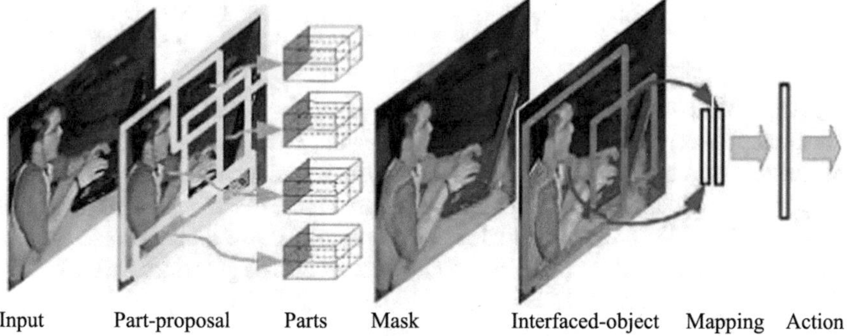

Fig. 1 Overview of the proposed scheme

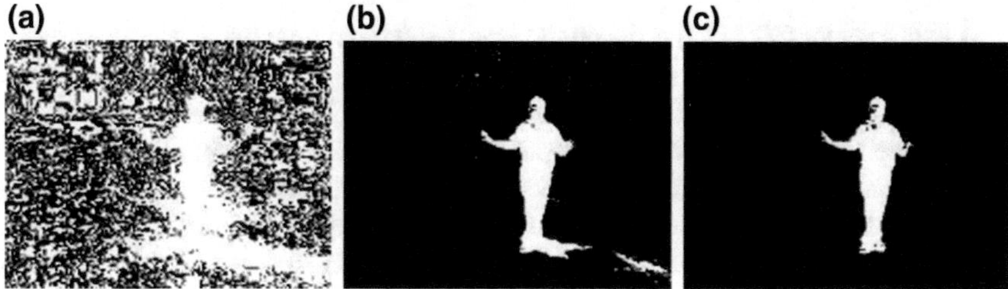

Fig. 2 Sample object delineation

such that,
s$1 \leq q \leq Q; 1 \leq r \leq R$
where, (p, q) are coordinates of top left corner of the objects bounding box, N is standard index of part object from where remaining parts are seeded, y is the image under consideration, Z is the resultant.

2.1 Computation of the Gesture-Mask

This computation is performed with the base assumption that the objects/parts involved in a particular gesture/action are neighbor to each other. An individual part is learned with respect to all other classes (remaining six). The visual coherence of all the gesture related parts (the pixel intensities) are considered meanwhile. They are isolated from the main input image thereby. So now, the task is now resembles a typical energy minimization problem. It can be formulated as a hidden Markov field. As derived from [18] it can be expressed Mathematically as,

$$\text{minimum}/(\alpha, \beta^{\mathrm{h}}, \beta^{\mathrm{l}}_{\mathrm{i}}, \gamma^{\mathrm{c}}) \sum_{\mathrm{i}}(\sum_{\mathrm{m}} \mathrm{V}(\alpha_{\mathrm{im}}, \mathrm{y}_{\mathrm{i,m}}, \beta^{\mathrm{h}}, \beta^{\mathrm{l}}_{\mathrm{i}}, \gamma^{\mathrm{c}})) + \sum_{\mathrm{i}}(\sum_{\mathrm{m,n}}(\mathrm{U}(\alpha_{\mathrm{i,n}}, \alpha_{\mathrm{i,m}}))) \quad (2)$$

Each of the parts is obtained by applying a clustering based on their pixels. This constitutes the part feature-space. Each of the gesture class is having few part-groups. As the numbers in this group is very less; hence, the action masks are obtained by the particular group sparsity which is a constant value. For the purpose of gesture representation, a feature representation is performed on the resultant. As each gesture is involving a human object, an action feature vector is to be generated for each of the seven action those have been considered in this work. Instead of generating feature vectors for all, only these seven gestures templates have been used.

The overall steps involved in generating the gesture masks are represented below in a stepwise manner.

1. A fisher vector (FV) [19] is generated from each input image that uses every part-based features. The Gaussian mixture model (GMM) is learned from these part features. The feature values are then normalized. The final feature vectors are computed by taking the mean and variance values from the Gaussians.
2. Individual gesture models are learned from their action models as computed in the above step. The feature vector corresponding to a specific model is just a row. Corresponding to this row is the label of the image. So, the images of corresponding classes are labeled, respectively.
3. The Gaussian components with nonzero values along with their centers are computed.
4. For all the images, the mask actions are computed. This is done by taking the parts which are closer to the Gaussian components.
5. The grab-cut is used to update the gesture marks into low-level profiles. The foreground and background GMM's are learned, and further refinement is done to the masks.
6. The global model is updated out of these gesture parts.

For the purpose of gesture recognition, a feature vector representation is needed here. For this, a pilot experiment is carried out to determine the best model of computing the feature vectors for the gestures. The very deep convolutional neural network (VD-CNN) has been applied directly to get the gesture vectors. This is shown in Fig. 3. The outputs of the last layer of the network are used to compute the parts. Let's take the output of a bounding box as X as a subset of $R^{s \times s \times d}$, s being the spatial dimensions and d being the filter. The pooling strategy used to extract the parts are given below (Fig. 4):

$$z^{M}_{r,s,d} = \mathrm{MAX}_{\mathrm{X(r',s',d),r \leq r' < r+m; s \leq s' < s+m}} \quad (3)$$

where (r, s) are relative coordinate values of the first corner of the part bounding box that is associated to its surrounding object bounding box, m is scale index of the parts. The Fig. 5 shows the difference between the only CNN approach and the VD-CNN approach where it is clear that the background scene is eliminated.

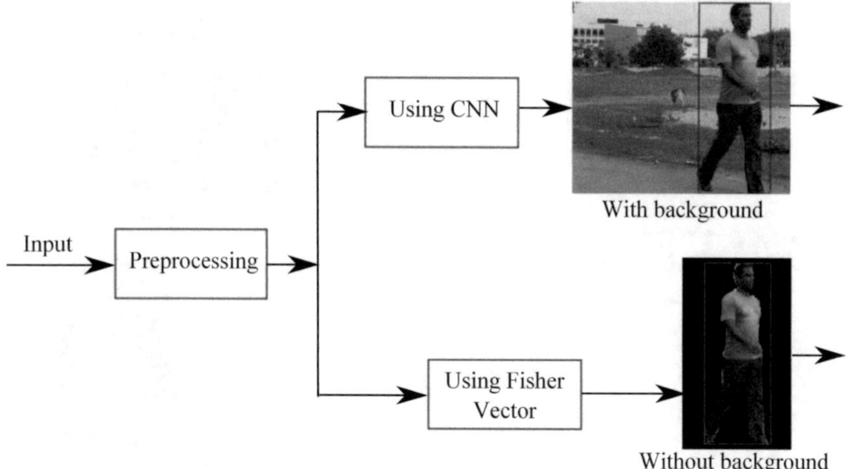

Fig. 3 Sample CNN model representation

Fig. 4 Sample fusion model representation

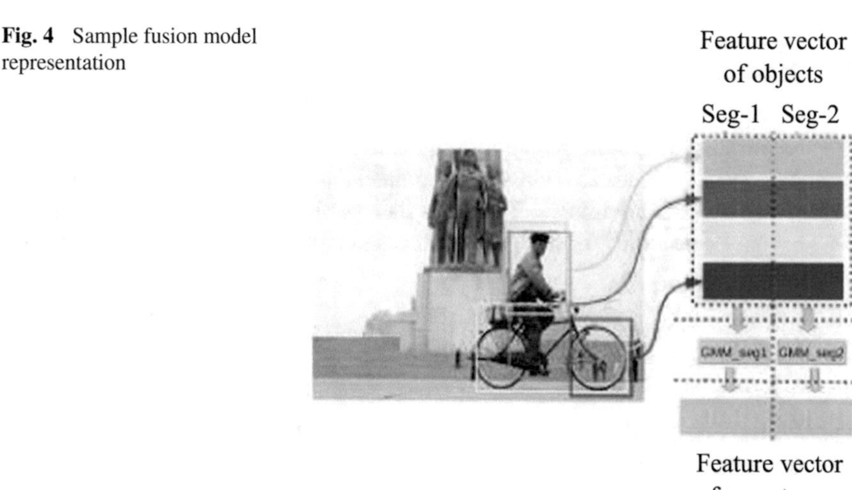

As of next approach, the objects are fused into gesture representation models. This leads to the formation of the gesture vector. This is shown in Fig. 4. This pilot experimentation favors the fusion model for generation of the feature vectors for gestures.

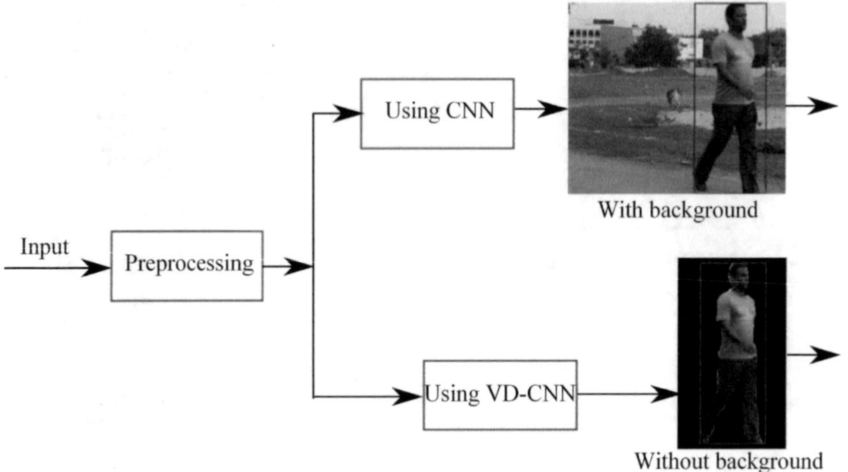

Fig. 5 Output comparison of VD-CNN with only CNN

3 Gesture Recognition by Classification

For the purpose of gesture recognition by classification, the linear support vector machine (SVM) is used. This classifier is trained out of the feature vectors obtained so far from the training images. Testing is then performed on suitable still images. The database considered for this purpose is explained in the subsequent section.

3.1 Input Database of Still Images

The Willow 7-action dataset [1] is considered for validating the effectiveness of the proposed scheme, this dataset consists of 108 images of each class. However, we have considered seven classes of gestures into account. These are *walking, standing, sitting, bending, biking, stretched-hands*, and *Holding-prop*. Fifty samples from each of the classes are taken as input for our experiment. Among these, the train/test split is 30/20, respectively. Thus, the total size of the dataset in our case is 3,500. A snapshot of the sample dataset with preprocessed outputs is shown in Fig. 6.

3.2 Experimental Evaluation

The proposed work was implemented for the dataset as discussed in the previous section. The results so obtained are quite satisfactory. A k-fold (k = 5)

Fig. 6 Sample dataset and preprocessed outputs

Fig. 7 Plot of comparison of rates of accuracy

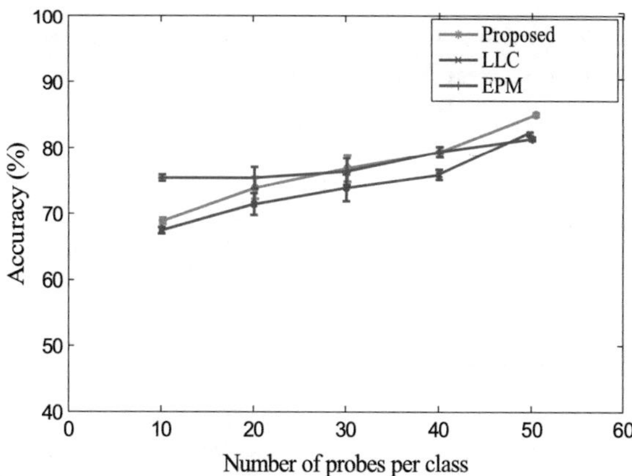

cross-validation strategy is adopted for determining the rate of accuracy. The overall accuracy obtained is 84.5%. The gesture-mask learning method thus performs better as compared to the existing bounding-box methods for gesture recognition. Proper labeling mechanism is also incorporated in the corresponding code of the proposed scheme. The rate of classifications has been presented in Table 1 in a form of confusion matrix. The out-performance of our method is presented in Table 1. Here, it has been compared with the state-of-the-art two other schemes. Plot of the k-fold cross-validation in a comparative manner has also been shown in Fig. 7. When the number of samples is increased, the accuracy for the scheme increases gradually and becomes stable afterward. This shows a good sign of the persistence of the proposed scheme (Table 2).

Table 1 Gesture-wise misclassification rate

Gestures	Walk	Stand	Sit	Bend	Bike	Stretch	Hold	Accuracy (%)
Walk	44	5	0	0	0	1	0	88
Stand	4	46	0	0	0	0	0	92
Sit	0	0	45	5	0	0	0	90
Bend	2	1	4	40	0	1	2	80
Bike	2	4	0	0	41	0	3	82
Stretch	1	2	1	1	2	40	3	80
Hold	3	3	1	0	1	2	40	80

Table 2 Gesture-wise comparison of rate of overall accuracy with other schemes

Mthods/Gestures	Walk (%)	Stand (%)	Sit (%)	Bend (%)	Bike (%)	Stretched (%)	Holding (%)	Overall (%)
EPM [13]	88	87	86	80	78	76	79	82
LLC [20]	89	82	80	80	81	78	79	81.3
Proposed	88	92	90	80	82	80	80	84.5

4 Conclusion

An efficient scheme has been proposed for recognition of six important gestures from still images. The scheme utilizes the very deep CNN for extraction of relativity between human parts with objects. Fisher vector approach has been used to generate feature vectors thereby. Gesture masks are computed with the help of GMM and grab-cut techniques. Finally, support vector machine has been used for proper classification of the feature vectors into six different gestures efficiently. For the whole purpose, Willow 7-action dataset has been used for validation. Along with this, local data captured at our end also have been used only for testing purpose. Future work includes recognition of more number of complex human gestures. Further, the study and analysis on forensic data will be part of the work.

References

1. Delaitre, V., Laptev, I., Sivic, J.: Recognizing human actions in stilll images: a study of bag-of-features and part-based representations. In: Proceedings British Machine Vision Conference (2010)
2. Poppe, R.: A survey on vision-based human action recognition. Image Vis. Comput. **28**(6), 97690 (2010)
3. Cheng, G., Wan, Y., Saudagar, A., Namuduri, K., Buckles, B.: Advances in human action recognition: a survey, 130 (2015)

4. Everingham, M., Gool, L.V., Williams, C., Winn, J., Zisserman, A.: The PASCAL visual object classes challenge 2012 (VOC2012) results. http://www.pascalnetwork.org/challenges/VOC/voc2012/workshop/index.html
5. Wu, J., Zhang, Y., Lin, W.: Towards good practices for action video encoding. In: Proceedings IEEE International Conference on Computer Vision and Pattern Recognition, pp. 2577–2584 (2014)
6. Zhang, T., Zhang, Y., Cai, J., Kot, A.: Efficient object feature selection for action recognition. In: International Conference on Acoustics, Speech and Signal Processing (2016)
7. Guo, G.-D., Lai, A.: A survey on stilll image based human action recognition. Pattern Recogn. **47**(10), 334–361 (2014)
8. Maji, S., Bourdev, L., Malik, J.: Action recognition from a distributed representation of pose and appearance. In: Proceedings IEEE International Conference on Computer Vision and Pattern Recognition, pp. 3177–3184 (2011)
9. Hoai, M.: Regularized max pooling for image categorization. In: Proceedings British Machine Vision Conference (2014)
10. Oquab, M., Bottou, L., Laptev, I., Sivic, J.: Learning and transferring mid-level image representations using convolutional neural networks. In: Proceedings IEEE International Conference on Computer Vision and Pattern Recognition, pp. 1717–1724 (2014)
11. Gupta, S., Malik, J.: Visual semantic role labeling (2015). arXiv:1505.0447
12. Gkioxari, G., Girshick, R., Malik, J.: Contextual action recognition with R*CNN. In: Proceedings IEEE International Conference on Computer Vision, pp. 1080–1088 (2015)
13. Sharma, G., Jurie, F., Schmid, C.: Expanded parts model for semantic description of humans in stilll images (2015). arXiv:1509.04186
14. Yang, H., Zhou, J.T., Zhang, Y., Gao, B.-B., Wu, J., Cai, J.: Exploit bounding box annotations for multi-label object recognition. In: Proceedings IEEE International Conference on Computer Vision and Pattern Recognition, pp. 280–288 (2016)
15. Gkioxari, G., Girshick, R., Malik, J.: Actions and attributes from wholes and parts. In: Proceedings IEEE International Conference on Computer Vision, pp. 2470–2478 (2015)
16. Prest, A., Schmid, C., Ferrari, V.: Weakly supervised learning of interactions between humans and objects. IEEE Trans. Pattern Anal. Mach. Intell. **34**(3), 601–614 (2012)
17. Mahapatra, A., Mishra, T.K., Sa, P.K., Majhi, B.: Human recognition system for outdoor videos using hidden markov model. AEU-Int. J. Electron. Commun. **68**(3), 227–236 (2014)
18. Rother, C., Kolmogorov, V., Blake, A.: GrabCutinteractive foreground extraction using iterated graph cuts. In: SIGGRAPH, pp. 309–314 (2004)
19. Sanchez, J., Perronnin, F., Mensink, T., Verbeek, J.: Image classifica- tion with the Fisher vector: theory and practice. Int. J. Comput. Vis. **105**(3), 222–245 (2013)
20. Wang, J., Yang, J., Yu, K., Lv, F., Huang, T., Gong, Y.: Localityconstrained linear coding for image classification. In: Proceedings IEEE International Conference on Computer Vision and Pattern Recognition, pp. 3360–3367 (2010)

Design of Rate 3/4, 16-States, 16-PSK, Trellis-Coded Modulation Code for AWGN Channel

Rajkumar Goswami, G. Rajeswara Rao and G. Sasi Bhushana Rao

Abstract Generally, 'encoding and decoding' and 'modulation and demodulation' are considered as two independent isolated activities, while designing any digital communication system. Implementation of 'encoding and decoding' such as block coding, TCM, turbo coding ensures that the errors introduced by the channel are mitigated, and the implementation of 'modulation and demodulation' such as BPSK, 8-PSK, 64-QAM at the baseband level enables the optimum data transfer depending on the channel conditions. In order to improve the BER, error correction schemes are implemented in the digital communication system under design, wherein data is encoded prior transmission; however, in this process bandwidth gets expanded. This bandwidth expansion, however, can be mitigated by suitability changing the modulation schemes, e.g. if the normal modulation scheme used is BPSK but if data is encoded at the rate ½, then changing modulation scheme to QPSK will preserve the bandwidth. Integration of convolutional code with the modulation type is known as trellis-coded modulation (TCM). The concept of TCM was though invented in 1970s; it is presently being utilised in many contemporary systems engaged in the field of data communication. A new TCM scheme having rate 3/4, 16-state, 16-PSK, in respect of additive white Gaussian noise (AWGN) channel has been proposed in this paper. Results have been quite encouraging and have indicated the coding gain of approximately 2 dB when compared to uncoded 8-PSK.

1 Introduction

Before discovery of TCM [1], the use of error correction coding techniques along with appropriate modulation methods was quite prevalent and had been regularly used for quite some time. Suppose a digital communication having certain

R. Goswami (✉) · G. R. Rao (✉) · G. S. B. Rao
College of Engineering, Andhra University, Visakhapatnam 530003, India
e-mail: rajkumargoswami@gmail.com

G. R. Rao
e-mail: grajesh197@gvpce.ac.in

V. Bhateja et al. (eds.), *Intelligent Engineering Informatics*, Advances in Intelligent Systems and Computing 695, https://doi.org/10.1007/978-981-10-7566-7_57

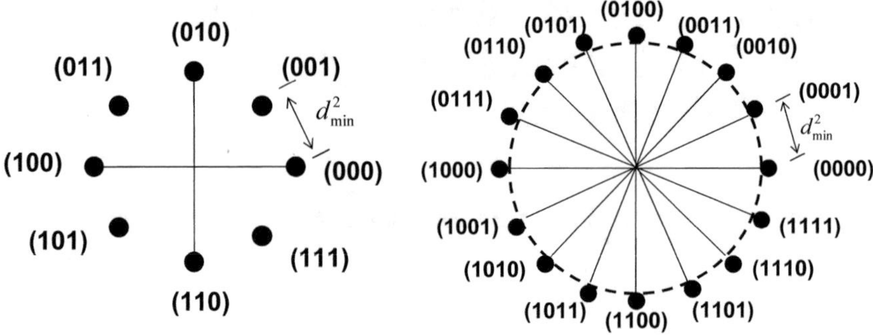

Fig. 1 Euclidean distance in case of 3 bits and 4 bits

bandwidth has been designed using 8-PSK modulation scheme and if redundant bits are added for mitigating the errors introduced by the channel, then it is obvious that the bandwidth will get affected, which will result in lower data rate. However, this expanded bandwidth can be compensated with the implementation of corresponding higher modulation scheme, while ensuring that the errors introduced by the channel are mitigated, which is the main concept of TCM. Further, Sinha and Shrivastava [2] have even shown that the TCM can also be utilised for reducing the peak-to-average power ratio (PAPR) in the OFDM. It can be observed from Fig. 1 that if the average energy is maintained constant, then the expanded signal set, which may be due to addition of redundant bits due to implementation of coding, decreases the Euclidian distance. This decrease in the Euclidian distance [3] implies that the signal points come closer to each other and, therefore, due to the effect of noise on these symbols in the channel, error may be caused while performing detection in the receiver. This increases the error rate, necessitating the incorporation of error correction coding techniques. For undertaking the reliable transfer of data, TCM is one such coding scheme, which preserves the bandwidth by selecting the appropriate modulation scheme for overcoming the effect of additional bits introduced during the coding.

This is highly apt for implementation through the channel those are band limited. The coding techniques used can be either block coding or convolutional coding; however, in case of real-time applications convolution codes are observed to have an edge over block codes [4]. It can be therefore deduced that in the TCM, convolutional encoding and the modulation are treated as a unique combined operation rather than considering them as separate activities in order to mitigate the errors caused by the channel by preserving the bandwidth.

Errors encountered during decoding process play important role in assessing the performance of designed code, and the most important type of error in convolutional codes is pair-wise error, which occurs when the decoder selects a sequence other than the one sent by the transmitter [5]. Therefore, constellation points whose Euclidean distance is as large as possible are assigned to the divergent and emergent branches.

The detector demodulates the signals based on the soft decision because irreversible loss of information occurs if hard-decision demodulation is employed [6]. Maximum likelihood (ML) criterion is applied for implementing soft decision, in which, Euclidean distance plays an important role because the received sequence will be closest to the one whose Euclidean distance is minimum [7].

The organisation of the paper is as follows. In Sect. 2, the system model has been explained in brief. The rules for designing the code for rate 3/4, 16-state, 16-PSK TCM scheme in respect of AWGN channel has been presented in Sect. 3. The construction of the code has been described in Sect. 4. The performance analysis has been presented in Sect. 5, and the conclusion has been presented in Sect. 6.

2 System Model

Basic block diagram of a TCM scheme [7, 8] over AWGN channel is shown in Fig. 2. Convolutional encoder encodes the input bits and produces a sequence of signals $s_l = (s_1, s_2, s_3, \ldots s_l)$ in which i indicates the time index and signal s_i denotes a vector of k-dimension, selected from the set of MPSK signal. Using the complex notation, a representation of signals s_l can also be made by a point in the complex (x, y) plane. Modulator appropriately modulates these signals which are subsequently transmitted through the channel. The channel corrupts these symbols, which are subsequently received by the receiver.

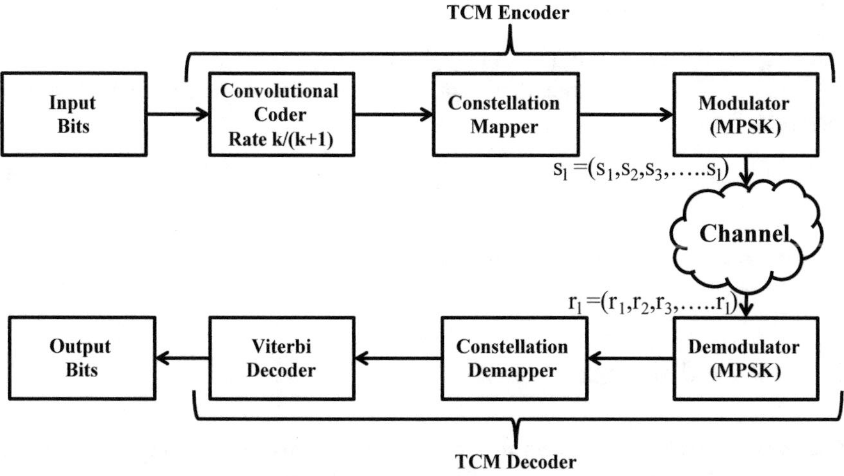

Fig. 2 Block diagram of system

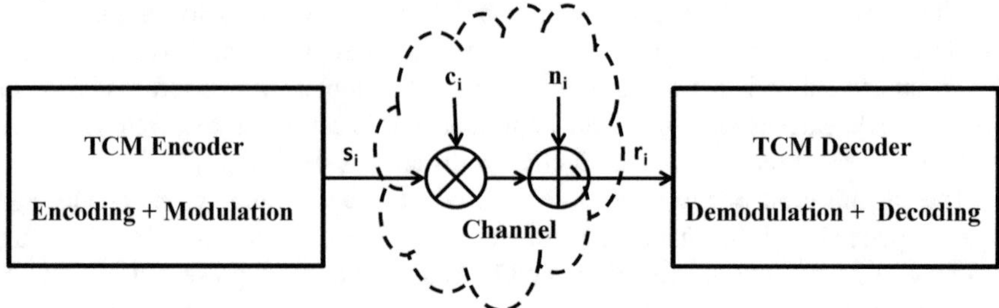

Fig. 3 Effect of channel conditions

Receiver demodulates these received signals, converts them to the digital signals and passed on to the decoder which based on the maximum likelihood (ML) principle decodes the received sequence.

The diagram in Fig. 3 indicates the effects of channel conditions (c_i, n_i) on the data symbols s_i for producing corrupted symbols r_i. The received signal at the *i*th instance, therefore, can be written as

$$r_i = c_i.s_i + n_i \tag{1}$$

where
n_i indicates zero-mean Gaussian noise having variance (σ_n^2) as $N_0/2$
c_i indicates complex channel gain, which is a Gaussian process having variance σ_c^2 and can also be expressed in phasor form as follows:

$$c_i = a_i \cdot e^{j\phi_i} \tag{2}$$

where a_i implies the amplitude and ϕ_i phase.

Assuming that the coherent detection is performed by the receiver Eq. (1) can be further written as

$$r_i = a_i.s_i + n_i \tag{3}$$

where a_i refers to the amplitude of noise.

3 Proposed Design

This section explains the rules that have been utilised in the design of rate 3/4, 16-state TCM codes in respect of AWGN channels. Rate 2/3, 8-state, 8-PSK TCM codes in respect of AWGN channel having an effective length, *L*, of 2 and

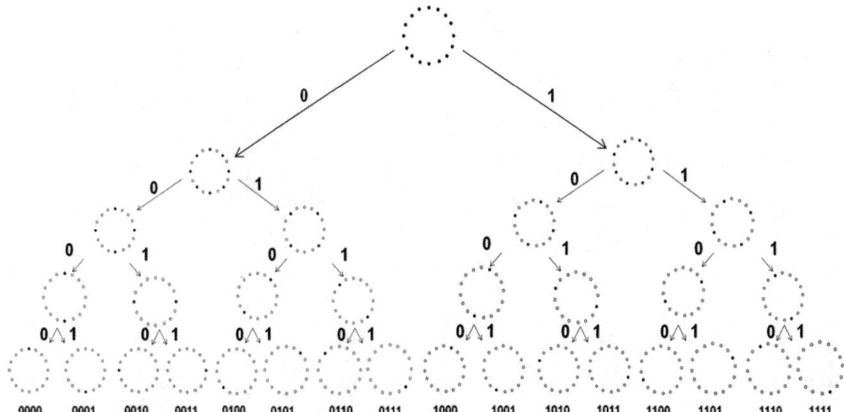

Fig. 4 Partitioning of 16-PSK signal set

$d^2_{free} = 4.586E_s$ has been designed by Ungerboeck [9]. The set of rules proposed in this section have been constructed on the same principle.

The proposed rules for designing the rate 3/4, 16-state, 16-PSK TCM schemes suitable for the AWGN channel are enumerated in subsequent paragraphs.

Partitioning of 16-PSK signal set (0000, 0001, ………1111) corresponding to each signal (s_0, s_1, ………s_{15}) into two subsets A_0 *and* A_1 consisting of {s_0, s_2, s_4, s_6, s_8, s_{10}, s_{12}, s_{14}} and {s_1, s_3, s_5, s_7, s_9, s_{11}, s_{13}, s_{15}}, respectively, having lowest intra-set Euclidean distance (δ_1) (Fig. 4).

In Fig. 5, constellation diagram pertaining to 16 signal points including the Euclidean distances amongst them has been shown.

State transition matrix plays a crucial role in designing the TCM. This matrix indicates the output of the encoder along with the next state of the encoder. In order to populate this matrix as a first step in designing the TCM encoder, following rules are to be followed:

Rule 'a': In any row/column, a signal should only occur once.

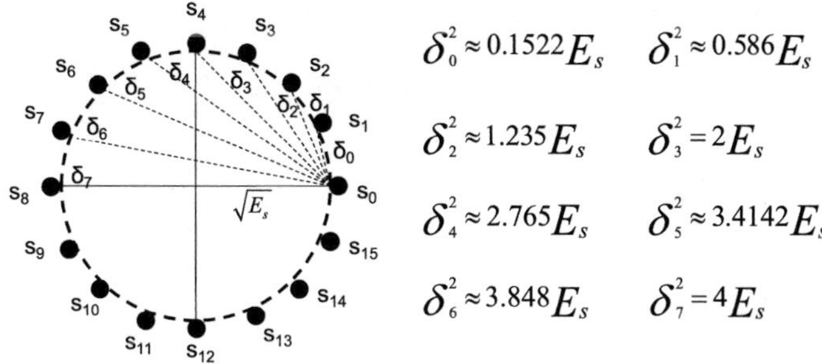

Fig. 5 Constellation diagram and Euclidean distances for 16-PSK signal set

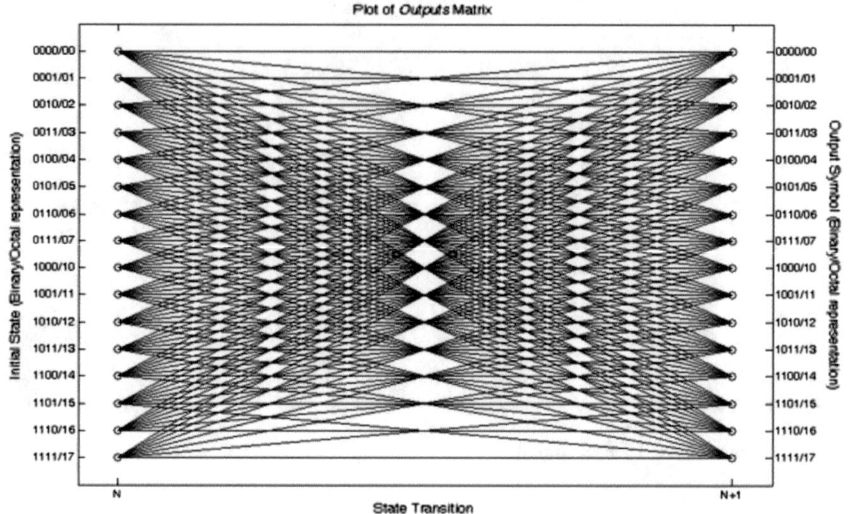

Fig. 6 Output matrix of proposed design scheme

Rule 'b': For a rate 3/4 code, paths emerging from a state are only 8; however, the states are 16, and therefore all transitions cannot happen. Accordingly, it is proposed to associate a signal with a transition path between two states based on the LSB of the initial state and the MSB of the destination state, which will ensure that for a given value of bit, i.e. 0 or 1, all associated signals will strictly be from one of the two sets, i.e. A_0 or A_1 as mentioned above. In Fig. 6, output matrix in respect of this design scheme has been shown.

Rule 'c': State transition matrix shown in Table 1 has a dimension of 16 × 16 because there are 16 states. In this matrix, *ij*th element indicates that signal is associated with the path coming out from state *i*, at *k*th stage, and going to state *j*, at $(k + 1)$th stage.

Minimum distance δ_1 amongst the each pairs of diverging and converging paths, i.e. from *i*th state to *j*th state has been ensured by rule 'a'. Rule 'b' combined with rule 'c' and by taking into consideration the total error events of length two has ensured the following squared product distance:

$$d_p^2(2) = \delta_7^2 \cdot \delta_5^2 \cdot \delta_3^2 \cdot \delta_1^2 = 16E_s^4 \tag{4}$$

where E_s is the energy.

Table 1 State transitions matrix

State	0000	0001	0010	0011	0100	0101	0110	0111	1000	1001	1010	1011	1100	1101	1110	1111
0000	b_0	b1	b2	b3	b4	b5	b6	b7	b8	b9	b_{0a}	b_{0b}	b_{0c}	b_{0d}	b_{0e}	b_{0f}
0001	b_{10}	b_{11}	b_{12}	b_{13}	b_{14}	b_{15}	b_{16}	b_{17}	b_{18}	b_{19}	b_{1a}	b_{1b}	b_{1c}	b_{1d}	b_{1e}	b_{1f}
0010	b_{20}	b_{21}	b_{22}	b_{23}	b_{24}	b_{25}	b_{26}	b_{27}	b_{28}	b_{29}	b_{2a}	b_{2b}	b_{2c}	b_{2d}	b_{2e}	b_{2f}
0011	b_{30}	b_{31}	b_{32}	b_{33}	b_{34}	b_{35}	b_{36}	b_{37}	b_{38}	b_{39}	b_{3a}	b_{3b}	b_{3c}	b_{3d}	b_{3e}	b_{3f}
0100	b_{40}	b_{41}	b_{42}	b_{43}	b_{44}	b_{45}	b_{46}	b_{47}	b_{48}	b_{49}	b_{4a}	b_{4b}	b_{4c}	b_{4d}	b_{4e}	b_{4f}
0101	b_{50}	b_{51}	b_{52}	b_{53}	b_{54}	b_{55}	b_{56}	b_{57}	b_{58}	b_{59}	b_{5a}	b_{5b}	b_{5c}	b_{5d}	b_{5e}	b_{5f}
0110	b_{60}	b_{61}	b_{62}	b_{63}	b_{64}	b_{65}	b_{66}	b_{67}	b_{68}	b_{69}	b_{6a}	b_{6b}	b_{6c}	b_{6d}	b_{6e}	b_{6f}
0111	b_{70}	b_{71}	b_{72}	b_{73}	b_{74}	b_{75}	b_{76}	b_{77}	b_{78}	b_{79}	b_{7a}	b_{7b}	b_{7c}	b_{7d}	b_{7e}	b_{7f}
1000	b_{80}	b_{81}	b_{82}	b_{83}	b_{84}	b_{85}	b_{86}	b_{87}	b_{88}	b_{89}	b_{8a}	b_{8b}	b_{8c}	b_{8d}	b_{8e}	b_{8f}
1001	b_{90}	b_{91}	b_{92}	b_{93}	b_{94}	b_{95}	b_{96}	b_{97}	b_{98}	b_{99}	b_{9a}	b_{9b}	b_{9c}	b_{9d}	b_{9e}	b_{9f}
1010	b_{a0}	b_{a1}	b_{a2}	b_{a3}	b_{a4}	b_{a5}	b_{a6}	b_{a7}	b_{a8}	b_{a9}	b_{aa}	b_{ab}	b_{ac}	b_{ad}	b_{ae}	b_{af}
1011	b_{b0}	b_{b1}	b_{b2}	b_{b3}	b_{b4}	b_{b5}	b_{b6}	b_{b7}	b_{b8}	b_{b9}	b_{ba}	b_{bb}	b_{bc}	b_{bd}	b_{be}	b_{bf}
1100	b_{c0}	b_{c1}	b_{c2}	b_{c3}	b_{c4}	b_{c5}	b_{c6}	b_{c7}	b_{c8}	b_{c9}	b_{ca}	b_{cb}	b_{cc}	b_{cd}	b_{ce}	b_{cf}
1101	b_{d0}	b_{d1}	b_{d2}	b_{d3}	b_{d4}	b_{d5}	b_{d6}	b_{d7}	b_{d8}	b_{d9}	b_{da}	b_{db}	b_{dc}	b_{dd}	b_{de}	b_{df}
1110	b_{e0}	b_{e1}	b_{e2}	b_{e3}	b_{e4}	b_{e5}	b_{e6}	b_{e7}	b_{e8}	b_{e9}	b_{ea}	b_{eb}	b_{ec}	b_{ed}	b_{ee}	b_{ef}
1111	b_{f0}	b_{f1}	b_{f2}	b_{f3}	b_{f4}	b_{f5}	b_{f6}	b_{f7}	b_{f8}	b_{f9}	b_{fa}	b_{fb}	b_{fc}	b_{fd}	b_{fe}	b_{ff}

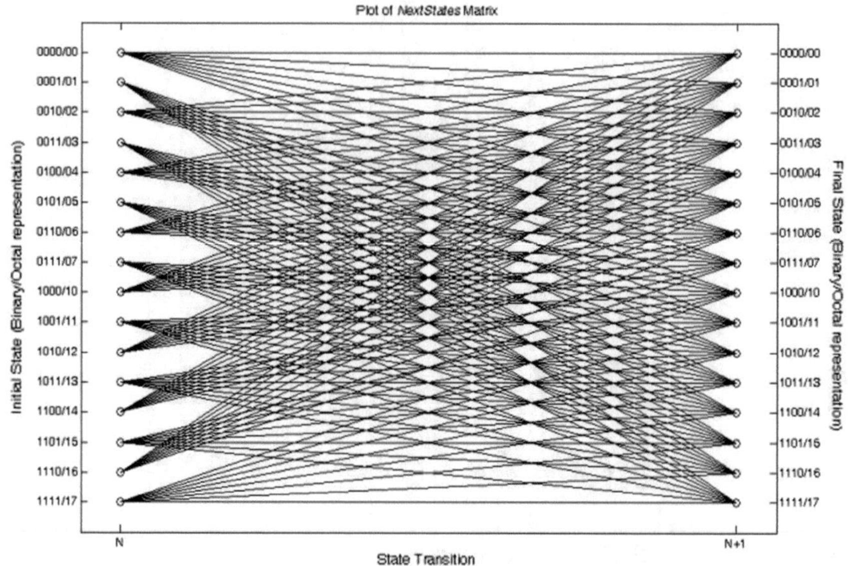

Fig. 7 Trellis diagram of the rate 3/4, 16-state, 16-PSK TCM

4 Construction of the Code

By taking the second rule into consideration, eliminate all those transitions which are not allowed and associate the A_0 subset with the word having LSB as 0 and A_1 subset with the ones having LSB as 1. This will ensure that A_0 subset will be associated with the even number rows, and A_1 subset will be associated with the odd number rows. In the next step, the signals are assigned to the first, second, third, and fourth row from A_0 and A_1 subset, correspondingly, by taking the second and third rules into consideration. This is followed by assigning the balance signal points in the similar manner. Trellis diagram of the code designed using this process has been shown in Fig. 7.

5 Performance Analysis

The performance analysis of the code is generally undertaken with the help of upper bound pertaining to the bit error probability, which is obtained by taking all possible error events of all lengths [10, 11] into consideration. It is to be noted that because of intractable form of the exact expression for $P_2(s_l, \hat{s}_l)$ generating function approach cannot be employed for enumerating such error events. However, by considering the dominant error events may be in small number, reasonable estimation of bit error probability can be established instead of considering error event

Fig. 8 BER versus SNR

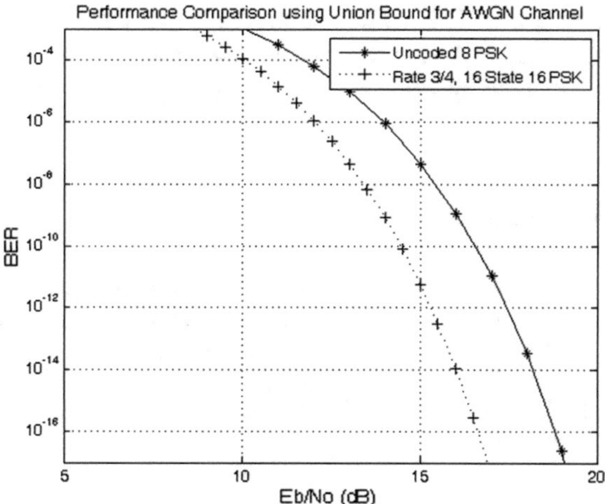

path of all length. Accordingly, the approximate bit error probability can be calculated as

$$P_b \approx \frac{1}{n} \sum_{l=1}^{\lambda} \sum_{s_l \neq \hat{s}_l} \bar{n}_l P_2(S_l, \hat{S}_l) \quad (5)$$

where

n_l indicates average bits in errors, encountered during error event $(s_l, \hat{s}_l)$.
n indicates bits/symbol and λ indicates the length pertaining to error events.

The performance bound was evaluated in respect of AWGN channel which was subsequently used for the evaluation of the pair-wise error event probability through the MATLAB simulation. Bit error rate (BER) versus SNR for the rate 3/4, 16-state 16-PSK TCM code designed in accordance with the rules proposed in this paper has been shown in Fig. 8. For the purpose of comparison, BER versus SNR in respect of the uncoded 8-PSK has also been plotted in the same figure.

6 Conclusion

In this paper, design of trellis-coded modulation scheme having rate 3/4, 16-state, 16-PSK, has been explored. The performance analysis has been undertaken in respect of AWGN environment. The simulation results indicate that in case of AWGN channel, the proposed coding scheme provides coding gain of approximately 2 dB in comparison to the uncoded 8-PSK modulation scheme. The conclusion is also drawn from the analysis that with the integration of the convolutional code and expanded bandwidth signal set, the significant performance can be improved albeit without bringing down the code rate. This work can be further extended to design the code having more number of states along with varying code rate including change of modulation technique from PSK to QAM.

References

1. Biglieri, E., Divsalar, D., Mclane, P.J., Simon, M.K.: Introduction to Trellis-Coded Modulation with Applications. Maxwell Macmillan International Editions (1991)
2. Sinha, S., Shrivastava, R.: OFDM PAPR reduction using trellis coded modulation. Infotech Educ. Soc. J. Sci. Technol. **1**(1) (2016)
3. Anderson, J.B., Svensson, A.: Coded Modulation System. Kluwer Academic/Plenum Publisher (2003)
4. Malhotra, J.: Investigation of channel coding techniques for high data rate mobile wireless systems. Int. J. Comput. Appl. **115**(3), 0975–8887 (2015)
5. Vallavaraj, A., et. al.: The effects of convolutional coding on BER performance of companded OFDM signals. In: International Conference on Communication, Computer & Power (ICCCP 2005) Muscat, pp. 14–16, Feb 2005
6. Jamali H.S., Ngoc, T.L.: Coded Modulation Techniques for Fading Channels. Kluwer Academic Publishers, Boston (1994)
7. Proakis, J.G., Manolakis, D.G.: Digital Signal Processing: Principles, Algorithms and Applications, 2nd edn. Prentice Hall of India, New Delhi (1995)
8. Schlegel, C.B., Pérez, L.C.: Trellis and Turbo Coding. Wiley (2004)
9. Ungerboeck: Trellis-coded modulation with redundant signal sets part II: state of the art. IEEE Commun. Mag. **25**(2), 12–21 (1987)
10. Kroll, J.M., Phamdo, N.C.: Analysis and design of trellis codes optimized for a binary symmetric markov source with MAP detection. IEEE Trans. Inf. Theor. **44**(7), 2977–2987 (1998)
11. Alamouti, S.M., Kallel, S.: Adaptive trellis-coded multiple-phase-shift keying for Rayleigh fading channels. IEEE Trans. Commun. **42**, 2305–2314 (1994)

Baseline Correction in EMG Signals Using Mathematical Morphology and Canonical Correlation Analysis

Vikrant Bhateja, Ashita Srivastava, Deepak Kumar Tiwari, Deeksha Anand, Suresh Chandra Satapathy, Nguyen Gia Nhu and Dac-Nhuong Le

Abstract Electromyogram (EMG) signal is a demonstration of muscular contraction. This being a non-stationary signal is distorted by Baseline Wander. Proper correction of Baseline Wander is a major issue while acquiring EMG signals as it may deteriorate the quality of the signal and make its diagnostic analysis difficult. This paper aims at proposing an effective method for Baseline Wander correction in the baseline-drifted EMG signals. Canonical correlation analysis (CCA) algorithm is first performed on the baseline-corrupted EMG signals to decompose them into various canonical components or variates. After that, morphological filtering deploying octagon-shaped structuring element is used to filter each canonical component. Finally, the results of the proposed technique are compared with the CCA-Gaussian- and CCA-thresholding-based techniques. Simulation results report that the Baseline Wander correction approach used in this work satisfyingly

V. Bhateja (✉) · A. Srivastava · D. K. Tiwari · D. Anand
Department of Electronics and Communication Engineering, Shri Ramswaroop Memorial Group of Professional Colleges (SRMGPC), Lucknow 226028, Uttar Pradesh, India
e-mail: bhateja.vikrant@gmail.com

A. Srivastava
e-mail: srivastavaashita95@gmail.com

D. K. Tiwari
e-mail: srmdeepaktiwari57@gmail.com

D. Anand
e-mail: deeksharoma2012@gmail.com

S. C. Satapathy
PVP Siddhartha Institute of Technology, Vijayawada, Andhra Pradesh, India
e-mail: sureshsatapathy@gmail.com

N. G. Nhu
Duytan University, Danang, Vietnam
e-mail: nguyengianhu@duytan.edu.vn

D.-N. Le
Haiphong University, Haiphong, Vietnam
e-mail: nhuongld@dhhp.edu.vn

V. Bhateja et al. (eds.), *Intelligent Engineering Informatics*, Advances in Intelligent Systems and Computing 695, https://doi.org/10.1007/978-981-10-7566-7_58

eliminates the Baseline Wander from EMG signals while distorting the original EMG signal to a minimum.

Keywords EMG · Baseline wander · CCA · Morphological filtering

1 Introduction

Electromyogram (EMG) signal is an electrical biomedical signal which is emanated whenever the muscles contract and are controlled by the nervous system. Its acquisition is done by the placement of electrodes on the surface of skin or by inserting needle deep inside the muscles [1]. EMG signal proves to be useful for diagnosis and treatment of different neuromuscular diseases and fatigue in muscles. However, due to reasons such as relative shaking of electrodes wires with respect to muscles, bad cable fixation and the electrical drifts found in equipments used for acquisition; the baseline shifts from its electrical zero which is termed as Baseline Wander [2]. These baseline fluctuations can degrade the signal quality which affects the analysis both qualitatively and quantitatively; therefore, its cancellation is required for effective diagnosis [3]. Various techniques have been implemented so far for the effective suppression of Baseline Wander from EMG signal. Fasano et al. [4] proposed a method of Quadratic Variation Reduction to baseline Wander estimation and removal for biomedical signals. This method produced satisfactory results but was computationally very intensive. Rodriguez et al. [5] employed discrete wavelet transform and high-pass filter to eliminate Baseline Wander from EMG signals. This method of BW elimination verified preferable over conventional methods but the presence of Motor Unit Action Potentials (MUAPs) that could not totally put aside from Baseline Segments altered the baseline estimation. Janani [6] analyzed wavelet families and found coif5 wavelet appropriate for elimination of Baseline Wander from ECG signals. This method proved effective for BW removal from EMG signal but to choose the Mother Wavelet according to signal was difficult task. Shin et al. [7] proposed an improved Detrending method for baseline removal in EMG signal. This method satisfactorily eliminates the BW from ECG signal but the requirement to assume a representative sampling frequency to establish the frequency interval makes it impractical to use. In the proposed work, an algorithm based on canonical correlation analysis (CCA) along with morphological filtering is used for the effective correction of baseline wander with minimum signal distortion. CCA has a major benefit that the estimated canonical components are ranked in order of their values of correlation coefficients and decomposes the EMG signal into various canonical components or variates. Morphological filters [8–10] are employed on the canonical variates to eliminate the baseline drift present and maintain the shape of native EMG signal. Signal-to-noise ratio (SNR) has been selected as the performance metric for the validation and effective analysis of the proposed approach. This performance metric also helps in comparing the proposed approach with other techniques of baseline wander removal. The organization of the rest of the paper is as follows. Section 2 gives a detailed description of the proposed approach. Section 3 discusses the results, and the conclusion has been described under Sect. 4.

2 Proposed Baseline Correction Algorithm

The proposed approach aims to combine the CCA algorithm followed by morphological filtering to eliminate the Baseline Wander from the corrupted EMG signal. Firstly, the EMG signal is fed to the Baseline Wander module where the signal decomposition via CCA takes place which converts the multicomponent frequency signal into various canonical variates. These canonical variates are then filtered by morphological filters. After that, the filtered components are added together to get the reconstructed baseline-corrected EMG signal. At the end, the signal fidelity assessment is done by performance metric SNR.

2.1 Canonical Correlation Analysis

Canonical Correlation Analysis (CCA) is a mathematical multivariate approach which makes use of the correlation coefficient to find about the relationship between two sets of linearly combined variables. CCA uses blind source separation (BSS) algorithm for the sake of separating corrupted signals. In this approach, firstly, the combination of variables from both sets is estimated and for governing the amount of resemblance between them, correlation coefficient is utilized. These pairs of combinations are called canonical variables or variates, and the analogue between them is called canonical correlation. The total amount of canonical variates is same as the minimum of variables in both sets [11].

We assume a multidimensional random variable $X = [X_1, X_2, \ldots\ldots.X_n]$ (n equals the quantity of sensors) is the outcome of combination of obscure source signals $S = [S_1, S_2, \ldots\ldots.S_n]$. X and S are related by matrix,

$$X = A.S \tag{1}$$

where A denotes mixing matrix and the goal here is to get the source matrix S by calculating the mixing matrix. It is done by bringing the demixing matrix W in a way that

$$Z = W.X \tag{2}$$

where the matrix Z closely equates the matrix S by a multiplying factor. Let Y be the lagged sketch of the matrix X in a way that $Y(k) = X(k-1)$, where k denotes the sample count. By removing the mean value of the variables X and Y, CCA acquires two basis vectors, one for X and other for Y in a way that the correlation between them is maximized. Consider the linear combinations given below:

$$x = (w_{x_1}x_1 + \ldots.. + w_{x_k}x_k) = w_x^T X \tag{3}$$

$$y = w_{y_1}y_1 + \ldots.. + w_{y_k}y_k = w_y^T Y \tag{4}$$

CCA gets the vectors w_x and w_y that maximize analogue ρ in x and y by the following expression:

$$\rho = \frac{w_x^T C_{xy} w_y}{\sqrt{(w_x^T C_{xx} w_x)(w_y^T C_{yy} w_y)}} \tag{5}$$

where C_{xx} and C_{yy} represent the auto-covariance matrices of X and Y, C_{xy} represents the cross covariance matrix of X and Y, w_x and w_y are the corresponding weight vectors. Initially, the first variates calculated using the CCA algorithm are found maximally correlated with each other. Similarly, second variates are calculated that are correlated exceptionally with each other but disassociated with the first pair of variates. So, the total variates are deducted iteratively [12]. As a result, all these variates are supposed as the sources which have maximum autocorrelation and are disassociated with one another. Thus, on employing CCA algorithm on the baseline-drifted EMG signal, the sources (deducted canonical components) are arranged in a way that the source at the top has the largest autocorrelation value and the source at bottom has the lowest autocorrelation value.

2.2 Morphological Filtering

Morphological filtering is a nonlinear transformation approach generally used for the purpose of regionally altering the geometrical properties of a signal. Morphological filters are type of set functions which adjust the linear representation of the signal to get the desired geometrical shape. In this approach, the whole examined signal is compared with a small probe called structuring element (SE), whose proper selection plays very vital role for the effective filtering of the signal [13]. The whole signal and structuring element are the two sets in mathematical morphology, onto which various operators are applied to change the structural shape of the signal. Various types of structuring elements are present in the mathematical morphology which enriches its applications. The shape and size of structuring element should be selected literally as its inaccurate choice may deteriorate the adjoining component of the EMG signal [14]. In the proposed approach, octagon-shaped structuring element of radius 3 as shown in Fig. 1 is chosen. Baseline Wander which is an artifact of low frequency is being eliminated by the help of morphological operators that consist of both high-pass and low-pass filter

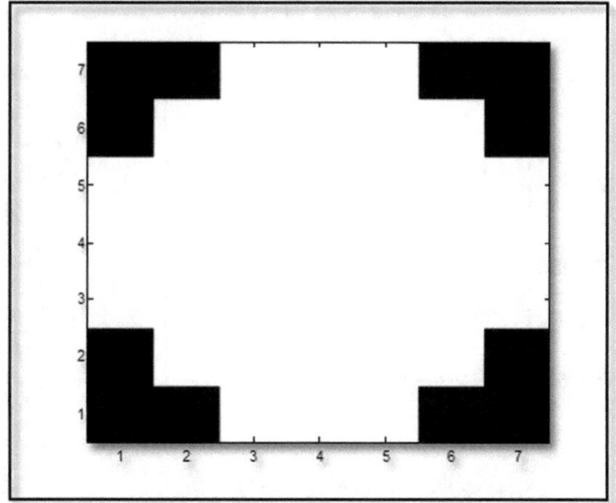

Fig. 1 Octagon structuring element of radius 3

characteristics. Therefore, the proposed approach uses top-hat filtering and bottom-hat filtering for the effective removal of baseline drift. The respective expressions of top-hat transform and bottom-hat transform are given by:

$$T_h = c_i - (c_i \circ b) \tag{6}$$

$$B_h = (c_i \bullet b) - c_i \tag{7}$$

where T_h and B_h represent top-hat and bottom-hat transforms, c_i represents the decomposed canonical variates, and b is the structuring element.

Top-hat filtering and bottom-hat filtering of the decomposed canonical components from the baseline distorted EMG signal are done which generate peaks and valleys, respectively, followed by subtraction of valleys from the peaks. This whole process is repeated for each variate, and after that the filtered variates are added to get the baseline-corrected EMG signal.

3 Results and Discussions

In the proposed approach, the EMG signals have been acquired from The UCI Machine Learning Repository [15] which includes 10 normal and 10 aggressive physical actions measuring the distinct human activities. In the simulation process, the EMG signals were firstly decomposed into their corresponding sources using the CCA algorithm. After the decomposition process, the decomposed canonical components were filtered using morphology. Proper selection of structuring element is a very necessary task in morphological operation so here 'octagon'-shaped structuring element of radius '3' has been used which is giving the correct output.

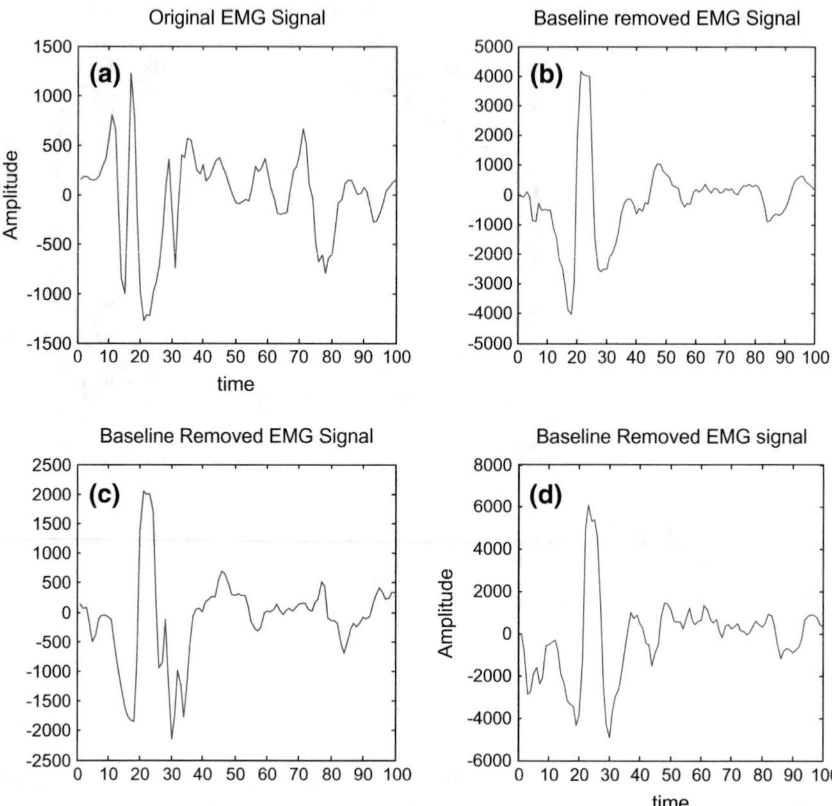

Fig. 2 **a** Original EMG signal (Pushing 3), **b** CCA-thresholding output signal, **c** CCA-Gaussian output signal, **d** Proposed approach output

Figure 2 demonstrates the results obtained using the proposed approach along with the outputs of the other approaches (CCA-thresholding and CCA-Gaussian). Figure 2a shows an original EMG signal (Pushing 3) in which AWGN is present. From Fig. 2b, it is observed that the baseline wander has been removed using CCA-thresholding algorithm but it is accompanied with change in the characteristic of the original signal. Figure 2c shows the output of the CCA-Gaussian approach in which some of the baseline drift has been removed but still it is present in the signal. But there is an improvement in the output of the proposed approach over the other two approaches. In this, baseline drift has been removed significantly from the original EMG signal which is clearly seen in Fig. 2d.

Furthermore, SNR has been used as performance metric to evaluate and validate the proposed approach. SNR values of the baseline-corrected EMG signal using the proposed approach have been compared with the CCA-thresholding and CCA-Gaussian approaches and shown in Table 1.

Table 1 Comparison of SNR (in dB) of other existing approaches used for baseline-corrected signal with the proposed approach

EMG signal ref. no.	Original	CCA-Thresholding [5]	CCA-Gaussian	Proposed approach
S1	–101.9355	–29.2235	–20.4157	–4.0793
S2	–65.1993	–18.2959	–17.7201	–4.5118
S3	–41.2400	–14.7545	–18.9656	–5.2869
S4	–50.3090	–17.1299	–17.2312	–5.5155
S5	–47.5911	–30.1842	–24.7958	–5.7964

Signal-to-noise ratio is a parameter which evaluates the improvement in the signal quality relative to noise. The SNR value of the approach which has been proposed is reported to be –4.0793 dB for signal (S1) which is much better as compared to CCA-thresholding and CCA-Gaussian. Similar results are obtained during simulations when tested with other signals. Therefore, the proposed approach certainly outshines over the other two approaches. The approach may be also extended for other applications of EEG as reported in works of [16–20].

4 Conclusion

This paper introduced a novel technique for the elimination of baseline drift from EMG signals based on CCA algorithm which is followed by morphological filtering. The CCA algorithm effectively decomposes the EMG signal into its source components while morphological filtering successfully removes the baseline wander present with minimum distortion of the original EMG signal [21, 22]. The performance of this approach was compared with CCA-thresholding and CCA-Gaussian filtering-based approaches. This approach achieved better results than the other two approaches. The drift present in the original EMG signal has been successfully corrected. Also, the proposed approach outperforms the existing CCA-thresholding- and CCA-Gaussian-based approaches in terms of SNR.

References

1. Luca, C.D.J., Adam, A., Wotiz, R., Gilmore, L.D., Nawab, S.H.: Decomposition of surface EMG signals. J. Neurophysiol. **96**, 1647–1654 (2006)
2. Ahsan, M.R., Ibrahimy, M., Khalifa, O.O.: EMG signal classification for human computer interaction: a review. J. Sci. Res. **33**, 480–501 (2009)
3. Canal, M.R.: Comparison of wavelet and short time fourier transform methods in the analysis of EMG signals. J. Med. Syst. **34**(1), 91–94 (2010) (Springer)

4. Fasano, A., Villani, V.: Baseline wander removal for bioelectrical signals by quadratic variation reduction, J. Signal Process. **99**, 48–57 (2014) (Elsevier)
5. Rodriguez, I., Gila, L., Malanda, A., Campos, C., Morales, G.: Baseline Removal from EMG Recordings. In: IEEE 23rd Annual International Conference on Engineering in Medicine and Biology Society, pp. 1–6 (2001)
6. Janani, R.: Analysis of wavelet families for baseline wander removal in ECG signals. Int. J. Adv. Res. Comput. Sci. Manag. Stud. **2**(2), 169–176 (2014)
7. Shin, S.W., Kim, K.S., Song, C.G., Lee, J.W., Kim, J.H., Jeung, G.W.: Removal of baseline wandering in ECG signal by improved detrending method. J. Bio-Med. Mater. Eng. **26**, 1087–1093 (2015)
8. Bhateja, V., Urooj, S., Mehrotra, R., Verma, R., Lay-Ekuakille, A., Verma, V.D.: A composite wavelets and morphology approach for ECG noise filtering. In: International Conference on Pattern Recognition and Machine Intelligence, pp. 361–366. Springer, Heidelberg (2013)
9. Bhateja, V., Verma, R., Mehrotra, R., Urooj, S.: A non-linear approach to ECG signal processing using morphological filters. Int. J. Meas. Technol. Instrum. Eng. (IJMTIE) **3**(3), 46–59 (2013)
10. Verma, R., Mehrotra, R., Bhateja, V.: A new morphological filtering algorithm for pre-processing of electrocardiographic signals. In: Proceedings of the Fourth International Conference on Signal and Image Processing 2012 (ICSIP 2012), pp. 193–201. Springer, India (2013)
11. Harrach, M.A., Boudaoud, S., Hassan, M., Ayachi, F.S., Gamet, D., Grosset, J.F., Marin, F.: Denoising of HD-sEMG signals using canonical correlation analysis. Med. Biol. Eng. Comput. **55**(3), 375–388 (2017)
12. Hassan, M., Boudaoud, S., Terrien, J., Marque, C.: Combination of canonical correlation analysis and empirical mode decomposition applied to denoising the labor electrohysterogram. IEEE Trans. Biomed. Eng. **58**, 2441–2447 (2011)
13. Verma, R., Mehrotra, R., Bhateja, V.: An improved algorithm for noise suppression and baseline correction of ECG signals, vol. 199, pp. 733–739. Springer, Heidelberg (2013)
14. Shrivastava, A., Alankrita, A.R., Bhateja, V.: Combination of wavelet transform and morphological filtering for enhancement of magnetic resonance images. In: International Conference on Digital Information Processing and Communications (ICDIPC), pp. 460–474 (2011)
15. Frank, A., Asuncion, A.: UCI Machine Learning Repository. University of California, School of Information and Computer Science, Irvine, USA (2010)
16. Lay-Ekuakille, A., Vergallo, P., Griffo, G., Urooj, S., Bhateja, V., Conversano, F., Casciaro, S., Trabacca, A.: Mutidimensional analysis of EEG features using advanced spectral estimates for diagnosis accuracy. In: 2013 IEEE International Symposium on Medical Measurements and Applications Proceedings (MeMeA), pp. 237–240. IEEE (2013)
17. Vergallo, P., Lay-Ekuakille, A., Giannoccaro, N.I., Trabacca, A., Labate, D., Morabito, F.C., Urooj, S., Bhateja, V.: Identification of visual evoked potentials in EEG detection by emprical mode decomposition. In: 2014 11th International Multi-Conference on Systems, Signals and Devices (SSD), pp. 1–5. IEEE (2014)
18. Lay-Ekuakille, A., Vergallo, P., Griffo, G., Conversano, F., Casciaro, S., Urooj, S., Bhateja, V., Trabacca, A.: Entropy index in quantitative EEG measurement for diagnosis accuracy. IEEE Trans. Instrum. Meas. **63**(6), 1440–1450 (2014)
19. Vergallo, P., Lay-Ekuakille, A., Urooj, S., Bhateja, V.: Spatial filtering to detect brain sources from EEG measurements. In: 2014 IEEE International Symposium on Medical Measurements and Applications (MeMeA), pp. 1–5. IEEE (2014)
20. Lay-Ekuakille, A., Griffo, G., Conversano, F., Casciaro, S., Massaro, A., Bhateja, V., Spano, F.: EEG signal processing and acquisition for detecting abnormalities via bio-implantable devices. In: 2016 IEEE International Symposium on Medical Measurements and Applications (MeMeA), pp. 1–5. IEEE (2016)

21. Qiang, L., Bo, L.: The muscle activity detection from surface EMG signal using the morphological filter. Appl. Mech. Mater. **195**, 1137–1141 (2012)
22. Tiwari, D.K., Bhateja, V., Anand, D., Srivastava, A., Omar, Z.: Combination of EEMD and morphological filtering for baseline wander correction in EMG signals. In: Proceedings of 2nd International Conference on Micro-Electronics, Electromagnetics and Telecommunications, pp. 365–373. Springer, Singapore (2018)

Measuring Discharge Using Back-Propagation Neural Network: A Case Study on Brahmani River Basin

Dillip K. Ghose

Abstract Prediction of discharge (runoff) is vital for flood control during peak periods of flow. The present work is focused on the prediction discharge using back-propagation neural network (BPNN) models. Parameters like stage (water level) have been collected on daily basis from Govindpur basins on River Brahmani to estimate discharge using BPNN model. Different architectures of models are trained and tested to predict the performance of models during June, July, and August of monsoon period for measuring discharges at the proposed station. The individual best performances for different models are found out to measure discharges during peak period of monsoon. Among June, July, and August, the model performance says the highest flow occurs during the month of July for the study period.

Keywords Stage · Discharge · Neural networks · Back-propagation

1 Introduction

The development of water resource engineering has closely followed the development in physical sciences. Very often it is difficult to develop models for water resources engineering problems due to its complex nature and uncertainty in water parameters with traditional methods of physical sciences. Development of new computational algorithms and their application to new areas cutting across various disciplines in science and engineering goes hand in hand. In recent years, such reports have increased phenomenally. Soft computing techniques such as artificial neural network (ANN) and fuzzy logic are being used in various engineering applications with high success rate.

D. K. Ghose (✉)
Department of Civil Engineering, National Institute of Technology,
Silchar 788010, Assam, India
e-mail: dillipghose2002@gmail.com

V. Bhateja et al. (eds.), *Intelligent Engineering Informatics*, Advances in Intelligent Systems and Computing 695, https://doi.org/10.1007/978-981-10-7566-7_59

Many researchers have investigated the potential of neural networks in modeling watershed runoff based on rainfall inputs as follows. Techniques of assessment of performance and the errors in the models were reported by Atiken [1] a holistic approach to natural resources management. Atiya et al. [2] applied neural networks to the problem of forecasting the flow of the River Nile in Egypt. They compared different methods of preprocessing the inputs and outputs including a method based on the discrete Fourier series. Carriere et al. [3] developed a virtual runoff hydrograph system that employed a recurrent back-propagation neural network to generate runoff hydrographs. He reported that neural network could predict runoff hydrographs accurately, with good agreement between the observed and predicted values. Fatima and Shaheen [4] estimated surface runoff for the Tarbela reservoir. The rational formula method was used to determine the surface runoff. French et al. [5] implemented a feed-forward back-propagation neural network to predict precipitation in Hanover, NH. They then compared these results with other algorithms including a radial basis function neural network and regression. Garrote and Bras [6] reported various applications of real-time flood forecasting using digital elevation models. Gautam et al. [7] reported that the groundwater table change before and after a bridge pier construction. Hino [8] analyzed daily maximum streamflow data of each month from three gauge stations for simulation using stochastic approaches. Mason et al. [9] verified clustering of rainfall–runoff patterns and modeling of each data set by different radial basis function neural networks. Minns and Hall [10] proved that improved performances are achieved with the extra hidden layer from many catchment areas. The primary objective of the study is to develop methodologies to solve the problems related to measurement of discharge at Govindpur, in the Brahmani river basin, India.

2 Study Area and Data

The Brahmani river basin about 480 km long, located in Eastern India as shown in Fig. 1. The Brahmani is the second longest river in Orissa after the river Mahanadi. It has a catchment area of about 39,033 square kilometers in Orissa alone. The Brahmani is formed by the confluence of the rivers South Koel and Sankh near the major industrial town of Rourkela at 22 15’ N and 84 47’ E. Govindpur is one of the major gauging stations which measures daily discharges for monsoon period. The data are collected from central water commission Bhubaneswar, Odisha, India, for Govindpur gauging station. The data spanning over two years from 2013–2105 during monsoon period is collected for model development. For developing model, the input data are the average stages of daily data and corresponding to the stages average discharge is considered as output.

Fig. 1 Study area

3 Methodology

Back-propagation is a systematic method for training artificial neural networks. The network is trained by supervised learning method to train the net to achieve a balance between the ability to respond correctly to the input patterns (Fig. 2).

Here batch learning or pre-epoch method is used in sequential learning, and the error is determined and back-propagated with updating of weights. The optimized hidden layers are within a range of two to seven layers. Here learning rate and momentum coefficient used are 0.001 and 0.01, respectively. Stages for present time one-day lag time and two-day lag time are $\mathrm{H_t}$, $\mathrm{H_{t-1}}$, and H_{t-2}, respectively, and considered as inputs in this architecture. Q_t is the predicted runoff considered as output of the architecture. Middle layer is the hidden layer which bridges between input and output.

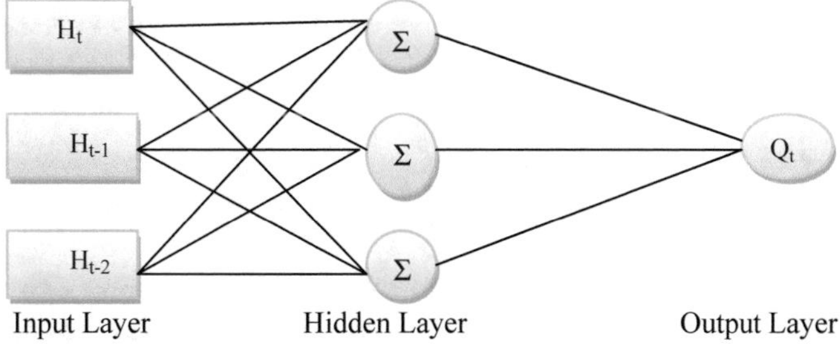

Fig. 2 Architecture of BPNN

4 Results and Discussions

In this study, BPNN models are developed for deriving runoff rating relationship using water levels (stage) as the input and the runoff (discharge) as the output. The daily runoff prediction models are developed using water levels and runoff with different lag periods as input and the present day runoff as output. A number of combinations of input data are experimented. The mean square error and model efficiency are computed for each of them. Lesser error and greater efficiency (approaching 1) are the satisfactory conditions for developing an efficient model. The results of the runoff prediction models for the months of June to August are presented in the following sections.

The scatter plot of actual versus modeled runoff showing on the plot coefficient of determination which represents the efficiency of model is shown in Fig. 3, and the linear scale plot of actual versus predicted daily runoff is shown in Fig. 4.

The simulated surface runoff is compared with their actual counterparts using the evaluation criteria earlier mentioned. Table 1 presents mean square error (MSE) for

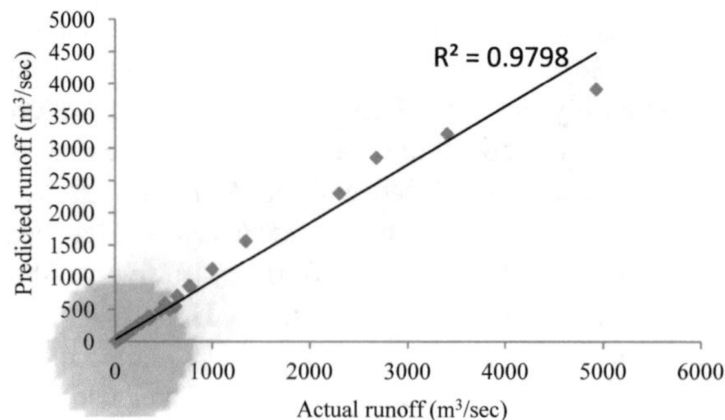

Fig. 3 Model efficiency of actual versus predicted runoff (June)

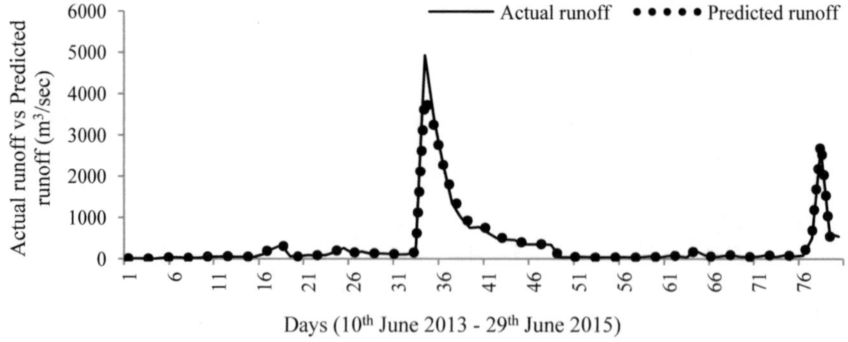

Fig. 4 Actual runoff and predicted runoff during testing phase of best model (June)

Table 1 Network architectures of BPNN runoff prediction model (June)

Month	Model inputs	Model architecture	Number of iteration	Mean square error		Model efficiency (%)
				Training	Testing	
June	H_t, H_{t-1}	2-3-1	3000	0.000499	0.000163	97.47
	H_t, H_{t-1}, H_{t-2}	**3-5-1**	**6856**	**0.000443**	**0.000138**	**95.88**
	H_t, H_{t-1},Q_{t-1}	3-3-1	9609	0.000351	0.000111	97.9

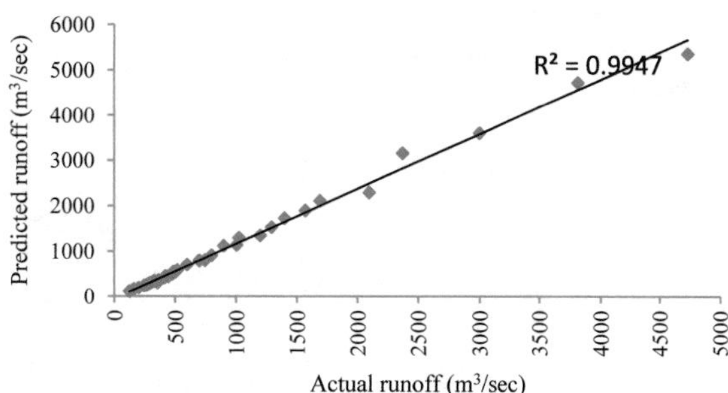

Fig. 5 Model efficiency of actual versus predicted runoff (July)

both the training and testing periods and model efficiency. In general, MSE decreased and efficiency increases with an increase in ANN architecture. Increase in ANN architecture may be read as increase in the number of hidden layers and/or the number of nodes in hidden layers. Out of different architectures presented in the table, the best model is 3-5-1 with inputs of H_{t-1} and H_{t-2}. This model has the highest efficiency, least errors in training and testing phases.

The scatter plot of actual versus modeled runoff showing on the plot coefficient of determination which represents the efficiency of model is shown in Fig. 5, and the linear scale plot of actual versus predicted daily runoff is shown in Fig. 6.

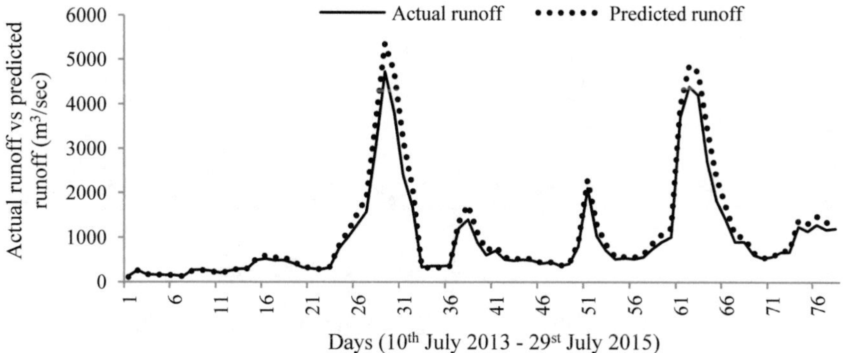

Fig. 6 Actual runoff and predicted runoff during testing phase of best model (July)

Table 2 Network architectures of BPNN and prediction of runoff for the month of July

Month	Model inputs	Model architecture	Number of iteration	Mean square error		Model efficiency (%)
				Training	Testing	
July	H_t, H_{t-1}	2-5-1	751	0.001436	0.000255	94.82
	H_t, H_{t-1}, H_{t-2}	**3-4-1**	**545**	**0.00153**	**0.000165**	**96.43**
	H_t, H_{t-1},Q_{t-1}	3-5-1	5615	0.001265	0.000179	95.25

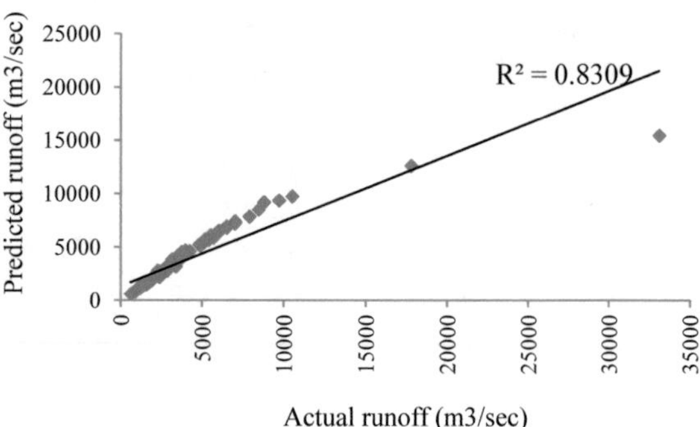

Fig. 7 Model efficiency of actual versus predicted runoff (August)

Mean square error (MSE) and model efficiency for both training and testing period of different architectures with different inputs are presented in Table 2. Model with architecture 3-4-1 having inputs H_t, H_{t-1}, H_{t-2} requires 545 iterations to achieve acceptable efficiency. Hence, in the month of July, two-day lag water levels used as input for prediction of runoff give better result as compared to one-day lag inputs and present-day inputs.

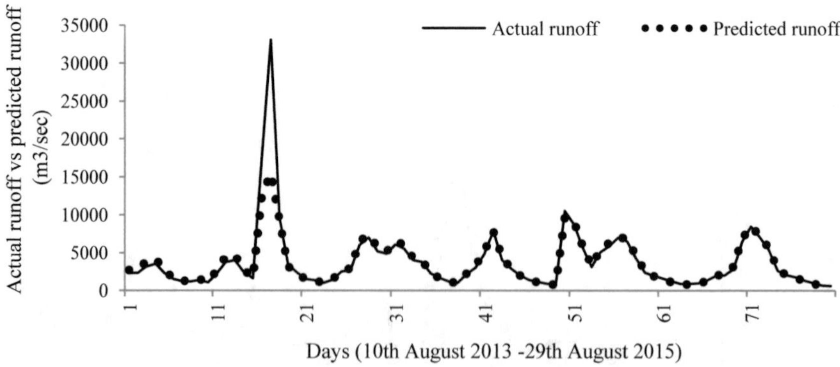

Fig. 8 Actual runoff and predicted runoff during testing phase of best model (August)

Table 3 Network architectures of BPNN and prediction of runoff for the month of August

Month	Model inputs	Model architecture	Number of iteration	Mean square error		Model efficiency (%)
				Training	Testing	
August	H_t, H_{t-1}	2-4-1	3005	0.000159	0.001756	82.65
	H_t, H_{t-1},H_{t-2}	**3-6-1**	**2232**	**0.000142**	**0.001648**	**94.02**
	H_t, H_{t-1},Q_{t-1}	3-5-1	5012	0.000162	0.001669	81.5

The scatter plot of actual versus modeled runoff showing on the plot coefficient of determination which represents the efficiency of model is shown in Fig. 7, and the linear scale plot of actual versus predicted daily runoff is shown in Fig. 8.

The results of three selected models 2-4-1 (H_t, H_{t-1}), 3-6-1 (H_t, H_{t-1},H_{t-2}), and 3-5-1 (H_t, H_{t-1},Q_{t-1}) are presented in Table 3. Out of all the models, the model in the second row is most efficient with efficiency of 94.02 (%).

5 Conclusion

Model development using BPNN found that 3-5-1 architecture shows best performances with coefficient of determination (model efficiency) as 0.9588 in June. During July, 3-4-1 architecture with model efficiency 96.43% is found to be best model among all, whereas 3-6-1 architecture with coefficient of determination 0.9402 is found to be suitable for predicting discharge during monsoon period. The predicted models are suitable for measuring discharges during June, July, and August for any gauging station of Brahmani river basin with different stages during different periods and recommended for measuring discharges with variation of stages in un-gauged stations by knowing the depths of flow in the river basin.

References

1. Atiken, A.P.: Assesing systematic errors in rainfall-runoff models. J. Hydrol. **20**, 121–136 (1973)
2. Atiya, A.F., El-Shoura, S.M., Shaheen, S.I., El-Sherif, M.S.: A comparison between neural-network forecasting techniques—case study, river flow forecasting. IEEE Trans. Neural Netw. **10**(2), 402–409 (1999)
3. Carriere, P., Mohaghegh, S., Gaskari, R.: Performance of a virtual runoff hydrograph system. J. Water Serv. (1993)
4. Fatima, K., Shaheen, A.: Estimation of surface runoff for Tarbela reservoir. In: ICAST Proceedings of 2nd International Conference on Advances in Space Technologies, Space in the Service of Mankind, 4747695, pp. 103–106. (2008)

5. French, M., Krajewski, W., Cuykendall, R.R.: Rainfall forecasting in space and time using a neural network. J. Hydrol. **137**, 1–31 (1992)
6. Garrote, L., Bras, R.L.: A distributed model for real—time flood forecasting using digital elevation models. J. Hydrol. **167**, 279–306 (1995)
7. Gautam, M.R., Watanabe, K., Saegusa, H.: Runoff analysis in humid forest catchments with artificial neural network. J. Hydrol. **235**, 117–136 (2004)
8. Hino, M.: Prediction of flood and stream flow by modern control and stochastic theories. In: Proceedings of the 2nd International IAHR, Symposium of Stochastic Hydraulic, 5-1-5-26. (1997)
9. Mason, J.C., Price, R.K.: Tem'me, A.: A neural network model of rainfall-runoff using radial basis functions. J. Hydraul. Res. **34**(4), 537–548 (1996)
10. Minns, A.W., Hall, M.J.: Artificial neural networks as rainfall-runoff models. Hydrol. Sci. J. **41**(3), 399–417 (1996)

Prediction of Suspended Sediment Load Using Radial Basis Neural Network

Dillip K. Ghose

Abstract Prediction of suspended sediment is vital for soil erosion during peak periods of inundation. The present work focused on development of modeling for suspended sediment concentration via Radial Basis Neural Network (RBNN) models. Parameters such as rainfall, discharge have been collected on daily basis from Rajghat gauging station of Subarnarekha basin to develop the model. Different architectures of models are fixed to envisage the performance of models during July, August, and September of monsoon period for measuring suspended sediment load. The individual best performances for different models are found out to measure sediment load during peak period of monsoon. Among July, August, and September, the model performance says the highest potential of erosion occurs during the month of September from the nearest zone of watersheds to the river basin during the peak period of flood. This work brings an idea for indirect measurement of sediment delivery ratio.

Keywords Rainfall · Runoff · Suspended sediment · Radial basis neural networks

1 Introduction

Sediment concentration in the river varies with several factors like sediment type, mode of sediment deposition, sediment transportation, and stream flow variability. The analysis of sedimentation data of Indian rivers shows that the annual siltation rate has been generally 1.5–3 times more than the designed rate and the reservoirs are losing capacity at the rate of 0.3–0.92% annually. The consequence of loss in storage due to sedimentation is precluding the intended usages. Though these problems are faced in all the river basins and are stated by many authors in their

D. K. Ghose (✉)
Department of Civil Engineering, National Institute of Technology,
Silchar 788010, Assam, India
e-mail: dillipghose2002@gmail.com

V. Bhateja et al. (eds.), *Intelligent Engineering Informatics*, Advances in Intelligent Systems and Computing 695, https://doi.org/10.1007/978-981-10-7566-7_60

research publications, the problems are very peculiar in the present study area. In the present work, effort is made to devise methods to solve these problems.

Achite and Ouillon [1] used regression relationships between discharge and suspended sediment concentration. Arabkhedri et al. [2] used two bias-corrected, rating curves to estimate sediment load. Ghose et al. [9] used artificial neural networks to predict suspended sediment load in the river Mahanadi. Some approaches are based on the power (work of the Flow) concept; Engelund and Hansen's approach [7], which depends on the power concept and similarity principle to obtain sediment transport function. Laursen [12] developed a functional relationship between the flow condition and the resulting sediment discharge; Jain [10] used three-layer feed-forward ANNs to establish an integrated stage-discharge-sediment concentration relation for two sites on the Mississippi river. Crowder et al. [6] identify the major hydrological and meteorological controls determining the dynamics of SSC during storm-runoff events and the magnitude of SSC in a headwater catchment in Luxembourg. Partal and Cigizoglu [14] used combined wavelet-ANN method to estimate and predict the suspended sediment load. Ganju et al. [8] used hydrologic proxies to develop a daily sediment load time series. Aytek and Kisi [3] used Genetic Programming (GP) for explicit formulation of daily suspended sediment–discharge relationship. Cobaner et al. [5] used neuro-fuzzy and ANM techniques namely generalized regression neural networks (GRNN), radial basis neural networks (RBNN), multilayer perceptron (MLP), and two different sediment rating curve (SRC). Baskaran et al. [4] used artificial neural networks (ANNs) to forecast and estimate sedimentation concentration values. Jothiprakash and Garg [11] used back-propagation neural network (BPNN) to estimate the volume of sediment retained in a reservoir. Melesse et al. [13] observed daily predictions were better compared to weekly predictions for all three rivers due to higher correlation within daily than weekly data. ANN predictions for most simulations were superior compared to predictions using MLR, MNLR, and ARIMA. The primary objective of the study is to develop methodologies to solve the problems related to sediment transport at Rajghat in the Subarnarekha river basin, India, during peak monsoon. The literature study brings an idea about different methods for estimating sediment load. The present study is an attempt for measuring suspended sediment concentration during peak period of monsoon.

2 Study Area and Data

The Subarnarekha basin (Fig. 1) flows over Odisha, Jharkhand, and West Bengal spanning over a total area of 29196 km^2 with length 297 km and breadth 119 km. The river basin lies between 85^0 8' to 87^0 32' east longitudes and 21^0 15' to 23^0 34' north latitudes, located in Peninsular India at northeast corner. It is surrounded by

Fig. 1 Study area: Rajghat gauging station

Chota Nagpur plateau. The river raises 600 m elevation near Nagri village of Ranchi district, Jharkhand. The river flow a length of 395 km before intermingling into the ocean, Bay of Bengal. Rajghat gauging station is located at a distance of 8 km nearest to Jaleswar town. The major part of the basin is covered with well-irrigated agriculture land accumulating to 60% approximately of the total watershed area. The area is usually flooded during monsoon period and soil erosion occurs in and around the region of watershed and flows into the river with runoff creating instability to the irrigation land. Primary data used for developing the model are daily precipitation and peak stages of water level. Secondary level of data used is the evaluation of discharge from the discharge rating curve projected by the corresponding stages of water level data. Daily data spanning over 2012–2016 are collected from Central Water Commission Bhubaneswar, (CWC) Odisha, India.

3 Methodology

Radial basis function network uses Gaussian transfer functions, which regularizes the networks. Gaussian transfer function enables describing the intensity distribution of sediment concentration produced by a point source basically radial basis

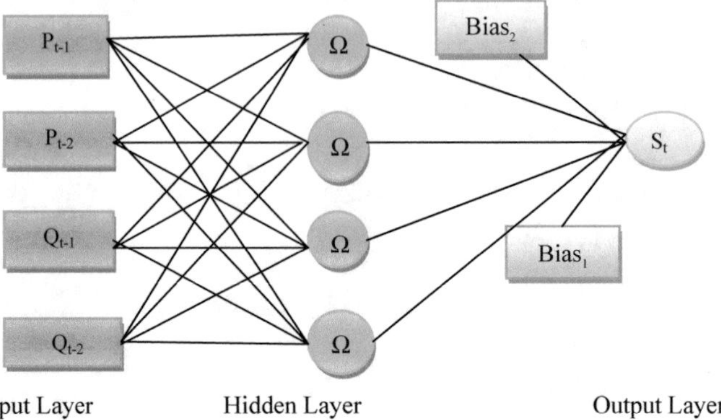

Fig. 2 Architecture of RBNN

neural network (RBNN) is composed of large number of simple and highly interconnected artificial neurons and can be organized into several layers, i.e., input layer, hidden layer, and output layer, and the architecture of RBNN is shown in Fig. 2.

Here unsupervised learning method is used in sequential learning and the error is determined with the updating of weights. The optimized hidden layers are within a range of two to nine layers. Here, the central vector is used within a range of 10–40. Here P_t, P_{t-1}, P_{t-2}, Q_{t-1}, Q_{t-2} are considered as inputs and S_t as output for measuring sediment load concentration.

4 Results and Discussions

Data of Rajghat station, which are used to develop RBFN models, mean square error in training and testing phases against the number of iteration are computed. Also the coefficient of efficiency or coefficient of determination is determined for each model. The best performing model is one which yields least errors (in training and testing phases) with the highest coefficient of efficiency. Number of center vectors are set for the models with several trials with different model parameters like learning rate(ŋ) and momentum coefficient (α). Learning rate and momentum coefficient pairs used for trial are (0.1, 0.3), (0.1, 0.5), (0.3,0.5), (0.5, 0.5), and (0.1, 0.7), respectively. Out of all the parametric pairs for radial basis function neural network, the best parameter pair used for sediment concentration rating models is (0.1 and 0.3).

The model architecture 3-20-1 of RBFN for the month of July is able to learn the process of variation of sediment concentration with the water level and runoff with the model efficiency of 95.21%. The scatter plot of actual versus modeled sediment concentration showing on the plot coefficient of determination which represents the

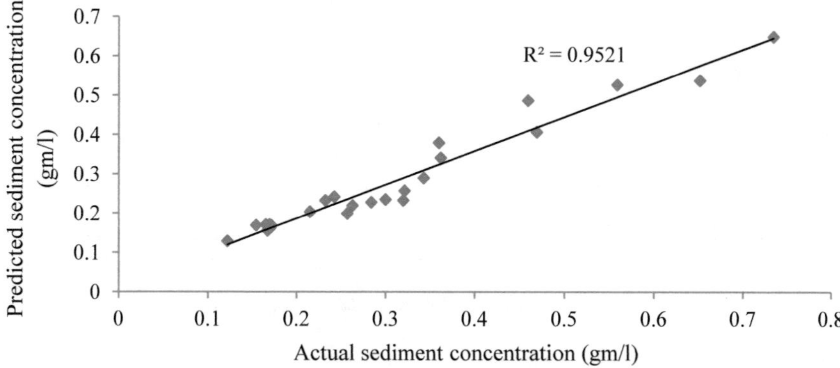

Fig. 3 Actual versus predicted sediment concentration (July)

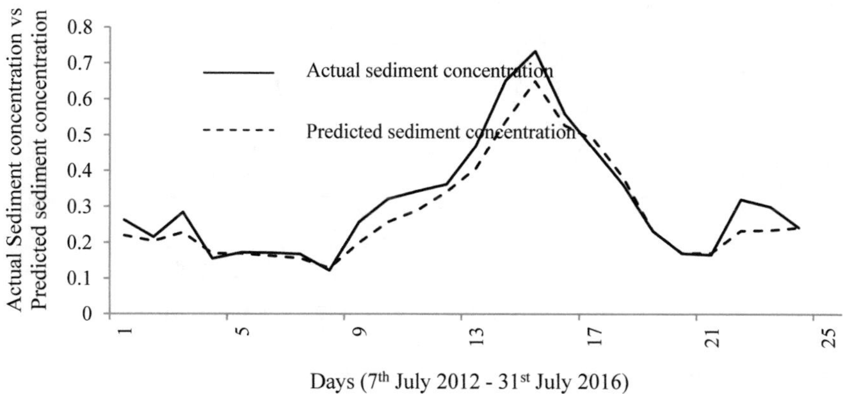

Fig. 4 Actual and predicted sediment concentration during testing phase (July)

efficiency of model is shown in Fig. 3 and the linear scale plot of actual versus predicted daily sediment concentration is shown in Fig. 4.

The scatter plot of actual versus modeled sediment concentration showing on the plot coefficient of determination which represents the efficiency of model is shown in Fig. 5 and the linear scale plot of actual versus predicted daily sediment concentration is shown in Fig. 6.

The scatter plot of actual versus modeled sediment concentration showing on the plot coefficient of determination which represents the efficiency of model is shown in Fig. 7 and the linear scale plot of actual versus predicted daily sediment concentration is shown in Fig. 8.

The architectures of different sediment concentration rating models and the final results of training and testing phases with the efficiencies of the model for different months (July, August, and September) are shown in Table 1.

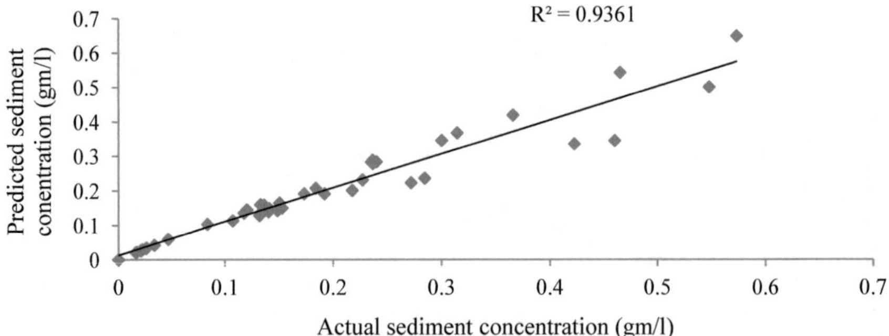

Fig. 5 Actual versus predicted sediment concentration (August)

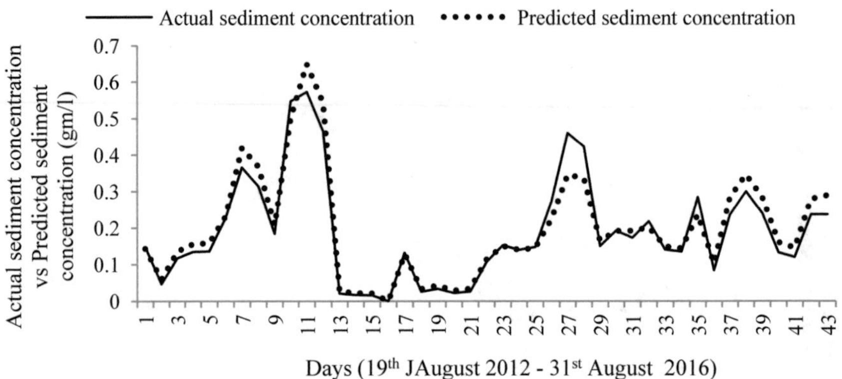

Fig. 6 Actual and predicted sediment concentration during testing phase (August)

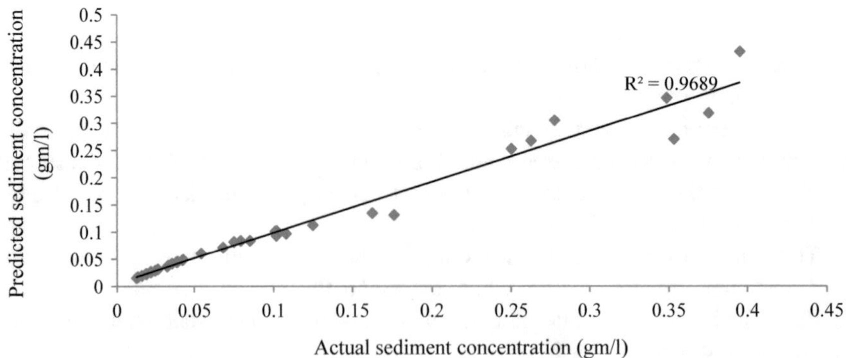

Fig. 7 Actual versus predicted sediment concentration (September)

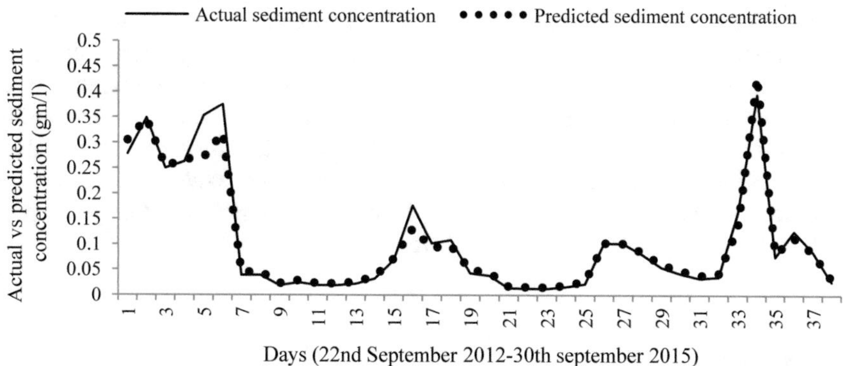

Fig. 8 Actual and predicted sediment concentration during testing phase (September)

Table 1 Details of RBFN used as sediment rating models

Month	Model inputs	Model architecture	Number of iteration	Mean square error		Coefficient of determination
				Training	Testing	
July	**P_{t-1}, Q_{t-1}, Q_{t-2}**	**3-20-1**	**220**	**0.005311**	**0.000605**	**0.9521**
	P_{t-1}, P_{t-2},, Q_{t-1}, Q_{t-2}	4-10-1	851	0.007318	0.000813	0.8852
	P_t, P_{t-1}, P_{t-2},, Q_{t-1}, Q_{t-2}	5-30-1	643	0.006324	0.000719	0.8973
August	**P_{t-1}, Q_{t-1}, Q_{t-2}**	**3-10-1**	**53**	**0.002718**	**0.000425**	**0.9361**
	P_{t-1}, P_{t-2},, Q_{t-1}, Q_{t-2}	4-20-1	202	0.006137	0.000519	0.8532
	P_t, P_{t-1}, P_{t-2},, Q_{t-1}, Q_{t-2}	5-30-1	158	0.007273	0.000493	0.9084
September	**P_{t-1}, Q_{t-1}, Q_{t-2}**	**3-20-1**	**1901**	**0.00276**	**0.000256**	**0.9689**
	P_{t-1}, P_{t-2},, Q_{t-1}, Q_{t-2}	4-10-1	2403	0.00589	0.000658	0.9136
	P_t, P_{t-1}, P_{t-2},, Q_{t-1}, Q_{t-2}	5-30-1	2268	0.00484	0.000586	0.8637

Similarly, August and September for sediment rating curves possess model efficiencies of 93.61 and 96.89%, respectively.

5 Conclusion

Model improvement using RBNN found that 3-20-1 architecture shows best performances with coefficient of determination as 0.9689 in the month of September. During September, 3-20-1 architecture with model efficiency 96.89% is found to be best model among all. The predicted models are suitable for measuring sediment concentration and an index for measuring soil erosion from watersheds of the river basin during peak monsoon period for any gauging station of Subarnarekha river basin with different discharges of different periods and recommended for measuring soil erosion characteristics from river banks with variation of suspended sediment concentration from the prescribed model. The developed models are capable of measuring sediment concentration and sediment delivery ratio at other gauging as well as un-gauging catchments of the river basin.

References

1. Achite, M., Ouillon, S.: Suspended sediment transport in a semiarid watershed. J. Hydrol. **343** (3–4), 187–202 (2007)
2. Arabkhedri, M., Lai, F.S., Ibrahim, N.A., Kasim, M.R.M.: The effect of adaptive cluster sampling design on accuracy of sediment rating curve estimation. J. Hydrol. Eng. **15**(2), 142–151 (2010)
3. Aytek, A., Kisi, O.: A genetic programming approach to suspended sediment modelling. J. Hydrol. **351**(3–4), 288–298 (2008)
4. Baskaran, T., Nagan, S., Rajamohan, S.: Prediction of sediment deposition in reservoirs using artificial neural network. Int. J. Earth Sci. Eng. **3**(4), 525–531 (2010)
5. Cobaner, M., Unal, B., Kisi, O.: Suspended sediment concentration estimation by an adaptive neuro-fuzzy and neural network approaches using hydro-meteorological data. J. Hydrol. **367** (1–2), 52–61 (2009)
6. Crowder, D.W., Demissie, M., Markus, M.: The accuracy of suspended sediments when log transformation produces nonlinear suspended sediment–discharge relationships. J. Hydrol. **336**(3–4), 250–268 (2007)
7. Engelund, F., Hansen, E.: A monograph for sediment transport and alluvial streams. Teknisk Forlag, Copenhagen (1972)
8. Ganju, N.K., Knowles, N., Schoellhamer, D.H.: Temporal downscaling of decadal suspended sediment estimates to a daily interval for use in hind cast simulations. J. Hydrol. **349**(3–4), 512–523 (2008)
9. Ghose, D.K., Panda, S.S., Swain, P.C.: Prediction and optimization of runoff via ANFIS and GA. Alex. Eng. J. **52**(7), 209–220 (2013)
10. Jain, S.K.: Development of integrated sediment rating curves using ANNs. J. Hydraul. Eng. ASCE. **1**, 30–37 (2001)
11. Jothiprakash, V., Garg, G.: Reservoir sedimentation estimation using artificial neural network. J. Hydraul. Eng. ASCE. **14**(9), 1035–1040 (2010)
12. Laursen, E.M.: The total sediment load of streams. J. Hydraul., Proc. Am. Soc. Civil Eng. **84**, 1530–1536 (1958)
13. Melesse, A.M., Ahmad, S., McClain, M.E., Wang, X., Lim, Y.H.: Suspended sediment load prediction of river systems, an artificial neural network approach. Agric. Water Manag. **98**(5), 855–866 (2011)
14. Partal, T., Cigizoglu, H.K.: Estimation and forecasting of daily suspended sediment data using wavelet–neural networks. J. Hydrol. **358**(3–4), 317–331 (2008)
15. Yitian, L., Gu, R.R.: Modelling flow and sediment transport in a river system using an artificial neural network. Env. Manag. **31**(1), 122–134 (2003)

A Hybrid Task Scheduling Approach Based on Genetic Algorithm and Particle Swarm Optimization Technique in Cloud Environment

Bappaditya Jana and Jayanta Poray

Abstract As per current trends, cloud services are becoming more popular day by day. Because these services satisfy several demands for heterogeneous resources without using dedicated IT infrastructure. In cloud platform, user gets shared pool of resources through Internet irrespective of any geographical location. But cloud services need to handle a gigantic amount of request, as the number of users increases exponentially. In order to manage a pool of requests till now no effective scheduling mechanism is available in practice. So to minimize the time delay and optimize the overall complexity, suitable scheduling methodology is very much required. In our study, we have presented a novel scheme for scheduling algorithm after the enhancement of genetic algorithm and particle swarm optimization technique. We have proposed a methodology that can provide a better response time from cloud provider and minimize the waiting time for particular clients in cloud environment.

Keywords Cloud computing · Genetic algorithm · Max–min scheduling Min execution time · Scheduling algorithm · Particle swarm optimization Hybrid task scheduling algorithm · Time hybrid task scheduling

1 Introduction

In recent advancement of information and communication technology, research on cloud computing becomes one of the most latest trends. Cloud computing appears as a fastest evolutionary virtual platform which promises for a good cost-effective environment including business model, education, personal intelligence, and many

B. Jana (✉) · J. Poray
Department of Computer Science & Engineering, Techno India University, Kolkata, West Bengal, India
e-mail: bappaditya.j.in@ieee.org

J. Poray
e-mail: jayanta.poray@gmail.com

V. Bhateja et al. (eds.), *Intelligent Engineering Informatics*, Advances in Intelligent Systems and Computing 695, https://doi.org/10.1007/978-981-10-7566-7_61

other diverse application for global users according to their need without investing software, hardware, or any other computing or networks infrastructure [1, 2]. The main purpose of this service to the user is to provide effective service with minimal cost [3]. As lots of advantages are coming from cloud services in our daily life, it is manually very difficult to assign or schedule a particular resource for a specific task [4]. So scheduling of task is one major challenging issue in cloud computing. Various methodologies have been proposed on this area. At present, some existing task scheduling approach using genetic algorithm and some cases modified form of genetic algorithm have been used in [5, 6, 7]. Some other existing algorithm such as FIFO scheduling, fair scheduling, and computing capacity scheduling is still facing some restrictions [8]. In cloud platform, redundant array of independence disks techniques is used to balance cost of the service and to maintain assurance of cloud data [9]. We have tried to propose a technique for task scheduling algorithm based on genetic algorithm (GA) and particle swarm optimization (PSO) technique and to optimize the existing issues in cloud platform. In our hybrid task scheduling algorithm, we have compared execution time with max–min scheduling execution and minimum scheduling time for a particular number of task and we have considered two times where execution time is less than other two cases. In our technique, we have shown that for a particular task, response time is comparatively better and waiting time is minimum than other two-mentioned scheduling approaches.

2 Overview on Cloud Services

Cloud system has two major components, one is cloud data distributor another is cloud service providers. In first cases, cloud collect data files from users or clients then splits into small clusters with respect to identical features and after splitting these are distributed according to the on-demand or requests of users. Memory distribution in cloud server is discussed in [10]. The distributed approach is maintained by redundant array of independent disk (RAID) technique. On the other hand, major goal of service providers is to store cluster of data in response according to the various requests showing availability of effective data and remove cluster of data as per requirement. Cloud service is a on-demand self-service. This service is generally divided into three categories, [11] such as Platform-as-a-Service (PaaS), Software-as-a-Service (SaaS), and Infrastructure-as-a-Service (IaaS). SaaS is on-demand software service and business model predicted by pay–per-use basis. In Saas, user can use data bases, hardware and software maintenance without any personal expenses. Again software is updated automatically without installing new software because in this service applications are hosted centrally. PaaS is one type of cloud computing in which cloud providers use an environment for developing applications and to provide various services through Internet. In Iaas, cloud provides an ideal computing platform to the client. Clients get virtual operating system and program execution environment, data bases, and Web server. Due to the

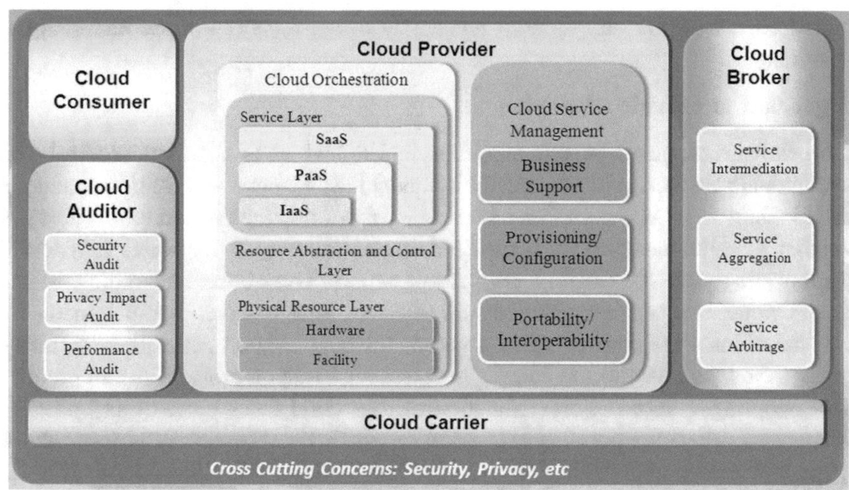

Fig. 1 Schematic diagram of cloud services

availability of central virtual architecture in cloud, many real-time operations are performed in cloud. The schematic diagram of overall cloud services is shown in Fig. 1, [12].

Clouds are classified into three types [13] such as public, private, and hybrid. Cloud behaves as virtual machine, and virtualizations of cloud are briefly described in [14].

3 Proposed Methodology

Genetic algorithm is based on biological evolution in population and concerned with Darwin's theory of evolution. Darwin proposed "Survival of the fittest" which implements method of scheduling in genetic algorithm [15, 16, 17]. In cloud computing, multiple resources have to be assigned according to the multiple requirements of jobs. So which resource will get preference for which task or job is a complex situation process. That's why absence of proper scheduling process, the time complexity will be high. Genetic algorithm is one type of evolutionary computing. Genetic algorithm was basically implemented on a set of chromosomes in ecological environment. At first, one group of chromosomes is considered as initial data set and then after generation, a new set known as offspring's is produced from parent set. This set of chromosomes is considered as a set of solutions in cloud computing and the set of solutions is known as population also. During generations, one population produces a newly updated population according to the expectation. A fitness function is used to find better solutions. Some genetic operator such as mutations, crossover, etc., is used repeatedly to get most optimized and simplest solution.

Algorithm 1: Genetic Algorithm

John Holland proposed genetic algorithm in 1962 which is a random search-based evolutionary method in biological environment [18]. Genetic algorithm includes six basics steps as **Initialization:** Generate population randomly from n number of solutions. **Use of fitness function:** f(x) is the fitness function of particular solution in existing population. **Selection:** According to the fitness value, two separate solutions are selected from the given population. These two individuals are considered as a parent for next coming generation. **Crossover:** Now two parents are coming in crossover based on probability and produce new offsprings. **Mutation:** Mutation of new child takes place at any position whenever the existing population tends to be consistent. **Accept and Replace:** After based on probabilistic mutation, the new solutions are generated and again replace new population as the present generation.

Algorithm 2: Particle swarm optimization technique

Kennelly and Eberhat proposed the concept of particle swarm optimization technique in 1995. It is one type of evolutionary search technique for a population in ecological environment where a particular set of random solution of existing population is considered as a particle. Every particle flies randomly with a velocity throughout whole search spaces and using a fit function they randomly fit according to swarm's historical behavior. This step is going repeatedly until getting better solutions.

4 Hybrid Approach

4.1 Evolutionary Algorithms as Optimizers

```
begin
solution representation
c: : = 0 ; // generation counter
initialize f(c)
evaluate f(c);
while(not terminate)
do
f1(c) :  selection(f(c));
f2(c) :  variation(f1(c));
Evaluate f2(c);
f(c+1) : = survivor (f(c). f2(c);
c : = c+1;
od
end
```

4.2 Traditional Evolutionary Approach in Binary-Coded Genetic Algorithms

4.2.1 Genetic Algorithms

- create a can for minimum cost having at most volume V.
- Objective function cost (d,h) : f(d,h) = Πdh + 2Πd2/4
- Constraint : Volume(d,h) : Πd2 h /4 ≥ V
- Representation in binary strings
- Fitness : Objective + penalty for constraint

4.2.2 PSO Approach

- Let us consider all particles fly around the whole search space and velocity of each particle are not static, they are updated dynamically.
- $x_i = x_i + v_i$ where x_i is the location and vi is the velocity
- $V_i = v_i + c_1$ rnd() (P_i, best − x_i) + c2 rnd() ($P_g - x_i$) Where Pi is the location with best condition for i-th particles and P_g is the actual location of the particle , selected as best and so far and c1, c2 in (0.2)
- According to the history, the momentum part is 1st term.
- The 2nd term is represented as thinking with private condition which is also known as cognitive part.
- The another remaining term is related with social components which means a collaboration.
- Inertia weight is Vi where $V_i = w\, V_i + \ldots$ (w in [0,4,0,9])
- Velocity restriction , V_i = min(Vmax,vi) and it prevents unbounded growth in velocity.
- Constricted PSO: $V_i = X(V_i + \ldots\ldots)$ (X = 0.7289) and it give better convergence.
- Bare bone PSO : Dropping V_i term
- Offspring are normally distributed around P_g and P_0
- Communication topologies :
- P_g is now replaced by I_g

Repeat the steps until a stopping criterion is meet to get a sufficient good fitness value.

Table 1 Average schedule length

Iteration number	Number of task	Max–Min scheduling time	Minimum execution time	Hybrid task scheduling algorithm
1	200	660	654	631
2	400	1350	1324	1286
3	600	2020	1996	1926
4	800	2772	2764	2692
5	1000	3446	3436	3376
6	1200	4164	4156	4116

5 Experimental Results

In our study, we have used a software for simulation known as cloudsim [18]. In [19] max–min algorithm maintains the task status that includes the updated time, execution time, and completion time to improve the performance. Our proposed algorithm is designed in cloudsim layer. Each service request from cloud user is called as task and which needs to be recorded by VMs. Each task has to be processed in VMs according to the processing capability of VMs. VMs have some memory and storage and controlled by millions of instructions. Associated band of number index entries found. Simulation number of task varied from 200 to 1200 with iterative increments by 200.

The data components give length of average scheduling and successful execution ratio which is shown in Table 1. According to Table 1, it is cleared that our presented methodology gives comparatively lower execution time than max–min scheduling time and minimum execution time. The hybrid-scheduling algorithm has decreased the scheduling length by 4.65% than max–min scheduling and by 4.15% than minimum execution time for 600 numbers of tasks which is shown in Fig. 2.

Similarly in Table 2, it is observed that our proposed methodology gives comparatively lower execution time than max–min scheduling time and minimum execution time. The hybrid-scheduling algorithm has decreased the scheduling length by 4.39% of max–min scheduling. On other hand by 5.61% than Minimum execution time for 400 numbers of task which is shown in Fig. 3.

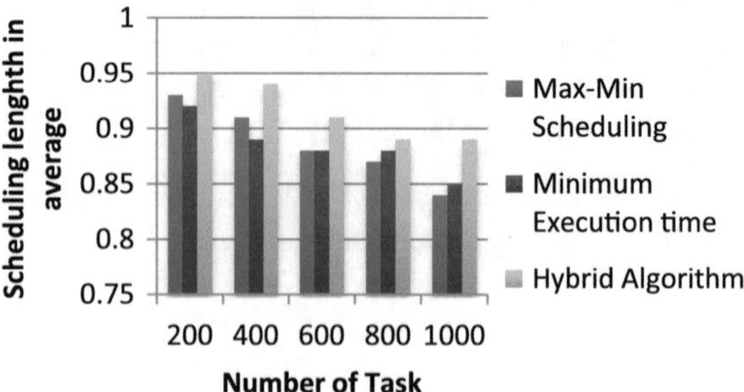

Fig. 2 Average schedule length

Table 2 Ratio of successful execution

Iteration number	Number of task	Max–Min scheduling time	Minimum execution time	Hybrid task scheduling algorithm
1	200	0.93	0.92	0.95
2	400	0.91	0.89	0.94
3	600	0.88	0.88	0.91
4	800	0.87	0.88	0.89
5	1000	0.84	0.85	0.89
6	1200	0.80	0.82	0.87

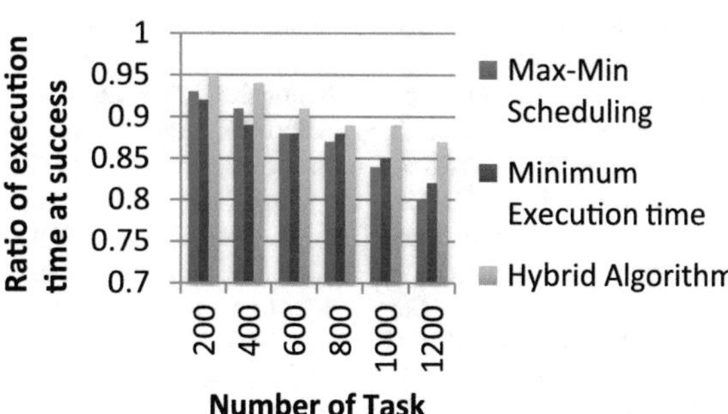

Fig. 3 Ratio of successful execution

6 Conclusions

We have presented a cost-effective hybrid task scheduling algorithm based on principle of genetic algorithm and coupled with particle swarm optimization technique which can provide a better response time from cloud provider and minimize the waiting time for a particular client in cloud computing environment. In future, we will focus on model implementation of dynamic task allocation in account of data center load factor, minimization of electric cost, and internal communications.

References

1. Zhu, K., Song, H., Liu, L., Gao, J., Cheng, G.: Hybrid genetic algorithm for cloud computing applications. In: 2011 IEEE Asia-Pacific Services Computing Conference (APSCC), pp. 182–187. IEEE (2011)
2. Paul, M., Sanyal, G.: Survey and analysis of optimal scheduling strategies in cloud environment. In: Proceedings of the IEEE International Conference on Information and Communication Technologies, Georgia, USA, pp. 789–792 (2012)

3. Lv, Z., Halawani, A., Feng, S., Li, H., Rehman, S.U.: Multimodal hand and foot gesture interaction for handheld devices. ACM Trans. Multimed. Comput. Commun. Appl. (TOMM) **11**(1s), Article 10, 19 pages (2014)
4. Tangang, Zhan, R., Shibo, Xindi: Comparatively analysis and simulation of load balancing scheduling algorithm based on cloud resource. J. Springer (2014)
5. Yin, H., Wu, H., Zhou, J.: An Improved Genetic Algorithm with Limited Iteration for Grid Scheduling. In: IEEE Sixth International Conference on Grid and Cooperative Comput ing, 2007. GCC 2007, Los Alamitos, pp. 221–227 (2007)
6. Guo, G., Ting-Iei, H., Shuai, G.: Genetic simulated annealing algorithm for task scheduling based on cloud computing environment. In: IEEE International Conference on Intelligent Computing and Integrated Systems (ICISS), 2010, Guilin, pp. 60–63 (2010)
7. Wang, J., Duan, Q., Jiang, Y., Zhu, X.: A new algorithm for grid independent task schedule genetic simulated annealing. In: IEEE World Automat ion Congress (WAC), 2010, Kobe, pp. 165–171 (2010)
8. Randles, M., et al.: Biased random walks on resource network graphs for load balancing. J. Springer (2009)
9. Qiyi, H., Tinglei, H.: An optimistic job scheduling strategy based on QoS for cloud computing. In: Proceedings of the IEEE International Conference on Intelligent Computing and Integrated Systems, Guilin, China, pp. 673–675 (2010)
10. Wei, Z., Pierre, G., Chi, C.: CloudTPS: scalable transactions for web applications in the cloud. IEEE Trans. Serv. Comput. **5**(4), 525–539 (2012)
11. Dean, J., Ghemawats, S.: MapReduce simplified data processing on large clusters[C]. In: Proceedings of the 6th Symposium on Operating System Design and Implementation, pp. 137–150. ACM, New York (2004)
12. Cryptographic Key Management Issues & Challenges in Cloud Services Ramaswamy Chandramouli Michaela Iorga Santosh Chokhani, NISTIR 7956, http://dx.doi.org/10.6028/NIST.IR.7956
13. Armbrust, M., Fox, A., Griffith, R., Joseph, A.D., Katz, R., Konwinski, A., Lee, G., et al.: A view of cloud computing. Commun. ACM **53**(4), 50–58 (2010)
14. Beloglazov, A., Buyya, R.: Managing overloaded hosts for dynamic consolidation of virtual machines in cloud data centers under quality of service constraints. IEEE Trans. Parallel Distrib. Syst. **24**, 1366–1379 (2013)
15. Michalewicz, Z.: Genetic Algorithms + Data Structures = Evolution Programs. Springer (1992)
16. Goldberg, E.: The existential pleasures of genetic algorithms. In: Winter, G. (ed.) Genetic Algorithms in Engineering and Computer Science, pp. 23–31. Wiley, New York (1995)
17. Yusoh, M., Izzah, Z., Maolin, T.: Clustering composite SaaS components in cloud computing using a grouping genetic algorithm. In: IEEE Congress on Evolutionary Computation, pp. 1–8 (2012)
18. Buyya, R., Ranjan, R., Calheiros, N.: Modeling and Simulation of scalable cloud computing environments and the cloudsim toolkit: challenges and opportunities. In: Proceedings of the 7th High Performance Computing and Simulation Conference, Leipzig, Germany, pp. 1–11 (2009)
19. Mao, Y., Chen, X., Li, X.: Max-Min task scheduling algorithm for load balancing in cloud computing. J. Springer (2014)

Segmentation of Edema in HGG MR Images Using Convolutional Neural Networks

S. Poornachandra, C. Naveena, Manjunath Aradhya and K. B. Nagasundara

Abstract In this paper, we present the segmentation of *edema* subregion in the high-grade gliomas (HGGs) MR images We use the T1, T2, FLAIR, and T1c MRI modalities in our work and employ convolutional neural network approach (CNN) architecture for the segmentation task. Preprocessing was done in each case by correcting the inhomogeneities in MR images, equalizing histogram, and applying Z-score normalization to all the volumes. Here, we experimented with the convolutional layers, activation functions, and max-pooling layers. The CNN was trained on 3D patches extracted from patient volumes. Our experimentation has given promising results with mean dice score of 0.68, positive predicted value of 0.63, and sensitivity of 0.65 for *edema* segmentation.

Keywords Gliomas · CNN · MRI · Dice score · PPV and sensitivity

1 Introduction

Brain tumors are the most aggressive and dreadful cancerous tissues which result in a mortality rate less than two years depending on the aggressiveness of the tumor (Grade I and II or Grade III and IV). Grade I and II are less aggressive, mortality

S. Poornachandra (✉)
Department of Computer Science and Applications, VTU RRC, Belgaum, India
e-mail: thisispoorna9@gmail.com

C. Naveena
Department of Computer Science and Engineering, SJB Institute of Technology, Bengaluru, India
e-mail: naveena.cse@gmail.com

M. Aradhya
Department of MCA, Sri Jayachamarajendra College of Engineering, Mysore, India
e-mail: aradhya.mysore@gmail.com

K. B. Nagasundara
Department of Computer Science and Engineering, JSS Academy of Technical Education, Bengaluru, India
e-mail: nagasundarakb@gmail.com

V. Bhateja et al. (eds.), *Intelligent Engineering Informatics*, Advances in Intelligent Systems and Computing 695, https://doi.org/10.1007/978-981-10-7566-7_62

rate is lower and is termed as low-grade gliomas (LGGs), whereas Grade III and Grade IV are more aggressive, mortality rate is higher and are termed as high-grade gliomas (HGGs). Tumors are bunch of cells that multiply and grow rapidly. Brain tumors can be classified into two types:

- In primary brain tumors, the cancerous tissues arise within the brain.
- Secondary brain tumors are those which spread into the brain via other tumor-affected regions like lung cancer, breast cancer they are also termed as *metastases*.

There are many medical imaging modalities, which are used for the diagnosis, treatment planning, and prognosis of various diseases, for example, CT is used for lung tumors, X-rays are used for diagnosing breast cancer, ultrasonography in cardiologic disorders. Magnetic resonance imaging is used for imaging soft tissue organs in the human body. And the gold standard to diagnose brain tumor is by using magnetic resonance imaging. Segmentation of brain tumors is necessary for monitoring the tumor volume measurement and tumor growth, further for surgical (resection) and radiotherapy planning and monitoring before and after treatment.

Manual segmentation of brain tumors is a labor intensive task and is subjected to inter- and intra-variation in observation. The MRI modalities used for segmentation of brain tumors are:

- T1-Weighted, used for distinguishing healthy tissues and anatomical analysis.
- T1-Weighted-contrast-enhanced, in this modality the active part of the tumor can be distinguished.
- T2-Weighted, in this modality the edema subregion appears bright.
- T2-FLAIR, in this modality the edema and CSF can be distinguished.

The different subregions of the brain tumor can be classified as follows:

- Necrosis—These cancerous tissues can be classified as dead part of the tumor.
- Edema—The swelling region of the complete which surrounds the tumor core.
- Enhancing—These cancerous tissues contribute as actively enhancing part of the tumor.
- Non-enhancing—These cancerous tissues are regarded as non-active part of the tumor.

Thus, it can be concluded that due to the rigid shape and complex structure of brain tumors, the manual segmentation of these tumors is a tedious job [1]. Not only the complex and fuzzy structure of brain tumors but also there are other challenges in brain tumor segmentation, which is discussed as below.

The two key challenges in brain tumor segmentation are as follows:

- Variation in Intensity: This is introduced with spatial intensity variations along each dimension and also due to the intensity inhomogeneity of homogeneous tissues.
- Non-standardization of intensity: This is introduced by the hardware specifications of the MRI machine (some MRI machines come with a magnetic field strength of 1.5 T and 3 T) and also by the variation in acquisition protocols like gradient strength, b0 value.

Hence, to overcome these drawbacks appropriate preprocessing steps should be employed for efficient segmentation and quantitative image analysis of brain tumors [2–5]. In this work, we propose our segmentation model, for edema (swelling region) subregion segmentation in high-grade glioma (HGG) MR images.

This paper is arranged as follows, Sect. 2 focusses on the literature survey in the chosen field of research work. Section 3 emphasizes on the methodology used for segmentation of *edema*, Sect. 3.3 focusses on the results and discussion. Finally, Sect. 4 concludes with the future work.

2 Literature Survey

In this section, we put forth the active researchers working in the research area of brain tumor segmentation and analysis. The research, development, and scientific publications in the field of computerized brain tumor segmentation has grown rapidly in the last two decades, and this also hints that the brain tumor segmentation research is still an ongoing work in progress. Rexilius et al. proposed histogram matching technique for initializing the segmentation by a tumor probability map. Eric Malmi et al. proposed a cascaded brain tumor segmentation approach in which two levels of random forests were used [6]. Mikael Agn et al. proposed a generative model with a prior tumor shape for segmentation of brain tumors and where healthy tissues were modeled by Gaussian mixture models [7].

Joana Festa et al. proposed on random decision forests using multi-sequence MRIs for automatic segmentation of brain tumor [8]. Tustison proposed on constructing a large feature space using Gaussian mixture models for segmentation of brain tumor using random decision forests [9]. Clark et al. proposed knowledge-based technique by using the prior knowledge from the properties of tumors and MRI modalities [10]. Marcel Prastawa et al. proposed on brain tumor segmentation by subject-specific modification and variation of atlas priors [11]. N K Subbanna et al. proposed on Bayesian techniques, probabilistic Gabor, and Markov random field-based segmentation of brain tumor in MRI sequences [12].

Hoo-chang Shin proposed a nonlinear decision boundary was tried to segment *edema* and tumor core using hybrid clustering and logistic regression for brain tumor segmentation in multi-sequence MRIs [13]. Bi Song et al. proposed brain tumor segmentation and classification by taking advantage of the brain anatomy; further they employed dilation for initial segmentation of tumor as the region of interest (ROI) and further proposed random forest classification model on the voxels, which come under the ROI for the multi-class (necrosis, edema, enhancing, non-enhancing) tumor segmentation [14]. Zikic et al. proposed on using context-aware spatial features a based on decision forests [15]. Abdel Maksoud and Eman A et al. proposed on 3D brain tumor segmentation based on hybrid clustering techniques using multi-views of MRI, in which the particle swarm optimization and k-means techniques were fused to segment the 2D slices of 3D brain MRI image [16]. Hence, these are

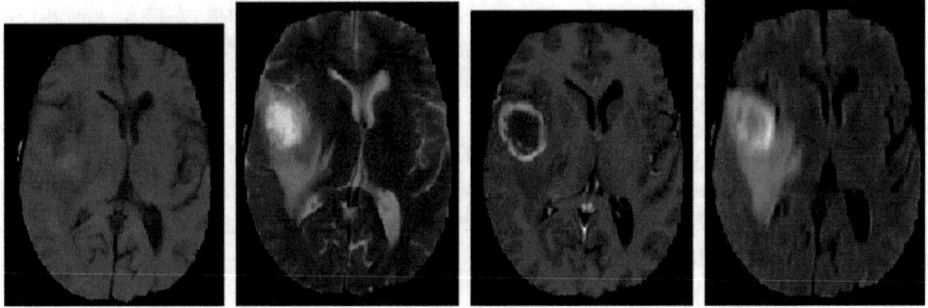

Fig. 1 Intensity inhomogenity corrected MR images

some of the active researchers who made essential contributions in the field of quantitative brain tumor image segmentation. This ongoing active research in the field of brain tumor segmentation motivated us to contribute our research in this field.

3 Methodology

In this section, we discuss the edema segmentation pipeline which consists of preprocessing step of MRI images and the segmentation process step of *edema*.

3.1 Preprocessing

In this subsection, we discuss the preprocessing steps applied to magnetic resonance images before quantitative image analysis is done. In MR images there are two artifacts, namely

- Inhomogeneity correction.
- Intensity standardization.

For inhomogeneity reduction (bias field correction) in MR images, we applied the nonparametric non-uniform normalization algorithm from the ITK package [17, 18]. Further, for the intensity standardization, we applied histogram matching technique followed by Z-score normalization.

Figure 1 shows the inhomogeneity corrected images using N4 algorithm, and Fig. 2 shows the intensity normalized MR images using histogram matching technique and Z-score normalization technique.

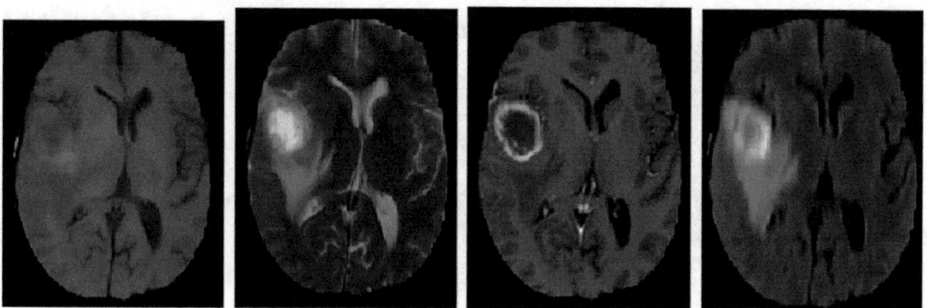

Fig. 2 Intensity normalized images using Z-score technique

3.2 Convolutional Neural Network

In this section, we explain the convolutional neural network (CNN) used in this paper. In the year 2012, breakthrough results were achieved in the field of computer vision for segmentation of natural scene images and this was achieved by Alex et al. by proposing a classification model "deep convolutional neural networks for ImageNet classification" in which convolutional neural networks (CNNs) were used to classify the 1.2 million high-resolution images with over 1000 categories. This state-of-the-art classification model marked a way for many of the computer vision applications in various fields like medical imaging, biometric applications, space imagery, self-driving cars [19]. Convolutional neural networks (CNNs) are the extension of traditional neural networks in which the main building block is the convolutional layer. The convolutional layers are stacked on top of each other for the automatic extraction of hierarchical features without the need of hand-crafted feature extraction process. Each layer extracts hierarchial features (low-level features like edges of an object, mid-level features like boundaries of an object, and high-level features like objects in the image) from its preceding layer of the hierarchy. A single convolutional layer takes input as a stack of input planes and yields output planes or feature maps. These feature maps are a map of responses of spatial nonlinear feature extractor, applied in a sliding window fashion. Each feature map is computed as follows:

$$O_i = b_i + \sum_k W_i k * X_k \tag{1}$$

where X_k is the kth input channel and $W_i k$ is the sub-kernel of the channel, b_i is the bias term and * is the convolution operation. The local deformations were overcome by introducing the max-pooling layer, which takes the max feature value for the sub-windows for each feature map. Each hidden unit in the fully connected layer is connected to all units in the preceding layer, and the weights in those connections are also specific to each neuron. Finally, the softmax layer is introduced which yields the probability distribution of classes.

$$softmax(x) = exp(x)/\sum exp(x) \tag{2}$$

The softmax layer is introduced at the last to get the class probability distribution for the segmented results.

3.3 Results and Discussion

In this section, we explain the experimental analysis and results obtained in our work. Firstly, we used the standard BRATS 2015 dataset for the experimentation purposes. This dataset consists of 220 multi-modal MRI scans (T1, T2, T1c, FLAIR) of patients with high-grade and 54 patients with low-grade gliomas. All the images are skull-stripped and were co-registered to T1c modality and were resampled to 1^3mm isotropic resolution with the 3D dimensions of each volume $240 \times 240 \times 155$. The ground truth segmentation includes four labels: (1) necrotic core, (2) edema, (3) non-enhancing tumor, (4) enhancing tumor core.

In this work, we segment the *edema (swelling region)* subregion of the brain tumor for HGG from the BRATS dataset. We implemented the CNN architecture for this purpose. We extracted 3D patches of size $32 \times 32 \times 32$ from all the four MRI modalities and to avoid class imbalance, 3D patches were extracted from the brain tumor-affected area only. The convolutional neural network architecture processed the input data, with four channels (T1, T1c, T2, FLAIR) in our dataset, hence, we inputted $4 \times 32 \times 32 \times 32$ patches into the convolutional neural network with convolutional layers, ReLu nonlinear activation functions, max-pooling layers, and finally, the softmax layer was applied to predict the class probability for the edema subregion and the background region. The classification model was implemented using the Python *Keras* deep learning library with the Theano [20] backend for symbolic tensor manipulation. And the model was trained on CUDA and cuDNN enabled NVIDIA Geforce GTX 970 GPU device. The performance evaluation of brain tumor segmentation is done using the following evaluation criteria [1].

- Dice score
- Positive predicted value (PPV)
- Sensitivity

The dice score is evaluated using the following expression

$$Dice = P_1 \bigcap T_1/(P_1 + T_1)/2 \tag{3}$$

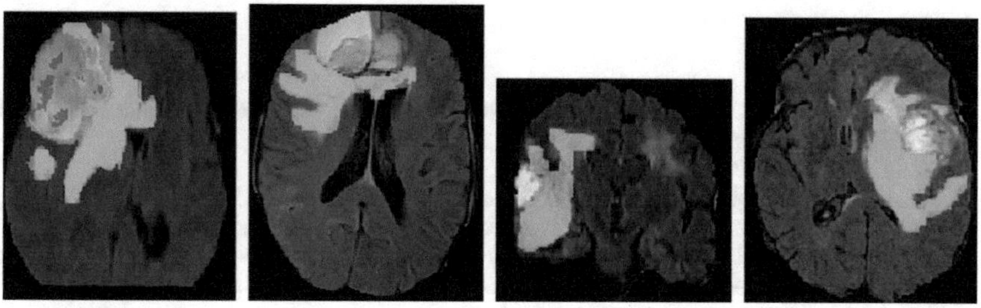

Fig. 3 *Edema* segmented MR images

The PPV is evaluate using the following expression

$$PPV = P_1 \bigcap T_1/P_1 \tag{4}$$

And, the sensitivity is evaluated using the following expression

$$Sensitivity = P_1 \bigcap T_1/(P_1 + T_1)/T1 \tag{5}$$

where T_1 is the true edema region and P_1 is an ensemble of voxels predicted as true positives for the *edema* subregion. Our model yielded a mean dice score of 0.68, PPV of 0.63 and sensitivity of 0.65 for the *edema* subregion of the BRATS test dataset. Our model was trained for HGG MR images, the segmentation model was not trained on LGG images, further the segmentation model yielded false positive for some of the test cases. Sergio Pierera et al. achieved dice score of 0.87, PPV of 0.89, and sensitivity of 0.86 for the complete tumor [21]. Konstantinos et al. attained dice score of 0.89, PPV of 0.89, and sensitivity of 0.90 for the whole tumor [22]. The network takes approximately 120 s per patient to train on an NVIDIA GeForce GTX 970 GPU. The *edema* segmentation results are shown in Fig. 3 for the BRATS test dataset where the green region indicates the *edema region* of the brain tumor.

4 Conclusion

Our work focusses on *edema* segmentation in the HGG MR images. Here, we consider binary classification in which we are making the other subregions of HGG other than *edema* as background itself. Comparatively, to other segmentation models, which consider the whole tumor region, our model has achieved nearly better results with dice score of 0.68, PPV of 0.63, and sensitivity of 0.65. In future, we extend our work for the segmentation of other tumor subregions *necrosis, enhancing, non-enhancing* using both the HGG and LGG MR images and evaluate the dice score, PPV, and sensitivity for the whole tumor and the tumor core [23].

References

1. Menze, B.H., et al.: The multimodal brain tumor image segmentation benchmark (BRATS). IEEE Trans. Med. Imaging **34**(10), 1993–2024 (2015)
2. Brain Tumors. N. Engl. J. Med. (2001)
3. De Angelis L.M.: Brain tumors, N. Engl. Med. **344**, 114–23 (2001)
4. Deimling, A.: Gliomas. Recent Results in Cancer Research. Springer, Berlin (2009)
5. Havaei, M., et al.: Deep learning trends for focal brain pathology segmentation in MRI. In: Machine Learning for Health Informatics: State-of-the-Art and Future Challenges, pp. 125–148. Springer International Publishing (2016)
6. Malmi, E., et al.: CaBS: a cascaded brain tumor segmentation approach. In: Proceedings MICCAI Brain, Tumor Segmentation(BRATS), pp. 42–47 (2015)
7. Mikael A. et al.: Brain tumor segmentation by a generative model with a prior on tumor shape. In: Proceedings MICCAI-BRATS, pp. 1–4 (2015)
8. Festa, J., et al.: Automatic brain tumor segmentation of multi-sequence MR images using random decision forests, pp. 23–26. In: Proceedings MICCAI Brain, Tumor Segmentation(BRATS) (2013)
9. Tustison, N., et al.: ANTs and Arboles. In: Proceedings MICCAI Brain Tumor Segmentation (BRATS) (2013)
10. Clark, M., et al.: Automatic tumor segmentation using knowledge-based techniques. IEEE Trans. Med. Imaging **17**, 187201 (1998)
11. Prastawa, M., et al.: Automatic brain tumor segmentation by subject specific modification of atlas priors. J. Acad. Radiol. (2003)
12. Subbanna, N., Arbel, T.: Probabilistic gabor and markov random fields segmentation of brain tumours in mri volumes. In: Proceedings MICCAI Brain Tumor Segmentation Challenge (BRATS), pp. 28–31 (2012)
13. Shin, H.-C.: Hybrid clustering and logistic regression for multi-modal brain tumor segmentation. In: Proceedings of Workshops and Challanges in Medical Image Computing and Computer-Assisted Intervention (MICCAI12) (2012)
14. Song, B., et al.: Anatomy-guided brain tumor segmentation and classification. In: International Workshop on Brainlesion: Glioma, Multiple Sclerosis, Stroke and Traumatic Brain Injuries. Springer, Cham (2016)
15. Zikic, D., et al.: Decision forests for tissue-specific segmentation of high-grade gliomas in multi-channel MR. In: International Conference on Medical Image Computing and Computer-Assisted Intervention. Springer, Berlin, Heidelberg (2012)
16. Abdel Maksoud, Eman A et al, 3D Brain tumor segmentation based on hybrid clustering techniques using multi-views of MRI. In: Medical Imaging in Clinical Applications: Algorithmic and Computer-Based Approaches. Springer Publications (2016)
17. Tustison, N.J., et al.: N4ITK: improved N3 bias correction. IEEE Trans. Med. Imaging **29**(6), 1310–1320 (2010)
18. Nyl, L.G., Udupa, J.K., Zhang, X.: New variants of a method of MRI scale standardization. IEEE Trans. Med. Imaging **19**(2) 143–150 (2000)
19. Krizhevsky, A., Sutskever, I., Hinton, G.E.: Imagenet classification with deep convolutional neural networks. In: Adv. Neural Inf. Process. Syst. (2012)
20. Bergstra, J., et al.: Theano: A CPU and GPU math compiler in Python. In: Proceedings 9th Python in Science Conference (2010)
21. Pereira, S., et al.: Deep convolutional neural networks for the segmentation of gliomas in multi-sequence MRI. In: International Workshop on Brainlesion: Glioma, Multiple Sclerosis, Stroke and Traumatic Brain Injuries. Springer, Cham (2015)
22. Kamnitsas, K., et al.: DeepMedic for brain tumor segmentation. In: International Workshop on Brainlesion: Glioma, Multiple Sclerosis, Stroke and Traumatic Brain Injuries. Springer, Cham (2016)
23. Rexilius, J., et al.: Multispectral brain tumor segmentation based on histogram model adaptation. In: Proceedings SPIE 6514, Medical Imaging (2007)

Real-Time Sign Language Gesture (Word) Recognition from Video Sequences Using CNN and RNN

Sarfaraz Masood, Adhyan Srivastava, Harish Chandra Thuwal and Musheer Ahmad

Abstract There is a need of a method or an application that can recognize sign language gestures so that the communication is possible even if someone does not understand sign language. With this work, we intend to take a basic step in bridging this communication gap using Sign Language Recognition. Video sequences contain both the temporal and the spatial features. To train the model on spatial features, we have used inception model which is a deep convolutional neural network (CNN) and we have used recurrent neural network (RNN) to train the model on temporal features. Our dataset consists of Argentinean Sign Language (LSA) gestures, belonging to 46 gesture categories. The proposed model was able to achieve a high accuracy of 95.2% over a large set of images.

1 Introduction

Sign language is a vision-based language which uses an amalgamation of variety of visuals like hand shapes and gestures, orientation, locality and movement of hand and body, lip movement and facial expressions. Like the spoken language, regional variants of sign language also exist, e.g., Indian Sign language (ISL), American Sign Language (ASL), and Portuguese Sign Language. There are three types of sign languages: spelling each alphabet using fingers, sign vocabulary for words, using hands and body movement, facial expressions, and lip movement. Sign language can also be isolated as well as continuous. In isolated sign language, people communicate using gestures of single word, while continuous sign language is a sequence of gestures that generate a meaningful sentence.

All the methods for recognizing hand gestures can be broadly classified as vision-based and based on measurements made by sensors in gloves. The vision-based method involves human and computer interaction for gesture recognition, while

S. Masood (✉) · A. Srivastava · H. C. Thuwal · M. Ahmad
Department of Computer Engineering, Jamia Millia Islamia, New Delhi 110025, India
e-mail: smasood@jmi.ac.in

V. Bhateja et al. (eds.), *Intelligent Engineering Informatics*, Advances in Intelligent Systems and Computing 695, https://doi.org/10.1007/978-981-10-7566-7_63

glove-based method depends on external hardware for gesture recognition. Significant works [1–6] in this field have been noted recently.

Ronchetti et al. [1] discussed an image processing based method for extraction of descriptor followed by a hand shape classification using ProbSom which is a supervised adaptation of self-organizing maps. Using this technique, they were able to achieve an accuracy of above 90% on Argentinean Sign Language.

Joyeeta and Karen [2] gave a method based on eigenvector. Skin filtering and histogram matching were performed in the preprocessing stage. The classification technique they used was based on eigenvalue-weighted Euclidean distance. They identified 24 different alphabets of Indian Sign Language with an accuracy of 96%.

Kumud and Neha [3] proposed a method for recognizing gestures from a video containing multiple gestures of Indian Sign Language. They extracted the key frame, based on gradient, to split the video to independent isolated gesture. The features were extracted from gestures by applying Orientation Histogram and Principal Component Analysis. Correlation, Manhattan, and Euclidean distance were used for classification and found that correlation and Euclidean distances gave better accuracies.

Anup et al. [4] demonstrated a statistical technique for recognizing the gestures of Indian Sign Language in real time. The authors created a video database for various signs. They used the direction histogram, which is invariant to illumination and orientation changes, as the feature for classification. Two approaches, Euclidean distance and K-nearest neighbor metrics, were used for gesture recognition.

Lionel et al. [5] proposed a system to recognize Italian sign language gestures. They used Microsoft Kinect and convolutional neural network (CNN) accelerated via graphic processing unit (GPU). They achieved a cross-validation accuracy of around 92% on a dataset consisting of 20 Italian gestures.

Rajat et al. [6] proposed a finely tuned portable device as a solution to alleviate this problem of minimizing the communication gap between normal and differently abled people. The architecture of the device, and its operations were discussed using three embedded algorithms which aimed for fast, easy, and efficient communication.

Another work by S. Masood et al. [7] have shown the use of Convolutional Neural networks for the purpose of character recognition for American Sign Language. In this work, the CNN based model was able achieve an overall accuracy of 96% on an image dataset of 2524 ASL gestures.

Also the creative works of the like of W. Vicars [8] have helped in general understanding in the field of American Sign Language recognition (Fig. 1).

In this work, we attempt to perform recognition on isolated sign language with the vision-based method. Unlike other works, we chose a dataset with larger gesture variants and significant video samples so that the resultant model having better generalization capabilities. In this work, we also attempt to explore the possibilities of exploiting the benefits of RNN in performing gesture recognition.

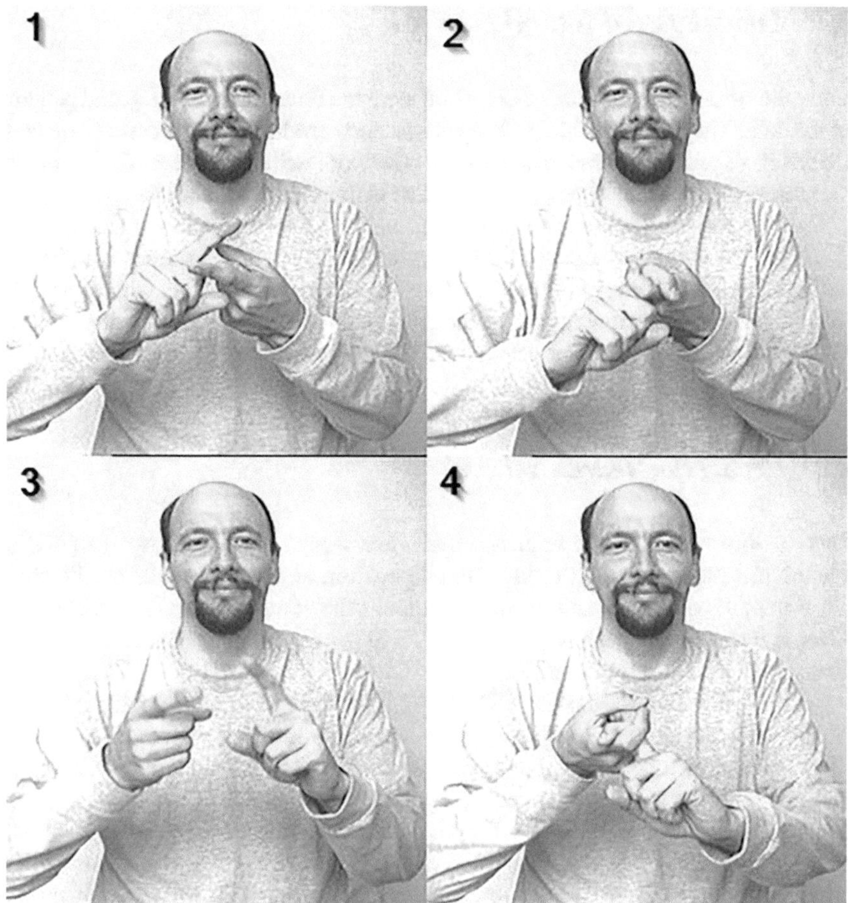

Fig. 1 ASL gesture for word "Friend" [8]

2 Algorithms Used

Video classification is a challenging problem as a video sequence contains both the temporal and the spatial features. Spatial features are extracted from the frames of the video, whereas the temporal features are extracted by relating the frames of video in a course of time. We have used two types of learning networks to train our model on each type of features. To train the model on spatial features, we have used CNN, and for the temporal features we have used recurrent neural network.

2.1 Convolutional Neural Network

Convolutional neural network or ConvNets are great at capturing local spatial patterns in the data. They are great at finding patterns and then use those to classify images. ConvNets explicitly assume that input to the network will be an image. CNNs, due to the presence of pooling layers, are insensitive to rotation or translation of two similar images; i.e., an image and its rotated image will be classified as the same image.

Due to the vast advantages of CNN in extracting the spatial features of an image, we have used Inception-v3 [9] model of the TensorFlow [10] library which is a deep ConvNet to extract spatial features from the frames of video sequences. Inception is a huge image classification model with millions of parameters for images to classify.

2.2 Recurrent Neural Network

There is information in the sequence itself, and recurrent neural networks (RNNs) use this for the recognition tasks. The output from an RNN depends on the combination of current input and previous output as they have loops. One drawback of RNN is that, in practice, RNNs are not able to learn long-term dependencies [11]. Hence, our model used Long Short-Term Memory (LSTM) [12], which is a variation of RNN with LSTM units. LSTMs can learn to bridge time intervals in excess of 1000 steps even in case of noisy, incompressible input sequences (Fig. 2).

The first layer is to feed input to the upcoming layers whose size is determined by the size of the input. Our model is a wide network consisting of single layer of 256 LSTM units. This layer is followed by a fully connected layer with softmax activation. Finally, a regression layer is applied to perform a regression to the provided input. We used Adaptive Moment Estimation (ADAM) [13] which is a

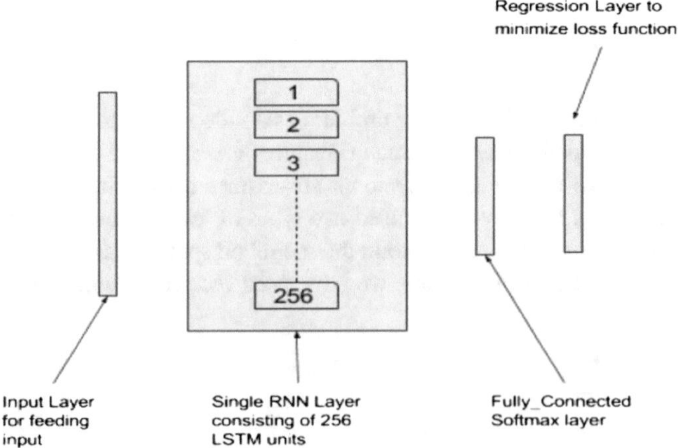

Fig. 2 Architecture of the proposed RNN model

stochastic optimizer, as a gradient descent optimizer to minimize the loss_function. Also, a wider RNN network was tried with 512 LSTM units and another deep RNN network with three layers of 64 LSTM units each. We tested these on a sample of the dataset and found that wide model with 256 LSTM units performed the best.

3 Methodology

Two approaches were used to train the model on the temporal and the spatial features, and both differ by the way inputs given to RNN to train it on the temporal features.

3.1 Prediction Approach

In this approach, spatial features for individual frames were extracted using inception model (CNN) and temporal features using RNN. Each video was then represented by a sequence of predictions made by CNN for each of their individual frames. This was given as input to the RNN. For every video corresponding to each gesture, frames were extracted and the background body parts other than hand were removed to get a grayscale image of hands which avoided color-specific learning of the model (Fig. 3).

Frames of the training set were given to the CNN model for training on the spatial features. The obtained model was then used to make and store predictions for the frames of the training and test data. The predictions corresponding to the frames of the training data were then given to the LSTM RNN model for training on the temporal features. Once the RNN model was trained, the predictions corresponding to the frames of the test data were fed to it for testing.

3.1.1 Train CNN (Spatial Features) and Prediction

Figure 4 depicts the role of CNN which is the inception model. From the training dataset, for each gesture “X”, all frames from each video corresponding to that gesture were labelled “X” and were given to the inception model for training.

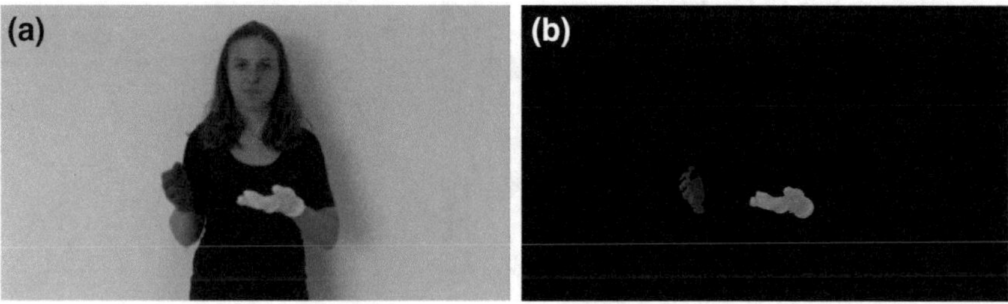

Fig. 3 **a** Sample frame from the dataset [14]. **b** Frame after background removal

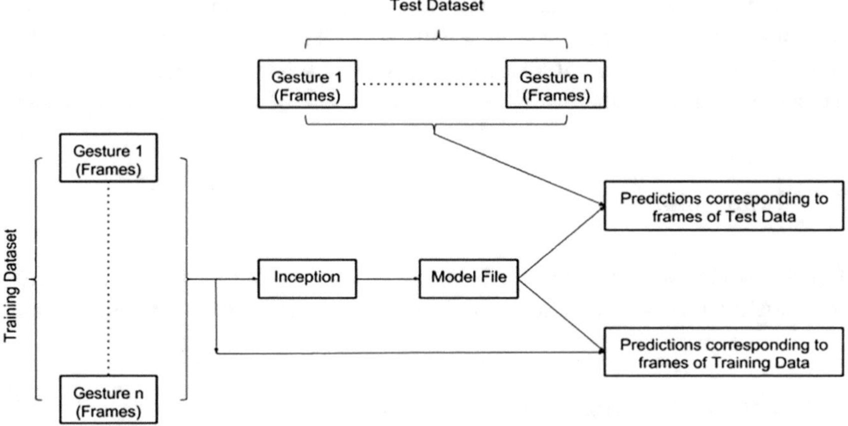

Fig. 4 Training CNN and saving prediction

The trained model was used to make predictions corresponding to frames of videos belonging to both the train and test set. This created two sets of predictions:

1. One corresponding to frames of training dataset—for training the RNN.
2. Another corresponding to frames of test dataset—for testing RNN.

Each gesture video was broken to a sequence of frames. Then after training CNN and making predictions, the video is represented as a sequence of predictions.

3.1.2 Training RNN (Temporal Features) and Testing

The videos for each gesture are fed to RNN as sequence of predictions of its constituent frames. The RNN learns to recognize each gesture as a sequence of predictions. After the Training of RNN completes a model file is created (Fig. 5).

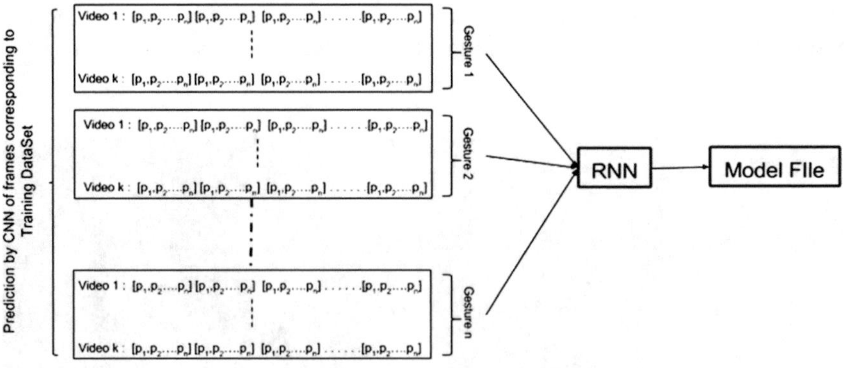

Fig. 5 Training RNN after getting the results from CNN

The predictions of CNN for frames of the test set were fed to the trained model for testing. The model was used to recognize the sequence of predictions in the test set.

3.2 Pool Layer Approach

In this approach, CNN was used to train the model on the spatial features and passed the pool layer output to the RNN before it is made into a prediction. The pool layer gives a 2048-dimensional vector that represents the convoluted features of the image, but not a class prediction. Rest of the steps of this approach are same as that of first approach. Both approaches only differ in terms of input given to the RNN.

The dataset used for both the approaches consists of Argentinean Sign Language (LSA) [14] gestures, with around 2300 videos for 46 gestures categories as shown in Table 1. Ten non-expert subjects executed the 5 repetitions of each gesture, i.e., 50 videos per gesture. Out of these 50 video samples, 40 videos per gesture were used for training, and 10 videos were used for testing. Thus, the training dataset had 1840 videos and the test dataset had 460 videos.

Table 1 Gestures used for training/testing

ID	Name	ID	Name	ID	Name
1	Son	17	Where	33	Barbeque
2	Food	18	Breakfast	34	Spaghetti
3	Trap	19	Catch	35	Patience
4	Accept	20	Name	36	Rice
5	Opaque	21	Yogurt	37	To-Land
6	Water	22	Man	38	Yellow
7	Colors	23	Drawer	39	Give
8	Perfume	24	Bathe	40	Away
9	Born	25	Country	41	Copy
10	Help	26	Red	42	Skimmer
11	None	27	Call	43	Sweet milk
12	Deaf	28	Run	44	Chewing gum
13	Enemy	29	Bitter	45	Photo
14	Dance	30	Map	46	Thanks
15	Green	31	Milk		
16	Coin	32	Uruguay		

4 Result

The prediction approach got an accuracy of 80.87% by recognizing 370 gestures correctly a test set of 460, while the pool layer approach scored 95.21% by recognizing 438 gestures.

It may be observed from Table 2 that all test videos for gesture "son" in prediction approach were incorrectly classified as they were misclassified as "colors".

It may be observed from Fig. 6a, b that the gestures, "Son" and "Colors", involve moving hands horizontally toward the right. The only difference is that in case of symbol "Son" it is done by holding two fingers up, while in case of "Colors" it is done with only one finger up. This extremely high level of similarity may be the reason behind misclassification of all "Son" gesture as "Color".

Table 2 Accuracy (in percent) per gesture using Approach 1 and Approach 2

ID	Gesture	App_1	App_2	ID	Gesture	App_1	App_2
1	Name	100	100	24	Spaghetti	70	100
2	Yogurt	100	100	25	Patience	70	100
3	Accept	30	90	26	Deaf	80	90
4	Man	70	100	27	Enemy	50	90
5	Drawer	100	100	28	Dance	100	90
6	Bathe	100	100	29	Rice	100	100
7	Opaque	70	90	30	To-Land	50	100
8	Country	100	100	31	Yellow	100	100
9	Water	60	90	32	Green	80	90
10	Red	100	100	33	Give	100	100
11	Call	90	100	34	Food	50	80
12	Colors	90	90	35	Away	80	100
13	Run	100	100	36	Copy	80	100
14	Bitter	100	100	37	Coin	100	90
15	Perfume	60	90	38	Where	100	90
16	Map	100	100	39	Skimmer	80	100
17	Born	80	90	40	Trap	100	80
18	Help	70	90	41	Sweet milk	100	100
19	Milk	100	100	42	Breakfast	100	90
20	None	80	90	43	Chewing gum	100	100
21	Uruguay	100	100	44	Photo	90	100
22	Son	0	80	45	Thanks	30	100
23	Barbeque	70	100	46	Catch	40	90

Fig. 6 **a** Set of few extracted frames of gesture "Son" [14]. **b** Set of few extracted frames of gesture "Color" [14]

4.1 Comparison Between the Two Approaches

The pool layer approach achieves a better accuracy than the prediction approach due to the increased size of the feature vector per frame being given to the RNN. In the prediction approach, for each frame the prediction of the CNN was given as input to the RNN. The prediction by the CNN was a list of values where the *i*th value denotes the probability of the frame belonging to the *i*th category. Hence, the size of the feature vector per frame was 46 in this work.

In the pool layer approach, for each frame the output of the pool layer, before it is made into a prediction, was given as input. The pool layer gives a 2048-dimensional vector that represents the convoluted features of the image. If each category is observed individually, it can be seen that for most of the categories the pool layer approach proves to be better. Though the high number of features have improved the correct classification rate for most of the gestures, but might have caused the RNN to discover random noise in the finite training set. Hence, it learnt some features that did not add any value to the judgement but may have led to overfitting.

5 Conclusion

In this work, we presented a vision-based system to interpret isolated hand gestures from the Argentinean Sign Language. This work used two different approaches to classify on the spatial and temporal features. CNN was used to classify on the spatial features, whereas RNN was used to classify on the temporal features. We obtained an accuracy of 95.217%. This shows that CNN along with RNN can be successfully used to learn spatial and temporal features and classify sign language gesture videos.

References

1. Ronchetti, F., Quiroga, F., Estrebou, C.A., Lanzarini, L.C.: Handshape recognition for Argentinian sign language using probsom. J. Comput. Sci. Technol. **16** (2016)
2. Singha, J., Das, K.: Automatic Indian Sign Language recognition for continuous video sequence. ADBU J. Eng. Technol. **2**(1) (2015)
3. Tripathi, K., Nandi, N.B.G.C.: Continuous Indian Sign Language gesture recognition and sentence formation. Procedia Comput. Sci. **54**, 523–531 (2015)
4. Nandy, A., Prasad, J.S., Mondal, S., Chakraborty, P., Nandi, G.C.: Recognition of isolated Indian Sign Language gesture in real time. Inf. Process. Manag., 102–107 (2010)
5. Pigou, L., Dieleman, S., Kindermans, P.-J., Schrauwen, B.: Sign language recognition using convolutional neural networks. In: Workshop at the European Conference on Computer Vision 2014, pp. 572–578. Springer International Publishing (2014)
6. Sharma, R., Bhateja, V., Satapathy, S.C., Gupta, S.: Communication device for differently abled people: a prototype model. In: Proceedings of the International Conference on Data Engineering and Communication Technology, pp. 565–575. Springer, Singapore (2017)
7. Masood, S., Thuwal, H.C., Srivastava, A.: American sign language character recognition using convolution neural network. In: Proceedings of Smart Computing and Informatics, pp. 403–412. Springer, Singapore (2018)
8. Vicars, W.: Sign language resources at LifePrint.com. http://www.lifeprint.com/asl101/pages-signs/f/friend.htm. Accessed 23 Sept 2017
9. Szegedy, C., Vanhoucke, V., Ioffe, S., Shlens, J., Wojna, Z.: Rethinking the inception architecture for computer vision. In: Proceedings of the IEEE Conference on Computer Vision and Pattern Recognition, pp. 2818–2826 (2016)
10. Abadi, M., Agarwal, A., Barham, P., Brevdo, E., Chen, Z., Citro, C., Corrado, G.S., et al.: Tensorflow: large-scale machine learning on heterogeneous distributed systems (2016). arXiv preprint arXiv:1603.04467
11. Bengio, Y., Simard, P., Frasconi, P.: Learning long-term dependencies with gradient descent is difficult. IEEE Trans. Neural Netw. **5**(2), 157–166 (1994)
12. Hochreiter, S., Schmidhuber, J.: Long short-term memory. Neural Comput. **9**(8), 1735–1780 (1997)
13. Kingma, D., Ba, J.: Adam: a method for stochastic optimization (2014). arXiv preprint arXiv:1412.6980
14. Ronchetti, F., Quiroga, F., Estrebou, C.A., Lanzarini, L.C., Rosete, A.: LSA64: an Argentinian sign language dataset. In: XXII Congreso Argentino de Ciencias de la Computación (CACIC 2016) (2016)

Application of Soft Computing in Crop Management

Prabira Kumar Sethy, Gyana Ranjan Panigrahi, Nalini Kanta Barpanda, Santi Kumari Behera and Amiya Kumar Rath

Abstract Indian agriculture is overwhelmed by numerous complications; some of them are usual, and some others are artificial like small and fragmented land-holdings, seeds, manures, crop selection, crop planning, fertilizers and biocides, irrigation, lack of mechanization, soil erosion, agricultural marketing, inadequate storage facilities, and so on. With the progression of different and specific outfits for the viability test of crop management are essential for providing reliable data observing to the performance of crop management. Valuable practical data can be collected by utilizing fuzzy logic-based scheme, in contrast with the intrinsic objectivity for collecting the data in gradual progression without any flaw. By dint of subject expertise and with the knowledge of scientific derivation, the approach should inspire to every corners of the country and management of cropping schemes. This paper analyzes the application of soft computing techniques in crop management in the field of farming and organic engineering is manifested. Upcoming progress and implementation using soft computing in the arena of farming and organic work to be think about.

Keywords Crop selection · Crop planning · Fuzzy Logic · Soft computing Crop management

P. K. Sethy (✉) · G. R. Panigrahi · N. K. Barpanda
Department of Electronics, Sambalpur University, Sambalpur, Odisha 768019, India
e-mail: prabirasethy@suniv.ac.in

G. R. Panigrahi
e-mail: gyan7420@gmail.com

N. K. Barpanda
e-mail: nkbarpanda@suniv.ac.in

S. K. Behera · A. K. Rath
Department of Computer Science and Engineering, VSSUT, Burla, Odisha 768018, India
e-mail: b.santibehera@gmail.com

A. K. Rath
e-mail: amiyaamiya@rediffmail.com

V. Bhateja et al. (eds.), *Intelligent Engineering Informatics*, Advances in Intelligent Systems and Computing 695, https://doi.org/10.1007/978-981-10-7566-7_64

1 Introduction

The crop management is foremost imperative research topic in farming, crop production, and vegetable, soil, and pest management. Crop scientists (or agronomists) are involved in improving food, feed, and fiber production. In agricultural systems, one of major challenges is crop selection and planning [1]. Analyzing and utilizing from the list of daily agricultural needful resources, the focus is to define the optimum cropping shapes. Optimal selection cum planning was projected to produce the maximum economic benefit [2]. This paper will analyze the development of advanced soft computing techniques, in view of crop management. With these concepts and methods, applications of soft computing techniques in crop management in the field of farming and organic engineering will be presented. It is a collection of "imprecise" calculating practices, which can have the exemplary degree to examine very multifarious problems. Hence for multifarious technical hitches, more predictable methods do not have the degree to yield worthwhile, methodical, and comprehensive kind of desired interpretations. But the soft computing has been considered and implemented deliberately which has proven itself and ran for last four decades successfully for technical research and engineering computing [3]. The primary aspects of soft computing techniques are fuzzy systems/logic, neural networks/computing, evolutionary algorithms/computation, genetic algorithms, pattern recognition and machine learning [4] and particle swarm optimization/intelligence combining techniques from these fields, through the final which are including acceptance systems, disorder theory and portions from book learning model results effective productivity in many areas like engineering applications, commercial, organization, finances, farming and organic engineering. Also great ability in solving compound problems in some inevitable field such as crop selection, crop planning, irrigation planning, water resources management, crop management, etc. It is having its own degree of improved assessment verification tool for an effective resolution for practical multifaceted amended complications. In farming and organic engineering, researchers and engineers have developed many methods and systems to study soil and water managements associated with crop growth and support management in precision farming [3]. Fresh approaches such as chaos computing and immune networks are also freshly considered to a part of soft computing. Undoubtedly, one can notice that quick derivation, precise in form, job-sharing, or fully automated decision-making systems of soft computing techniques can be employed in order to decrease costs, reduce wrong decisions, avoid human failures and save time, which can be more profitable, or cost cutting in trade, and even in many different extents to complete the intended assignment efficiently in a well-defined manner.

Crop management problem: Apropos selection cum planning together depends on land types, reaping rate, affairs of meteorological, farming inputs, crop demand, capital obtainability, and the cost of production [5]. Few are quantifiable and can be enumerated. Perhaps, it is very difficult to forecast some of the erratic factors like precipitation, meteorological conditions, flood, hurricane, and other natural catastrophes [3]. Resources scarcity makes the problem generally as a constrained optimization problem. One of the vital issues in those difficulties is the optimization objective(s). Based on one or multi-objective optimization, one or more goal is to be considered. Maximization of net return, maximization of gross margin, minimization of water use and soil conservation are some of goals that are desired.

Optimization Methods: By Soft computing optimization techniques which have been outstretched to solve problems in the field of farming schemes such as crop selection, crop planning, crop management, irrigation planning, water resources management, vegetable production, beef production, wildlife and livestock management, sugarcane transportation and many more.

2 Techniques of Soft Computing

2.1 *Soft Computing*

Soft computing which creates taxonomies to run each tiny element is a degree of the unit of likeness for every class. It offers more edifying figures and hypothetically precise result, especially for uneven three-dimensional resolution data classifications. Artificial neural networks, fuzzy logic, decision tree, genetic algorithms, pattern recognition, and so on come under soft classification methods.

2.2 *Fuzzy Logic*

In 1965, in exigency with scientific model by the dint of proposed uncertain logics, fuzzy logic was being presented using Boolean logics by an acumen people Lotfi Zadeh. According to him, this is a direct interpretation which has imprinted from conventional set theory. Looking to the different degree of indeterminate situations, this can be arranged for different state of valued cognitive kind of easy and well going solutions specifically meant for the imprecisions and doubts. The preeminent part of this logic is that it uses very simplest and regular kind of protocols and languages which can apply immediately for its cognitive sort of human endorsed solutions in addition to prevent ambiguity. In accordance to create an allegory, the conventional system of set logic is a part and parcel of fuzzy sets, as shown in Fig. 1.

More than two valued fuzzy logic (FL) which permits transitional values to demarcate among predictable estimations from expected values. Ideas like somewhat high otherwise very quick can be expressed arithmetically and performed as to implement the real going thought process which is very much significant as like human by the help of software design. This can be substituted to usual concepts of set of reason and relationship which is its root as per bygone Greek perception. FL

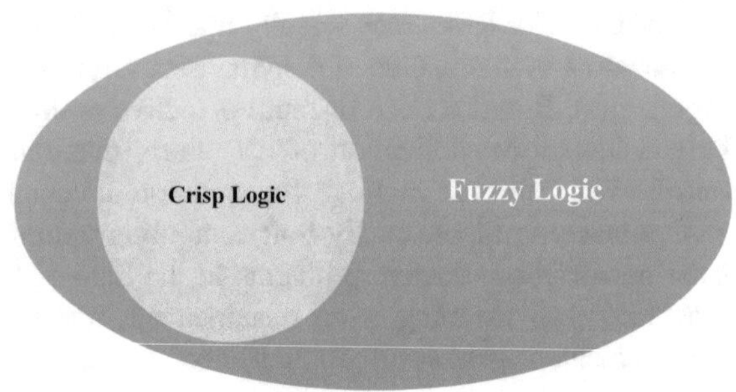

Fig. 1 Set theory of the subordinate of fuzzy logic

arose as a commercial device for monitoring cum navigation for the modern day intricate professional engineering systems like in the field of farming and organic engineering.

2.3 *Artificial Neural Networks*

A method of computing system has stimulated by the construction, dispensation technique, and learning aptitude of an organic brain. Faces of artificial neural networks (ANN)—a huge amount of very unassuming processing neuron-like dispensation elements, a different ratio of networks through the components, dispersed depiction data over the networks, and records are acquired through a learning process as Fig. 2 illustrates.

It is inevitable because of its immense throughput rate, disseminated illustration, learning capability, simplification ability, error tolerance, etc. Components of ANN are:

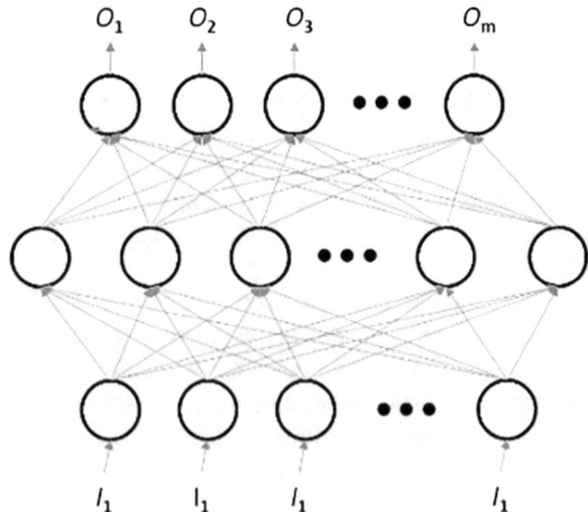

Fig. 2 Artificial neural network

- Execution parts
- Geometrical representation
- Learning process

Applications are pattern arrangement, bunching/classification, utility guesstimate, forecast/predicting, result oriented, addressable index memory management.

2.4 *Genetic Algorithms and Evolutionary Algorithms*

The field of evolutionary computation is relatively new (1970s), and its nomenclature is still a little bit fuzzy: as far as concern, the term "evolutionary algorithms" encompasses all stochastic algorithms made by "imitation" + "assortment" with concrete solutions, slackly imitating upon the evolvement of instinctive populaces in natural surroundings. Inside this big group, there are many notable subsets, such as: genetic algorithms (GA), characterized by a binary string representation of the candidate solutions.

Evolution strategies (ES) use vectors of real-valued numbers as representation. One of the most famous ES is the covariance matrix adaptation ES (CMA-ES).

Genetic programming (GP), one of the most recent techniques (1990s), uses binary trees (or other complex structures, such as directed graphs) as internal representation of candidate solutions. This type of representation is incredibly powerful, as it allows for evolving almost everything, from Bayesian networks to equations, from assembly programs to gene regulatory networks.

One of the most famous success stories for GP is Eureqa, by Nutonian, Inc. And there are many other interesting subsets, such as evolutionary programming, memetic set of rules, linear inherited user interface design, and so on. A genetic or evolutionary algorithm applies the principles of evolution found in nature to the problem of finding an optimal solution to a solver problem. To solve the bit string encodation, an "evolutionary algorithm" can be employed using the choice variables and problematic functions. Most of saleable yields are extensively relied on evolutionary algorithm.

An optimize differentiation between an evolutionary algorithm versus "classical" optimization methods is:

- Arbitrary versus resolute action
- Inhabitants versus uniform resolution
- Making innovative way out by alteration
- Linking results by inserting delimiter.
- Opting results by means of "reproductive success"
- Downsides of evolutionary algorithms

Various applications are groundwater monitoring networks, image processing-dense pixel matching, learning ambiguous regulation using hereditary processes, plant floor layout, learning ambiguous regulation using genetic algorithms, and so on.

2.5 Pattern Recognition and Machine Learning

The difference is the features that are take out. Simply put, In Pattern recognition, Machine Learning involves extracting useful features from data and building a statistical model. Those features usually are extracted using advanced filtering, signal processing, time–frequency analysis, etc., whereas other applications of machine learning such as natural language processing might use word counts, bag-of-words, etc. In the end, you are using features to create a model.

Data acquisition, usually not related to machine learning, is ultra-important for the success of a pattern recognition system. Traditional machine learning often focuses on the abstract model learning part, which usually uses pre-extracted feature vectors as its input, and seldom pays care to data acquisition. Instead, machine learning algorithms assume the feature vectors that satisfy some mathematical or statistical properties and constraints and learn machine learning models based on the feature vectors and their assumptions.

Pattern recognition and machine learning are two closely related subjects, and the differences may gradually disappear. For example, the recent deep learning trend in machine learning emphasizes end-to-end learning: The input of a deep learning method is the raw input data (rather than feature vectors), and its output is the desired prediction.

Various applications for the proof of identity cum validation: e.g., certificate plate recognition, biometric impressions, and medical diagnosis: e.g., different cancer screening and even for farming and organic engineering like crop selection, crop planning, crop disease identification, and so on.

2.6 Particle Swarm Optimization

It is a populace kind of speculative improving technique given by Dr. Eberhart and Dr. Kennedy in 1995. From the world of computer, in view to a given degree of class, PSO is a method of reckoning which solves a problem constantly by attempting to develop a clear solution. It is having many similarities with metamorphic estimation techniques such as genetic algorithms (GA). It has originated with a populace of arbitrary results and detects for best target by modernizing peers. It does not have any evolution operators such as crossover and mutation. The budding solutions, which hereupon stated as elements, sail into the problematic space using the existing finest elements. From many years, PSO has ran and implemented satisfactorily in many research areas. In comparison to other approaches, it itself has proven satisfactorily for its better accurate quicker results and which is inexpensive as well. With the least parameters, adjustment can also achieve better to improved result. It can be used in broad variety of applications.

There are many areas that it covers like machine learning, operations research, optimization algorithms, detection of periodic orbits, neural networks training for bioinformatics, fuzzy cognitive maps learning, crop pattern detection, and so on.

2.7 Crop Management

For crop pest management, by Pydipati et al. (2005), color co-occurrence method has been used for the textural examination to govern whether classification algorithms would be used to recognize diseased and usual citrus leaves. ANN classifiers based on the radial basis function (RBF) networks are one of the classifying approaches investigated with fuzzy outputs that designate a degree of strength. Rani, A.S. (2017) [1] reported by, henceforth, expectation of horticultural generation on a yearly premise is intense employment, however deserving of the exertion since it will give the agriculturists a premonition of what they could expect in the following year. Late advances are presently a day's ready to give a considerable measure of data on horticultural-related exercises, which can then be broke down with a specific end goal to discover imperative data and to gather pertinent data. The monetary commitment of horticulture to India's GDP is relentlessly declining with the nation's expansive-based financial development. Arrangements followed in the nation and nature of innovation that got to be distinctly accessible after some time have strengthened a portion of the varieties coming about because of normal variables. The agricultural yield principally relies on upon climate conditions, maladies and bugs, arranging of reap operation. Successful administration of these components is important to assess the likelihood of such negative circumstance and to minimize the outcomes. Precise and dependable data about chronicled trim yield are along these lines essential for choices identifying with rural product administration. The findings and temporal data of information mining procedures are very much significant and interesting to the farmer by the time of execution. In result, prediction has really influenced on agriculturist's long-term happenstance for specific yield, crops, and so on. The rural harvest totally depends upon climate

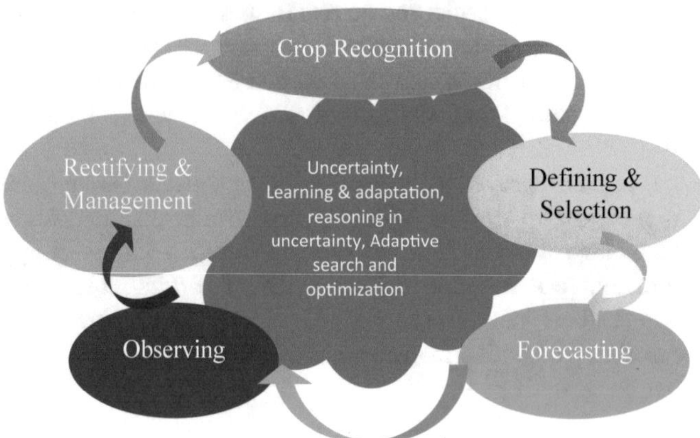

Fig. 3 Flow diagram of fuzzy logic in crop management

conditions. Understanding the relative importance of these climate elements to product yield variety could give significant data about harvest planting and administration under environmental change condition for ranchers. Development in information estimate requires computerized strategy to concentrate vital information. By applying information mining procedure, it is conceivable to remove helpful learning and patterns. Learning picked up in this way might be connected to build work proficiency and enhance basic leadership quality. Data innovation has turned into a necessary piece of our everyday life. Procedures for overseeing information have turned out to be essential and regular in industry and administrations. This is particularly valid for agribusiness, keeping in mind the end goal to modernize and better apply GPS innovation. Investigation is required for finding the future climatic states of specific locale where the information mining assumes an imperative part to analyze chronicled climate information and locate the required arrangement. The information can be broke down in a social database, an information stockroom, a Web server log, or a straightforward content record. The aim of this paper is to give insights into various information mining strategies of farming area so analysts can get insights into proper information mining systems in setting of harvest administration in light of climate conditions.

Time to time, many researchers have researched and contributed many proposed results for farmer of India by doing experiment using different kind of prediction logics and shown the future in this. Step-by-step flow diagram using the application of soft computing in the field of crop management is stated in Fig. 3.

3 Analysis of Various Soft Computing Schemes

See Table 1.

Table 1 Summarization for crop selection, planning and management-related problems, and solving techniques

Sl.	Author name and year	Title	Technique	Area
1	Pravin Kumar et al. (2017) [6]	Efficiency measurement of fertilizer manufacturing organizations using fuzzy data envelopment analysis	Fuzzy data envelopment analysis	Rank and efficiency of fertilizer
2	N. Sundaravall, Dr. A. Geetha (2016) [7]	A study and survey on rainfall prediction and production of crops using data mining techniques	Fuzzy logic k-mean Neuro-fuzzy with genetic algorithm	Prediction of rainfall and crop production
3	Asghar Mahmoudi et al. (2016) [3]	Simulation of control system in environment of mushroom-growing rooms using fuzzy logic control	Fuzzy logic Simulink	Temperature and humidity in mushroom production
4	Fahim Jawad et al. (2016) [8]	Examination of best harvest farming using fuzzy system	Fuzzy logic	Optimum crop cultivation
5	Sneha Murmu Sujata Biswas (2015) [9]	Application of fuzzy logic and neural network in crop classification	Fuzzy logic Neural network	Crop mapping Estimating crop water requirement
6	Miss. Snehal S. Dahikar et al. (2014) [10]	Agricultural crop yield prediction using artificial neural network approach	Artificial neural network	Crop yield prediction
7	Siti Khairunniza-Bejo et al. (2014) [12]	Application of artificial neural network in predicting crop yield	Artificial neural network	Prediction of crop yield
8	P. Lavanya Kumari et al. (2014) [2]	Optimum allocation of agricultural land to the vegetable crops under n certain profits using fuzzy multi-objective linear programming	FMOLP	Best cropping pattern

(continued)

Table 1 (continued)

Sl.	Author name and year	Title	Technique	Area
9	Harsimranjit Singh Narinder Sharma (2014) [11]	A review of fuzzy-based expert system in agriculture	Fuzzy logic	Soil preparation, seed selection, pesticide management, water scheduling
10	Ingole Kartik et al. (2013) [3]	Estimation and recognition of crop by fuzzy logic in MATLAB.	Fuzzy logic MATLAB	Crop detection
11	Mohammad Mansourifaretal (2013) [12]	Optimization crops pattern in variable field ownership	Genetic algorithm	Crop planning Crop pattern
12	Moussa waongo et al. (2013) [13]	A crop model and fuzzy rule-based approach for optimizing maize planting dates in Burkina Faso, West Africa	Fuzzy logic	Optimize crop planting date
13	Ehsan Houshyar et al. (2012) [14]	Sustainable and efficient energy consumption of corn production in southwest Iran: combination of multi-fuzzy and DEA modeling	Fuzzy logic Data envelopment analysis (DEA)	Efficiency of corn production
14	Leila Naderloo et al. (2012) [15]	Application of ANFIS to predict crop yield based on different energy inputs	Adaptive neuro-fuzzy inference system (ANFIS)	Grain yield of wheat
15	Dattatray angaram Regulwar et al. (2010) [5]	Fuzzy approach-based management model for irrigation planning	Multi-objective fuzzy linear programming) (MOFLP)	Crop planning Optimal cropping pattern
16	Yanbo Huang et al. (2010) [4]	Development of soft computing and applications in agricultural and biological engineering	Soft computing techniques	Crop management Precision agriculture

4 Conclusions of Various Crop Management Schemes

See Table 2.

Table 2 Analysis report on conclusions from different schemes

Sl.	Author name and year	Schemes	Conclusions
1	Pravin Kumar et al. (2017)	Fuzzy data envelopment analysis	The uniqueness of this approach is that it incorporates the minimum, average, and maximum values of feeding and taking the data in the form of a triangular fuzzy number (TFN) in the analysis
2	N. Sundaravall, Dr. A. Geetha (2016)	Fuzzy logic k-mean Neuro-fuzzy with genetic algorithm	On this approach, the device used here can be used to predict coproduction of any place definitely uploading the facts in accordance to users
3	Asghar Mahmoudi et al. (2016)	Fuzzy logic Simulink	Inputs and outputs data were collected from target HVAC and growing hall and controlling systems were developed based on fuzzy control system
4	Fahim Jawad et al. (2016)	Fuzzy logic	Future scheme is very helpful to take correct choice of harvesting the right crop at right time
5	Sneha Murmu Sujata Biswas (2015)	Fuzzy logic Neural network	Neural networks or neuro-fuzzy models have also been applied in few studies where the performance was better than the other classification studies
6	Miss. Snehal S. Dahikar et al. (2014)	Artificial neural network	ANN is beneficial tool for crop prediction. Analyze by using feed-forward back-propagation ANN (MATLAB)
7	Siti Khairunniza-Bejo et al. (2014)	Artificial neural network	The results of this study indicate that the conventional color digital camera could be employed for fast and accurate, non-destructive measurement, and determination of the status of N and chlorophyll content in rice plant
8	P. Lavanya Kumari et al. (2014)	FMOLP	The application of fuzzy concept in cropping pattern for high economic expectations has successfully tackled the uncertainty and imprecision in profit
9	Harsimranjit Singh Narinder Sharma (2014)	Fuzzy logic	Including fuzzy logic in expert system to handle imprecise information agriculture field is been useful and has shown good results
10	Ingole Kartik et al. (2013)	Fuzzy logic MATLAB	Resulted corrosion model by means of spatial statistics using soils, usage land and crop shield, climate and geometrical arrangement features was technologically advanced using GIS methods

(continued)

Table 2 (continued)

Sl.	Author name and year	Schemes	Conclusions
11	Mohammad Mansourifaretal (2013)	Genetic algorithm	Genetic algorithm gave acceptable results to optimize crops planning. The study demonstrated that crops production in study area could be increased by using optimization planning model
12	Moussa waongo et al. (2013)	Fuzzy logic	The use of fuzzy logic to estimate planting rules instead of binary logic gives further flexibility in estimating reliable planting dates where strict thresholds may fail
13	Ehsan Houshyar et al. (2012)	Fuzzy logic Data envelopment analysis (DEA)	By using a combination of fuzzy modeling and DEA technique, agricultural sustainability would better be estimated and could improve up to level of sustainability which is being done by the efficient farmers
14	Leila Naderloo et al. (2012)	ANFIS	ANFIS was employed in order to predict the irrigated wheat grain yield based on the input energy
15	Dattatray angaram Regulwar et al. (2010)	MOFLP	The objective of the study is to develop an optimal cropping pattern that maximizes net benefits, crop production, employment generation/labor requirement, and manure utilization simultaneously
16	Yanbo Huang et al. (2010)	Soft computing techniques	The powerful adaptive learning is an instrumental tool for the implementation and venturing of possible collaboration among different approaches

5 Advantages of Employing Soft Computing in Crop Management

Soft computing applications have two verified key advantages: (1) By solving nonlinear problems, it can be prepared easily, where possibly mathematical representations are not existing and (2) it has presented the human acquaintance such as perception, credit, empathetic, culture, and others in the field of computing.

It yields the opportunity of establishing smart and brainy schemes like independent at hand adjust systems and computerized designed systems.

In brief, the advantages of using soft computing are its ability to comprehend fuzziness, ambiguity, and fractional truth to attain manipulability and sturdiness on pretending human decision-making conduct with minimal cost. In other words, soft computing delivers an occasion to represent vagueness in human thinking with the doubt in real life.

6 Conclusions and Future Scope

From the current review of crop management using different techniques, the following can be emphasized. In the present situation of farming and organic management system, decision making toward certain areas is to consider as compromise solution, not necessarily always a leading solution. The advanced model has high precision and is able to control parameters precisely and close to set point values. Fuzzy logic takes advantage from created protocols and image sorting which results in very less or equal time intense than the other predictable approaches. The application of fuzzy concept in cropping management for high commercial expectations has successfully tackled the uncertainty and imprecision in profits. Crop management is yet the one and only incredible agronomic doings across India. In association, least input to the agriculture can yield extraordinary results for the nation. Thus, the advanced fuzzy set-based measurable policy is capable in integrating the ambiguity for planning model. As a whole, soft computing is efficient, intelligent, and user-friendly and with the quality study, the soft computing improves its performance in various fields.

References

1. Rani, A.S.: The Impact of Data Analytics in Crop Management based on Weather Conditions (2017)
2. Kumari, P.L., Reddy, G.K., Krisna, T.G.: Optimum allocation of agriculture land to the vegetable crops under uncertain profits using fuzzy multiobjective linear programming. IOSR J. Agric. Vet. Sci. **7**(12), 19–28
3. Ingole, K., et al.: Crop prediction and detection using fuzzy logic in Matlab. Int. J. Adv. Eng. Technol. **6**(5), p. 2006 (2013)
4. Huang, Y., et al.: Development of soft computing and applications in agricultural and biological engineering. Comput. Electron. Agric. **71**(2), 107–127 (2010)
5. Regulwar, D.G., Gurav, J.B.: Fuzzy approach based management model for irrigation planning. J. Water Resour. Prot. **2**(06), p. 545 (2010)
6. Kumar, P., Singh, R.K., Shankar, R.: Efficiency measurement of fertilizer-manufacturing organizations using Fuzzy data envelopment analysis. J. Manag. Anal. (2017)
7. Sundaravalli, N., Geetha, A.: A Study & Survey on Rainfall Prediction and Production of Crops Using Data Mining Techniques (2016)
8. Jawad, F., et al.: Analysis of Optimum Crop Cultivation using Fuzzy System. In: 2016 IEEE/ACIS 15th International Conference on Computer and Information Science (ICIS). IEEE (2016)
9. Murmu, S., Biswas, S.: Application of fuzzy logic and neural network in crop classification: A review. Aquati. Procedia **4**, 1203–1210 (2015)
10. Dahikar, S.S., Rode, S.V.: Agricultural crop yield prediction using artificial neural network approach. Int. J. Innov. Res. Electr. Electron. Instrum. Control Eng. **2**(1), 683–686 (2014)
11. Singh, H., Sharma, N.: A Review of Fuzzy Based Expert System in Agriculture. Int. J. Eng. Sci. Res. Technol.
12. Mansourifar, M., et al.: Optimization crops pattern in variable field ownership. World Appl. Sci. J. **21**(4), 492–497 (2013)

13. Waongo, M., et al.: A crop model and fuzzy rule based approach for optimizing maize planting dates in Burkina Faso, West Africa. J.Appl. Meteorol. Climatol. **53**(3), 598–613 (2014)
14. Houshyar, E., et al.: Sustainable and efficient energy consumption of corn production in Southwest Iran: combination of multi-fuzzy and DEA modeling. Energy **44**(1), 672–681 (2012)
15. Naderloo, L., et al.: Application of ANFIS to predict crop yield based on different energy inputs. Measurement **45**(6), 1406–1413 (2012)

A Hybrid Shadow Removal Algorithm for Vehicle Classification in Traffic Surveillance System

Long Hoang Pham, Hung Ngoc Phan, Duong Hai Le and Synh Viet-Uyen Ha

Abstract Shadow is one of the common parts in the natural scenes and has become an important topic in the field of computer vision. In many vision-based traffic surveillance systems, shadows interfere with fundamental tasks such as vehicle detection, classification, and tracking. Thus, it is necessary to suppress the effect of shadows. A difficult part of the shadow removal problem is how to accurately detect and remove shadow regions and recover the boundaries of the vehicles, while still achieving real-time processing performance. Many powerful methods have been proposed to solve this dilemma; however, instabilities at the boundaries of moving vehicles are still challenges. In this paper, an improved algorithm to remove shadow regions, and quickly recovering the boundaries of moving vehicles is presented in a detailed manner. The proposed method applies edge information with background subtraction to handle daytime traffic scenes. Our approach has demonstrated more accurate results than previous approaches regardless of lighting luminance levels or shadow orientations.

Keywords Traffic surveillance system · Shadow removal · Edge detection Vehicle recovery · Daytime detection · Vietnam

1 Introduction

The past decade has seen increasingly rapid advances in the field of computer vision which in turn has led to a renewed interest in traffic surveillance systems. Vision-based traffic surveillance systems have the capability to provide fast and reliable information that is necessary for traffic management and congestion mitigation.

L. H. Pham (✉) · H. N. Phan · D. H. Le · S. Viet-Uyen Ha (✉)
School of Computer Science and Engineering, International University, Vietnam National University HCMC, Block 6, Linh Trung, Thu Duc District Ho Chi Minh City, Vietnam
e-mail: phlong@hcmiu.edu.vn

S. Viet-Uyen Ha
e-mail: hvusynh@hcmiu.edu.vn

V. Bhateja et al. (eds.), *Intelligent Engineering Informatics*, Advances in Intelligent Systems and Computing 695, https://doi.org/10.1007/978-981-10-7566-7_65

The main objective is to detect interesting objects (moving vehicles, people, and so on). Other targets include classifying objects based on their features and appearance (shape, color, texture, and area), counting and tracking vehicles (trajectory, motion), assessing the traffic situation (congestion, accident). While later processes are dependent on specific application requirements, the initial step of object detection must be robust and application-independent.

However, a major problem with this kind of application is the appearance of shadows in daytime scenes. In many traffic surveillance systems, shadows interfere with fundamental tasks such as moving vehicle detection, classification, and tracking. Firstly, cast shadows that appear next to the conveyances distort vehicles' shapes and confound vehicle classification process. Secondly, many vehicles are connected by shadows and thus are detected as one big vehicle, which affects the procedure of counting and tracking.

Many powerful methods have been proposed to solve the problem of shadow removal; however, instabilities at the boundaries of moving vehicles are still challenges. In this paper, we present a simple, effective, and robust algorithm which can remove shadow regions and recover vehicles' boundaries in real-time. Our approach includes three steps: (1) detecting moving vehicles; (2) subtracting shadow regions; (3) recovering vehicles' boundaries. The main contribution of this paper is that we successfully recover the moving vehicle boundaries by combining edge information from the input frame and the lightness component with better performance in the experimental results than related methods.

The rest of the paper is organized as follows. In Sect. 2, our method and relevant theories will be discussed in detail. Section 3 describes the experiments and results obtained. The conclusion is presented in Sect. 4.

2 The Proposed Method

2.1 Outline of the Algorithm

The outline of the proposed method is described as follows. Given a new frame captured by the camera (*I*), a pair of background (*BG*) and foreground (*FG*) images are obtained. Then, *I* and *FG* are used in the moving object extraction process. The output, *MO*, which is an RGB image containing moving objects over a black background, is then converted to grayscale (MO^{Gr}) and HSV color model from which the lightness component (MO^{V}) is extracted and threshold. The Canny edge maps of *BG*, *FG*, MO^{Gr}, MO^{V}, denoted by E_{BG}, E_{FG}, E_{Gr}, E_{V}, respectively, are generated. With E_{BG}, E_{FG}, and E_{Gr}, the edge pixels of shadows can be removed (E_S). Next, E_V is combined with E_S to refine the vehicle edge maps, RE_S. Finally, the post-processing process is performed to denoise and construct the final binary mask, *MV*.

2.2 Moving Object Detection

It is a common task in traffic surveillance systems to utilize background modeling to construct the background, *BG*, and to detect moving objects in traffic scenes. The effectiveness of the background model is evaluated through a binary foreground mask, *FG*, in which moving objects and their cast shadows are marked as white blobs. However, in outdoor scenes, a real background is not always available and can be affected by extrinsic factors including slow-moving or stationary objects, and camera vibration (e.g., strong wind, heavy vehicles). To account for these problems, we adopt the background subtraction algorithm proposed by Nguyen et al. [1]. The outcomes of this model acquire good precision and real-time performance, which are critical factors in real-world applications. Figure 1 shows the examples of *BG* and *FG*.

The next step is to refine *FG* to remove unwanted objects and noises. Regarding this issue, we adopt the observation zone technique presented in both [2, 3]. In this approach, the observation zones which are a region where vehicles traveling through have steady changing rates in their appearances are automatically defined on the camera angle. Particularly, this procedure is crucial to enhance the performance of the system through improving the quality of vehicle classification and reducing the computational efforts by focusing on a smaller subset of moving objects. Figure 1d illustrates the refined foreground mask, *RFG*, after applying observation zone technique on *FG*:

$$RFG(p) = \begin{cases} 255 & \text{if } FG(p) \in OZ, \\ 0 & \text{otherwise.} \end{cases} \quad \text{where } OZ \text{ is the observation zone.} \tag{1}$$

Then, *RFG* is combined with *I* to create the moving object mask, *MO*, by applying an AND operation: $MO(p) = RFG(p) \wedge I(p)$. Figure 2c illustrates examples of *MO*. Also, we convert *MO* to gray-level, MO^{Gr}, and HSV color model from which the lightness component, MO^{V}, is extracted.

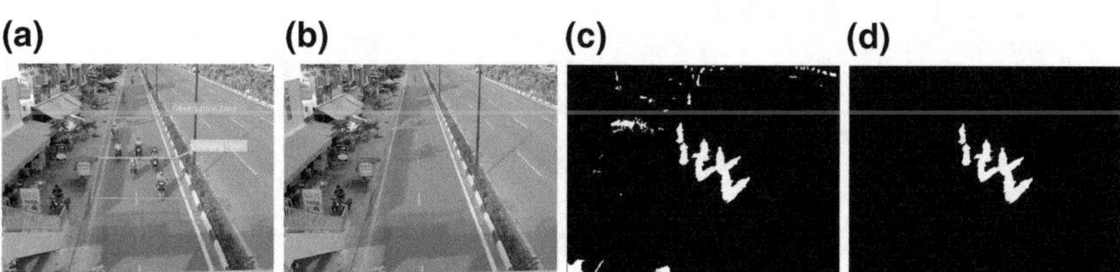

Fig. 1 **a** Observation zone and counting line. **b** Background model, *BG*. **c** Initial foreground mask, *FG*. **d** Refined foreground mask, *RFG*

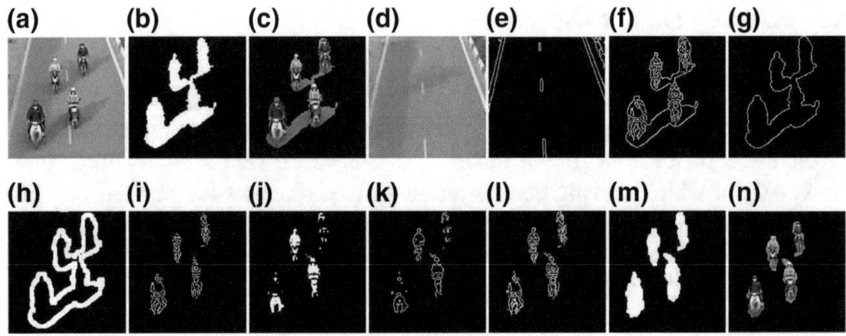

Fig. 2 Main steps in our algorithm with frame 451 of dataset PVD02. Note that the frame is cropped and scaled up for easier interpretation. **a** I. **b** RFG. **c** MO. **d** BG. **e** E_{BG}. **f** E_{Gr}. **g** E_{FG}. **h** DE_{FG}. **i** E_S. **j** MO^V. **k** E_V. **l** RE_S. **m** MV. **n** MV in RGB

2.3 Shadow Region Subtraction

Shadow region subtraction is the kernel of our contribution which comprises of two sections: (1) edge map generation; (2) shadow edge subtraction.

(1) Edge Map Generation. With respect to this issue, we make use of Canny edge detection. During the process of Canny edge detector, two threshold values (TH_1 and TH_2) are used to filter out the edge pixels with weak gradient values, which are caused by noise and color variation, and preserve the edge pixels with high gradient values. If the pixel value is smaller than the lower threshold (TH_1), it will be suppressed. If the edge pixel value is higher than the upper boundary (TH_2), it is marked as strong edge pixels. If the edge pixel value is between TH_1 and TH_2, it is marked as weak edge pixels. The two threshold values are empirically determined values, which will need to be defined when applying to different images. The problem becomes determining the optimal values for the thresholds when processing multiple frames captured under varying lighting conditions. We solve this problem by taking the median of I and then construct the upper and lower thresholds based on a percentage of this median:

$$\begin{cases} TH_1 = \max\{0, (1.0 - s) * M(I)\} \\ TH_2 = \min\{255, (1.0 + s) * M(I)\} \end{cases} \tag{2}$$

where $M(I)$ denotes the median of I and s is an optimal value used to vary the percentages. Typically, a lower value of s indicates tighter threshold, whereas a larger value of s gives wider threshold. In practice, $s = 0.52$ tends to give good results on the datasets we are working with. Table 1 summarizes the lower and upper thresholds calculated in our experiments with different lighting conditions.

The results, E_{BG}, E_{FG}, E_{Gr}, E_V, are the binary masks of values 0 and 255, as shown in Fig. 2e, f, g, k.

Table 1 Summary of TH_1 and TH_2 values calculated using Eq. 2

Video sequences	Conditions	TH_1	TH_2
Highway3	Strong shadows	66	216
VVK01	Morning, faint shadows	80	255
PVD01	Afternoon, strong shadows	69	228
PVD02	Afternoon, cloudy, varied shadows	65	215

(2) Shadow Edge Subtraction. From Fig. 2f, three observations can be made: (1) The cast shadows present sharp edges because the illumination source is far from the objects; (2) the vehicle has significant edges; however, the corresponding shadow is edgeless; (3) the edge of the cast shadow fastens on the boundary region of the moving foreground mask.

In order to remove the edges of shadows, we first compute the boundary of foreground mask, E_{FG}, which represents the outline of both the vehicle and its shadow. Then, the dilated boundary of the foreground mask which is more than one pixel thick is acquired:

$$DE_{FG} = E_{FG} \oplus E \tag{3}$$

where E is a 5×5 dilated structure element and $\oplus$ denotes morphological dilation.

Also, the edge map of the background model, E_{BG}, is computed to remove the edges created by background textures, such as road markings and pedestrian crossing pavements. Finally, the edge pixels of moving vehicles, E_S, are the interior edges of E_{Gr} (as shown in Fig. 2i), that is:

$$E_S(p) = \begin{cases} 255 & E_{Gr}(p) = 255 \text{ and} \\ & DE_{FG}(p) = 0 \text{ and} \\ & E_{BG}(p) = 0, \\ 0 & \text{otherwise.} \end{cases} \tag{4}$$

2.4 *Vehicle Boundary Recovery*

After the operator of Sect. 2.3, almost all the edge pixels in cast shadow areas are eliminated. However, the edges near the boundary of the vehicle have also been removed in the process, as shown in Fig. 2i. The recovery process consists of refining the vehicle boundary and recovering the vehicle mask.

To obtain additional edge information for the vehicle boundary, we look at the color aspect of shadow. We can use the value channel to acquire additional information to refine the vehicle boundary. The value mask, V, is established by applying a binary threshold with a value of 130 on the moving object mask, MO, which is shown in Fig. 2j. This threshold value is chosen by some initial experiments, and then they are fixed for all experiments. Then, the edge map E_V is obtained as in Fig. 2k. Then, we combine E_S with E_V to create the refined edge map RE_S, that is, $RE_S(p) = E_S(p) \vee E_V(p)$ as shown in Fig. 2l.

Then, the recovery of vehicle shape is performed. First, a morphological erosion is applied using the square structure element E with the size of 5×5. Second, the contour of each vehicle is extracted by removing small connected regions and repainting the vehicle shapes. Then, a morphological dilation operation with the same square structure element, E, is used to compensate the effect of the morphological erosion. The final extraction of moving vehicles is denoted by MV, which is shown in Fig. 2m, n. Illustrations of MV show that our algorithm can remove all shadow areas while still preserving the vehicle features.

3 Experiments and Discussion

To evaluate the proposed algorithm, we performed some experiments using the Highway3 video from dataset ATON [4], which is often used in recent publications regarding traffic surveillance system, and our datasets captured in Ho Chi Minh City, Vietnam. The video sequences were captured at daytime under a different lighting condition with the resolution of 640×480 and at the frame rate of 30 fps. We also compare our approach with related methods: Chromacity [5], Geometry [6], Edge [7], Physical [8], SR Texture [9], and LR Texture [10] on a system having a configuration of Intel Core i7 2630QM and 8GB of RAM.

3.1 Subjective Evaluation

Experiments were conducted on four traffic datasets. We have also compared our results with other methods. For all frames in these video sequences, our algorithm can achieve satisfactory results of removing shadows from moving vehicles.

On the first row of Fig. 3, we performed experiments on dataset Highway3. The result of the LR Texture method still contains some noises on the right of the vehicle which are created by the textures on the road. Moreover, the Edge method could not recover the left side of the vehicle after subtracting the shadow areas. In our method, because the road textures are removed during the shadow edge subtraction process, the vehicle boundary is refined using additional information. Hence, we could obtain a cleaner vehicle extraction. The second row of Fig. 3 illustrated the experiments on dataset VVK01, which was captured in the morning with faint shadows under low lighting conditions. In this scenario, all methods, except for our and LR Texture, fail to deliver satisfying results and misclassify parts of the bus as shadows. Compar-

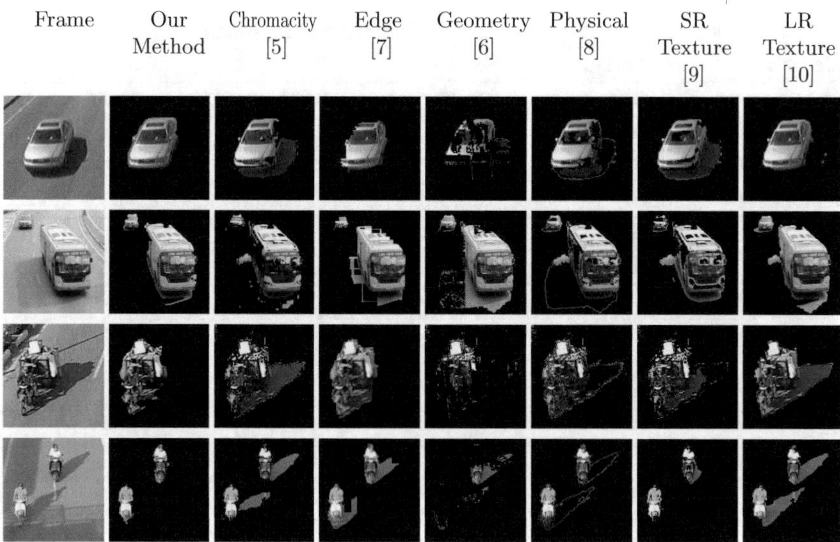

Fig. 3 Comparison results of shadow detection and removal. First row: the car from frame 1171 of dataset Highway3. Second row: the bus from frame 3750 of dataset VVK01. Third row: the unconventional vehicle from frame 1059 of dataset PVD01. Fourth row: the motorbikes from frame 948 of dataset PVD02

ing with LR Texture results, our method presents a better result. The third row of Fig. 3 is an experiment carried out on dataset PVD01, which was captured in the afternoon with strong shadows casting alongside the vehicles. We demonstrate the effectiveness of our shadow removal method under a special case with an unconventional vehicle. Although the Edge method can remove the shadows on the right side, it fails to detect the shadows under the vehicle, but our method can remove all the shadows. These results show that our algorithm can dynamically adapt to the real-world traffic scenes and therefore outperforms other methods.

3.2 *Objective Evaluation*

The error rates ε, proposed by [11], of the binary mask of moving vehicles, *MV*, are used to show the effectiveness of our algorithm. It is defined as follows:

$$\varepsilon = \frac{N_e}{N_I} \tag{5}$$

where N_I is the frame size and N_e is the number of pixels in *MV* that are different from the reference alpha plane.

Table 2 Average error rates of shadow detection and removal algorithms

Methods	Highway3	VVK01	PVD01	PVD02
Chromacity [5]	0.0072	0.0025	0.0017	0.0055
Edge [7]	0.0145	0.0044	0.0008	0.0049
Geometry [6]	0.0022	0.0069	0.0005	0.0046
Physical [8]	0.0049	0.0015	0.0009	0.0021
SR Texture [9]	0.0095	0.0106	0.0011	0.0027
LR Texture [10]	0.0094	0.0227	0.0019	0.0068
Our method	0.0021	0.0024	0.0006	0.0004

Table 3 Average processing time

Step	VVK01 (ms)	PVD01 (ms)	PVD02 (ms)
Background subtraction	12.40	15.62	15.93
Foreground object extraction	6.53	6.20	5.49
Canny edge detection	3.42	3.45	3.29
Shadow edge subtraction	0.22	0.28	0.28
Vehicle edge refinement and post-processing	1.81	1.85	1.78
Total	24.38	27.40	26.77

Table 2 reports the error rates for all methods used in these experiments. We can see that our method does not always have the lowest error rates in all video sequences. However, they are relatively equivalent to the best results. Also, the average error rates of each method are: 0.138% (our method), 0.443% (Chromacity [5]), 0.355% (Geometry [6]), 0.615% (Edge [7]), 0.235% (Physical [8]), 0.598% (SR Texture [9]), and 1.020% (LR Texture [10]). It shows that our algorithm can robustly detect shadows with minimum error rates in different scenarios.

The average processing time in each step of our algorithm can be found in Table 3. The average time needed to process one frame is ranging from 24.38 to 28.61 ms. Our algorithm has been optimized to run in the parallel fashion so that it only introduces a small delay to the overall processing time. Therefore, we are confident that after incorporating with our algorithm, the surveillance traffic system can achieve real-time processing capability (fps ≥ 25).

4 Conclusion

In this paper, we have proposed a novel shadow removal algorithm in daytime traffic scenes. The algorithm is based on edge information from both the input frame and the lightness component of HSV color model. The advantages of our method are: (1) The algorithm is robust to a variety of shadow orientations, shapes, and appearances under different lighting conditions; (2) the algorithm precisely removes shadows from both smooth and textured backgrounds. Experiments show that our algorithm performs better than previous method as it can run in real-time speed when processing single or multiple traffic sequences.

Acknowledgements The study was supported by Science and Technology Incubator Youth Program, managed by the Center for Science and Technology Development, Ho Chi Minh Communist Youth Union, the contract number is "20/2017/ HĐ-KHCN-VU".

References

1. Nguyen, T.P., Tran, D.N.-N., Huynh, T.K., Ha, S.V.-U.H.: Disorder detection approach to background modeling in traffic surveillance system. J. Sci. Technol. Vietnamese Acad. Sci. Technol. **52**(4A), 140149 (2014)
2. Pham, L.H., Duong, T.T., Tran, H.M., Ha, S.V.U.: Vision-based approach for urban vehicle detection and classification. In: 2013 Third World Congress on Information and Communication Technologies (WICT 2013), pp. 305–310 (2013)
3. Ha, S.V.-U., Pham, L.H., Tran, H.M., Ho-Thanh, P.: Improved vehicles detection and classification algorithm for traffic surveillance system. J. Inf. Assurance Secur. **9**(5), 268277 (2014)
4. ATON datasets by UCSD. http://cvrr.ucsd.edu/aton/shadow/
5. Cucchiara, R., Grana, C., Piccardi, M., Prati, A.: Detecting moving objects, ghosts, and shadows in video streams. IEEE Trans. Pattern Anal. Mach. Intell. **25**(10), 1337–1342 (2003)
6. Hsieh, J.-W., Wen-Fong, H., Chang, C.-J., Chen, Y.-S.: Shadow elimination for effective moving object detection by Gaussian shadow modeling. Image Vis. Comput. **21**(6), 505–516 (2003)
7. Xiao, M., Han, C.-Z., Zhang, L.: Moving shadow detection and removal for traffic sequences. Int. J. Autom. Comput. **4**(1), 3846 (2007)
8. Huang, J.B., Chen, C.S.: Moving cast shadow detection using Physics-based features. In: 2009 IEEE Computer Society Conference on Computer Vision and Pattern Recognition Workshops, CVPR Workshops 2009, pp. 2310–2317 (2009)
9. Leone, A., Distante, C.: Shadow detection for moving objects based on texture analysis. Pattern Recognit. **40**4 (2007)
10. Sanin, A., Sanderson, C., Lovell, B.C.: Improved shadow removal for robust person tracking in surveillance scenarios. In: 2010 20th International Conference on Pattern Recognition, pp. 141–144 (2010)
11. Chien, S., Ma, S., Chen, L.: Efficient moving object segmentation algorithm using background registration technique. IEEE Trans. Circuits Syst. Video Technol. **12**(7), 577586 (2002)

Real-Time Analysis of a Nonlinear State Observer with Improved Disturbance Compensation for Sensorless Induction Motor

S. Mohan Krishna, J. L. Febin Daya and C. Kamal Basha

Abstract This paper presents a comparison and real-time analysis of sliding mode disturbance observers for speed sensorless induction motor drives. The rotor speed tracking bandwidth and the load rejection capability are improved by altering the profile of the sliding hyperplane used in the state observer. The entire system is built in Simulink environment, and real-time RT-lab blocksets are integrated into the same and tested in a new Processor-in-Loop-based test bench. The Processor-in-Loop test bench uses the OP4500 real-time target and loop back adapters for signal routing. This ensures that there is a real-world signal transfer between the plant and the controller.

Keywords Sensorless · State estimation · Adaptive control
Model-based design · Processor-in-loop (PIL) · RT-lab

Nomenclature

$i^s_{ds}, i^s_{qs}, i^r_{dr}, i^r_{qr}, i^*_{as}, i^*_{bs}, i^*_{cs}$	D-, q-, and three-phase reference current components
v^s_{ds}, v^s_{qs}	D- and q-axis stator voltage components
T_r, R_s, R_r	Rotor time constant, stator, and rotor resistance
Σ, L_r, L_m, L_s	Leakage reactance, rotor, magnetizing, and stator self-inductance
$L_{ls}, L_{lr},$	Stator and rotor leakage inductances
$\psi^s_{ds}, \psi^s_{qs}, \psi^s_{dr}, \psi^s_{qr}$	D-axis and q-axis flux linkages
$\theta_f, \theta_{sl}, \theta_r T^*_e$	Field, slip, and rotor angle, reference electromagnetic torque

S. M. Krishna (✉) · C. Kamal Basha
Department of Electrical and Electronics Engineering, MITS (Madanapalle Institute of Technology and Science), Madanapalle, Andhra Pradesh, India
e-mail: smk87.genx@gmail.com

J. L. F. Daya
School of Electrical Engineering, VIT University, Chennai campus, Chennai, India

V. Bhateja et al. (eds.), *Intelligent Engineering Informatics*, Advances in Intelligent Systems and Computing 695, https://doi.org/10.1007/978-981-10-7566-7_66

1 Introduction

Variable frequency drives play a major role in almost all industrial sectors, some of which are process control, motion control, industrial automation, etc. Induction motors are generally employed owing to their rugged construction, robustness, less maintenance, and reliability. Speed sensors or encoders mounted on the shaft of these motors give information about the rotor speed, but the sensor is highly sensitive to hostile environment and also occupies additional mounting space. Estimation of speed without the use of speed sensor was a topic of interest and occupied considerable research space over the last two decades. Of several speed estimation algorithms, adaptive control-based parameter estimation schemes were easy to implement and popular. Besides, there was flexibility in changing the configuration to adapt more parameters. The model reference adaptive system (MRAS)-based observers are sought after in model-based design and have a high rate of adaptation [1–4]. MRAS used for state estimation is explained briefly in the next section. As the induction motor is highly nonlinear in nature, the vector control strategy decouples the torque and flux components to produce a linearized torque control strategy. For the indirect vector control algorithm, the rotor speed is essential. Thus, the estimation is generally performed in stationary frame and the control is done in synchronously rotating reference frame.

The performance of two MRAS state observers with conventional and modified disturbance rejection capabilities is studied and analyzed. The improved sliding mode disturbance observer gives better performance in terms of speed range, torque levels, etc., and is verified in a real-time Processor-in-Loop environment which adds more credibility to the findings.

2 Mathematical Modeling of the Conventional and Proposed State Observers and Current-Regulated Vector Controller

The model reference adaptive systems form an important component of adaptive control-based parameter estimation scheme. In this paper, the MRAS-based observer scheme takes inputs from the machine terminals and estimates parameters by means of an algorithm derived from the model quantities. The adaptive mechanism comprises the control law which is designed in such a way that the state of the adaptive model (desired performance) is constrained to that of the reference model (process). The sliding hyperplane is used for faster convergence of the states of both the models by restraining the stator current error dynamics. The observer is further illustrated and explained in detail with appropriate mathematical foundation. Luenberger was a scientist actively involved in the theory of state observers.

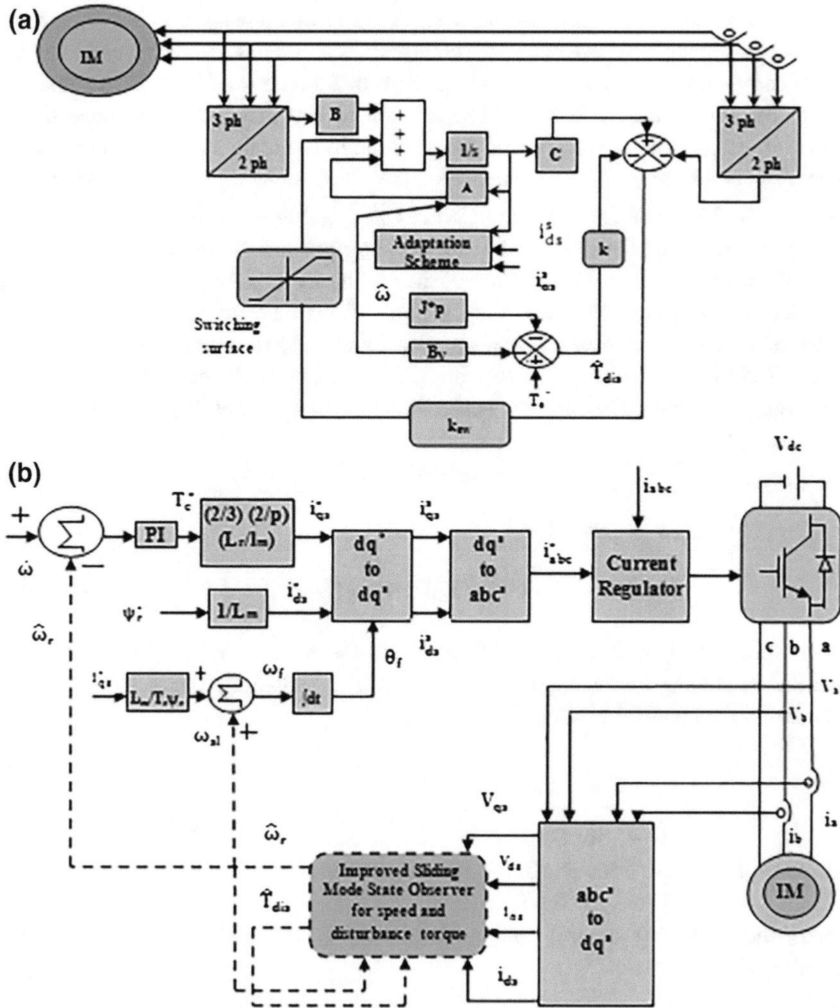

Fig. 1 **a** Proposed observer with modified disturbance rejection **b** Observer-based sensorless drive system

These observers made use of an additional correction term in the observer state dynamic equation for faster convergence [5]. The integration of the switching surface or sliding hyperplane into the Luenberger observer structure gave rise to sliding mode Luenberger observers [6, 7]. The observer with the disturbance torque estimate included in the sliding hyperplane is shown in Fig. 1a.

The labels are elaborated as follows. 'A' is the parameter matrix, '^' symbol indicates estimated parameters, 'X' comprises the state variables which are the direct and quadrature axes stator currents and rotor fluxes, and 'k_{sw}' is the observer switching gain, which can be chosen in such a way that the eigenvalues of the observer are proportional to that of the machine to maintain stability under normal operating conditions. 'J', 'p', and 'B_V' are the moment of inertia, differential operator, and viscous friction coefficient. 'T_e^*', '$\widehat{T}_{dis}$' are the motor model electromagnetic torque and the estimated disturbance torque, and 'k' being a positive gain. Care should be taken in the selection of the sliding hyperplane as it is significant in stability and observer dynamics. The Lyapunov function candidate V is used for the derivation of the adaptive mechanism, and its derivative should satisfy the Lyapunov stability criterion [8]. The sliding hyperplane is denoted by 'S' and the Lyapunov function candidate is a scalar function of the same [6].

$$\dot{V}(S) = S(x)\dot{S}(x) \tag{1}$$

The control law is given by:

$$u(t) = u_{eq}(t) + u_{sw}(t) \tag{2}$$

where the control vector, u(t), is resolved into the equivalent control vector $u_{eq}(t)$ and the switching vector $u_{sw}(t)$, respectively. The switching vector satisfying the conditions for stability is shown below [9]:

$$u_{sw}(t) = \eta \, \mathrm{sign}(S(x, t)) \tag{3}$$

where $\mathrm{sign}(S) = \begin{cases} -1 \text{ for } S < 0 \\ 0 \text{ for } S = 0 \\ +1 \text{ for } S > 0 \end{cases}$

where the switching control gain η should be chosen such that (1) is negative definite. This means:

$$S(x)\dot{S}(x) < 0$$

This ensures the disturbance is constrained. The nonlinear high-frequency switching also leads to a phenomenon known as chattering. In order to eliminate this, a saturation function is defined having a boundary layer of width (Φ) which replaces sign(S) with sat(S/Φ):

$$\mathrm{sat}(S/\Phi) = \begin{cases} \mathrm{sign}\left(\frac{S}{\Phi}\right) \text{if} \left|\left(\frac{S}{\Phi}\right)\right| \geq 1 \\ \left(\frac{S}{\Phi}\right) \text{if} \left|\left(\frac{S}{\Phi}\right)\right| < 1 \end{cases} \tag{4}$$

The structure of the observer using state space formulation is given below:

2.1 Motor Model (Reference)

$$\frac{dx}{dt} = [A]x + [B]u \tag{5}$$

$$y = [C]x \tag{6}$$

where

$$x = \left[i_{ds}^{s},\ i_{qs}^{s},\ \psi_{dr}^{s},\ \psi_{qr}^{s}\right]^{T},\ A = \begin{bmatrix} A_{11} & A_{12} \\ A_{21} & A_{22} \end{bmatrix},$$
$$B = \left[\tfrac{1}{\sigma L_s} I \quad 0\right]^{T}, C = [I, 0],\ u = \left[v_{ds}^{s} \quad v_{qs}^{s}\right]^{T},$$
$$I = \begin{bmatrix} 1 & 0 \\ 0 & 1 \end{bmatrix},\ J = \begin{bmatrix} 0 & -1 \\ 1 & 0 \end{bmatrix},$$
$$A_{11} = -\left[\frac{R_s}{\sigma L_s} + \frac{1-\sigma}{\sigma T_r}\right] I = a_{r11} I, A_{12} = \frac{L_m}{\sigma L_s L_r}\left[\frac{1}{T_r} I - \omega_r J\right] = a_{r12} I + a_{i12} J,$$
$$A_{21} = \frac{L_m}{T_r} I = a_{r21} I\ ,\ A_{22} = \frac{-1}{T_r} I + \omega_r J = a_{r22} I + a_{i22} J,$$

2.2 Disturbance Torque Estimation

The inertia and viscous friction coefficients are considered as constant, and the coulomb friction is not taken into account:

$$\widehat{T}_{dis} = T_e^{*} - J\frac{d\,\widehat{\omega}}{dt} - B_V\widehat{\omega} \tag{7}$$

2.3 Sliding Mode Luenberger Observer with Conventional Disturbance Rejection—Observer 1

$$\frac{d\widehat{x}}{dt} = \left[\widehat{A}\right]\widehat{x} + [B]u + k_{sw}\mathrm{sat}\left(\widehat{i}_s - i_s\right) + \widehat{d} \tag{8}$$

Sliding surface or hyperplane is $s = \widehat{i}_s - i_s$ and $\widehat{d} = k\widehat{T}_{dis}$ and

$$\widehat{y} = [C]\widehat{x} \tag{9}$$

where $\widehat{i}_s$, i_s = estimated and actual stator currents.

$$A = \begin{bmatrix} A_{11} & \widehat{A}_{12} \\ A_{21} & \widehat{A}_{22} \end{bmatrix},\ \widehat{A}_{12} = \frac{L_m}{\sigma L_s L_r}\left[\frac{1}{T_r}I - \widehat{\omega}_r J\right] = a_{r12}I + \widehat{a}_{i12}J,$$
$$\widehat{A}_{22} = \frac{-1}{T_r}I + \widehat{\omega}_r J = a_{r22}I + \widehat{a}_{i22}J$$

The switching gain is obtained from the following reduced order matrix:

$$k_{sw} = \begin{bmatrix} k_1 & k_2 \\ -k_2 & k_1 \end{bmatrix}^T \tag{10}$$

To ensure stability of (2) by means of pole placement, the switching gain is chosen. Therefore,

$$k_1 = (m-1)a_{r11} \tag{11}$$

$$k_2 = k_p,\ k_p \geq -1 \tag{12}$$

2.4 *Sliding Mode Luenberger Observer with Improved Disturbance Rejection—Observer 2*

The model disturbance and the stator current error are constrained within the sliding hyperplane. Therefore, the profile of the sliding hyperplane is changed.

$$\frac{d\widehat{x}}{dt} = \left[\widehat{A}\right]\widehat{x} + [B]u + k_{sw}\mathrm{sat}\left(\widehat{i}_s - i_s - \widehat{d}\right) \tag{13}$$

where $s = \widehat{i}_s - i_s - \widehat{d}$ and $\widehat{d} = k\,\widehat{T}_{dis}$

$$\widehat{y} = [C]\widehat{x} \tag{14}$$

2.5 *Adaptive Mechanism with Lyapunov Function Candidate*

The function candidate is denoted by M and is expressed as:

$$M = e^T e + \frac{(\widehat{\omega}_r - \omega_r)^2}{\lambda} \tag{15}$$

'λ' is positive. Therefore,

$$\frac{dM}{dt} = e^T\left[(A + k_{sw}C)^T + (A + k_{sw}C)\right]e - \frac{2\Delta\,\omega_r\left(e_{ids}\widehat{\varphi}^s_{qr} - e_{iqs}\widehat{\varphi}^s_{dr}\right)}{c} + \frac{2\Delta\omega_r}{\lambda}\frac{d\widehat{\omega}_r}{dt} \tag{16}$$

where $e_{ids} = i^s_{ds} - \widehat{i}^s_{ds}$, $e_{iqs} = i^s_{qs} - \widehat{i}^s_{qs}$

On equalizing the second and the third term of (16),

$$\frac{d\widehat{\omega}_r}{dt} = \frac{\lambda}{c}\left(e_{ids}\widehat{\varphi}^s_{qr} - e_{iqs}\widehat{\varphi}^s_{dr}\right) \tag{17}$$

'c' being an arbitrary positive constant.

2.6 *Structure of Indirect Vector Controller with Hysteresis Band Current Regulation*

The indirect rotor flux-oriented indirect vector control strategy is used as the control algorithm. The current regulation is employed because of its inherent short-circuit current protection capability, independence on load parameters, and fast torque response [10].

$$E_c = \widehat{\omega}_r - \omega^* \tag{18}$$

$$T^*_e = e_c\left[k_p + (k_i/s)^*T_s\right] \tag{19}$$

where e_c, k_p, k_i, and T_s are the speed error, the controller gains, and the sampling time. The rotor flux assumptions for different sets of speeds are given below:

$$\Psi_r = 0.96,\ \text{for}\ \widehat{\omega}_r < \omega_{bsync} \tag{20}$$

$$\psi_r = 0.96^*\left(\frac{\widehat{\omega}_r}{\omega_{bsync}}\right),\ \text{for}\ \widehat{\omega}_r > \omega_{bsync} \tag{21}$$

The torque and flux requests together form the inputs for the vector controller. The orthogonally oriented torque-producing and flux-producing components are obtained as shown:

$$i_{ds}^{*} = \left(\frac{\psi_r}{L_m}\right)\left[1 + \frac{dT_r}{dT_s}\right] \tag{22}$$

$$i_{qs}^{*} = \left(\frac{2}{3}\right)\left(\frac{2}{P}\right)\left(\frac{L_r}{L_m}\right)\left(\frac{T_{ref}}{\psi_r}\right) \tag{23}$$

It is indirectly imposed by means of the slip speed. The field angle is used for the inverse transformation of components, and therefore, it is obtained from the slip speed as shown:

$$\theta_f = \theta_{sl} + \theta_r \tag{24}$$

$$i_{as}^{*} = i_{ds} \sin\theta + i_{qs} \cos\theta \tag{25}$$

$$i_{bs}^{*} = \left(\frac{1}{2}\right)\left\{-i_{ds}\cos\theta + \sqrt{3}i_{ds}\sin\theta\right\} + \left(\frac{1}{2}\right)\left\{i_{qs}\sin\theta + \sqrt{3}i_{qs}\cos\theta\right\} \tag{26}$$

$$\mathbf{i}_{cs}^{*} = -\left(\mathbf{i}_{as}^{*} + \mathbf{i}_{bs}^{*}\right) \tag{27}$$

The current errors are processed through tolerance or the hysteresis band regulator to give rise to gating signals for the switches of the inverter. The entire setup is shown in Fig. 1b.

3 Processor-in-Loop Environment for Real-Time Verification

Here, the interaction between the RT-lab integrated plant (inverter, motor, and observer) and the vector controller is by means of loopback cables and not by Simulink wires. Two loop-back adapters and a 40 pin flat ribbon cable are used for real-time signal routing from the plant to the controller and vice versa. The output of the plant, which is, the estimated speed and the three phase currents, are transferred via analog output channels and captured by the analog input channels in the controller subsystem.

The pulses are transferred by means of the digital output channels of the controller subsystem and captured by the digital input channels of the plant subsystem. The output and input channels are configured accordingly. Both the analog and digital loopbacks are stand-alone hardware equipments; however, the digital loopback needs a +5 V or +12 V source for both Vsource and Vref.

The OP4500 real-time target has a single processor core activated, where the plant and the controller rest in the same subsystem. The entire model is present in the workstation or PC which is then connected to the real-time target through TCP/IP protocol. It is shown in Fig. 2.

(a)

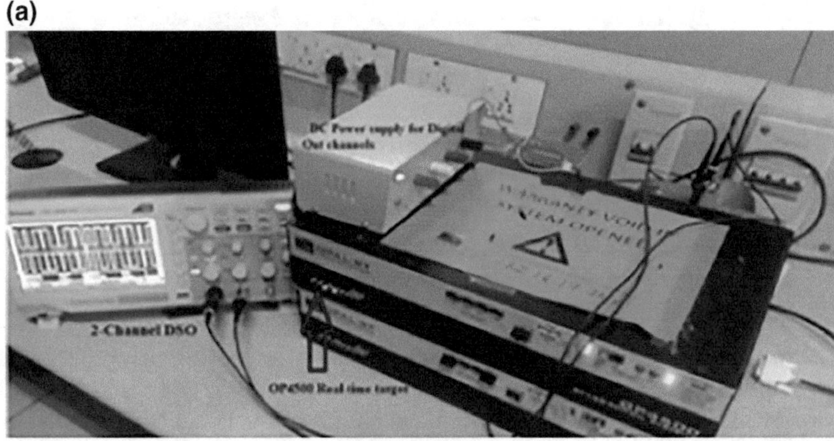

(b)

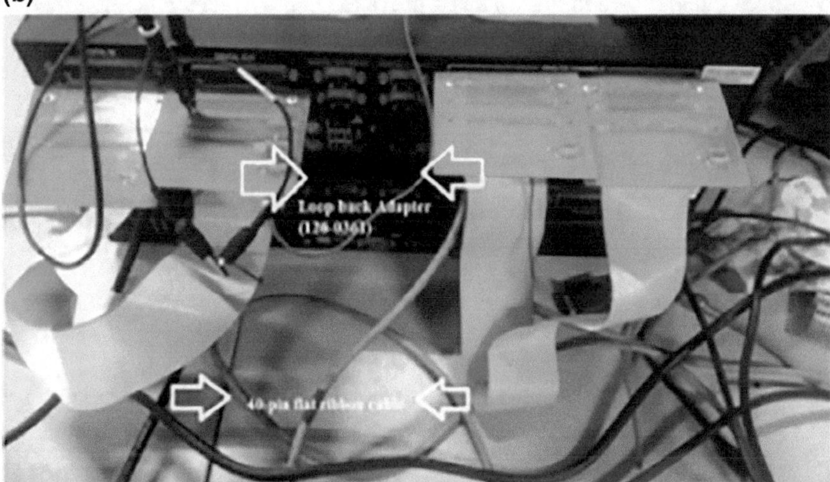

Fig. 2 **a** Front view of the OP4500 real-time target **b** Rear view of the OP4500 real-time target with loopback cables and adapters

4 Real-Time Results and Analysis

The simulation configuration is changed to fixed step discrete, and the time step used is 50μs. Along with the model disturbance, the measurement noise in the form of Gaussian noise is introduced at the terminals of the machine while measuring the terminal voltages and currents. The Simulink and RT-lab integrated model is executed in the real-time target. The disturbance observers are tested for the test cases.

4.1 Low-Speed Range—at Constant Load Perturbation of 100 Nm

See (Figs. 3 and 4).

4.2 Constant Speed Range of 100 Rad/S—at Step Load (Initially at 5 Nm, After 15 S, 200 Nm)

The motor model parameters and ratings considered for the study are provided in the Appendix. The two observers are compared for low speeds and changes in load perturbation. The changes in the speed and load commands are treated as model disturbances. The testing is carried out in motoring mode. The results in the low-speed range shown in Figs. 3 and 4 prove the improvement in Observer 2 over

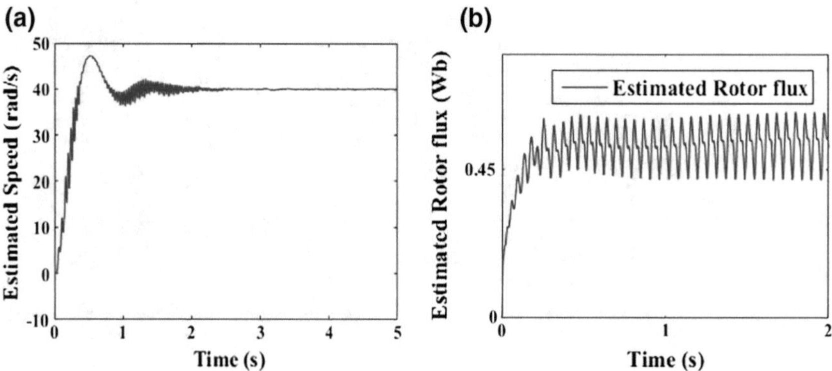

Fig. 3 **a** Estimated rotor speed and **b** estimated rotor flux of Observer 1

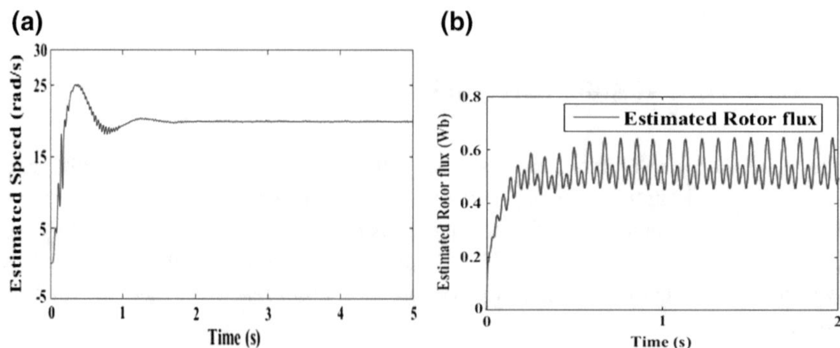

Fig. 4 **a** Estimated rotor speed and **b** estimated rotor flux of Observer 2

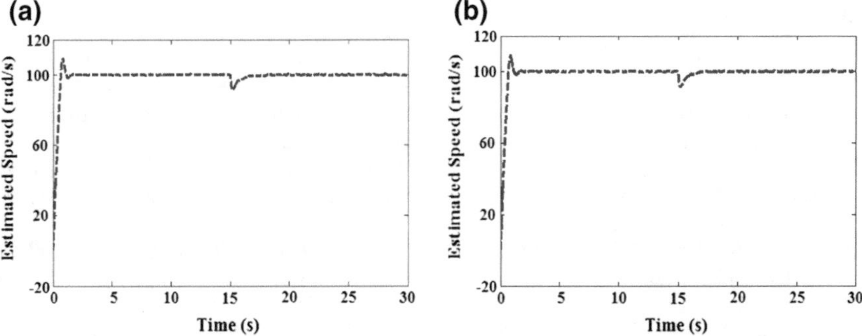

Fig. 5 **a** Estimated rotor speed of Observer 1 and **b** Observer 2

Observer 1. The Observer 2 tracks at lower speed of 20 rad/s with considerably lesser oscillations than that of Observer 1 at 40 rad/s (below which the observer goes out of bounds) with more oscillations. This is due to superior state convergence of the sliding hyperplane. But, the medium speed tracking profiles of both the observers are similar as shown in Fig. 5. This infers that the change in the profile of the sliding hyperplane affects the lower speed regions more than the medium speed regions.

5 Conclusion

The RT-lab-based PIL verification is used for testing the observers with the proposed and the conventional disturbance compensation mechanism. The PIL testing forms one of the most important model-based design paradigms. The virtual plant and controller are executed in a real-time target OP4500, and the signal exchange is also real world. The improved observer displays better performance in terms of tracking and disturbance rejection at low-speed ranges, thereby increasing its speed bandwidth. It occupies lesser computational space. Also, the change in the profile of the sliding hyperplane makes the state observer comparatively more robust to disturbances and nonlinearities.

Appendix

Model ratings and parameters: A 50 HP, three-phase, 415 V, 50 Hz, star-connected, four-pole induction motor with equivalent parameters: $R_s = 0.087$ Ω, $R_r = 0.228$ Ω, $L_{ls} = L_{lr} = 0.8$ mH, $L_m = 34.7$ mH, inertia, J = 1.662 kgm^2, friction factor = 0.1.

References

1. Krishna, M., Daya, J.L.F.: Dynamic performance analysis of mras based speed estimators for speed sensorless induction motor drives. In: IEEE International Conference on Advances in Electronics, Computers and Communication, pp 1–6 (2014)
2. Krishna, M., Daya, J.L.F.: MRAS speed estimator with fuzzy and PI stator resistance adaptation for sesnorless induction motor drives using RT-Lab. Perspectives in Science, pp. 121–126. Elsevier (2016)
3. Krishna, M., Daya, J.L.F.: An improved stator resistance adaptation mechanism in MRAS estimator for sensorless induction motor drives. In: Advances in Intelligent Systems and Computing, vol. 458, pp. 371–385. Springer-Verlag (2016)
4. Krishna, M., Daya, J.L.F., Sanjeevikumar, P., Mihet-Popa, L.: Real-time performance and analysis of a modified state observer for sensorless induction motor drives using RT-Lab for electric vehicle applications. Energies, MDPI, pp. 1–23 (2017)
5. Krishna, M., Daya, J.L.F.: Effect of Parametric variations and Voltage unbalance on adaptive speed estimation schemes for speed sensorless induction motor drives, pp. 77–85. IAES, Indonesia, Int. J. Power. Elect. Drive. Sys (2015)
6. Krishna, M., Daya, J.L.F.: A modified disturbance rejection mechanism in sliding mode state observer for sensorless induction motor drive. Arab. J. Sci. Eng. 3571–3586 (2016) (Springer-Verlag)
7. Krishna, M., Daya, J.L.F.: Adaptive speed observer with disturbance torque compensation for sensorless induction motor drives using RT-lab. Turk. J. Elect. Eng. Comp. Sci. Tubitak 3792–3806 (2016)
8. Slotine, J.J.E., Li, W.: Applied Non Linear Control. Prentice-Hall (1991)
9. Comanescu, M.: Design and analysis of a sensorless sliding mode flux observer for induction motor drives. In: IEEE International Electric Machines & Drives Conference, pp. 569–574 (2011)
10. Bose, B.K.: Power Electronics and Variable Frequency Drives—Technology and Applications. Wiley (2013)

Correction to: Cloud Adoption: A Future Road Map for Indian SMEs

Biswajit Biswas, Sajal Bhadra, Manas Kumar Sanyal and Sourav Das

Correction to: Chapter "Cloud Adoption: A Future Road Map for Indian SMEs" in: V. Bhateja et al. (eds.), *Intelligent Engineering Informatics*, Advances in Intelligent Systems and Computing 695, https://doi.org/10.1007/978-981-10-7566-7_51

In the original version of the book, the author names in reference 9 of chapter "Cloud Adoption: A Future Road Map for Indian SMEs" have been corrected, which is a belated correction. The correction chapter and the book have been updated with the change.

The updated version of this chapter can be found at
https://doi.org/10.1007/978-981-10-7566-7_51

V. Bhateja et al. (eds.), *Intelligent Engineering Informatics*, Advances in Intelligent Systems and Computing 695, https://doi.org/10.1007/978-981-10-7566-7_67

Author Index

V. Bhateja et al. (eds.), *Intelligent Engineering Informatics*, Advances in Intelligent Systems and Computing 695, https://doi.org/10.1007/978-981-10-7566-7